Configuration Management

Configuration Management

Theory and Application for Engineers, Managers, and Practitioners

Second Edition

Jon M. Quigley and Kim L. Robertson

CRC Press is an imprint of the
Taylor & Francis Group, an **informa** business
AN AUERBACH BOOK

CRC Press
Taylor & Francis Group
6000 Broken Sound Parkway NW, Suite 300
Boca Raton, FL 33487-2742

Printed on acid-free paper

International Standard Book Number-13: 978-0-367-13725-0 (Paperback)

Library of Congress Cataloging-in-Publication Data

Names: Quigley, Jon M., author. | Robertson, Kim L., author.
Title: Configuration management : theory and application for engineers, managers, and practitioners / Jon M. Quigley, Kim L. Robertson.
Description: Second edition. | Boca Raton : Taylor & Francis, a CRC title, part of the Taylor & Francis imprint, a member of the Taylor & Francis Group, the academic division of T&F Informa, plc, 2019. | Includes bibliographical references.
Identifiers: LCCN 2019021294 | ISBN 9780367137250 (paperback : acid-free paper)
Subjects: LCSH: Configuration management. | Production management.
Classification: LCC QA76.76.C69 Q54 2019 | DDC 004/.0688—dc23
LC record available at https://lccn.loc.gov/2019021294

Visit the Taylor & Francis Web site at
http://www.taylorandfrancis.com

and the CRC Press Web site at
http://www.crcpress.com

Dedication

This second edition is for Tina O'Dell, CM'er par excellence, who is an inspiration to everyone who knows her.

It is also for Steven Easterbrook cofounder of the Configuration Management Process Improvement Center whose untimely passing on April 17, 2019 deeply affected the thousands of lives he touched with his largesse, photos, music, mentoring, and deep understanding of CM and how best to teach it.

Contents

Equations

Figures

Tables

Tables

Foreword to the Second Edition

The first edition published in 2015 was the culmination of a project that I started over 25 years earlier. Jon Quigley, Value Transformations, LLC, invited me to co-author when his regular writing partner, Kim H. Pries, dropped out due to other commitments. John Wyzalek/Taylor & Francis was instrumental in seeing it published. That initial effort has been well received. So why do we need a second edition a scant four years later? Several key events have taken place since publication that made the first edition information dated and a second edition necessary.

1. The SAE G-33 configuration management committee completed crafting of Revision C to the SAE/EIA 649-C,[1] *Configuration Management Standard.* A. Larry Gurule's able leadership ensured that the new revision reflected an international consensus for CM of supplies and services in government and commercial sectors.
2. I joyfully completed the task of writing a set of 24 articles tracing the history of CM from antiquity to the nuclear age for the CMPIC CM Trends eZine. That effort paved the way to my taking the first four CMPIC certification classes, giving me new insights.
3. Prof. Dr. A. (Ton) G. de Kok of the Technical University of Eindhoven (TU/e) chose the first edition as the textbook for a class in CM for students seeking a graduate degree and invited Jon and me to present the week of February 3, 2019. As the course materials evolved, it became apparent that reorganizing the chapters would make it a better book.
4. First edition comments regarding minor typographical errors, wordiness, and the need to expand some areas reached critical mass and were taking over my home office.
5. Technologies such as model based engineering (MBE), and 3D printing have come online or evolved and impact how CM implementation needs to advance in the coming decades.
6. Dee and I have taken several international trips ferreting out the origins of past technologies that bear on the history of managing configurations.

I am truly humbled by the knowledge trove residing in the professionals I have had the good fortune to become acquainted with through G-33, CMPIC, TU/e, and our travels. Their generosity in time spent clarifying industry specific CM implementation and one-on-one mentoring was the catalyst needed to correctly weave the multitudes of digital and non-digital threads we gathered into the fabric of this second edition. Many are listed in the Acknowledgments section of this book. Others we are indebted to remain nameless, like the docent at the Philips Museum in Eindhoven who spent an hour answering questions and dozens of museum and historic site curators just like him across Great Britain, Europe, and Turkey.

Daniel K. Christensen – NAVAIR, Martijn Dullaart – ASML, and Stein L. Cass, Jr. PhD – Ball Aerospace & Technologies Corp. were kind enough to review the second edition text and their comments were invaluable.

I hope you enjoy it,

Kim L. Robertson
Erie, Colorado

[1] SAE. (2019). SAE/EIA 649-C-2019-02, Configuration Management Standard. SAE International. Warrendale, Pennsylvania.

Preface

Peter F. Drucker[2] said that anyone is a manager, if, by virtue of their position or knowledge, they are responsible for contributions to the organization that materially affects the capacity of the organization to bring new product to market, contribute to market share, or to reduce cost and schedule while maintaining safety, quality, and user satisfaction. We combined that with the premise that "an engineer solves problems" and set off down the road of discovery.

Books about the theory and practice of any subject must consider that examples cited are no more than snapshots in time. Statements about the future need to be couched in terms like, "this is probably something that you need to pay attention to because it is going to hurt if you aren't prepared for it." Another issue with books like this one is that unless the author(s) are careful the text will become outdated before it is published. With these cautions firmly in mind we have done our best to provide a second edition that addresses configuration management (CM) theory and practice in a meaningful way.

This is not a volume that tells you how to do CM. It will hopefully provide you the back story regarding why configurations need to be managed and some insight in what is headed your way as you contemplate the future of your business. The business environment is so interconnected, and technologies are evolving so fast that you can no longer afford to partition responsibilities. A single department is incapable of managing everything associated with a configuration. The effort must be distributed. You need a basic grasp of that interconnectedness to succeed.

Some aspect of managing configurations includes every individual working for the company. It is part of the company's DNA and without understanding what it is or why it is necessary you put the safety, quality, producibility, reliability and sales of the services you are providing or the product you are selling at risk.

Management of a configuration starts with the first discussion that a market niche exists presenting an opportunity for a new service or product. It continues as designs are crafted, products or services are produced, tested, marketed, sold, and maintained through to eventual disposal at end of life.

CM applies to everything!

[2] Drucker, P. F. (1967): *The Effective Executive*, Harper/Business, Harper/Collins Publishers, Inc., 10 East 53rd Street, New York, NY

Preface

Acknowledgments

To our families we express love and appreciation. Jon would like to thank his wife Nancy and his son Jackson Quigley for their support. Kim would like to thank his wife Dee Robertson. Without Dee's encouragement, critiques, understanding and acceptance of long nights and weekends crafting this edition and her Scottish tenacity in completing the final sanity check of its contents the update never would have been completed on time.

We would all so like to shout out a big thank-you to everyone who previously contributed to the first edition that formed the base we departed from for this second edition, and to everyone who contributed directly or indirectly to this second edition. Their sage guidance, critique, and feedback has steered the updates and corrections contained herein. Contributions came in the form of CM Trends articles; via email and teleconferences; during CMPIC annual SWAT events, quarterly SAE G-33 meetings, G-33 SAE/EIA 649 Revision C Sprints, our meeting with ASML, Philips, and DAF Trucks; preparing for the master's level course in CM being taught at the Technical University of Eindhoven; and through LinkedIn. Individual contributions are noted in the book text by source.

The influencers on this second edition are, in alphabetical order:

Bonnie Alexander – Navy (U.S.); David Armstrong – Bentley Systems; Peter Bilello – CIMData; Authur Bingcang – Boeing; Mike Bostelman – Cummins; Andre Brosnan – INDEC; Cornita Bullock –DOD (U.S.); Steven Buskermolen – DAF Trucks N.V.; Daniel K. Christensen – Navy (U.S.); David Clark – Lord; Lea Clark – Navy (U.S.); Leo Clark – CMPIC LLC; Shari Councell – Northrop Grumman; Mary Covington – Intuitive Research & Technology; Marcellin Dagicour – Airbus; A. "Ton" G. de Kok – Technical University of Eindhoven; Robert Del Valle – Perspecta; Michael Denis – CapGemini; Jaqui Dow – Honda Aircraft Company; Kristina Ducas – Navy (CAN); Martijn Dullaart – ASML;

Steven and Kathy Easterbrook – CMPIC; Todd Egan – IPX; Lisa Fenwick – CMstat; Tim Fitch – L3 Com; John Gellios – FAA; A. Larry Gurule – i-infusion, Inc.; Kelly Hansen – Raytheon; Gregg Hendry – Ball Aerospace & Technologies Corp. (Retired); Thad Henry – NASA (Retired); Tyra "Ty" Holt – NSWC (U.S.); Reginald Hughes – USMEP; Paul Irwin – Feature[23]; Mitch Kaarlela – Lockheed Martin; Jeff Klein – Boeing; Leisa Lemaster – USAF; Brynt Lindsey – BAE Systems; Carol Maguire – Boeing; Jonathan Maloof – Lockheed Martin; Dan McCurry – Boeing; Rich McFall – CMstat; Cynthia Mendez – Lockheed Martin; James Newman – Bentley Systems; Tina O'Dell – Cummins; Moe Parker – PSA; Mike Potts – Docio; Crystal Reed – SPARWAR (U.S.); Jasper Rodenhuis – ASML; Mary Salisbury – ManTech International; Stacy Speer – Raytheon; Paul Steinback – Defence (AU); Rick St. Germain – CMPIC Canada; Guy Stoot – DAF Trucks N.V.; Alejandra Sudbury – VENCORE; Theresa Sullivan – USAF; Tom Tesmer – CMstat; Michael Treadwell – Configuration Management Advocates; Bas van den Berg – Philips; Jos van Lieshout – Philips; Leon Waldron – Boeing (Retired); Ken Wallace – Sikorsky Aircraft Corporation; Joel Ward – L3 Com; Wolfgang Weiss – IBW; Donna A. Williams – U.S. Army RDECOM; and Jack Wasson – FAA Program Office (Retired).

Finally, we would like to say thank-you to our editors - John Wyzalek, Glenon Butler, Madhulika Jain, Laurel Sparrow, and all the fine people at Taylor & Francis Group/CRC Press, for believing in us and bringing the book to press.

Any errors due to translating and synergizing information from a multitude of source data, discussions, and personal experience are our own.

Acknowledgments

To our families we express love and appreciation. [illegible] would like to thank his wife Nancy and his son Jackson Quigley for their support. [illegible] Robitaille thanks his wife [illegible] Allison [illegible] for [illegible] encouragement, patience, understanding, and acceptance of long nights and weekends creating this edition and [illegible] completing the final supply [illegible] update [illegible] would have been [illegible].

We would [illegible] to take [illegible] out a big thank you to everyone who previously contributed to the first edition that formed the basis [illegible] for this second edition and to everyone who contributed directly [illegible] to this second edition. [illegible] and feedback [illegible] the updates and corrections contained herein. Contributions came [illegible] CM [illegible] email and [illegible] CMPIC [illegible] events, [illegible] CM [illegible] GEIA/SAE [illegible] Revision C [illegible] meeting with [illegible] for [illegible] CM [illegible] for the second [illegible] Additional contributions [illegible] the [illegible]

The contributors on this second edition are, in alphabetical order:

Bruno Amerdeo [illegible] David Armstrong [illegible] Boeing [illegible] Mike [illegible] Steve [illegible] Daniel K[illegible] Chris Clark [illegible] CM [illegible] David Garcia [illegible] University of [illegible] Robert [illegible]

Steven and Kelly Eisenbrook [illegible] Todd Goin [illegible] Lisa Hammock [illegible] Hauser [illegible] Paul [illegible] Technologies [illegible] Tom [illegible] Carol Magraine – Boeing [illegible] Cynthia Moody [illegible] Lockheed Martin [illegible] O'Dell [illegible] Pena [illegible] Jasper Koonmans – ASML; Mary Salisbury – ManTech [illegible] Paul Saniquet [illegible] Defence [illegible] NAVAIR [illegible] Teresa Sullivan [illegible] Configuration Management Associates [illegible] Phillips [illegible] Waldron [illegible] Retired [illegible] Wallace [illegible] Ward [illegible] Westcott [illegible] Retired

Finally, we would like to say thank you to our editors [illegible] Maddison [illegible] Sparrow [illegible] and all the fine people at Taylor & Francis Group CRC Press for believing in [illegible] of bringing the book to press.

Any errors due to translating and adapting the material from a more limited source base [illegible] our personal experience are our own.

Authors

Jon M. Quigley, Value Transformations, LLC, developed a deep affinity and appreciation for science and technology at a very early age. His fascination with the biodiversity and geologic history around him engendered his developing a talent in illustration. His detailed renderings reflect a keen eye and amazing fidelity as he translates what he sees onto whatever media he had at hand to share this wonder with others. He leveraged this curiosity into the study of biology and engineering topics that eventually resulted in a BSc in electronics engineering.

A believer in continuous learning and avid reader with Renaissance tendencies, he constantly explores those things that pique his curiosity. He has devoted his professional life to understanding as much of the product development process as possible. Jon has been awarded 7 US patents and earned two master's degrees along the way. His MBA in marketing is uniquely paired with an MSc in project management with an emphasis on informal communications.

Jon often states that, "The product team's ability to produce exceptional products—under budget and on schedule—must be tempered with sound engineering quality and safety processes, knowledge, skills, and business acumen if it is to succeed. Simply identifying and coordinating the actions required to make something does not ensure that it will meet requirements or be reproducible."

Jon has been a member of SAE International for more than 20 years and has been part of developing environmental, electrical system, communications, and architecture standards, as well as a member of the risk management task force. He is also a sought-after instructor in teaching risk management and the TIEMPO technique he developed.

In addition to holding the PMP and CTFL certifications, Jon has collaborated on multiple books as well as articles for more than 40 magazines and podcasts on a variety of product development, configuration management, testing, quality, and project management topics. He is a principal and founding member of Value Transformation LLC, a product development training and cost improvement organization. Value Transformation LLC was started so others will not have to experience the mistakes he has observed over a very long and productive career. Jon lives in North Carolina with his wife Nancy.

Kim L. Robertson has always been fascinated by the interconnectedness of things around him. He spent 10 years as a wildlife artist specializing in marquetry of the fauna of the western United States. His inlaid wood pictures contained thousands of individual pieces and hung in galleries in Park City, Utah; Jackson Hole, Wyoming; Scottsdale, Arizona; and Denver, Colorado. It was during this phase of his life that he began to explore product to marketplace interdependencies from an end-to-end supply chain perspective.

A BSc in mathematics paired with a second major in physical sciences led to a position as specification writer at Martin Marietta. This evolved into positions in change and then contract management. While at Martin Marietta, Kim earned an MA in organizational management with a subspecialty of contracts administration. His grasp of program infrastructure fast tracked him into the position of contracts policy and review manager for the Martin Marietta Launch Systems division. He was instrumental in migrating department policies and procedures from the paper world to a full digital format on the company's first intranet rollout. His master's thesis was on the effectiveness of Martin Marietta's configuration and contracts training program.

During an aerospace market slump, Kim leveraged his DOD, NASA, and commercial experience to become a contracting officer at the FDIC/RTC Denver office. He served in the source selection branch assisting with cleaning up the financial fallout resulting from the U.S. savings and loan failures. As the FDIC/RTC work ended, he transitioned to a position at the Rocky Flats Environmental Test Site in supply chain management assisting U.S. Department of Energy (DOE) prime contractors with the environmental remediation effort. When offered a full-time position on commercial and NASA aerospace programs at Ball Aerospace & Technologies Corp., Kim accepted. He is currently the C&DM department policy and review chair and C&DM manager for various NASA instrument programs.

Kim holds a CMPIC Configuration Management Principles and Implementation certification and is a National Defense Industrial Association (NDIA) certified configuration data manager (CCDM). He is a member of the SAE International G-33 committee for CM Standards, National Defense Industrial Association (NDIA), and the Association for Configuration and Data Management (ACDM). He provides group training on a variety of CM related subjects and does assessments of CM implementation. He can be reached via LinkedIn.

Kim lives in Colorado with his wife, Dee, and dogs Rufus and Grizzly.

1 Overview of the Product Life Cycle

1.1 QUESTIONS ANSWERED IN THIS CHAPTER

- What are the five phases of product development?
  - What are the key elements of product development?
  - What are concerns associated with these areas?
- What are the input–output constraints and facilitators of the configuration management process?
  - What are the input–output key elements?
- What are the differences between product market adaptation and product performance adaptation?
- How are the seven deadly sins used in marketing?
- How does cost and the influence of critical design decisions change over time?
- What are configuration market adaptations?
- What examples of product adaptive ration exist?
- What is the difference between design reuse and common design?
- What are the design reuse and common design key elements?
- What criteria are involved in defining a numbering schema?
- What is meant by a cost objective?

1.2 INTRODUCTION

Many separate pieces of information relative to configuration management (CM) exist. They are contained in specifications and handbooks from multiple nations, industries, and groups with a focus on one market sector or another. Each can be referenced for the specifics of this thing known as CM. The current guidance documentation describing configuration management consists of the following seven elements.

1. CM planning
2. Management of configurations
3. Configuration identification
4. Configuration change management
5. Configuration status accounting
6. Configuration verification
7. Configuration audit

These seven elements are based on what we have found across all market segments. The list is not consistent with guidance put forth by the International Organization for Standards [ISO] or Electronic Industries Alliance [EIA], who only list five elements These standards combine CM planning and management and configuration verification and audit. Our seven elements are listed for clarity.

Practitioners of CM have been so well trained and the definitions for the terms used in their art have such specific meanings that at times they lose focus on the management theory that resulted in the documentation they know so well and cite chapter and verse. CM and its sub-element data management (DM) did not spring fully formed from the military minds of the post–World War II era around 1960. They evolved over time and all have interesting back stories. Reproducibility of products refined to a level where part interchangeability was possible existed as early as 210 BCE. Specifications, drawings, and ledgers existed as far back as 2200 BCE.

At other times, the definitions so methodically drafted, vetted, and released to the public over the last 40–60 years are not an exact fit with the issues faced today. Words and concepts become muddled in the current context. They were originally created to define not the *how* or *why* something is done but rather *what* had to be done in a primarily paper-based engineering environment. Companies have been creative about leveraging the words in the specifications and requirements to have a broader—albeit still limited—scope. It becomes more complicated because word meanings morph over time. The word *bald* is an example; today it is taken to mean "without hair." Historically it meant "white"; thus, we have bald eagle, piebald goat, or dog with a bald face. Taken out of context the statement, "He appeared like a Norse god as he stepped before the gathered assembly … having gone bald with the passage of time; his thick flowing hair cascaded over his shoulders like drifting snow" appears to be inconsistent. Yet, at the time it was written there was no dichotomy.

The CM requirements documentation suite has been revised mostly as a matter of necessity. It has been accomplished with delays due to the internal checks and balances required to administer specifications and standards that have application over multiple branches of a country's military, space, nuclear regulatory, or other government *influenced* infrastructure. Some standards like SAE/EIA 649, *Configuration Management Standard*[1] try to refresh every five years. Others like MIL-STD-973, *Military Standard Configuration Management*,[2] canceled 30 September 2000, still find their way into contracts as requirements documents or are referenced in some versions of defense documents like the DD Form-250, *Material Inspection and Receiving Report (MIRR).*

The practitioners of CM we conversed with were found to be dedicated, productive, competent, and passionate. They know the definition of every acronym, word, and form specified for their management specialty. They perform CM activities to the best of their abilities given the woefully inadequate software tools at their disposal. These take the form of everything from home grown databases running on Unix or Linux, server-based databases like dBase or Microsoft Access, CM modules attached as an afterthought to engineering requirements planning systems, and in some cases spreadsheets and ledgers.

Unfortunately, few CM practitioners interviewed had made the leap from a market sector specific implementation to a total life cycle view of all market sectors. CM implementation is inclusive of all phases of a product or service's life cycle from the end user, contractor, and subcontractor perspectives. The myopic view of market sector specific CM has led to a limited understanding of the implications of poor CM implementation as well as a lack of understanding of how it applies to the world marketplace. It has also led to a tendency for practitioners of CM in industries specializing in sales to government to discount CM practices, as they apply to all other market sectors. This is unfortunate as companies in the non-government sales market sector have developed CM tools and processes that are decades ahead of those used in the government sales market sector.

Few people alive today were part of the committees of experts gathered together to generate the requirements or can explain their thinking when the requirements were first created. It is time to look back to where the elements of CM came from, see where we are today, and find how this contextual history can help plot the course forward. Despite extensive research, neither author found any existing text that answered these basic questions pertaining to CM:

- What is the history?
- Why are we where we are now?

---

[1] SAE. (2019). SAE/EIA 649-C-2019-02, Configuration Management Standard. SAE International. Warrendale, Pennsylvania.

[2] DOD, U.S. (1992). *MIL-STD-973, Configuration Management.* U.S. Government.

- Where are we going?
- Why is product and service standardization an outcome of CM?

We started on a quest to find out the answers to these questions. It has directed us down through the ages to answers very different from those we anticipated and rooted in antiquity. Unexpected sources, twists, and turns made the gathering of source data more of an archeology expedition, leading us to a degree of self-discovery along the journey. Our research was aided by many who were involved in CM communities on LinkedIn, friends, colleagues, subject matter experts, deep introspection, discussions, visits to 23 countries, and reading of related materials.

We have attempted to sift through the material and present it in phases to make it easier to digest. Some may be eager to delve directly into what they see as the meat of CM, and may find that our presentation of the subject based on its evolution misses the point of having a book with CM in its title. Our intent is first to ground the reader in management theory from a configuration-centric perspective before discussing the specifics of CM practices.

One of the principal roles of CM is to protect us from ourselves. In so doing, we also protect our suppliers and our customers. When high school or college students first encounter bills of materials or find that the fast food industry is controlled by processes, they experience shock and awe. An automotive wire harness with 368 leads must have a bill of materials with at least 368 components (Figure 1.1; some will be common, but we must still account for their presence), associated connectors, solder, anti-fretting compound, and so on. Every possible component and process falls under CM.

For those unfamiliar with processes, Anne Mette Jonassen Hass[3] says this: "An analogy is sometimes made between a process description and a recipe. That analogy is not far off. A recipe includes a list of ingredients (inputs), workflow, and a description—maybe even a picture—of the expected result. The recipe may also include information about the skills required to follow the instructions, suggestions for tools to use, expected time to produce the result, and the expected calorie count." Processes and procedures exhibit many of these same characteristics, with inputs, workflow, and outputs.

Regulations impacting CM come from extraordinary sources. The evolutionary context of the well-known U.S. Federal Acquisitions Regulations clause 52.237-2, Protection of Government Buildings, Equipment, and Vegetation, is an example. Veteran aerospace leader Marvin "Marv" D. VanderWeg, past vice-president of contracts at United Launch Alliance and later SpaceX, relayed

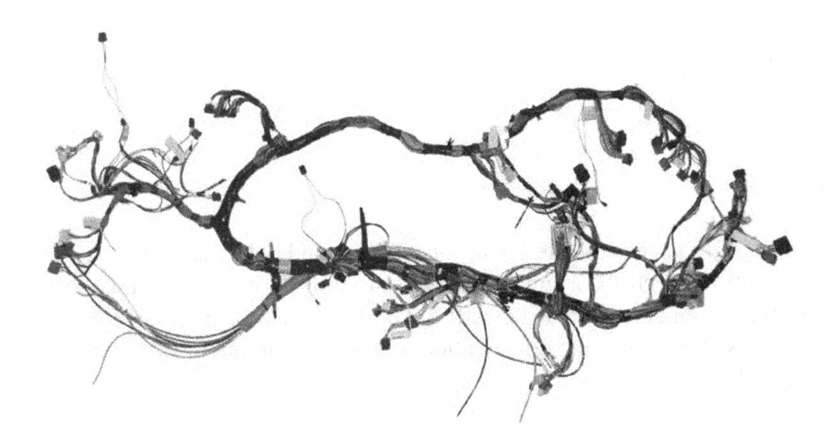

FIGURE 1.1 Wire harness—typical.

[3] Hass, A. M. J. (2003), *Configuration Management Principles and Practice*, Addison-Wesley Pearson Education, Inc. 75 Arlington Street, Suite 300, Boston, MA 02216.

the story to Kim Robertson. According to Marv, an intoxicated trash removal truck driver ran over a base commander's wife's flower bed. The base commander, whom Marv knew personally, found that he had no legal recourse against the trash removal company for the damage. As a result, he fought to get 52.237-2 written. It is now a standard clause in U.S. government contracts.

The ultimate result of CM is product and service standardization, such as the acceptance of video home system (VHS) over Betamax for home video in the 1980s. Standardization through CM has been going on for centuries. Screw threads are a prime example. Starting with intracompany standardization by pioneers such as Henry Maudslay in the early 1800s, it eventually spread to industry worldwide.

CM implementation is mandated in every business improvement, product safety, and quality process at a low level before going to higher levels such as Capability Maturity Model® Integration®, (CMMI), Information Technology Infrastructure Library, ISO series (i.e., ISO 20000), systems engineering, Six Sigma, and total quality management.

What is difficult is to try and develop items and services that are not ultimately CM intensive. Perhaps the ephemera we call *thoughts*, *feelings*, and *emotions* fall into this category. Of course, CM implementation does not necessarily track everything, because not everything has equal importance. A readily replaceable item such as printer paper is probably very low priority (the recipe for making the paper would be under DM and intellectual property controls); however, any item that is part of a product or service or related to it must fall under CM.

We have seen companies deliver products with unknown software, unknown hardware, and unknown features to customers for evaluation. We have also seen services performed in such a way that there is little consistency in the end results. Every time we see such rushed events occur, it has resulted in a marketing fiasco. So, what is the moral of the CM story? It is the necessity to determine those things that are important and establish realistic controls, so that we always know what we are working with and can reliably and consistently provide a service or build/make the product—the key point is that it must be done safely, with repeatability, and with the highest quality.

We begin the odyssey with the product life cycle, which is critical to understanding how CM activities change over time. Contractual documentation under DM and change control, as well as each of the seven elements that make up CM of a product, are directly influenced by the product life cycle phases.

1.3 PHASES

The five product phases are:

- Development
- Introduction
- Growth
- Maturity
- Decline

Each of these areas has an associated range of activities and risks. Each requires the entire organization to accomplish objectives or supply inputs to ensure the success of that phase and maximize the benefits to the organization. Later we will break down each of these areas and discuss what happens, what types of activities the organization will perform to maximize market potential, and how to mitigate risk.

1.3.1 Development

In the development phase of the product development life cycle, you are generating ideas for the product or service you wish to develop. You see opportunities in the marketplace and wish to explore whether you can capitalize (not always a vulgar word) upon these opportunities. Perhaps innovative technology has become available that facilitates product improvements or cost reduction. You may want to investigate the application of that technology within your existing businesses. Maybe you see a new market developing. Perhaps you are evaluating the product or service from an entirely new application perspective. At any rate, this phase starts off by idea generation and rigorous critique of those ideas to guide the development of the product or service (or to eliminate the idea for the product or service).

This critique is facilitated via initial prototyping and reviews of the possible design solutions. Concurrently with the development of the product, you will define your manufacturing capabilities or production flows and review that development work as well. All of this will guide your design or final configuration and manufacturing processes.

You will also reach out to your prospective customers for their comments, concerns, and acceptance. Their input will make it possible to generate a business case that moves beyond mythical to a tangible need. The business case helps you understand the design constraints, scope, and schedule needed to achieve the desired profitability margin. This profit calculation is based on the following:

- The volume of customer mandated testing you can expect.
- The cost to make and deliver the product (includes manufacturing, warehousing, and delivery).
- The date the product or service can be brought to market.
- The selling price of the product.

The selling price is also derived from the volume calculation as there will be some volume implication due to the customer sale price (higher price can reduce number of customers). You will garner as much data as you can via market testing to arrive at this optimum selling price. Ultimately, at the end of development, you have a product that you will move on to the next step—introducing the product to the marketplace in earnest (as opposed to learning what the customers think and the price the market will tolerate).

In the development phase, you explore the variety of product adaptations or permutations that meet the market objective. As this definition process continues, you begin to understand the scope of each variation and how they drive the design, costs, and return on investment. Some examples of the things you may want to control during the development phase are:

- Concepts that are under consideration for the product
- Product hardware variation needs as defined by the market (customer)
- Product software variation as defined by the market (customer)
    - Parameters
- Design failure modes effects analysis
- Design documentation (including changes to)
    - Producibility
    - Safety
    - Software
    - Hardware
    - Test cases
- Prototype material throughout the development process
- Prototype software throughout the development process

Neglecting to control the configuration and associated data during the development process will cause a litany of problems. Without CM implementation at this phase, you may not know exactly what you have built up until the day of production launch. You may find it difficult to learn and make definitive statements about the product without a chain of controlling events to confirm the truth of those statements.

1.3.2 Introduction

You have developed a product or service that was based on market research and perceived opportunity. You have tested that product both from the market perspective and from the safety, quality, and customer expectations and your own business requirements. Now it is time to introduce the resultant product to your clientele.

Because you now have something to sell, your plans for introduction will now be executed. If this product or service has some similarities to the rest of the items in your product line, then you will likely rely on those existing distribution mechanisms.

You may discover the product being used in conditions you did not envision. This requires you to remediate safety concerns through taking direct action. As an extreme and a bit flippant example, say you were not aware that customers would use your hair dryer while bathing.[4] You are required by law to attach a product safety warning that such use could result in electrocution and death.

You have not finished learning about the product even in this phase of the life cycle. Your product development process may have called for such manufacturing development activities to be under configuration control.

- Process studies
- Measurement system assessment
- Process failure mode effects analysis
- Control plans

However, these are experientially based and theoretical exercises. You have not likely produced a product at the rate you forecast in your projected delivery volumes. Even if you have previously produced product quantities or services you believe represent the expected volume, you have not done it recently. As you ramp up to full production volumes, process improvement opportunities and constraints will surface. These improvements can come from any part of the supply chain.

1.3.3 Growth

You have developed a product or service that is useful to your clients. Your production volumes grow, allowing you to improve the product or service cost and pricing structures. Besides cost improvements due to volume purchasing, you can work with your service providers and manufacturing line to reduce rework costs. You may choose to steadily drive your price down also, to increase the number of consumers. It is possible to improve quality and safety while reducing the price of the product or service to improve the profit. This new market ownership situation is unlikely to last as other potential suppliers enter the market with their own offerings.

During the growth phase of the product or service, you may find other adaptations for the product or service and CM considerations will once again take center stage. You may elect to reinvest some of the profits into the product or service to meet consumer demands that cannot be realized through price reduction. Small adaptations to the product or service may also allow you to enter untapped market segments.

Requirements and even product concept of operations should begin with the documentation process so that product development and refinement is known, quantified, and verified. One of the biggest issues we found was uncontrolled changes to requirements leading to loss of any tie between the technical requirements and allocation of those requirements to services, systems, and sub-systems. Product or service development and refinement must be planned, and configuration planning is the most important aspect of a successful CM implementation.

[4] http://answers.google.com/answers/threadview/id/738460.html accessed 2019-01-07

CM planning should describe interdepartmental responsibilities, what data is to be recorded and controlled, what contractual constraints and opportunities are in play, what each department gives and receives from other departments, who makes up the configuration control board, what falls under change management, how decisions are made, CM department roles and responsibilities for each product and, at the enterprise level, how and where data is stored/managed, the CM standards used, how interface control decisions are accomplished, data disaster recovery, and much more. Without proper CM planning, product reproducibility, quality, and safety are at risk.

Managing the configuration is essential to determining how comparable products or services are differentiated and brought to the marketplace. If the product has software components, you may be able to radically change the performance or the features to alter product performance instead of incurring hardware costs. For example, you can have a platform that operates at one speed, and then through software parameter assignments increase the speed of performance and market the faster model at a higher price.

1.3.4 Maturity

Eventually, product sales will start to decline. Though sales growth is still positive, it is at a lower rate of growth (less steep) as the product enters the maturity phase in its product life cycle. The market is becoming saturated; perhaps there are newer versions or competing products or services from other suppliers. Projections show your product is nearing the point in the life cycle where it is no longer commercially viable. You should be planning a product refresh or replacement. You can extend the maturity phase, perhaps, with some form of "What is old is new again" marketing scheme. You may be lucky enough to be featured in a movie (like the VW Beetle was with *Herbie the Love Bug*) or be trending as a new retro look in Los Angeles and New York's SoHo districts (like the Hush Puppies Shoes®). Though your production and distribution processes can be quite mature, there is always the possibility of stemming the rate of decline via cost-improvement techniques.

- Design to produce
- Lean manufacturing
- Total quality management
- Six Sigma
- Product tear downs

All these activities can help you understand where your costs can be better controlled, thus retaining your profit margins in the face of sales volume reductions. These techniques have a range of sophistication or complexity that any organization can use to ascertain and improve its cost structure. Nevertheless, you will eventually see product sales steadily decline until you reach the point where there is no growth, the harbinger of the decline phase.

1.3.5 Decline

Sales continue to decline, and you are losing customers and market share. You still may be able to eke out a profit, but you will have to work especially hard doing so. If you were forward thinking, you likely have a replacement ready to market, which will restart the entire process all over again. You may maintain some replacement parts for the product already sold, but your manufacturing processes will be ramping down. If you did not believe you could produce a new widget that meets customer demands or needs, you may ultimately discontinue the product model line. This was seen recently in the U.S. automotive sector with the discontinuance of Eagle, Hummer, Mercury, Oldsmobile, Plymouth, Pontiac, Saturn, and Scion. Another alternative is to sell the division producing the product, as was done in 2016 when General Electric sold its appliance division to Haier for €4,700,000,000.

1.3.6 After the Decline

If you have been developing a replacement product to be introduced in parallel prior to your present product's demise, you may find opportunities for improving the associated development costs based on your aged product. Perhaps your replacement product will have the same envelope and enclosure geometry. The new iteration may fall in the adaptive radiation category (discussed in Section 1.5.2.1). In that case, you can effectively employ some of your existing tools for the next generation of the product.

1.4 CM AND THE DYNAMICS OF CHANGE

CM, as practiced in different enterprises, may concentrate on a single aspect of the product life cycle or the product life cycle itself as viewed from the perspective of one piece of the total market segment. This results in narrowing down the understanding of how configurations are managed as well as why they are managed. The enterprise may be the customer, or it may play the role of a vendor as part of the supply chain. The focus of CM implementation at each lower supply chain level becomes more removed from the top-level CM implementation for the end item or service. It is critical that the configuration management approach encompass the entire scope of the life cycle, as well as the scope of the place in the supply chain the enterprise supports, and the dynamics that are inherently involved throughout a product life cycle. Providers of components, mechanisms, assemblies, and other items delivered to the end user all play a critical part in the overall system.

The more each provider of services or products understands about the CM process and product planning, designing, and manufacture, or configuration architecture of the whole, the better the product or service becomes. To truly understand CM, it must be viewed in relationship to the business infrastructure. A rudimentary grasp of the functional resources within an organization and their role in CM implementation, as well as basic concepts regarding primary functions of those roles, provides managers and engineers the perspective required to succeed. The following discussions as well as those in the next two sections pertain to CM implementation as applied to the entire enterprise infrastructure rather than to DM, records management, and release management.

The CM department in many enterprises performs only a subset of the entirety of CM due to variations in organizational structure. Many who make their careers in CM also define it along this same narrow subset perspective. This is like one chemistry lab assistant's explanation to Kim Robertson that chemistry is taking chemicals from large drums, placing them into smaller containers so students can mix them together, and then disposing of the new mixture in a safe manner.

The CM process, as seen by the U.S. Department of Defense (DOD), starts with the identification of a new need followed by mission analysis. This flow, shown in Figure 1.2, is typical of the government market segment. It is, however, only part of the story as defined by the U.S. DOD and is written as guidance for government personnel. It is a minimalistic view of what the government must do and what the government expects the contractor to do. Viewed another way, it is nothing more than a choreographed script for buyers and sellers in a specific market segment, where the government controls the environment the sellers exist in. This artificial environment rarely exists in the non-government sales market sectors.

The commercial world is governed not only by market change dynamics that greatly influence CM considerations, but also the viability and sustainability of products and the infrastructure required to market and support them. This is a much broader view of goods and services driven by logistical necessity or by response to international events than CM strictly tied to government considerations. It is critical that CM be implemented with a firm grasp of market stability, what drives it, and how it relates to the product life cycles.

Some may find it easier to grasp many aspects of change dynamics from an ecological viewpoint. In many cases, there is a direct corollary between market sectors and the predator–prey relationships. Ecological interactions occur between organisms and their environment. They also

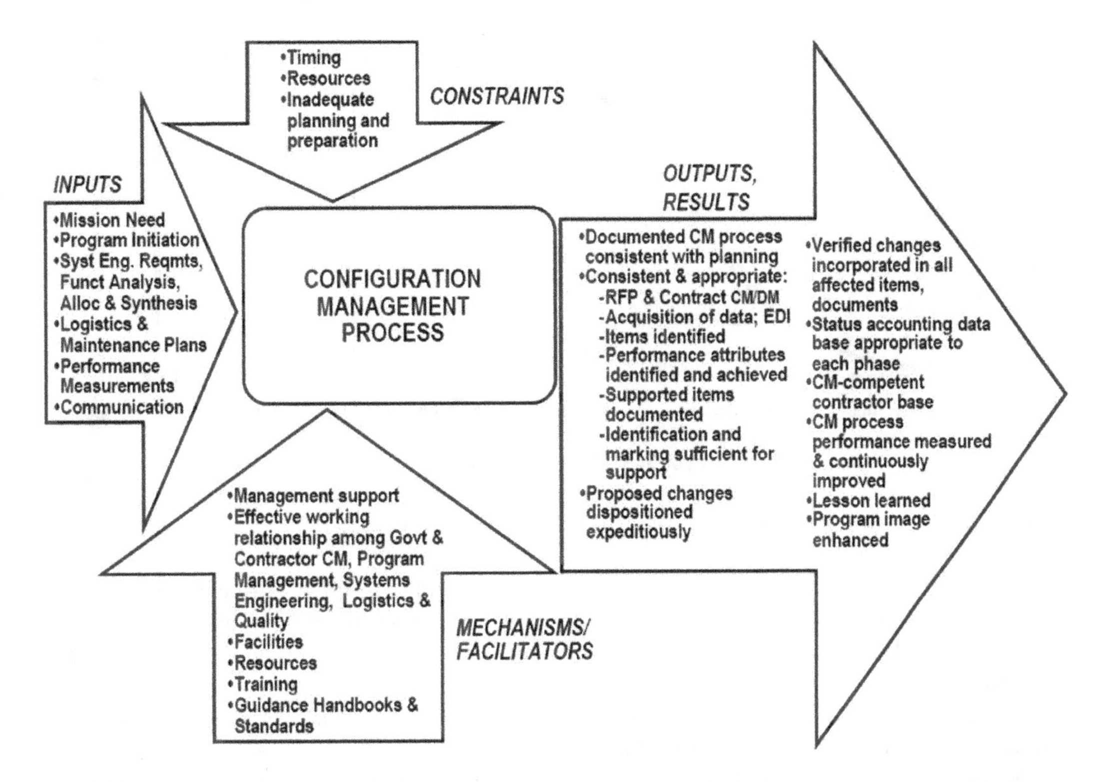

FIGURE 1.2 U.S. DOD configuration management process. (Data from DOD, U.S., MIL-HDBK-61a, Military Handbook: Configuration Management Guidance, Federal Government, U.S. Figure 1.2, 2013.)

occur between species trying to exist in the same environmental niche. Allelopathic trees like black walnut and eucalyptus release chemicals that inhibit the growth of other species. Predator–prey relationships can also develop that are beneficial to both species. The weaker species members of the prey are not allowed to reproduce or forage and the weaker predators starve. Both are symbiotic within their ecological niche. Unique defensive and offensive strategies develop with each species.

Symbiotic relationships develop as well where both species benefit. This creates a case of logistics growth plus an increase in population growth of one species because of the presence of greater numbers of another species. The Lotka–Volterra[5] equations (Equation 1.1) form a simple framework for modeling this kind of species relationship. Richard Goodwin utilized similar equations in 1965 to model the dynamics of various industries.[6] Assume that N and M are population densities, R = intrinsic growth rate, K = carrying capacity of the local environment, and β = coefficient converting encounters of one species with a new member of another.

$$\frac{dN}{dt} = r_1 N\left(1 - \frac{N}{K1} + \beta_{12}\left(\frac{M}{K_1}\right)\right)$$

$$\frac{dM}{dt} = r_2 M\left(1 - \frac{M}{K2} + \beta_{21}\left(\frac{N}{K_2}\right)\right)$$

EQUATION 1.1 The Lotka–Volterra equation for species dependency.

[5] Lotka, A. J. (1925). *Elements of Physical Biology.* Baltimore, MD: Lippincott Williams & Wilkins.

[6] Goodwin, R. M. (1967). *A Growth Cycle: Socialism, Capitalism and Economic Growth*, Feinstein, C.H. (ed.). Cambridge: Cambridge University Press.

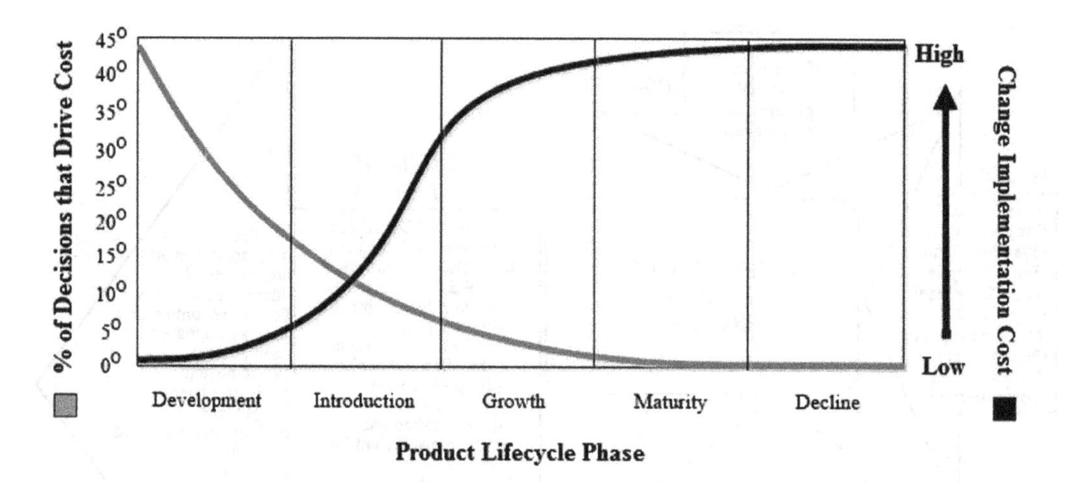

FIGURE 1.3 Cost and influence of critical design decisions plotted over time.

Sayings like, *We're going to eat their lunch!* and *They're eating us alive!* have existed in boardrooms for a long time. Both pay tribute to the ecological similarities of predator–prey and business survivability conditions. Wherever possible, the ecological as well as the systematic view will be discussed.

Each phase of the product life cycle brings different dynamics into play as the product interacts with the marketplace. Each phase will influence the management of the product's configuration. A sound grasp of CM theory and practice is required to minimize the product to market churn to reflect better earnings before interest and taxes (EBIT). Understanding the financial aspects requires some formulas and we promise to keep them to a minimum.

CM activities shift as a product matures from guiding decisions relative to design and ease of logistics activities to activities that assure production repeatability. Thirty years of experience on over 25 different programs ranging in value from €21,466,900 to €17,470,000,000 and 270 projects ranging in size from €4,366 to €78,601 has shown that 80% of the decisions driving cost are made in the product development phase of the product life cycle. It has also shown that the further along a product is in the life cycle, the more expensive design changes become. There was less than a 1.5% variance over the sample size. Figure 1.3 graphically depicts these findings.

The environment in which an organism or a product finds itself changes over time. Most think of the Sahara as a dry, foreboding expanse with life crowded along the Nile River. This has not always been the case. Stefan Kröpelin of the University of Cologne in Germany observed, "The climate change at (10,500 years ago) which turned most of the (8.4 million square meter) large Sahara into a savannah-type environment happened within a few hundred years only, certainly within less than 500 years."[7] Organisms suited for the savannah-type Sahara were not as successful in the dry periods that bracketed it.

Environmental climate change is not the only factor that influences the types and density of organisms within the environment. A single organism may thrive in a relatively unchanging environment only to be displaced by another that is better suited to thrive in that ecological niche.

Manufactured items that are viable today can become non-viable in the future for similar reasons. They are dependent on the economic climate change and competition for the same market niche. To remain viable, an enterprise must be able to adapt and modify its products and services, marketing strategy, and management of its approach using CM of its product and production infrastructure. As innovative technologies emerge, older ones cease to be supported, in much the same way as wolves

[7] Bjorn Carey, 2006. *Sahara Desert Was Once Lush and Populated.* July 20. http://www.livescience.com/4180-sahara-desert-lush-populated.html (accessed September 20, 2014).

survive in the tundra. The expenditure of energy for the nutrition gained is greater during the migration of caribou as a teaming effort with other wolves. Yet, the wolf is content in summer months to rely on a diet, in some cases, comprised predominantly of voles and other small prey.

Cyclic demand for products and product features also ebbs and flows with economic changes. Business plans that relied heavily on one external dependency, such as transportation of goods from one location, may find the model less viable than one relying on regional distribution or in some cases co-location with the end user or user community.

1.4.1 Survivability Conditions

Organisms and products/services can survive only if several conditions exist.

- Nature
  - The physical environment is suitable (food, climate, and predators).
  - The competitive environment is suitable (few or no competitors).
  - Reproduction rate exceeds lifespan (its reproduction rate exceeds its mortality rate).
  - High dispersal rate (its dispersal rate allows it to exploit the entire niche).
  - Adaptability (it has an ability to adapt to changing conditions, giving it advantages over the current niche holder).
- Products
  - The physical environment is suitable (raw materials, technology, fabrication capability, and competitors).
  - The competitive environment is suitable (few or no competitors and it is meeting a perceived or real need).
  - Reproduction rate exceeds lifespan (purchases exceed product end of life).
  - High dispersal rate (marketing and distribution channels push out competitors).
  - Adaptability (it has an ability to adapt to changing market needs, giving it advantages over the current niche holder).

Two main survival strategies come into play when making the decision to enter a market segment: product market adaptation and product performance adaptation. Both require an understanding and synergy of the regional, national, international, and cultural norms that influence how and where a product will be used. Both achieve the net result of increased market share in that segment.

Marketing can often make or break a product. In its simplest form, it relies on captivating the customer's attention using appeals to one or more of the seven deadly sins mentioned by Dante Alighieri.[8] Any evening spent watching commercial television in the U.S. will result in one or more "shouty" men or women selling products or services through one of these means.

1. Luxuria (lechery/lust)
   a. If you use this product, you'll get the guy/girl/house/vacation of your dreams!
2. Gula (gluttony)
   a. Now offered in six exciting colors … collect them all. Just pay separate shipping and handling fees.
3. Avaritia (avarice/greed)
   a. I made enough using this app to pay for my new car and had more than I started with when I was done!

[8] Alighieri, D. (d. 1321). *The Divine Comedy: Complete the Vision of Paradise, Purgatory and Hell*, translated by Rev. H. F. Cary. Salt Lake City, UT: Project Gutenberg eBook. November 30, 2012. http://www.gutenberg.org/files/8800/8800-h/8800-h.htm.

4. Acedia (sloth/discouragement)
    a. Tired of using the remote or trying to find your favorite radio channel? Just tell your smart device what you are looking for and let it find it for you!
5. Ira (wrath)
    a. Hate weeds? Get rid of them now with Zap 'Em now!
6. Invidia (envy)
    a. The beauty secrets known only by the stars are now yours!
    b. You'll have the nicest landscaping in your neighborhood!
7. Superbia (pride)
    a. You deserve it!
    b. My dog is better than yours!

Each decision made regarding the product's configuration allows it to be marketed in one of these seven ways. Each decision made regarding the product's configuration also allows your competitors to develop a counter marketing strategy to combat against it. There is an additional marketing sin not considered by Dante that must be added.

8. Metus (Fear)
    a. All your appliances will eventually die, so buy appliance insurance now!

The following case study concerns the opening of a new market segment known as the "horseless carriage." Here, you will find every survivability condition required for creation of a market sector.

1.4.2 Evolution of the Horseless Carriage

New market segments generally open due to the need to solve a problem. Such problems may be real (as in the case of the environmental crisis solved by the automobile) or the need may be contrived. New markets and products are rarely developed through the inspiration of a single individual. The automotive market came about through a synergy of the existing body of knowledge and other *environmental* conditions both in the marketplace and in nature.

One topic of discussion at the world's first international urban planning conference in 1898 was growing health concerns due to horse excretions and the death of creatures that accompanied them (animals that died in harness were often left in the street where they expired). As the primary means of locomotion for wagons and other forms of transport, horse populations exceeded human population in cities.[9]

Manufacturers of "horseless carriages" using steam, electric, and internal combustion engines as the motive force attempted to meet the challenge. Gasoline was found unsuitable at first as a fuel for combustion engines, so Henry Ford's Model T was designed to use ethanol. Steam-powered and electric-powered vehicles held an early advantage but were eclipsed in the market by the internal combustion engine. Nicolas Joseph Cugnot built the first steam tractor in 1769. By 1881, Amédée-Ernest Bollée's "La Rapide" achieved the impressive speed of 62 km/h. Electric cars improved to the point that in 1899 Camille Jénatzy's electric automobile "La Jamais Contente" set a world record for land speed of 100 km/h.

The success of the internal combustion-powered vehicle is no different than three species competing for the same ecosystem, where one develops an advantage over the other two. Eventually, the species with the greatest advantage gains a dominant position in the ecological environment. A similar struggle in the automotive environment is going on today. Due to changes in the Earth's environment, ethanol and electric are again in competition or being combined with the internal combustion engine as viable competitors to gas-powered motors in the automotive market segment.

[9] Morris, E. (2007). *From Horse Power to Horsepower.* Berkeley, CA: University of California Transportation Center—ACCESS number 30.

1.4.2.1 A Look at the Converging Technologies

The synergy of converging technologies, processes, and components leading up to Henry Ford's moving production line facilitating the internal combustion engine automobile's dominance has a long and often surprising history. Metalwork and metallurgy were critical components of the success of the horseless carriage. Forging of metals requires heating the metal to a malleable temperature, placing it on a solid object, and striking it with another solid object. Finding a way to do this rapidly while allowing the smith to move the metal being forged was critical to rapid production of tools and weapons. Animal and water power were first harnessed to do this and led to the development of the helve hammer mill.

Evidence suggests that the helve hammer mill may have existed in China as early as 1050 BCE. If proven out, this predates the Roman use by 1,600 years or more. Hammer mills of this kind are thought to have been used in 210 BCE to manufacture the weaponry for the estimated 8,000 terracotta warriors found at the burial site of Qin Shi Huang, the first emperor of China.

Yang Fuxi, a 10th-generation maker of ancient crossbows, states that a crossbow trigger found at the site contained parts interchangeable with one found 3218.69 km away. This supports statements by archeologist Liu Zhangcheng that 30,000 arrowheads in the pits were tested and the difference in their sizes was less than 1 mm. Some evidenced a difference of 0.22 mm. We know from this that weapon manufacturing was standardized.[10] A select sample of these arrowheads is shown in Figure 1.4.

Strict control of the configuration, manufacturing process, and quality management were required to achieve such a high degree of uniformity. It is unknown if production methods included checking the arrowheads against a master gauge for uniformity, perhaps ceramic as a half-mold from a master arrowhead. Evidence also exists that metallurgy was known and practiced using a blend of copper, tin, nickel, magnesium, and cobalt.[11] Bronze swords found at the site were also found to be coated with a layer of chromium oxide that prevented the weapons from oxidizing for 2000 years.[12] This led to speculation that sophisticated surface coating technology existed at the time. It is now believed the chromium oxide coating is residue from painted wooden weapon handles long since decomposed.

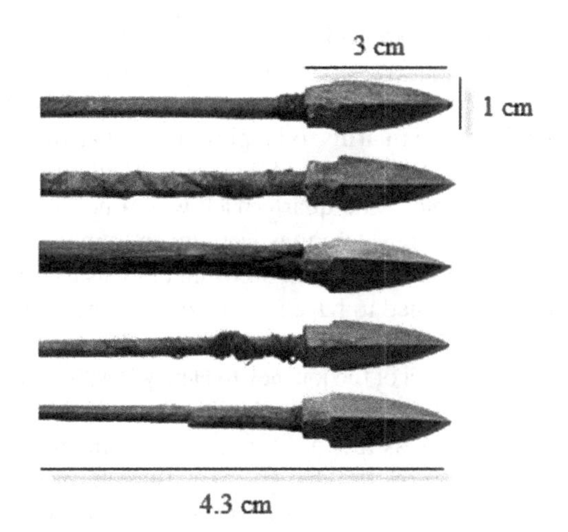

FIGURE 1.4 Bronze arrowheads 210 BCE, China.

[10] Talley, S. (Director). (2011). *Secrets of the Dead: China's Terracotta Warriors*. [Motion Picture]. PBS.org.

[11] *New Terra-Cotta Warriors Dig Begins in China. (2009). National Geographic (AP).* http://news.nationalgeographic.com/news/2009/06/090615-terracotta-excavate-video-ap.html

[12] Zhewen, L. (1993). *China's Imperial Tombs and Mausoleums*. Beijing, China: Foreign Languages Press. p. 216.

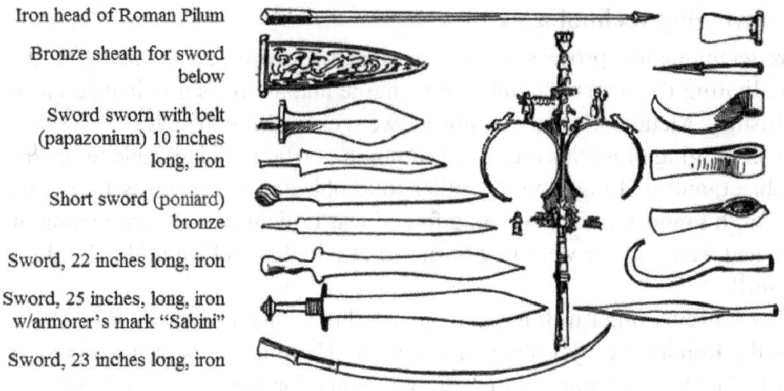

FIGURE 1.5 Roman weapons and descriptions. (From Demmin, A., *An Illustrated History of Arms and Armour: From the Earliest Period to the Present Time*, translated by C. C. Black. M. A. G. Bell and Sons, London, 1911. The figure is a composite with each of the illustrations taken from a different page of the book. https://archive.org/details/illustratedhisto00demmrich. The book was accessed on June 22, 2014. With permission.)

Moving forward in time to Rome, we find specifications for iron spearheads (Figure 1.5). Consistency in spearhead size, weight, and material would have been deemed invaluable in the Roman empire. Without them, it would have been very hard to have the efficient fighting force that Rome required to manage its provinces.

Each legionnaire was equipped with a *gladius* (a short sword about 46 cm long), a *pugio* (a dagger 18–28 cm long), and one or more *pila* (a 2 m long javelin). Of these three items, the gladius-like swords may have varied in weight and perhaps balance depending on the reach of arm and arm strength versus age and personal preference of the legionnaire. The pugio may have varied in length and configuration also based on the preference and perhaps birth origin of the legionnaire. It is conceivable that the pilum was controlled to very strict standards for efficiency in battle if nothing else.

The pilum was designed to be thrown and stick into an enemy's shield, making the shield too unwieldy to use for defense. Since this weapon was used by every legionnaire, standardizing length, weight, and center of gravity would have held distinct advantages. Standardization would have allowed the pilum of a fallen comrade in arms to be picked up by anyone and thrown. Imagine the difficulties involved if today's gladiatorial sport of American football used balls of differing size, weight, and center of gravity each play. The quarterback would never be able to properly train or stay "in the zone." Plausibly, similar considerations were given to manufacture of the pilum for the same reasons. The mass of the light pilum of the Augustan era was about 2.2 kg.[13] The weight and center of gravity of the pilum is assumed to have remained constant enough to allow it to meet the objective of effectively engaging the opponent at a range of 10–15 m.

The forging of metal is but one aspect of the journey to Henry Ford's production line. Development of fuels was another. In the period 1800–1900, the United States relied on coal, natural gas, camphene, and white naphtha (also known as fuel oil) for heating. White naphtha was also used as fuel for lamps as a less expensive alternative to whale oil. Distillation of naphtha from oil was documented in the *Book of Secrets* in the ninth century.[14] By-products of distillation included gasoline (Figure 1.6) and others, which all went unused until the 1900s. Product entries into an existing market segment at times are based on utilization of by-products that no one else has found a use for. Often, by-products that had no known use were disposed of in local rivers and lakes. At times,

[13] Walbank, F. W. (1957). *A Historical Commentary on Polybius*, 3 vols. Oxford: 1999 Special Edition Clarendon Press.

[14] Taylor, G. B. (2008). *Al-Razi's "Book of Secrets": The Practical Laboratory in the Medieval Islamic World.* California State University, Fullerton, CA.

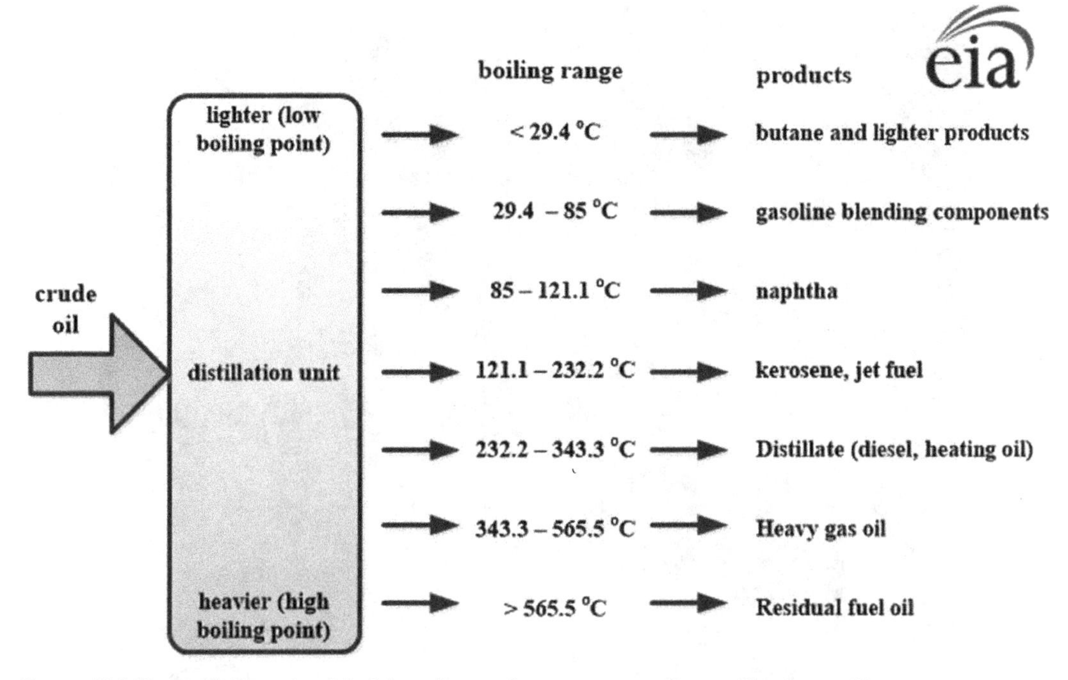

FIGURE 1.6 Crude oil by-products and distillation temperatures. (Data from U.S. Energy Information Administration—redrawn to convert degrees Fahrenheit to degrees Celsius.)

people would see if the pollutants would burn and were taken by surprise when the lake caught fire; in his youth, Earl Ray "E. R." Miller was one such experimenter.[15]

Each of the other components utilized in the final vehicle has a similar evolutionary history. Henry Ford built on a long history of innovation. His 1913 moving production line reduced the time to manufacture an automobile from 12 hours to 2 hours and 30 minutes and the associated labor costs by 95.2%. A photo of the Ford production line is provided in Figure 1.7.

It was predated in 1903 by Michael Joseph Owens of Owens Bottle Machine Company, where glass bottles were produced at a rate of 240 per minute at a reduced labor cost of 80% (Figure 1.8). Henry Ford and Michael Joseph Owens used different CM implementation approaches to achieve their production needs. Michael Joseph Owens automated the complete set of manufacturing operations into one machine. Henry Ford broke the manufacturing operations into manageable steps and placed them sequentially, improving unit manufacturability. CM decisions relative to the ease of production, assembly, repair, accessibility to things that need to be repaired or replaced, transport, product options, parts sparing, tooling, components, materials, and unit life were critical to the success or failure of each competitor.

Sometimes configuration changes due to adaptive radiation are so significant that they impact an entire market segment. This is certainly true in the automotive world. The first automobile with a vehicle control system layout like those used today was the Cadillac Type 53. The instrument cluster location behind the steering wheel combined with the gas pedal on the right, the brake pedal in the middle, and the clutch on the left were much more intuitive to use than the layout in Ford's Model T. Austin copied the Cadillac Type 53 layout and employed it in its Austin 7. It then licensed the

[15] Earl Ray "E. R." Miller Sr., test driver for Ford and Hupmobile, who helped lay the brick track used in the first Indianapolis 500, mechanic for Cadillac on the first U.S. transcontinental race, Hollywood stuntman, and entrepreneur in mechanical innovations and an early Portland Cement franchisee.

FIGURE 1.7 Assembly line at the Ford Motor Company's Highland Park plant. (Reproduction Number: LC-USZ62-19261 [b&w film copy neg.], 1913. Library of Congress Prints and Photographs Division Washington, DC, http://www.loc.gov/pictures/item/2011661021/.)

FIGURE 1.8 Ten-arm Owens automatic bottle machine. (Hine, L.W., 1874–1940 photographer, Reproduction Number: LC-DIG-nclc-05512 [b&w digital file from original glass negative] Library of Congress Prints and Photographs Division, Washington, DC, http://www.loc.gov/pictures/ collection/nclc/item/ncl2004001184/PP/.)

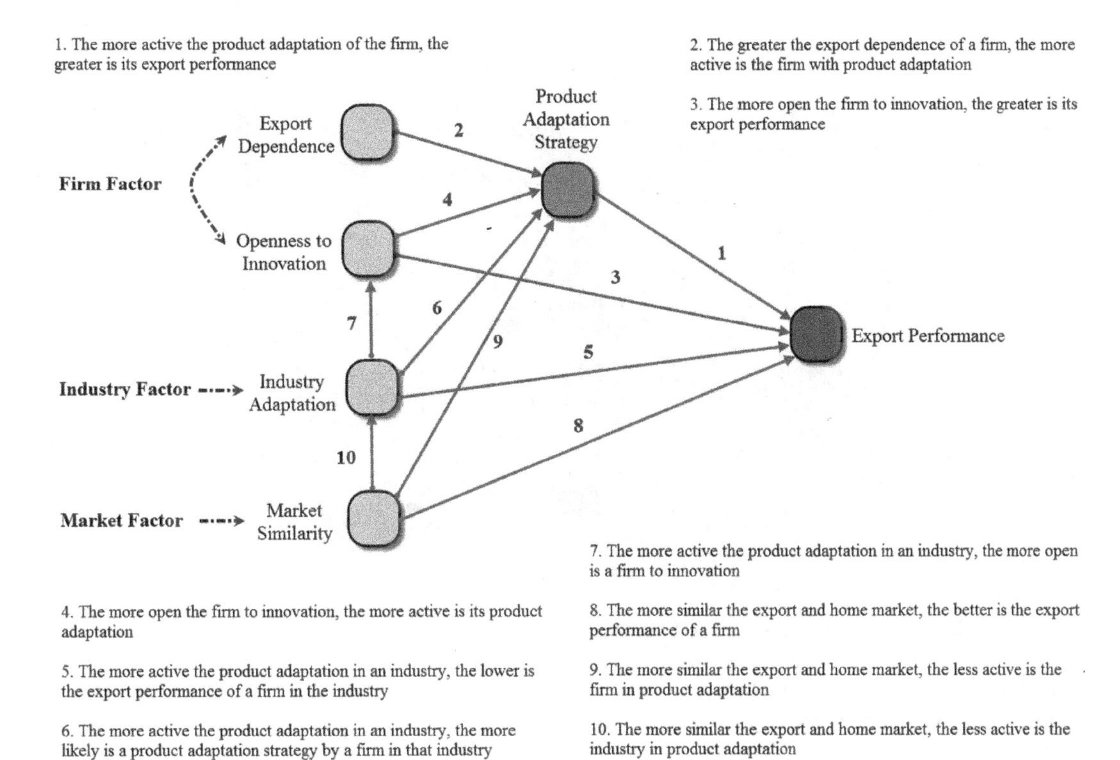

1. The more active the product adaptation of the firm, the greater is its export performance
2. The greater the export dependence of a firm, the more active is the firm with product adaptation
3. The more open the firm to innovation, the greater is its export performance
Product Adaptation Strategy
Export Dependence
Firm Factor
Openness to Innovation
Export Performance
Industry Factor
Industry Adaptation
Market Factor
Market Similarity
1
2
3
4
5
6
7
8
9
10
7. The more active the product adaptation in an industry, the more open is a firm to innovation
4. The more open the firm to innovation, the more active is its product adaptation
8. The more similar the export and home market, the better is the export performance of a firm
5. The more active the product adaptation in an industry, the lower is the export performance of a firm in the industry
9. The more similar the export and home market, the less active is the firm in product adaptation
6. The more active the product adaptation in an industry, the more likely is a product adaptation strategy by a firm in that industry
10. The more similar the export and home market, the less active is the industry in product adaptation

FIGURE 1.9 Market performance. (Adapted from Calantonea, R. J., et al., J. Bus. Res., 59, 2, 176–185, 2006.)

Austin 7 design to BMW and Datsun (now Nissan). It was not long before the configuration change was standardized worldwide. Similar adaptive radiation of configuration changes is being made today, from standardization of map symbols in North Atlantic Treaty Organization (NATO) countries to refined hardware interfaces for computing devices.

Market performance is based on the two strategies of product market adaptation and product performance adaptation. It is graphically shown in Figure 1.9. The market performance construct identified earlier is sound and forms the basis of the discussions that follow.

1.4.3 Product Market Adaptation

Product market adaptation is the strategy of introducing new products into an existing market based on either a modification of a competitor's existing product or an improvement of that product. It is not based on an enterprise investment in research and development to pioneer innovations or to open a new market segment. It is rather an attempt to enter a market segment previously established by someone else.

Product market adaptation is seen in the history of the rental car business. Joe Saunders in Nebraska affixed a mileage meter to the left front wheel of his Model T Ford in 1916 and rented it out at €0.088 per mile. By 1925, he had a monopoly in the new rental car market segment he created and was operating in 21 states. Others saw the profitability potential of entering the market segment and went into direct competition, including Walter L. Jacobs, who entered the market segment in 1918. In 1923, he sold his company to John Hertz, president of Yellow Cab and Yellow Truck and Coach Manufacturing Company. By 1925, Hertz had established the first coast-to-coast rental network using franchises and advertised heavily (Figure 1.10).

The Hertz® Drive-Ur-Self System was acquired in 1926 by the General Motors Corporation when it bought Yellow Truck from John Hertz. Hertz rental cars dominated the market until Warren

FIGURE 1.10 Hertz® Drive-Ur-Self System. (Advertisement circa 1925 provided courtesy of the Hertz® Corporation. With permission.)

Avis entered the market segment in 1946, specializing in car rentals at airports. At that time most of the Hertz rentals took place in cities.

Historically, product market adaptation also took the form of acquisition of or merger with an enterprise in the market sector and agreements between competitors to sell for the same price. This was taken to extremes in the United States during the 1800s with the forming of trusts (monopolies) in the oil (Standard Oil), sugar, steel (U.S. Steel), and railroad industries. The result was control of quantity, quality, and price. Passing of the Sherman Act in 1890 against price fixing and illegal activities and the Clinton Act in 1914 to limit mergers gave small companies the ability to truly compete in the market. Compliance was overseen by the U.S. Federal Trade Commission (FTC) established in 1914.

Today, many mergers are simply to acquire technologies the enterprise does not currently possess, so that it can build rather than buy components used in a product. In other cases, mergers are used to diversify so that an enterprise can survive a broader set of market-specific fluctuations. Glenn L. Martin Company (aerospace and launch/missile systems) did this when it merged with American-Marietta Corporation (paints, dyes, metallurgical products, and construction materials) to form the Martin Marietta Corporation in 1961.

Market adaptation can be observed in the ebb and flow of firms manufacturing smart devices (phones, tablets, etc.) and the integration of their capabilities across many aspects of daily life. This integration includes seeing what the nanny is doing in real time; managing personal finances; and locating a vehicle in a parking lot, turning it on, and pre-warming the seats. They are also manifesting themselves in reaching a balance of standardization and market adaptation while doing a corresponding dance with consumer privacy.

While not all international products are designed with this objective, the proportion of international products that balance standardization and adaptation appears to be growing, and anecdotes justifying the merits of this come from a spectrum of industries. These include food and cosmetics (where adaptation would appear to be important) along with copiers and televisions (where standardization of interfaces would appear to be critical). The logic is that through standardization and adaptation, multinational corporations (MNCs) can simultaneously harness the benefits of efficiency and responsiveness and combine inputs of both headquarters and foreign subsidiaries into their products, for a greater competitive advantage.[16] One area where this is noticeably lacking is in the safety governance of the U.S. automobile industry. Should the United States adopt European Union vehicle safety standards, it would reduce the weight and cost of U.S. vehicles while increasing fuel efficiency, making it less expensive to buy domestically as well sell as outside of the U.S. marketplace.

1.4.4 Product Performance Adaptation

Product performance adaptation differs from product market adaptation. It is not a strategy for entering a new market by copying and perhaps improving on a competitor's product. It is rather the process of understanding the needs and norms of a market area and adapting your introduction of a product or the product itself to increase the sales performance in that market segment.

Consider the case of adapting a vehicle for the world market. About 65% of countries drive vehicles on the left-hand side of the road. These countries have the steering wheel on the right-hand side of the vehicle (known as right-hand drive). The 35% that drive on the right-hand side of the road have the steering wheel on the left-hand side of the vehicle (left-hand drive). This identification of a cultural norm could be considered one of the keys to product performance adaptation. Documentation indicates that left-hand drive Model T Fords were marketed in Australia, Great Britain, and elsewhere. The Model T was built for an international market with few paved roads and was a boon to those in rural communities. Not all product performance adaptations evidencing consideration for cultural norms were initially successful.

An example of this was the first introduction of the Subaru 360 in 1968 by Malcolm Bricklin. Subaru is manufactured by Fuji Heavy Industries. It was named the 360 (Figure 1.11) due to its 356 cm^3 two-stroke engine size that produced 25 hp. The 360—affectionately nicknamed "the Ladybug"—was diminutive in size, measuring a scant 2990 cm long × 1380 cm high × 1300 cm wide and weighing in at 419.573 kg. This made the Ladybug 26% smaller than the 1969 VW

FIGURE 1.11 Subaru 360.

[16] Subramaniam, M., and Hewett, K. (2004). Balancing standardization and adaptation for product performance in international markets: Testing the influence of headquarters-subsidiary contact and cooperation. *Management International Review*, Vol. 44, No. 2, 171–194. Reprinted with permission of Springer Science + Business Media, Heidelberg, Germany. pp. 171–194.

Beetle, which measured 4070 cm long × 1500 cm high × 1540 cm wide. Although the Ladybug was technically sophisticated at the time, with unibody construction and torsion bar trailing arm suspension, it was rated Not Acceptable by *Consumer Reports* in 1969 after a crash test against an American automobile. By 2017, Subaru had resolved such issues and held 3.8% of the U.S. automotive market share.[17]

Subhash C. Jain makes the point that different markets for a product may exist in various stages of product life cycle development. If a product's foreign market is in a different stage of market development than its native market, some changes in the design may be required to make an adequate product/market match in the foreign market.[18] When viewed in this context, the Subaru 360 was entering a foreign market with a product that was not an excellent product to market match. Jain goes on to state, "Culture influences every aspect of marketing. The products people buy, the attributes they value, and the principals whose opinions they accept are all culture-based choices." This aspect of product performance adaptation was not considered and made the *Consumer Reports* rating worse than it would have been for a more familiar product.

Henry Ford was forced to discontinue the production of the Model T due to dated vehicle design, in order for his company to remain viable. Other innovative manufacturers were modifying body styles and vehicle capability and Model T sales were rapidly declining. In 1927, fewer than 500,000 Fords were sold compared to 1,215,826 Chevrolet sales in 1926 for General Motors Corporation.[19] Ford decided to shut down production, and the last Model T was produced on May 26, 1927. Approximately 60,000 workers were laid off and Ford dealers were forced to purchase all remaining Model T inventory and parts if they wished to remain dealers. Ford factories retooled for a competitive re-entry into the market and the new Model A Ford was released in December 1927.[20] Ford had previously called a vehicle the Model A (Figure 1.12) and started production in 1903, pre-dating the start of Model T production in 1908 by five years.

1903 Ford Model A

1930 Ford Model A

FIGURE 1.12 1903 vs. 1930 Ford Model A.

[17] https://www.statista.com/statistics/249375/us-market-share-of-selected-automobile-manufacturers/. Accessed January 9, 2019.

[18] Jain, S. C. (1989). Standardization of international marketing strategy: Some research hypotheses. *Journal of Marketing*, Vol. 53, No. 1 (American Marketing Association), 70–79.

[19] General Motors Corporation (1926). Eighteenth Annual Report of the General Motors Corporation Year Ended December 31, 1926.

[20] Earl Ray "E. R." Miller Sr. former Ford employee.

1.5 CM AND MARKET ADAPTATIONS

Although firms modify their market strategies to include both product market adaptation and product performance adaptation, it is important to consider the effect of consumers having an affinity for the product itself. This is accomplished not only through such things as brand loyalty or the need the product meets, but also through how the culture it is marketed to values both strategies.

1.5.1 Mass Customization

In his 1922 autobiography, *My Life and Work*, Henry Ford stated, "Any customer can have a car painted any color that he wants so long as it is black."[21] Earl Ray "E. R." Miller[22] of Ely, Nevada, worked for Henry Ford during the Model T era. He states,

> Henry Ford was a master in the art of common design. He attempted to standardize everything he could to make his production of the Model T more profitable. The crates that parts were shipped to the factory in were made to his specifications and after being unpacked we took them apart and used the parts in the assembly line. Not a board, screw, nut or bolt on those crates was thrown away. They all ended up in one of the cars. The wood scraps were made into charcoal and sold under the brand name "Kingsford."
>
> He preferred to paint the Model T black simply because it dried faster, and he could put more vehicles out the door. The only production year that black was the only color option was 1914. From 1908 through 1914 black wasn't even an option. I remember grey, green, blue, and red ones. He called me and Louie Chevrolet up to his office one day and he was late, so we were sitting on his desk smoking cigarettes. I never found out what he wanted because he fired us on the spot. He didn't like anyone smoking.
>
> Well Louie started his own company and I went to work for Hupmobile as a test driver. My job was to take the cars out on the road and try to break them. Every night they would take the body of my test car off the automobile chassis and put it on a new chassis that incorporated the fixes from the day before. Off I'd go the next day trying to break it again while they looked at what I'd done to the automobile the day before and figure out how to fix those. Lots of folks think that engineers know what they are doing all along in the design process. They don't. They design, build, test and if it breaks go back to design. You remember that.

Today, mass customization surrounds us and drives not only the product offerings but also product marketing, manufacturing, and sales. Sales offices and websites offer versions of what is referred to as a product configurator—a software program that makes it possible to add and/or change functionalities of a core product or to build a fully custom product for purchase. It uses simple and complex rules to determine the options mix, as some options do not interface well with other options, and certain options also require other options. The U.S. public sees product configurators associated with automobile and computer sales websites where a potential customer can design a vehicle or computer to meet their needs based on available options.

They exist in other industries as well for product upgrades during operations and maintenance. As with most operations and maintenance activities, when the product has not been maintained by the equipment manufacturer or an authorized agent, the as-maintained configuration may not be known. Several companies we talked to stated that this required them to take 150% more parts to the field for upgrades as they had to perform any out-of-date maintenance to the entire system before the upgrade was installed to ensure that the upgrade would perform properly. ASML, for example, tracks the configuration of well over 1,000 delivered units as part of its operations and maintenance activities.

---

[21] Ford, H., and Crowther, S. (2005). *My Life and Work.* Salt Lake City, UT: Project Gutenberg eBook. p. 72, Chapter IV.

[22] Earl Ray "E. R." Miller Sr. describing to his grandson Kim Robertson design and supply chain innovations he assisted with at Ford during Model T production cost reduction.

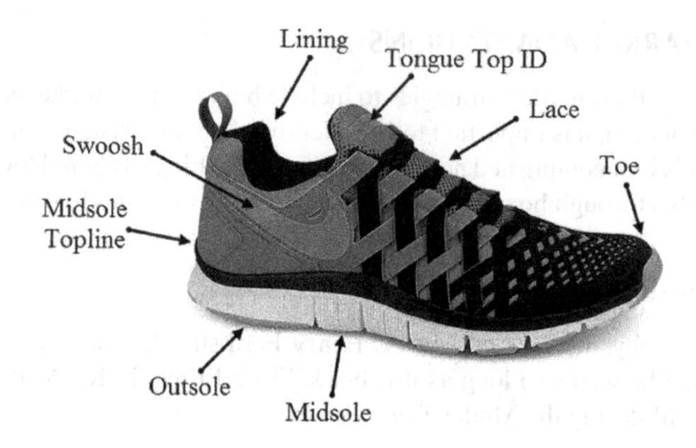

FIGURE 1.13 Components of the Nike Free Trainer 5.0 ID.

A visit to the Nike website in 2014 provided an excellent example of a product configurator, the Nike Free Trainer 5.0 identification (ID) (Figure 1.13). Launched in 2013, it gives the customer multiple color options (Equation 1.2 and Table 1.1) for each shoe component that is selected by the consumer before placing an order. Based on the numbers of colors available and assuming the use of the Nike logo, for this product alone Nike is offering the consumer 758,160,000 unique color variants of the product.

$$9 \times 8 \times 15 \times 15 \times 13 \times 15 \times 4 \times 6 \times 10 = 758{,}160{,}000$$

EQUATION 1.2 Nike Free Trainer 5.0 identification (ID) color variants.

The consumer's customization choices are logged, manufactured, and delivered within four weeks. This ability plays a substantial part in the firm's market share, brand loyalty, and customer satisfaction.

Another short-lived online version of mass customization was introduced by LEGO®. LEGO® has built the company around reuse of stock components to produce a variety of *new* products. To facilitate or showcase this capability, it introduced a tool called LEGO® Digital Designer (Figure 1.14). The tool was essentially a self-contained computer-aided design package that made it possible to develop a product without the use of actual materials. Once the product was designed, the bill of materials, assembly instructions, and packaging for the product could be generated, giving each package a custom appearance through the LEGO® Design byME web application. Until January 16, 2012, these designs could be uploaded, along with instructions and a box design, to the LEGO® Design byME website and then be ordered for delivery as a real, packaged set. With LEGO's expanding component base, robotic creations are commonplace and limited only by an individual's creativity.

Volkswagenwerk GmbH took a simpler approach to configuration and market adaptation. Detroit in the 1950s was experiencing excessive parts inventory costs to maintain the wide option offerings for any given model and passing the costs on to the consumer. In 1956, the air-cooled VW vehicle was featured by *Popular Mechanics* in its article "Owning a VW Is Like Being in Love." This article set the stage not only for public acceptance of the unconventional mechanical configuration, but equally for the unconventional marketing approach. That year, General Motors had a U.S. advertising budget of €141,917,900 compared to VW's €524,007.[23] Volkswagen limited exterior colors for the standard sedan to seven, with a small choice of pre-matched interior colors, and 16 accessories. This kept dealer and production inventories low. A keen attention to CM also enabled dealers

[23] Rieger, B. (2013). *The People's Car: A Global History of the Volkswagen Beetle*. Cambridge, MA: Harvard University Press.

TABLE 1.1
Nike Free Trainer 5.0 ID Customization Options. XDR, Xtra Durable Rubber

Shoe Part	Upper	Toe	Swoosh	Lace	Lining	Midsole Topline	Midsole	Outside			Tongue Top ID	
Sub Choices								Solid	Translucent	XDR Outdoor	Nike Logo	ID Max of 5 Characters
Black												
Wolf Grey												
White												
Flash Lime												
Atomic Teal												
Volt												
Electric Green												
George Green												
Vivid Sulpher												
Midnight Navy												
Blue Glow												
Game Royal												
Solar Red												
University Red												
Team Red												
Team Orange												
Total # Choices	9	8	15	15	13	15	4	6	NA	NA	10	NA

FIGURE 1.14 LEGO® Digital Designer.

to easily stock accessories and vehicles separately, facilitating dealer stock turnover. Accessories could be added at any time and were not part of the vehicle fabrication process. In 1963, 209,747 Volkswagen 1200 Sedans (Beetles) were exported to the United States.

Walt Disney Productions' release of *The Love Bug* in 1968 provided an unlooked-for marketing boost for Volkswagenwerk GmbH. The movie was based on the 1961 book *Car, Boy, Girl* by Gordon Buford. Buddy Hackett, who played Tennessee Steinmetz in the film, stated that the pearl white Volkswagen was named "Herbie" after one of his comedy skits about a ski instructor.

The Beetle was chosen due to the fact that, of all the cars Walt Disney lined up in his casting call, it was the only one petted by passersby at the Disney studio.[24] Use of the vehicle in the movie was not sponsored by the manufacturer, yet the release of the movie and its sequels played a part in Volkswagenwerk GmbH's ability to maintain a sound market share despite an aging technology. The movie also facilitated expansion of movie-related product markets by other manufacturers. In February 1972, Volkswagen celebrated the sales of over 15 million Beetles by February 1970, surpassing the Ford Model T. Today, it holds the record for the most popular vehicle ever produced.[25]

1.5.2 Performance Adaptations

Changing the product itself to either increase or adapt its performance appears to have a counterpart in the natural world. CM again plays a key role, not only in determining how a product is designed, but in all aspects of pertaining to producibility and maintainability. Charles Darwin published his journals of the second survey expedition of HMS Beagle (from December 27, 1831, to October 2, 1836) with the title *Journal of Remarks.* Due to its popularity, it was reissued in May 1839 as the *Journal of Researches.* The *Journal of Remarks* was a reprint of Darwin's account of the voyage from volume three of *The Narrative of the Voyages of H.M. Ships Adventure and Beagle.* He spent 39 months of the voyage on land as the expedition's gentleman geologist. This experience

[24] Stevenson, R. (Director). (1968). *The Love Bug Special Edition DVD (2003).* [Motion Picture]. Walt Disney Productions, Burbank, CA.

[25] *VOLKSWAGEN Brand History.* (n.d.). Retrieved July 23, 2014, from *Autoevolution*: http://www.autoevolution.com/volkswagen/history/.

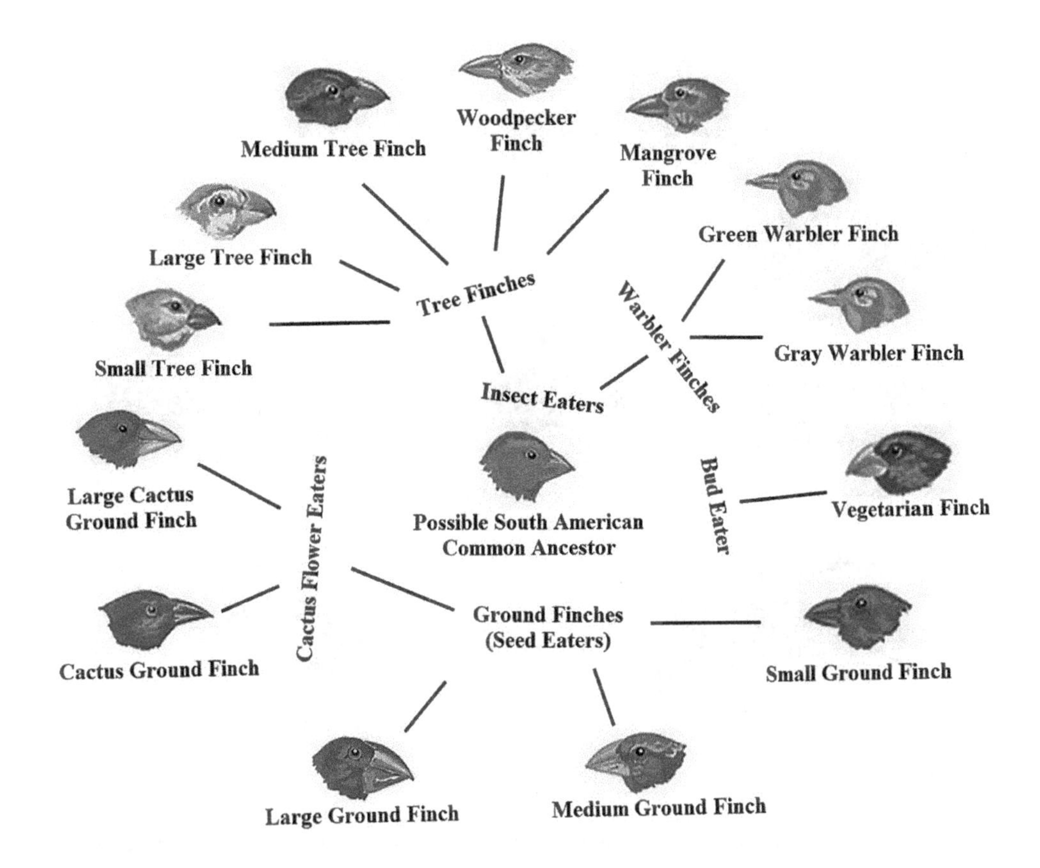

FIGURE 1.15 Galápagos finch species.

led to his theory of evolution. He postulated that the diversity of finch species he observed on the Archipiélago de Galápagos was possibly due to a diversification, later termed "evolution," from a common South American ancestor (Figure 1.15).

1.5.2.1 Adaptive Radiation

Such diversification allowed multiple similar species to exist in greater densities as they did not compete for the same food source. The scientific term for this is niche partitioning. Niche partitioning allowed the diversity of sauropods (e.g., lizard-hipped like the brontosaurus) and theropods (e.g., bird-hipped like the allosaurus) exploiting the same biomass during the Jurassic. A new species can develop in as little as two generations[26] like the most recent Galápagos finch species "Big Bird" (not shown in Figure 1.15).

Niche partitioning can also be seen in industry and is the basis for much that has transpired since the theory was postulated. If the same observations are made with respect to the evolution of the Model A from the Ford Quadracycle, adaptive radiation moves from theory to the practical in the automotive world (Figure 1.16). Ford's approach to maximizing common platforms for the 1929–1931 Model A variants was not unique (1903 Model A Ford is not shown to avoid confusion).

[26] Lamichhaney, S., Han, F., Webster, M. T., Andersson, L. B., Rosemary Grant, B. R., and Grant, P. R. (*Science* 23 Nov 2016) Report: Rapid hybrid speciation in Darwin's finches http://science.sciencemag.org/content/early/2017/11/20/science.aao4593

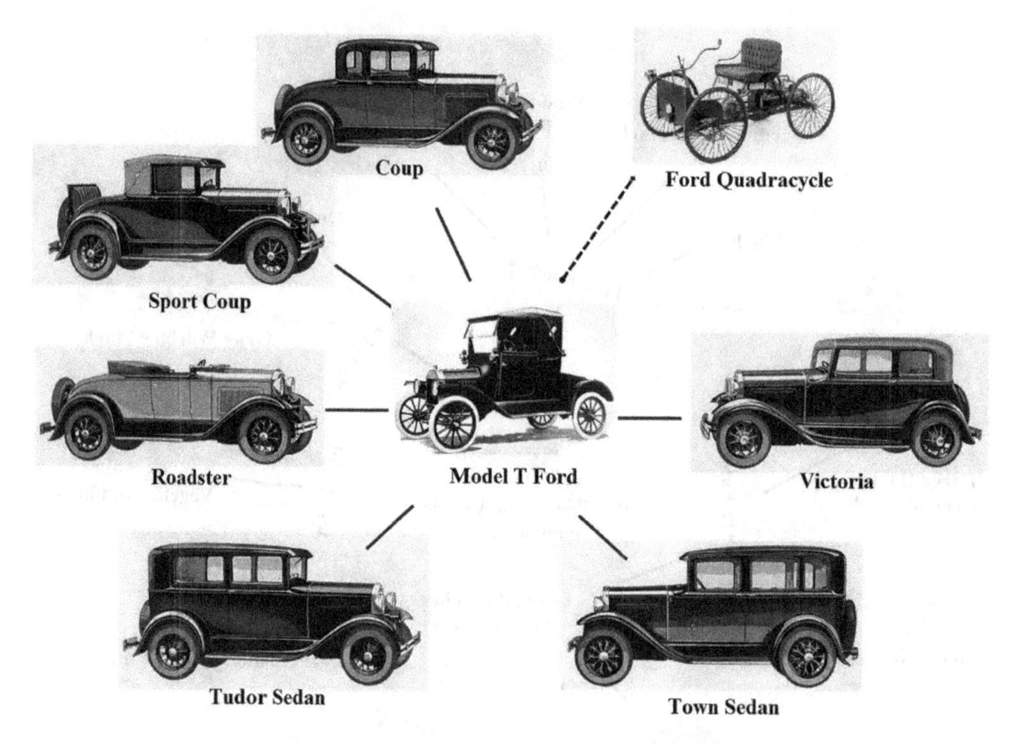

FIGURE 1.16 Adaptive radiation in engineering.

It has remained in practice in many industries until the present. Automotive niche partitioning means that you can buy tires, fan belts, sparkplugs, engine oil, and other products from someone other than the company that manufactured the vehicle. A timeline of automotive technologies is provided in Figure 1.17.

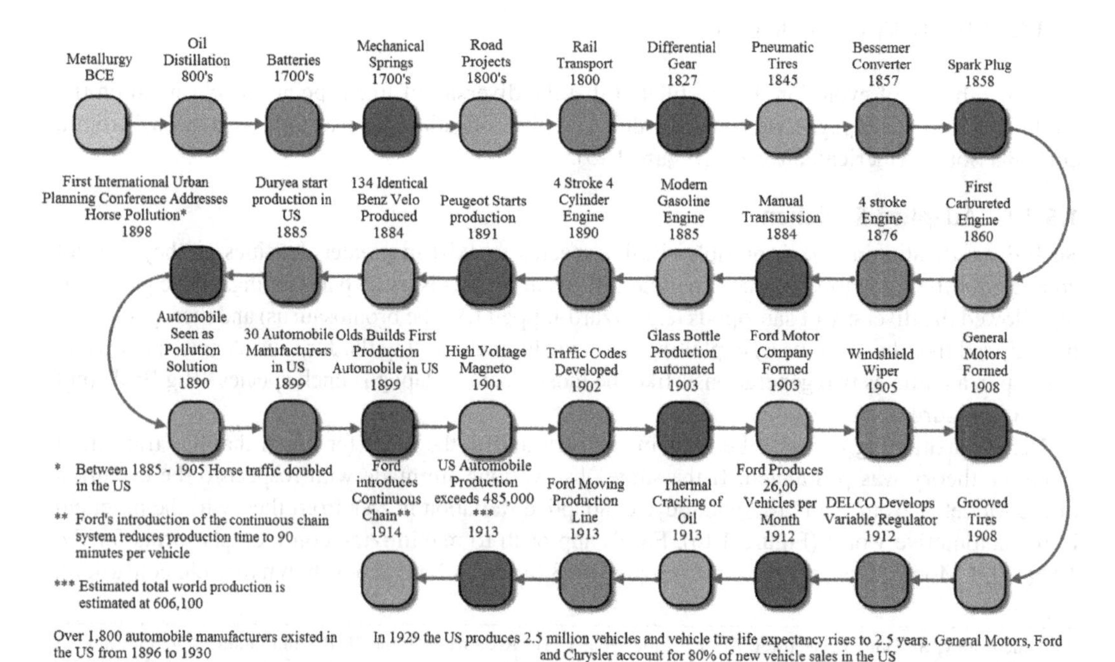

FIGURE 1.17 Timeline of critical automotive technologies and influencers from 1800s to 1930s.

Hardware, software, and firmware development can be traced in much the same way as experienced breeders of botanicals and animals trace the influence of the parents to offspring. This pedigree is evident in the world of space launch systems and the competition to capture market share now that governments can no longer afford the great development expense due to changes in world politics. Although some contenders are funded in whole or in part by their respective governments, several launch vehicles are being developed independently. Space Exploration Technologies (SpaceX) is one such entry and has effectively used the concept of leveraged innovation in the development of the Falcon 9 launch booster. SpaceX has an aggressive and managed configuration approach for developing its market share (Figure 1.18) that includes a project launch of the Starship timed to meet the 2022 Mars conjunction window.

One advantage of planned radiation is the ability to design in contingency profiles with predetermined and managed configuration capabilities. On May 19, 2012, during launch processing of

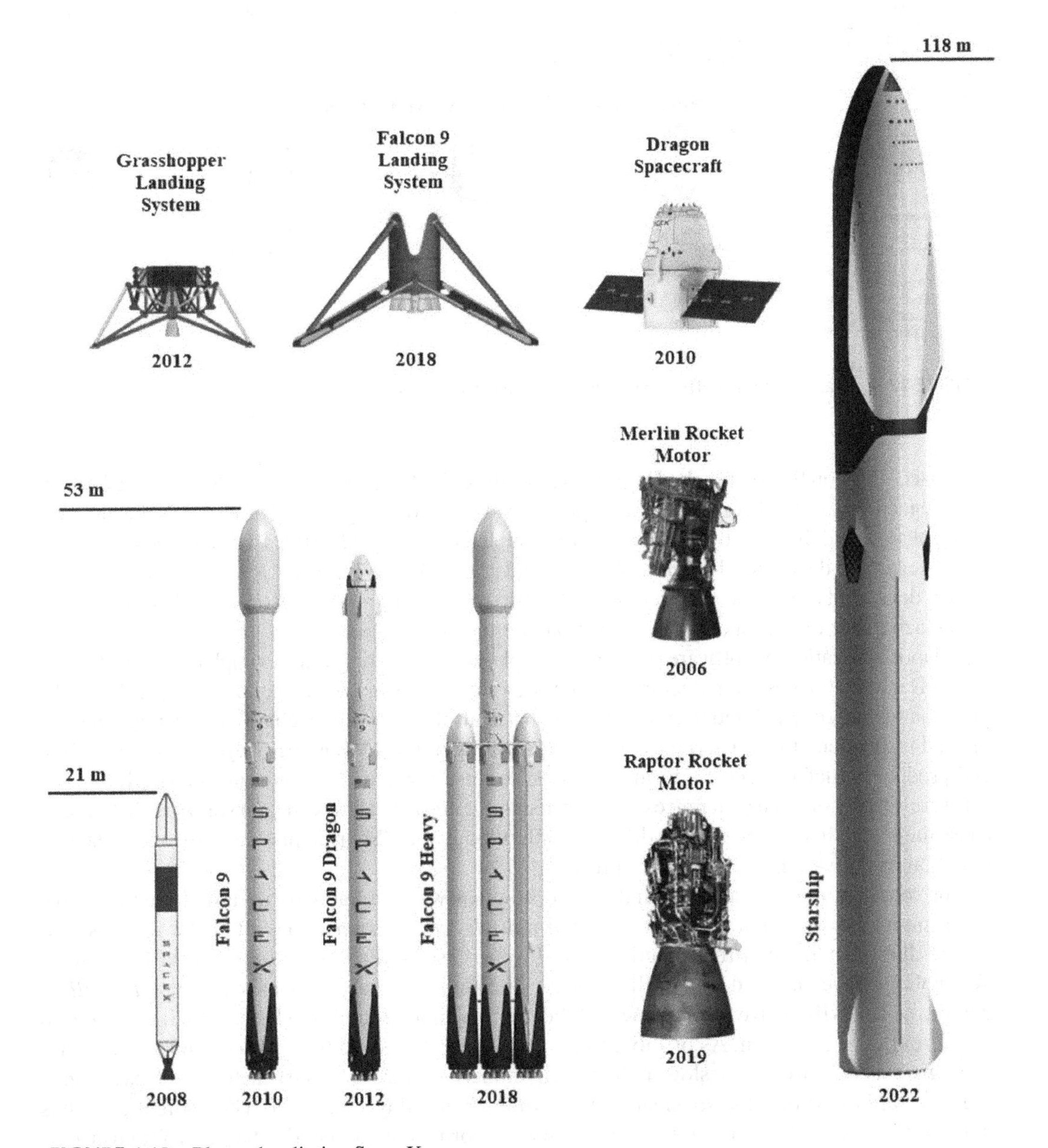

FIGURE 1.18 Planned radiation SpaceX.

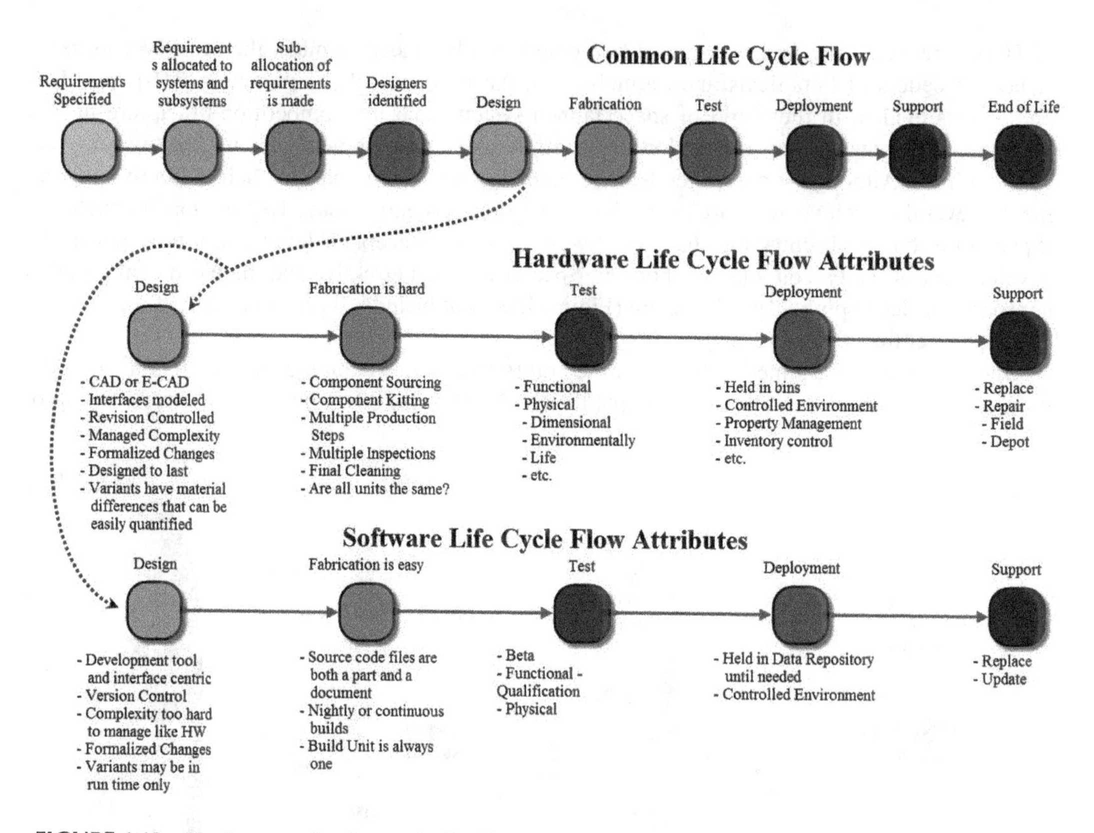

FIGURE 1.19 Hardware and software similarities and differences.

the SpaceX Falcon 9 carrying the Dragon spacecraft to the International Space Station, the booster experienced an event where the engine 5 pressure started trending high. The engine's check valve was replaced, and the launch was successfully completed on May 22, 2012. Issues of this kind under the government-developed launch systems would normally have caused a six-week to ten-week launch delay and a possible de-stack of the launch vehicle and re-manifesting of other vehicles in the launch queue until root cause investigations were completed.

Adaptive radiation of software can best be explained by relating it to a simpler case study than that of Darwin's finches. Manufacturing and design changes in hardware take time simply due to the manufacturing process. It requires sourcing of materials through supply chain management, receiving and inspection of the materials, performing forming or machining operations, inspection of the end-product for defects, and conformance to the engineering and test requirements. Hardware and software similarities and differences are depicted in Figure 1.19. Hardware and software differences cause them to radiate along slightly different paths though the CM principles are the same. Adaptive radiation in software is depicted in Figure 1.20.

For some, it may be easier to understand how software radiates using a different example from nature. In the laboratory, *Escherichia coli* (*E. coli*) bacteria are often the organism of choice due to their rapid reproduction rate. In one case, scientists at the Georgia Institute of Technology have inserted 500 million-year-old Paleozoic-era bacteria genes into *E. coli* to determine if it will evolve the same way it did the first time around or whether it will evolve into a different, new organism. As of July 2012, 1,000 generations had been observed. The new bacteria grew about two times slower than their modern-day counterpart at first and have since mutated rapidly to become stronger and healthier than today's bacteria. The ancient gene has not yet mutated to become more like its modern form, but rather, the bacteria found a new

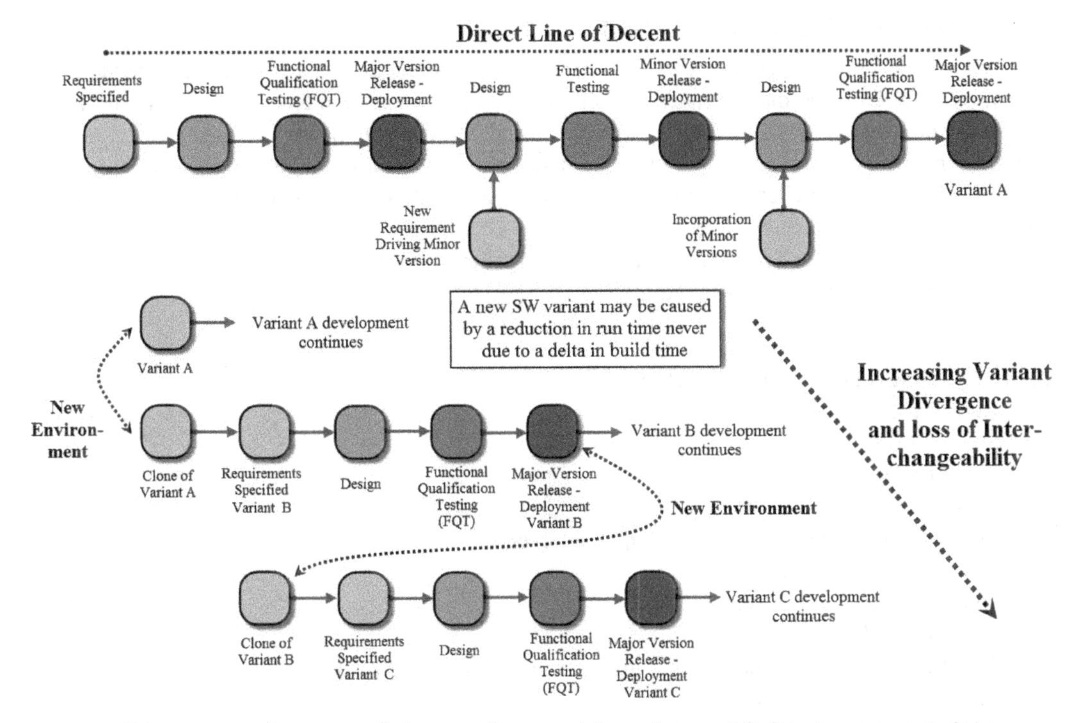

FIGURE 1.20 Adaptive radiation in software.

evolutionary trajectory.[27] Software follows a similar *evolutionary* path to that described by Betul Kacar and Eric Gaucher for the modified *E. coli* bacteria.

When an enterprise capitalizes on the capabilities of its infrastructure to create production tooling and shop aids, and includes the users and fabricators of the tools and aids in the process, it results in items that are better designed, more easily produced, and more rapidly tested. Leveraging design collaboration across the entire enterprise results in not only better designs but designs that are less expensive to make. This aspect of management of the item's configuration is often overlooked. Designers create items that may cause expenditure of fabrication times that could be cut drastically if production is involved up front with configuration decisions.

When innovation is done with suppliers as equal partners in the design and profitability, it becomes "leveraged innovation."[28] This brings the talents of all involved to fill a market niche they have not been in before or to capitalize on a market they are already a player in. This approach to the supplier–client relationship is becoming increasingly necessary to support a more stable form of market share based on the principle of one firm knowing what it does best partnering with another firm with different capabilities to leverage a synergy of knowledge and expertise to the benefit of both, manifesting in a greater market share and profit distribution. Leveraged innovation can also take place between different subsidiaries under a single enterprise.

---

[27] Cristoph Adami, David M. Bryson, Charles Ofria, and Robert T. Pennock. (eds.) (2012). *Artificial Life XIII: Proceedings of the 13th International Conference on the Synthesis and Simulation of Living Systems*. July. MIT Press. pp. 11–18. http://mitpress.mit.edu/sites/default/files/titles/free_download/9780262310505_Artificial_Life_13.pdf

[28] Francis Bidault, C. D. (1998). *Leveraged Innovation: Unlocking the Innovation Potential of Strategic Supply*. Houndmills: Macmillan Press Ltd.

1.5.2.2 New Drawing versus New Dash Number

The idea of product elements changing over time or in response to a new need is of importance to many aspects of CM, the market, and infrastructure support. The elements of form, fit, and function are often discussed in CM circles as reason to re-identify one item from another. Unfortunately form, fit, and function are often not clearly understood. Just as the Galápagos finches filled the unique environmental niches, the configuration of a single item such as a screw is purposefully designed to fill a specific need (Figure 1.21).

Screw heads take on different forms depending on their use. Other elements of the screw design also contribute to form, fit, and function.

- Head diameter
- Drive type (hex, Phillips, flat, etc.)
- Head undercut
- Shoulder diameter
- Shoulder length
- Shoulder chamfer
- Thread form (square, triangular, trapezoidal, etc.)
- Lead, pitch, and start
- Tolerance
- Thread length
- Thread and thread standard (there are over 30 thread standards)
- Material and material pedigree
- Strength (hardness level)
- Right or left handedness
- Surface coatings and finish

Yet, form, fit, and function are more than this. Three seemingly identical machine screws may have to be marked differently. They may have been cleaned to different cleanliness levels. Two screws cleaned to the same cleanliness level may be different because one was lubricated after cleaning. Two screws cleaned to the same cleanliness level and lubricated differently after cleaning may have to be marked differently due to the lubrication used.

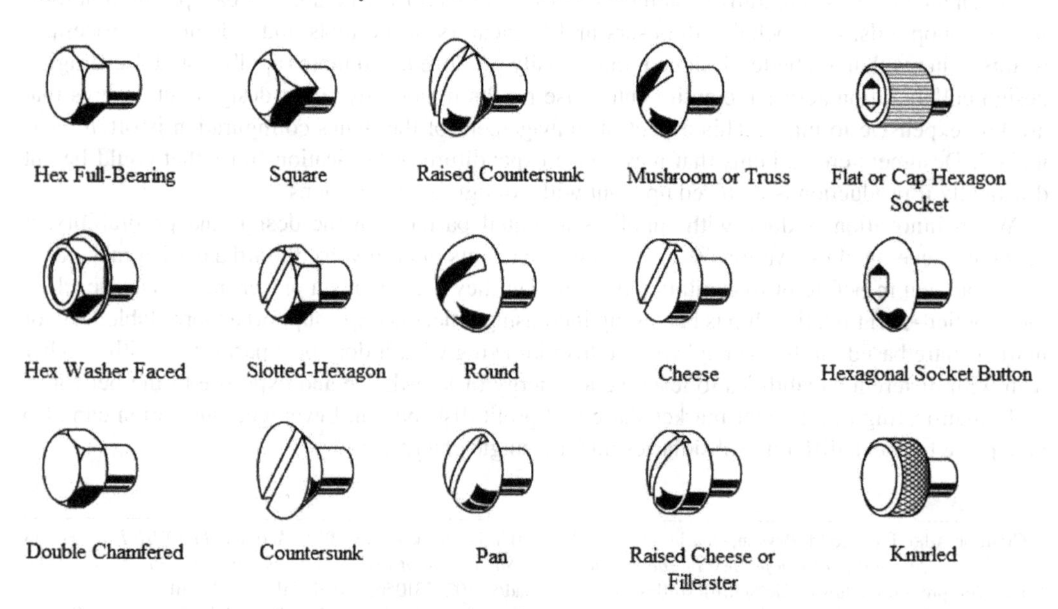

FIGURE 1.21 Screw heads forms.

TABLE 1.2
ISO Cleanliness Level

	ISO cleanliness level	
ISO	Can be used on ISO level	Can receive parts from ISO level
1	1	1
2	2	1, 2
3	3	1, 2, 3
4	4	1, 2, 3, 4
5	5	1, 2, 3, 4, 5
6	6	1, 2, 3, 4, 5, 6
7	7	1, 2, 3, 4, 5, 6, 7
8	8	1, 2, 3, 4, 5, 6, 7, 8
9	9	1, 2, 3, 4, 5, 6, 7, 8, 9

Taking the surface form, fit, and function criteria down to the particulate level is critical in many applications. It is even more critical if the item is lubricated or made of specific materials that outgas (also known as off-gas). Car dashboards outgas, causing car windows to become smoky over time due to the outgassed substances.

Outgassing is a concern for any electronic equipment intended for use in high-vacuum environments. It refers to the release of gas trapped within a solid, such as a high-frequency circuit-board material. The effects of outgassing can impact a wide range of application areas in electronics, from satellites and space-based equipment to medical systems and equipment. In space-based equipment, released gas can condense on such materials as camera lenses and detectors, rendering them inoperative. Hospitals and medical facilities must eliminate materials that can suffer outgassing to maintain a sterile environment.[29]

ISO cleanliness level 1 assemblies can only contain ISO cleanliness level 1 parts but can be used on any other ISO level assembly (Table 1.2). This is referred to as one-way part substitution. ISO particle size is shown in Table 1.3. Again, there is a biological similarity in human blood types. Blood typing is concerned with antigens, antibodies, and rhesus (Rh) factors. People with blood type O Rh– can only receive blood from other O Rh– donors, but they can donate blood to any other blood group (O Rh– is known as the universal donor). People with blood type AB Rh+ can only donate blood to other AB Rh+ recipients, but they can receive blood from any other blood group (AB Rh+ is known as the universal recipient). This concept is depicted in Table 1.4.

CM application is tailored to the product, and re-identification is one aspect of CM. Part re-identification due to form, fit, and function criteria in a medical, aerospace, drug manufacture, or other market segment, where cleanliness levels and outgassing are a concern, needs to be more rigorously applied than in other industries. All aspects of the product and the components that make up the item's product structure are taken into consideration. U.S. government DOD-STD-100[30] provides a sound definition of an interchangeable item.

> When two or more items possess such functional and physical characteristic as to be equivalent in performance and durability and capable of being exchanged one for the other without alteration of the items themselves or of adjoining items except for adjustment, and without selection for fit or performance, the items are interchangeable.

[29] Coonrod, J. (2010, November 19). *What Is Outgassing and When Does It Matter?* Retrieved July 22, 2014, from Rog Blog: http://mwexpert.typepad.com/rog_blog/2010/11/19/, reprinted with permission of Rogers Corporation.

[30] DOD, U.S. (1987). *DOD-STD-100, Engineering Drawing Practices, Revision D.* U.S. Government, Pentagon, Washington, DC. pp. 700–5.

TABLE 1.3
Certification Particle Size (μm)

ISO	Certification Particle Size (μm)						FS209
Class	0.1	0.2	0.3	0.5	1	5	Class
1	10	2	–	–	–	–	–
2	100	24	10	4	–	–	–
3	1,000	237	102	35	8	–	1
4	10,000	2,370	1,020	352	83	–	10
5	100,000	23,700	10,200	3,520	832	29	100
6	1,000,000	237,000	102,000	35,200	8,320	293	1,000
7	–	–	–	352,000	83,200	2,930	10,000
8	–	–	–	3,520,000	832,000	29,300	100,000
9	–	–	–	35,200,000	8,320,000	293,000	–

Source: ISO, International Organization for Standards.

Often, the determination regarding how part re-identification is managed is based on whether the item remains in family or starts a new family (is out of family). The notion of something being in family or out of family can be interpreted in several ways depending on the product line and markets of the enterprise. One way to look at families is to study the managed configuration of the product line segment from Eppinger®, a manufacturer of fishing lures in its Original Dardevle® Spoons. The spoons come in a set number of standardized sizes based on a lure profile originally established in 1906 by the firm's founder, Lou Eppinger. Lures are marketed in a variety of finishes, patterns, and pattern color combinations, including the option for consumers to customize some aspects of the design (Figure 1.22).

Each pattern could be considered as a distinct family with color variants being configuration controlled as sub-families. In the image, 15 patterns have been shown. This is a small subset of the entire Eppinger® offerings for this lure's line. Another approach could be to consider the Dardevle as a family rather than a product type with individual families.

Decisions regarding individual family members often will depend on how the design is viewed within the enterprise. If designs are created from the designer's viewpoint, they take one form. If designs are produced from a manufacturing viewpoint, they may appear another way. Both

TABLE 1.4
Blood Compatibility

Blood Compatibility			
Type	Rh Factor	Can Donate Blood to	Can Receive Blood from
A	+	A Rh+, AB Rh+	A Rh+, A Rh–, O Rh+, O Rh–
A	–	A Rh+, A Rh–, AB Rh+, AB Rh–	A Rh–, O Rh–
B	+	B Rh+, AB Rh+	B Rh+, B Rh–, O Rh+, O Rh–
B	–	B Rh+, B Rh–, AB Rh+, AB Rh–	B Rh–, O Rh–
AB	+	AB Rh+–	A Rh+, A Rh–, B Rh+, B Rh–, AB Rh+, AB Rh–, O Rh+, O Rh–
AB	–	AB Rh+, AB Rh–	A Rh–, B Rh–, AB Rh–, O Rh–
O	+	A Rh+, B Rh+, AB Rh+, O Rh+	O Rh+, O Rh–
O	–	A Rh+, A Rh–, B Rh+, B Rh–, AB Rh+, AB Rh–, O Rh+, O Rh–	O Rh–

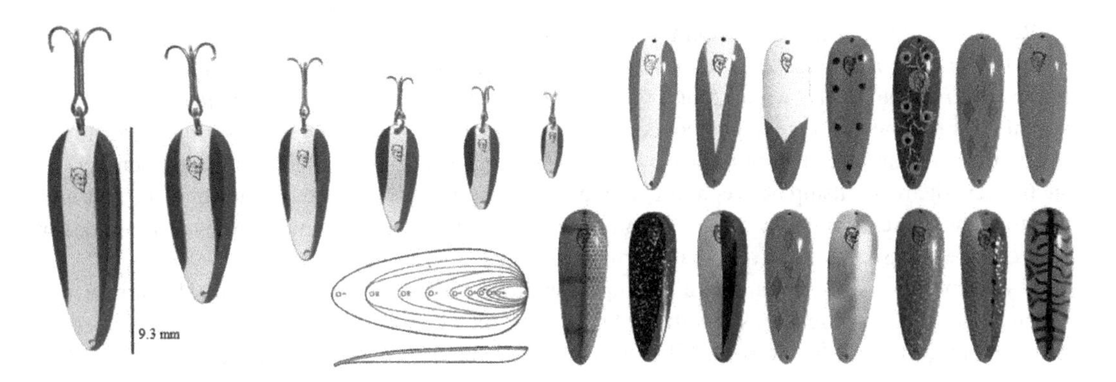

FIGURE 1.22 Eppinger® Original Dardevle® Spoons.

views rely on what is known as a "product structure." Each object manufactured consists of parts, sub-assemblies, assemblies, and so on; each of these has a unique part identifier to distinguish it from something else. Identifiers may be numeric (111, 222, 333, 444, etc.), alphanumeric (A12bc6, A12cd6, etc.) or alpha (aaa, bbb, ccc, etc.) depending on the approach taken by the enterprise itself. In some cases, family relationships are tied back to the identifier established for the drawing itself as a prefix to the root drawing number in a relational scheme.

A practical part numbering system design should account for the limits of short-term memory. The *magic limit* is typically considered to be 7 (+ or – 2 items). Many years of academic study, verified by real-world experience, proves that data entry errors increase as the number of characters increases. After a certain length, errors increase at an increasing rate: at 15 characters, the error probability is close to 100%.[31] Several numbering schemes are shown in Figure 1.23.

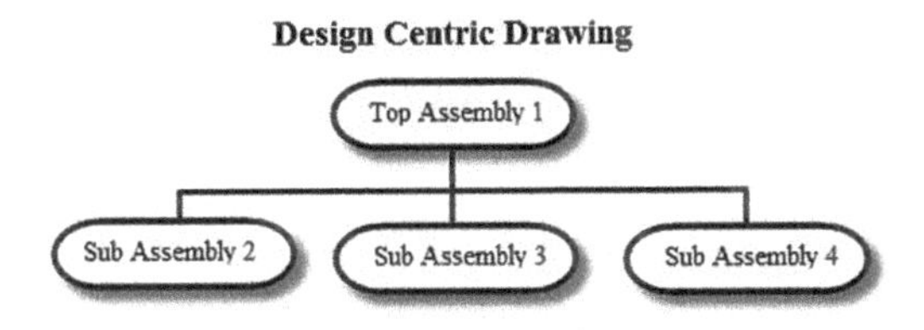

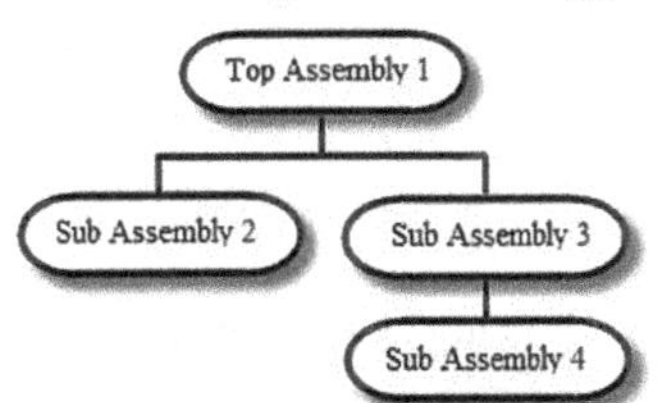

NAS Washers
Significant Number Scheme*

Item #	ID mm	OD mm	Thick mm
NAS620-0	1.600	2.515	0.406
NAS620-10	4.953	8.763	1.600
NAS620-10L	4.953	8.763	0.813
NAS620-2	2.261	4.039	0.406
NAS620-3	2.591	4.572	0.813
NAS620-3L	2.591	4.572	0.406
NAS620-4	2.921	5.309	0.813
NAS620-416	6.477	11.887	1.600

* Numbering Scheme is limited and fragile if new numbers must be added into an existing sequence. The more criteria required the longer the part number

Hewlett Packard
Relational Number Scheme**

Item #	ID
123456	Drawing
123456-001	First Detail
123456-002	Second Detail
123456-200	First Software Executable
123456-201	Second Software Executable
123456-500	First Hardware Assembly
123456-501	Second Hardware Assembly
123456-700	First EDU

** 3 Digit suffix does not support items such as High precision Resistors. A 7 digit numbering system provides 9 Million drawings / document numbers if Zero is not used as the leading digits

Commodity Code
Semi Relational Number Scheme***

Item #	ID
1	First Commodity Family
1000001	First Family Member
1000002	Second Family Member
1000003	Third Family Member
2	Second Commodity Family
2000001	First Family Member
2000002	Second Family Member
2000003	Third Family Member

*** 3 Never start a Family with Zero. 000001 may be written as 1 or interpreted by the SW system as 1

FIGURE 1.23 Drawing design and part numbering scheme.

[31] Miller, G. A. (1956, March). The magical number seven, plus or minus two: Some limits on our capacity for processing information. *The Psychological Review*, Vol. 63, 81–97.

Relational drawing number schemes are common across some manufacturing organizations. The suffix is often referred to as a "dash number." Under this kind of a system, creation of a new addition to *in-family* parts would require a new dash number. Creation of a new *out-of-family* part would require a new drawing. The concept of new dash number versus new drawing is meaningless in firms that assign random numbers to their drawings and to the items the drawing denotes. In such cases, the relationship between drawing and part is tracked in the system used for records and release. These are mistakenly called "configuration management systems." CM is much broader than change release, tracking, and management of data.

Use of numbering schemata is also important for revision and version control. Numbering schema and hierarchal relationships differ between parts and software. Branching is described in an excellent post, Introducing Branching by Abstraction.[32] A trunk may look as follows.

```
<root>
trunk/
foo-components/
foo-api/
foo-beans/
foo-impl/
  build.xml
  src/
  java/
  test/
  cruisecontrol-config-snippet.xml
remote-foo/
bar-services/
bar/
  build.xml
  src/
  java/
  test/
  cruisecontrol-config-snippet.xml
  bar-web-service/
```

It is not a new concept and promotes branching from releases only.

```
<root>
trunk/
releases/
rel-1.0/
rel-1.1/
rel-1.2.x/
```

A software versioning system is <Unique Asset Name> <Major _ Version _ #>.<Medium _ Version _ #>.<Minor _ Version _ #>, where Major, Medium, and Minor could be used to communicate the complexity of the changes.[33]

---

[32] Hammant, P. (2007, April 27). Introducing Branching by Abstraction. Retrieved July 23, 2014, from Paul Hammant's blog: http://paulhammant.com/blog/branch_by_abstraction.html#!

[33] Guerino, F., Chairman for The International Foundation for Information Technology (IF4IT) from a reply to a post on release naming conventions downloaded November 14, 2013.

The one outgrowth of form, fit, and function considerations is the belief in some industry sectors relying on DOD-STD-100 or the American Society of Mechanical Engineers root documents that parts have revisions.

This is at odds not only with standard industry practice but also with the U.S. government requirements as defined in DOD-STD-100[34] where the practice is often found. The standard states:

> The term *revision* refers to any change after that drawing has been released for use. (pp. 700–707)
> Part numbers shall not include the drawing revision (See paragraph 402.5. e). (pp. 400–404)

It may be due to carrying forward misconceptions from paper-drafting systems, where parts defined on the drawing were perceived to have the same revision as the drawing. This concept arose because they were identified on the drawing. This misconception is also seen in the U.S. government end item data package requirements defined by some government agencies.

We looked at over 40 manufacturing firms using part revisions and found no consistency in the reasoning. We conjecture that if your company puts revisions on parts, perhaps you use it to segregate parts when rework, retest, or re-inspect is cited as the part disposition. Items that haven't been modified are at the previous revision and items that have been modified are at the new revision.

If a part is not form, fit, and function interchangeable with any previous design iteration, it must be given a new identifier. In a production environment, parts in a single storage location or bin must always be interchangeable. As a result, the concept of part number revisions is meaningless. Interchangeability of parts must also apply to alternate parts from a second source of supply. As discussed previously, it does not apply to parts of a higher quality level as higher-grade parts are often substituted for lower grade parts despite a difference in part numbers. Tracking and reporting mandatory, latest, and as-built revision on parts spends valuable resources and is not required if the CM change implementation process is properly followed.

As parts go through steps toward design maturity, decisions are made regarding disposition of previously manufactured parts. Such dispositions are to scrap all existing inventory as the design is unsound or to rework the inventory to the new revision. Either decision equates to the rule that all parts contained in a single parts bin are 100% upward and downward compatible. Dispositions of *use as is* are irrelevant. The old iteration and new iteration are form, fit, and function interchangeable and there is no disposition of these parts associated with the change (e.g., the disposition is "Not Applicable"). The cost of maintaining any relationship between a part and its revision is not trivial. It impacts all phases of design and production. A simplified design/production flow is provided in Figure 1.24.

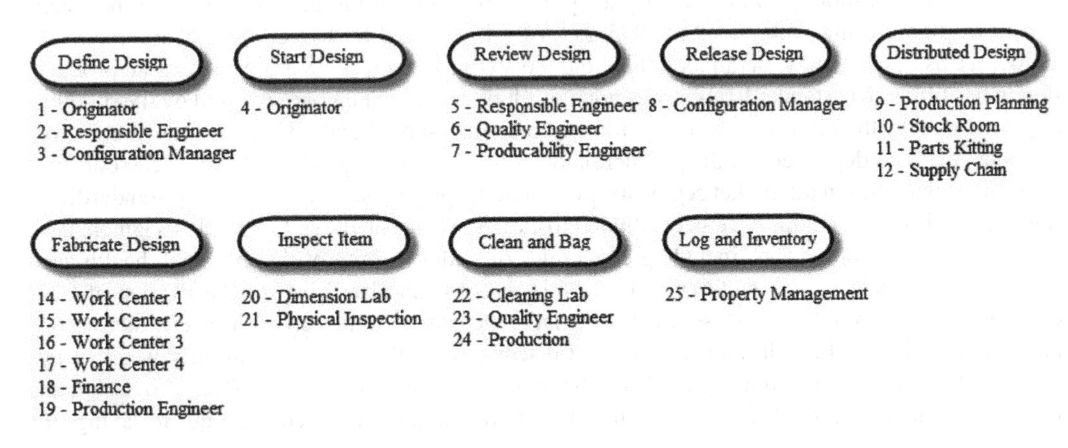

FIGURE 1.24 Simplified design/production flow with eight steps and 25 functions.

---

[34] DOD, U.S., *DOD-STD-100, Engineering Drawing Practices, Revision D.*

From the early 1900s to today, in many sectors a single drawing specified the configuration of 40 distinct assemblies and details. In one case, such a drawing was revised 37 times, with most revisions adding new configurations. Within the product data/life cycle management (PD/LM) system, each distinct revision level had been traced throughout all the revisions. Dispositions of Not Applicable had been duly entered and tracked by quality assurance, checked by production engineering, monitored by the property management system, and cross-verified when parts to make the items were consumed on fabrication orders. Assume that 25 people are identified in the simplified design/production flow. Each person involved spent no more than 20 minutes for each drawing revision, but this equates to over 308 hours, as shown in Equation 1.3.

$$20 \text{ minutes} \times 37 \text{ revisions} \times 25 \text{ people} = 18{,}500 \text{ minutes}$$
$$18{,}500 \text{ minutes} / 60 \text{ minutes per hour} = 308.3 \text{ hours}$$

EQUATION 1.3 Simple change review time calculation.

The number of people and the time spent was purposely kept low. Using a separate set of initial parameters, we can evaluate cost impact on a medium-sized product. Let's assume that 500 parts are manufactured by the firm and that there are three revisions per drawing (initial release, revision 1, and revision 2). This gives us a feel for total project cost impact, as shown in Equation 1.4.

$$20 \text{ minutes} \times 500 \text{ parts} \times 25 \text{ people} \times 3 \text{ revisions} = 750{,}000 \text{ minutes}$$
$$750{,}000 \text{ minutes} / 60 \text{ minutes per hour} = 12{,}500 \text{ hours}$$

EQUATION 1.4 Medium program change review time calculation.

Extrapolate this across the entire enterprise to all products after scaling the assumptions to the market segment, and the impact of a poor CM decision regarding parts with revision control influences negative EBIT as well as market share in this market segment.

1.5.2.3 Design Reuse versus Common Design

As a product type matures, standardization follows if the item is utilized across a broad enough market. One result of standardization is a benefit to the consumer. Another result is the loss of market segment for one or more providers as they adapt from one standard to the other. The war between Sony's Betamax format introduced in 1975 and VHS introduced by Japan Vector Corporation in 1976 is a classic case study. Betamax fell behind the VHS format in 1978 and Sony ceased production in the 1980s. The standardization was not a result of agreement on one standard by the producers, but driven by the marketplace. The market segment measured hundreds of millions of units for a product that was dependent on the production of a complementary product, the videocassette.[35]

Similar dynamics in all market segments are still being played out in other areas of standardization. Screw thread standards have been alluded to earlier. Standardization in a market segment can occur for a variety of reasons, including economies of scale, scarcity of raw material, health and safety concerns, and market preference. Economic reasons can also drive standardization of CM decisions within a single enterprise. Such internal standardization takes two forms: design reuse and common design. Each has advantages and disadvantages that must be carefully weighed by CM and the design team. Both capitalize on the premise that it is more cost effective to design an item once than to have it redesigned each time an item or similar item is required, due to savings in

[35] Cusumano, M.A., Mylonadis, Y., and Rosenbloom, R.S. (Draft March 25, 1991). *Strategic Maneuvering and Mass-Market Dynamics: The Triumph of VHS over Beta*, WP# BPS-3266-91. http://dspace.mit.edu/bitstream/handle/1721.1/2343/SWP-3266-23735195.pdf.

recurring costs (costs to produce) and non-recurring costs (costs to design plus tooling and fixtures). Returning to a portion of E. R. Miller's statement about Henry Ford's production methods will set the stage for examining this area of CM.

> Henry Ford was a master in the art of common design. He attempted to standardize everything he could to make his production of the Model T more profitable. The crates that parts were shipped to the factory in were made to his specifications and after being unpacked we took them apart and used the parts in the assembly line. Not a board, screw, nut or bolt on those crates was thrown away. They all ended up in one of the cars. The wood scraps were made into charcoal and sold under the brand name "Kingsford."

Miller refers to common design, not design reuse. In the case of the Model T, some elements of the product line were used on all the Model T variants.

1.5.2.3.1 Design Reuse

Design reuse is a concept founded in antiquity. The apprentice learned how to make certain items pertaining to the trade from the master and in turn became a journeyman, master, and perhaps grand master with apprentices. Shoes for horses could be scaled up or down in size depending on the animal being shod. Wagons, waterwheels, harnesses, and other products followed the same general pattern with the necessary modifications to size and shape to accommodate the end need. As production methods for making mechanical movements became refined, allowing greater tolerance control, they were documented. Designs were gathered and published in books such as *Five Hundred and Seven Mechanical Movements*.[36]

Design reuse is a different concept than common design. Design reuse is the leveraging forward of technologies to create the next level of technological advancement. Figure 1.25 is an example of this concept as manifested in the Apple iPhone through 2014. Design reuse can be traced back to design inception regardless of iteration. The benefit of this is to establish the provenance of the design and clearly establish the original ownership of the technological evolution. It is an example of CM principles and application assisting in defending against copyright, patent, and other infringements.

In *Red Dwarf, Series IV*, "Meltdown" episode,[37] Pythagoras is insistent that the solution to the war somehow involves triangles. Einstein tells him that not all problems in life can be solved by triangles. This is not unlike the modern design-decision process, where engineering talent has become

FIGURE 1.25 Apple® iPhone® models.

[36] Brown, H. T. (2010). *Five Hundred and Seven Mechanical Movements*, 18th ed. Mendham, NJ: Brown and Seward republished the Astragal Press.

[37] Bye, E. (Director). (1991). *Red Dwarf, Series IV*, "Meltdown." [Motion Picture].

so ingrained into a single way of perceiving solutions to engineering challenges that a single methodology is always applied. It is also directly applicable to the determination of part and assembly candidates to be produced through design reuse in a production environment, as well as the manufacturing infrastructure necessary to produce them.

Standard manufacturing processes involve shaping of materials in a distinct variety of ways. Products can be milled, formed on a lathe, cast (includes injection molding and roto-molding), forged, and, in certain instances, three-dimensionally printed. If an item has traditionally always been milled for a bell housing, and it is not on the critical path as far as lead time, then forging it may not be considered due to perceived economies of scale, despite historical data indicating that if 10 units are produced, the cost for milling and up-front investment to create a forged, injected, or stamped housing reaches the breakeven point.

In a limited-need scenario, such as aerospace applications, determining what can and should be considered as a candidate for a shared design can be key not only to keeping a competitive edge but also to program-by-program profitability. This is a one-sigma, or at best, a two-sigma solution set despite Six Sigma approaches elsewhere in the actual manufacturing process. In addition to the normal quality, cost, and schedule constraints, baselining the product design prior to evaluating options and impacts can drastically affect any exploration of anticipated performance and market development.

Discussions of *what sigma something is* are meaningless at this point and this has become an overused catchphrase for everything from process errors to manufacturing yield. Concentrating on potential future application of a static design element and maximized efficiency of production, as well as growing application of the design element, should outweigh sigma discussions.

Premature lock-in of a design or artificially imposing design constraints to force shared design is abandoning the idea that form follows function. In the natural world, having a strong skeletal framework is the first necessity. Just as some natural designs occur in varied species across time (e.g., ceratopsian and rhinoceros head to neck configuration), superior design stems from a sound construct. Putting it another way, do not put skin on the animal before you know what the bones look like and how they are articulated. Then add the muscle, organs, and—once you know if it is a dog, cow, bird, or fish—decide how to package it. The *Ceratopsoidea, Pachycephalosauria* and *Rhinocerotidae,* in all their variations, shared a similar skull to spine construct for supporting a head that could be used as a battering ram but were decidedly different in outward form.

In the day-to-day design-manufacturing cycle (including services design), design gives way to production, which in turn gives way to test. As tasks are completed, design staff move on to the next project and the eventual designs move through the product life cycle. Our research indicates that over a period of 20 years, the cumulative knowledge of design decisions' soundness is no longer held in the memory of the design staff. This 20-year refresh rate where the learning curve starts over can be mitigated if the CM implementation is sound and designed to facilitate keeping the "secret sauce" recipe from passing out of memory. Issues still crop up, however, such as Green Gunge (degraded di-isoctyl phthalate) in older PVC insulated cables despite its no longer being commonplace in electronics manufacturing.

1.5.2.3.2 Common Design

Common design is a single design element (part or assembly) that is used across multiple applications. A solitary product group owns the design and controls its evolution. A product line family may have multiple common design parts in it. A single model in a product line sub-family may also have common design parts that are common to that model only. Common design initially costs more to create because the company is designing to produce components key to its survival on company money. Once the components have been developed, however, the cost payback is enormous. It is even greater if these common designs are sold commercially and not only as part of the company's key products. These items (e.g., software, firmware, parts, sub-assemblies, and assemblies) make up part of the collective intellectual property (IP) of the enterprise.

In her book, *Configuration Management Principles and Practice*,[38] Anne Mette Jonassen Hass speaks to common design although she does not call it by that name. She states, "Components may be created in various ways—as the result of a conscious management decision or in a specific project during the development work. Either way, the responsibility and ownership should be placed centrally in the company, outside the projects where they are used. The collection of components should be regarded as an independent project, on an equal footing with other products, with all this entails in support functions (project management, quality assurance, configuration management)."

In an environment where the design is used on domestic as well as government programs, common design is treated much as commercial off-the-shelf parts are treated from an IP standpoint. The design was not produced or modified using government funds, so the government has no intellectual rights to the data. Individual programs also have no control over the design and cannot modify it for their use if it is not suitable as is. Modifications to the design to add a new *in-family* or *out-of-family* variant are controlled by the product group that developed it. Common design is a single design element either (1) developed at private expense or sold to the commercial sector, or (2) if developed using government funds, the contractor retains unrestricted rights and the government receives information only in copies with no rights. Under the U.S. Federal Acquisition Regulations (FAR) Part 12, Acquisition of Commercial Items, FAR 12.211, Technical Data:

> Except as provided by agency-specific statutes, the Government shall acquire only the technical data and the rights in that data customarily provided to the public with a commercial item or process. The contracting officer shall presume that data delivered under a contract for commercial items was developed exclusively at private expense. When a contract for commercial items requires the delivery of technical data, the contracting officer shall include appropriate provisions and clauses delineating the rights in the technical data in addenda to the solicitation and contract (see Part 27 or agency FAR supplements).

Experience on multiple FAR Part 12 procurements has shown that provisions and clauses are rarely included in either the solicitation or the resulting contract. This leaves the data unencumbered on FAR Part 12 firm fixed price (FFP) contracts (e.g., all rights in data reside with the contractor).

Shared design is a precursor to common design parts in instances where the design development is shared by two or more cost objectives (e.g., there is more than one program charge number). Multiple cost objectives are of great concern when an enterprise sells the same items to commercial as well as government markets. The shared design utilizes one set of engineering drawings and may be required to provide copies to each funding source. Shared design can be converted to a common design item if all cost objectives have been met and enterprise funds are used to make the conversion. Recurring and non-recurring costs must be properly allocated to all cost objectives. Do not assume that excess parts can be ordered and then purchased by a second cost objective. In the case of the U.S. government programs, an FFP program under FAR Part 15, "Contracting by Negotiation," may include 52.232-32 "Performance-Based Payments."

> (f) Title.
>
> (1) Title to the property described in this paragraph (f) shall vest in the Government. Vestiture shall be immediately upon the date of the first performance-based payment under this contract, for property acquired or produced before that date. Otherwise, vestiture shall occur when the property is or should have been allocable or properly chargeable to this contract.

Title constraints can be avoided through proper planning of cost allocations to cost objectives. Under this arrangement, CM considerations include treating shared design as a separate project not

[38] Hass, A. M. J. (2003), *Configuration Management Principles and Practice*, Addison-Wesley Pearson Education, Inc. 75 Arlington Street, Suite 300, Boston, MA 02216.

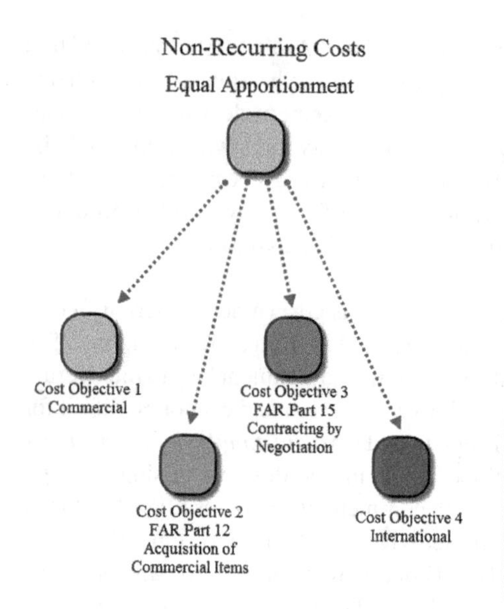
Non-Recurring Costs
Equal Apportionment
Cost Objective 1
Commercial
Cost Objective 2
FAR Part 12
Acquisition of
Commercial Items
Cost Objective 3
FAR Part 15
Contracting by
Negotiation
Cost Objective 4
International

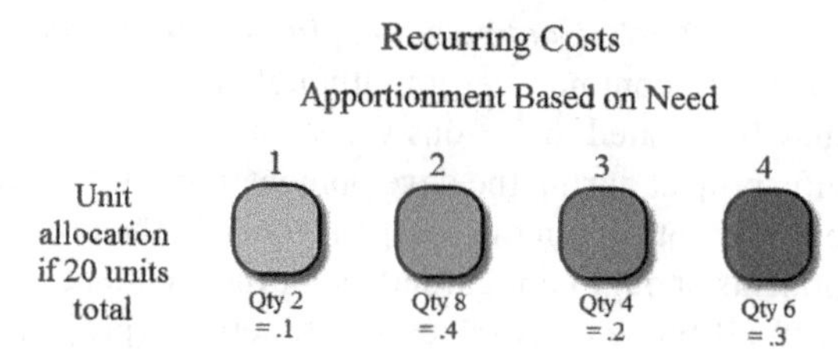
Recurring Costs
Apportionment Based on Need
Unit
allocation
if 20 units
total
1
2
3
4
Qty 2
= .1
Qty 8
= .4
Qty 4
= .2
Qty 6
= .3
- Purchasing allocates costs based on unit consumption required by program
- Work Orders are packaged together from all cost objectives to minimize fabrication set-up charges
- Set up charges are equally allocated to each cost objective
- Item fabrication, inspection and acceptance costs are allocated based on unit quantity
- e.g. if 20 units are produced each unit receives equal allocation with cost apportionment based on cost objective quantity

FIGURE 1.26 Shared design non-recurring and recurring cost allocation.

tied to any single cost objective. As it is utilized across multiple programs, each shared design item would have a distinct configuration identifier. In some cases, the configuration identifier could be a standard number for a family of items with sub-differentiation by part number. Individual part numbers are incorporated into the product's structure. Typical allocations for recurring and non-recurring costs for multiple programs with shared design are shown in Figure 1.26.

In all shared design scenarios, the data markings are applied after release instead of being embedded in the document. This may be accomplished by modifying metadata fields in the PD/LM system not subject to engineering (e.g., contract number, charge number, program name, configuration item identifier, etc.). Proper CM of the PD/LM system would include a metadata field that allows the document legends to be chosen from a pull-down menu or the legends to always be a predetermined choice. Ability to modify the metadata fields should be given to a finite number of properly trained individuals performing the record and release function who are closely connected to the product area functional management. A shared design for four different cost objectives is shown in Figure 1.27.

In the case of a firm doing business with domestic commercial, international, and U.S. government customers, cost accounting as well as IP and International Traffic in Arms Technical Assistance Agreement considerations must be addressed prior to data being distributed outside the enterprise. CM implementation includes document format, when data markings can be applied, charging of recurring and non-recurring cost elements, and data delivery constraints.

Maximum effectiveness of CM application is achieved using a single drawing in both cases. This requires document formats configured to support the application of all document markings after release. PD/LM systems should be configured to apply such markings when documents are output from the system and must be able to export single, multiple, or all documents associated with any individual product with a single request in an un-editable format such as portable data file (.pdf). Documents may include drawings, analysis, or supporting documentation such as reports, minutes, action items, specifications, and vendor statements of work.

Digital artifacts are artifacts produced or generated within an engineering environment. They describe any aspect of the product executed or used in execution during production. Digital artifacts include data, algorithms, processes, models, and 2D and 3D representations. Digital artifacts when combined with drawings, analysis, and supporting documentation are known as digital information.

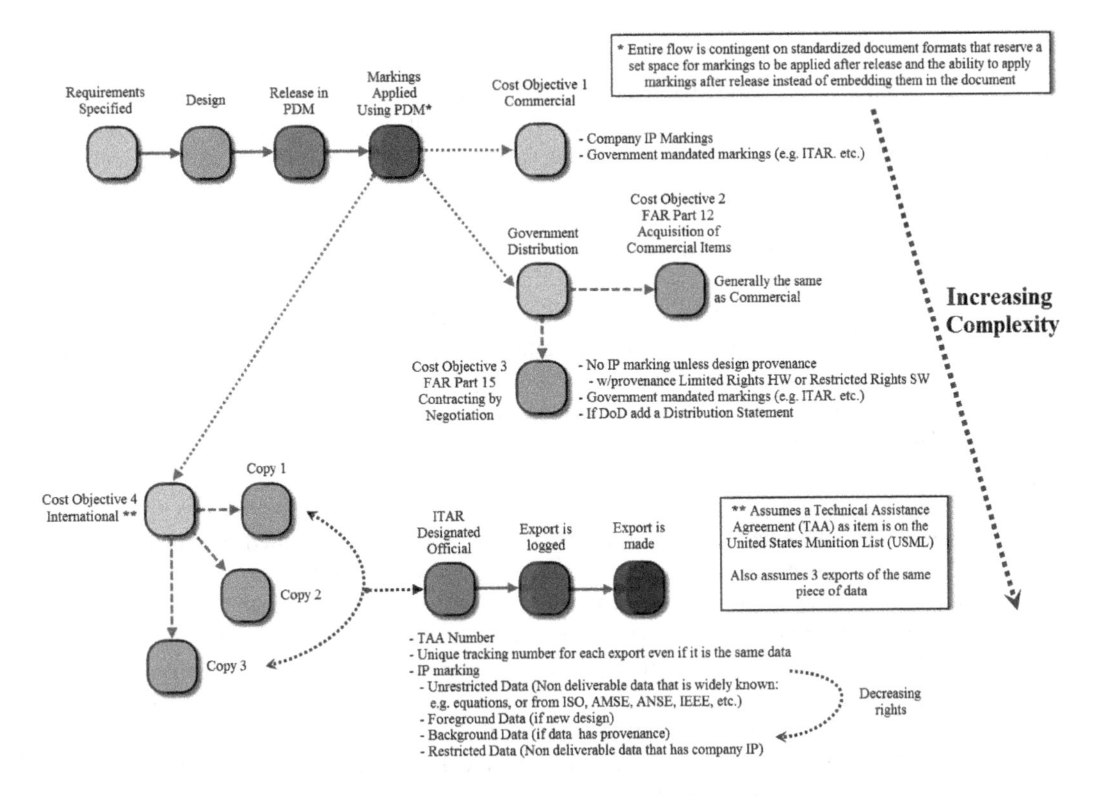

FIGURE 1.27 Typical shared design with four different cost objectives.

All digital information should be under some form of configuration management. During e-mail discussions with Kim L. Robertson on March 14, 2018, pertaining to Revision C of SAE/EIA 649-C, Rick St. Germain, managing director, CMPIC Canada, kindly provided the following insight for this edition:

> The engineering process generates vast amounts of knowledge, originating in the heads of engineers. To be useful to others, this knowledge must be codified (rendered into a portable form using symbolic representations) into a medium. I think about this as "content deposited into a container" (e.g., Engineers render design intent knowledge into a 3D model format held in a computer file or data store). We humans do it this way because we haven't found a way yet to handle knowledge directly (or even measure it, for that matter), so we handle it by its container. The most valuable bit is the content. The container is merely used to hold it. (e.g., When I buy a nice bottle of wine, I'm paying (mostly) for the wine, not the bottle.)
>
> The codification (rendering) process translates knowledge into a symbolic format that falls into two broad classes: (1) Book form whose purpose is to inform the reader, and (2) Software form that actively executes on a computer to create a tangible effect, beyond merely informing. The main point here is that we need to distinguish the content from the container. A Word file (the container) holds text-based content that informs—it does not itself execute, even though it is stored on a "digital" medium. It's the Word app that executes and allows us to modify that content (a tangible effect), so the Word file content is still Book form—a document—even though it is on a digital (not paper-based) medium.

Rich media pdf files known as 3di pdf or smart pdf files have interactive content embedded within them and are projected to be the norm for drawing data deliverables by 2026. Many CAD software programs today can export 3di pdf formats. The 3di pdf model image can be rotated and contains the ability to pull up the BOM for any portion of the model selected within the pdf.

1.6 SHARE DESIGN DECISION

During a visit to the Proton expendable rocket-manufacturing facilities in the former U.S.S.R. by a U.S. based launch vehicle provider, it was discovered that manufacturing tolerances for mechanical parts were only held to ±0.00025 cm if the part interfaced with another part (a critical interface). For non-critical interfaces, manufacturing tolerances for mechanical parts were only held to ±0.0025 cm. After a debrief of the market segment, the president and several vice presidents were tasked with evaluating the firm's commercial booster design to determine how much could be saved by relaxing manufacturing tolerances.

They were also tasked to determine which items represented the greatest savings if a shared design relationship existed between its U.S. government and commercial launch booster business. At the time, the company was also designing a new heavy booster. It was estimated that total customer need would not exceed 50 boosters when the decision was made.

Because of the internal evaluation, the following recommendations were made.

1. The configuration manager recommended that new design efforts and any change to existing designs should include relaxed tolerances. This should be extended to include any shared design elements between the commercial and government booster lines if weight and center of gravity were not diminished below the advertised booster load lift limits. CM estimated that all drawings for the existing design's shared design items could be converted in less than one month due to the amount of design activity at the time with a unit savings on every part manufactured.
   a. Mission assurance concurred with the CM recommendation.
2. The vice president for the launch booster product line discovered that one fabrication process took 200 hours to complete for a part that was critical to each launch vehicle, and recommended that the company invest €1,570,000 to automate the production process. At the time the fully burdened rate for production engineers was €87.33 per hour, resulting in a net savings of €17,466.90 per unit.
3. The vice president of production was concerned about part quality if all machining was not held to the same standard due to the retraining involved, and recommended against relaxing any tolerances. Instead he recommended investment of €1,050,000 in an automated welding fixture that would reduce the cost to weld launch booster segment skins from €6,986.76 per seam to €873.34 per weld seam. There were five seams on each stage and three stages for a net savings of €13,100 per launch booster.
   a. Systems engineering was also against relaxing tolerances on manufactured parts due to the time it would take to retrain the engineering design staff and marketing considerations regarding the firm's stringent manufacturing processes.

The firm decided to proceed with recommendation 2 due to the total cost savings per unit. Only 30 more boosters were ever produced before the launch vehicle was phased out.

Here are some additional pieces of information regarding the breakeven point for each recommendation.

1. Recommendation 1 was not implemented. The review committee determined the savings not quantifiable as the implementation would be phased into existing design and implemented on all new design. Estimated but not verifiable savings were 10% of the total manufacturing hours and a reduction in rejected parts of 90%. Drawing design time averaged 20 hours per drawing, and 1,400 drawings were estimated for the new design of the heavy booster. Design impact for existing designs was estimated at €4.366.72 total.

2. Recommendation 2 was implemented. It later turned out that the breakeven point was the 90th rocket as each rocket required only one part produced by the automated production process, as shown in Equation 1.5.

$$€1,570,000 / €17,466.90 = 89.88$$

EQUATION 1.5 Breakeven point calculation for recommendation 2.

As only 30 units were made, a scant €524,007 of the €1.57 million implementation cost was recouped after installation. The implementation was 33% effective (30 units made/90 units to break even = 0.3333). Implementing this project was a misguided effort that expended valuable resources with negative payback.

3. Recommendation 3 was not implemented as the review committee determined the breakeven point to be unit 80, as shown in Equation 1.6.

$$€1,050,000 / €13,100.17 = 80.15$$

EQUATION 1.6 Breakeven point calculation for recommendation 3.

2 Overview of the Supporting Enterprise Infrastructure

2.1 QUESTIONS ANSWERED IN THIS CHAPTER

- What enterprise elements does configuration management (CM) directly interface with?
- Is there a clean process interface with release and change management?
- What is the process to hand off to operations in a clean and consistent manner?
- How is profitability associated with CM implementation?
- What is the connection between innovation and CM implementation?
- Describe an incremental and iterative development model and the role of CM in that method.
  - What are their key elements?
  - What concerns are associated with the key elements?
- What is the global technology solutions (GTS) backbone and how does it affect CM? (GTS and IT will be used interchangeably in this book.)
- Is there a sane way to manage applications and tool upgrades?
- What are the challenges faced in the implementation of GTS systems security?
- How are platform transitions managed from a configuration perspective?
- What is the importance of CM of modular systems?
- What elements of cost comprise general and administrative (G&A) activities?
- What elements of cost comprise overhead (OH) activities?
- What internal and external forces drive intellectual property (IP) decisions?
- How do decisions regarding shared design, design reuse, common design, and leveraged innovation impact IP?
- What is the International Traffic in Arms Regulation (ITAR) and how does it impact CM domestically and internationally?

2.2 INTRODUCTION

CM activities influence and are influenced by many elements and organizations within the enterprise. Each speaks in a language that is often profession-centric. All are working toward the same goals of survivability, market share, and profitability. They aim for environmental sustainability and cultural diversity, while remaining compliant with local, national, and international laws and meeting contractual requirements. The CM implementation must become multilingual within the enterprise to mitigate profession-centric "job speak." The implementation must be based on an understanding of not only the language and purpose of each element, but also how decisions regarding managing a configuration influence and are influenced by each element. It is very important to understand the responsibilities of each of these elements and organizations and the CM activities each performs. At the very least, data management and change management are involved in every case.

Without an understanding of each element, it is difficult to gauge the effectiveness of CM implementation or to know if the proposed improvements to processes result in cost savings and an increase in earnings before interest and taxes (EBIT). Everyone is responsible for EBIT and the calculations are made using Equations 2.1–2.4. Operating expenses, COGS, depreciation, and amortization needs from a product line perspective will change over the product life cycle as well as by market segment. Decisions regarding managing the configuration impact each of the cost elements at the product and enterprise levels.

$$\text{Expenses} = \text{Operating expenses} + \text{Cost of goods sold} + \text{Depreciation} + \text{Amortization}$$

EQUATION 2.1 Expenses.

$$\text{Operating expenses} = \text{OH} + \text{G\&A} + \text{Administrative expenses}$$

EQUATION 2.2 Operating expenses.

$$\text{COGS} = \text{Inventory} + \text{Nonrecurring design labor} + \text{Recurring production labor}$$

EQUATION 2.3 Cost of goods sold (COGS).

$$\text{EBIT} = \text{Revenues} - \text{Expenses}$$

EQUATION 2.4 Earnings before interest and taxes.

If we step back and look at all available markets as an ecological system inhabited by companies behaving in predator–prey and symbiotic relationships, then a market segment is like an ecological biome. Within it you find interrelated service providers and manufacturing and user communities within a specialized social formation. The biome is subject to variable environmental conditions acting on a loosely integrated system. Communication is one factor in this technocenosis. As it evolves, it both adds and reduces complexity to the overall biome. Naturally occurring biomes are shown in Figure 2.1.

In biocenosis, humidity, altitude, temperature, and territorial configurations play a part in the biome. In a technocenosis-based ecology, financial gain, environmental impact (both internal infrastructure and biological overall ecological impact), perceived need, legal, ITAR, and other imposed regulations as well as profitability and growth potential form a similar stressor set relative to survival in the biome. Under technocenosis, the need for reduction of information to a portable form requires a specialized data solution. Only comparatively recently has GTS come to mean "electronic" GTS and that computer has supplanted electronic computing.

Cave painting, quipus, word pictures, cuneiform, tokens in clay envelopes, tattooing messages on shaved heads, quills and ink, block printing, movable type, teletype, strike-based bar, wheel and

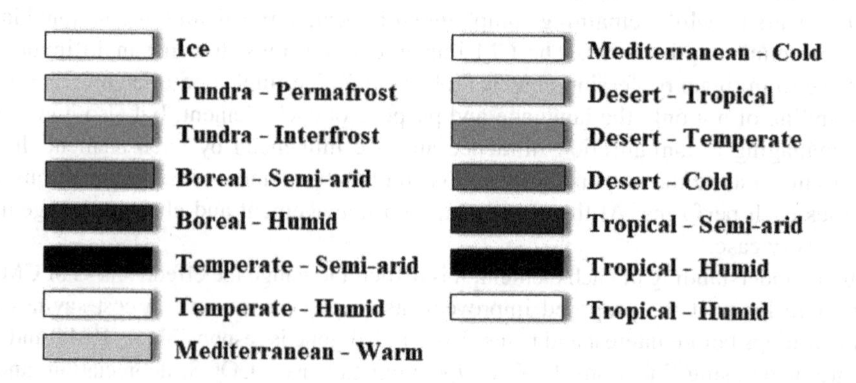

FIGURE 2.1 Biomes in nature.

ball, and dot matrix printers were supplanted in turn by inkjet, laser, and portable data format files. Intranet, Internet, and the cloud interrelate and follow in much the same way. Elements of the cloud exist internally to many companies today, taking the form of distributive virtual servers that are easily scaled to need. Aspects of the cloud are also being implemented externally to many technology companies as well. As with any product reaching a tipping point due to complexity, most of the financial stream both expended and acquired is not subject to the initial design, development, test, evaluation, and asset initial sale, but rather to the maintenance and data reduction of the item and the subcomponents over its operational life cycle. In a spacecraft, this secondary aspect is often as high as 90% of total mission cost. With aircraft and missile defense systems, it is more simply due to how long they remain operational.

Increasingly, profitability is dependent on infrastructure security. How we view security, IP, ITAR, and other imposed requirements is being challenged. Soon we will be forced to change simply because it will be too painful to continue doing things the way we have been. The cloud, 3D printing, distributed manufacturing, integrated design/development, and MBE are gaining a foothold in every technology-based biome. They are supplanting existing technologies and evolving rapidly.

Our current definitions of *secure* will also change. Data in the cloud is no more or less safe than any other enterprise GTS solution; it is simply different, and each solution has its place. The only real differentiator is the capability of someone to break security protocols protecting the data. The technological landscape is in constant flux, and as the ecological landscape changes, so must those companies that live in it. It is a simple case of recognize, accept, adapt, integrate (where it makes sense), and survive. Those who do this well will survive longer.

2.3 INFORMATION TECHNOLOGY

GTS relies on a highly sophisticated integrated level of CM activities. This is contrary to the trend of many companies that have moved GTS out from under the overall CM infrastructure and treat it as if it were a separate service provider. Of all the firms evaluated, only three were found to have a released CM plan for GTS.

The growth of the importance of GTS follows a natural progression when its history is examined. Jeremy Butler[1] highlighted four developmental periods for solving the communications, input, processing, and output cornerstones of GTS. Since 2003, Jeremy Butler's vision must be expanded to add the information age.

1. Pre-mechanical age: 3000 BCE–1450 CE
2. Mechanical age: 1450–1840
3. Electromechanical age: 1840–1940
4. Electronic age: 1940–1990
5. Information age: 1990 to present

2.3.1 A Brief History of Information and Information Security

Development of a symbolic form of verbal communication (so that data could be recorded, transferred or stored, retrieved, and recommunicated) was critical to the development of GTS as we understand it today.

We live in a world where crypto mining and Internet of Things (IoT) malware are becoming more automated. This necessitates equally strong automated countermeasures where companies are always doing too little, too late. In the third quarter of 2018, cyber criminals released an average of

[1] Butler, J. G. (1997). *Concepts in New Media*. University of Arizona, Tucson, AZ.

480 new threats per minute,[2] a 70% jump from the previous quarter. The idea of trust is now entering the lexicon and raising the following questions:

- How can I trust that you are who you say you are, when you access your computer and the systems interfacing with it?
- Is a two-factor authentication system good enough for access to the intranet?
- If a two-factor authentication system is good enough, then what form should it take?
- Are fingerprints and a password enough?
- If not, are biometric scans of an employee's facial features analyzed against a master biometric scan plus a password enough?
- Do you need to consider implementing some form of triple authentication?

We can't give you an answer. The world is changing too rapidly. What we can give you is a look at the history of information and information security. Let's start the journey in the Tigris and Euphrates valley where one pressing concern was how to document a contractual agreement.

Clay tokens bearing wedge-shaped markings known as *cuneiform* found in Turkey, Syria, Israel, Jordan, Iraq, and Iran were used for record keeping as early as 8000 BCE. (Any form of writing that is composed of wedge-shaped marks is known as cuneiform.) Hollow clay envelopes (known as *bullae*) marked to correspond with the clay tokens captured inside have been found in Sumer from around 3500 BCE (Figure 2.2). Denise Schmandt-Besserat, who researched the nature of information management, made a compelling case that the tokens continued within the clay envelopes recorded grain, livestock, oil, and perhaps manufactured goods. This was the equivalent of the accounting spreadsheet combined with a security system. Once a bulla was broken, the contents were traded for the goods they represented. The tokens inside the bullae were then re-used. Eventually, clay tokens were replaced by documenting transactions on clay tablets with wedge-shaped (cuneiform) symbols imprinted in the wet clay as an early pictographic writing system.

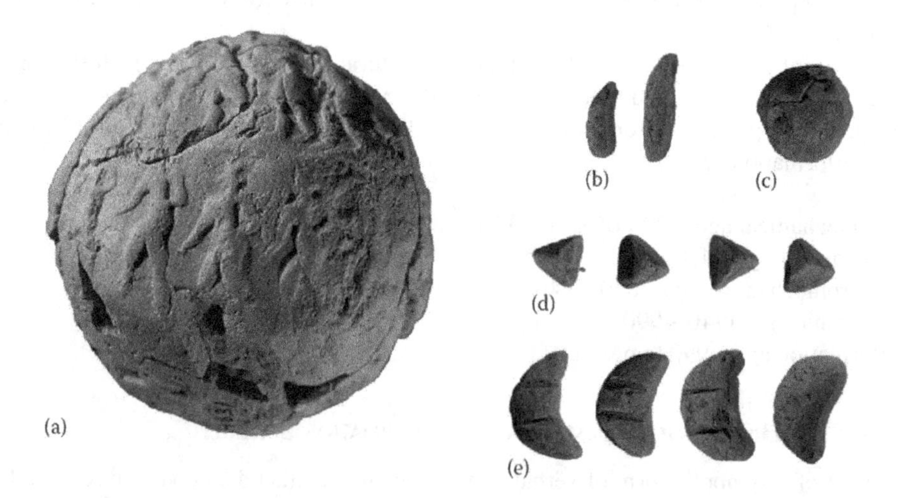

FIGURE 2.2 Bullae envelope with 11 plain and complex tokens inside from Near East, around 3700–3200 BCE: (a) 6.5 cm bullae with row of men walking and a predator attacking a deer representing wages for four days' work; (b) two small measures of an unknown commodity; (c) one large measure of barley; (d) four days; and (e) four measures of metal.

[2] McAfee (2018). *McAfee Global Threat Intelligence*. McAfee: McAfee Labs Threat Report (p. 24). https://www.mcafee.com/enterprise/en-us/assets/reports/rp-quarterly-threats-dec-2018.pdf

Other forms of security were utilized until information could be retrieved by the intended recipient. The fictional accounts contained in "The Gold-Bug" and "The Purloined Letter,"[3] "The Adventure of the Dancing Men,"[4] and the lanterns in the Old North Church during "Paul Revere's Ride"[5] were ingenious but the historical record provides even more ingenious methods. In 49 BCE[6] during the Persian wars, the Greek Histiaeus is said to have sent a message to Aristagoras tattooed on the shaved scalp of a slave. Once his hair grew back, the slave started walking and found his way to Aristagoras, who shaved the slave's head and retrieved the message. Due to this intelligence, he attacked and burned Sardis with help from the Athenians and Eretrians. Steganography also took the form of messages etched on wooden tablets that were subsequently covered with wax. The wax was incised with something mundane like accounting data and the hidden message was passed on to its intended source, who melted off the wax to read the message.

One area of information security that has already derived some benefit from steganography is digital rights management.[7] Cryptography also has important connections to steganography. Since its inception, digital aspects of information security have impacted the ethics surrounding both steganography and cryptography. Today, steganography can hide data in redundant pixels, characters or sounds as computers treat sound, image, and text files as equal parts of the bit stream.

Jeremy Butler[8] traces its beginnings to the development of the Sumerian cuneiform script by 3000 BCE. Thousands of clay tablets have been found and translated; most of them document business transactions. Sumerian cuneiform script was superseded by the Phoenician alphabet in the Neo-Assyrian Empire between 934 BCE and 609 BCE. Yet this may be an oversimplification, as the precursor of the quipus existed as early as 3000 BCE in Andean South America. Quipus of up to 2,000 individual cords made from dyed llama or alpaca hair strings recorded numeric and other values in a base 10 positional system.

Quipus (Figure 2.3) as a communication and information storage medium offered several advantages over cuneiform script for storing massive quantities of data. Quipus utilized to full advantage the textile-making technologies of the time. Strings could be made with a right-to-left spin or a left-to-right spin. Three types of knots were typically used:

- Single—simple overhand or Flemish knot
- Long cord wrapped around itself two or more times—double, treble, fourfold, fivefold … etc.
- Figure-eight or e-knot (the cord forms the shape of the number 8), and the ends of the string exit through the holes of the 8.

Knots could be made in two different orientations, resulting in different slants to the knot axis. Information (pendant) cords could be attached to the main cord in two ways and cords were also made in differing colors and color combinations, with the overall result of multiple data parameter choices that could be reconfigured as needs changed over time.

Yet the capturing and sharing of technical information predates even this. The first writing system was believed to have been invented around 4000 BCE in the late Neolithic. Petroglyphs have been found from as early as 10,000 BCE.[9] Evidence exists that may prove to be a calendar in

---

[3] Poe, E. A. (1904). *The Purloined Letter (from The Works of Edgar Allan Poe—Cameo Edition).* Funk & Wagnalls Company.

[4] Doyle, A. C. (1903, December). The adventure of the dancing men. *The Strand Magazine.*

[5] Longfellow, H. W. (2013, September 28). *Paul Revere's Ride.* Retrieved September 6, 2014, from http://poetry.eserver.org/paul-revere.html.

[6] Herodotus. (2003). *The Histories, Book 5,* Translated by A. D. Selincourt. London: Penguin Group.

[7] Grodzinsky, Frances S. (n.d.). *The Ethical Implications of the Messenger's Haircut: Steganography in the Digital Age.* Retrieved September 20, 2013, from http://bibliotecavirtual.clacso.org.ar/ar/libros/raec/ethicomp5/docs/htm_papers/25Grodzinskypercent20Frances percent20S.htm.

[8] Butler, J. G. (1998). *A History of Information Technology and Systems.* University of Arizona, Tucson, AZ.

[9] Diringer, D., and Minns, E. (2010). *The Alphabet: A Key to the History of Mankind.* Kessinger Publishing.

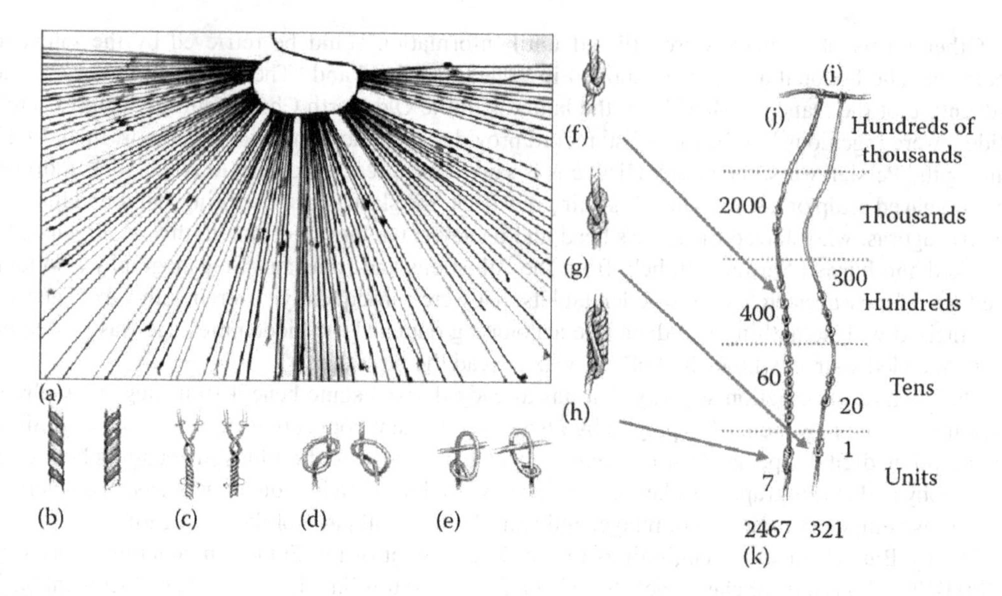

FIGURE 2.3 Quipus: (a) typical completed quipus; (b) cord color; (c) twist—right or left; (d) knot direction; (e) method of attachment; (f) single knot; (g) figure-eight knot; (h) long knot; (i) main cord; (j) two hanging cords; and (k) cord totals.

15,000 BCE,[10] and a symbol-based communication has been found from as far back as 30,000 BCE in Chauvet-Pont-d'Arc Cave in the Ardèche region of southern France. A simplified timeline of the development of information history is provided in Figure 2.4. The recently discovered cave paintings in three different sites in Spain are considered older than 61,982 BCE and may push the date for symbol-based communication back an additional 30,000 years, implying Neanderthal authorship.[11]

The evolution of related technologies in the pre-mechanical age concerned the decipherable recording of information that was critical for survival—invocation of deities for a good hunt or good harvest, calendars for marking the seasons based on astronomical observations, tracking elements of business, and contracts for goods and services. Basic design parameters for engineering purposes and other exigencies of the moment were the primary concern. City-state infrastructures grew and merged. As a result, new ways of capturing information in a recognizable form evolved to support gathering of data to be recorded. James Burke[12] traces the evolution and connections between key technologies such as electronic computers that gave rise to the information age.

Although not specifically addressed by James Burke, the evolution of the software programming capability based on the I Chin or Zhouyi traces to 3 BCE. Within it, 64 hexagrams are described that formed the base for the binary number system.[13] This led to George Boole's creation of Boolean algebra[14] and its application to digital logic and computer science. Prior to the advent of electro-mechanical devices capable of operating in a strictly yes/no, on/off mode often associated with computers, there is a rich history of analog devices reaching back to the Antikythera mechanism in

[10] Jones, D. (2013). *The Mystery of Göbekli Tepe and Its Message to Us.* Retrieved April 15, 2014, from http://www.newdawnmagazine.com/articles/the-mystery-of-gobekli-tepe-and-its-message-to-us.

[11] Seglie, D., Menicucci, M., and Collado, H. (2018, February 23) *Neanderthal cave art, Science,* 23 February 2018, Vol. 359, Issue 6378, pp. 912–915 http://cesmap.it/neanderart2018-international-conference-report-from-science-23-feb-2018-vol-359-issue-6378-pp-912-915-u-th-dating-of-carbonate-crusts-reveals-neandertal-origin-of-iberian-cave-art/

[12] James Burke (born December 22, 1936) is a British broadcaster, science historian, author, and television producer.

[13] Leibniz, G. W. (1703). *Explication de l'Arithmétique Binaire VII.223.* Retrieved August 22, 2014, from www.leibniz-translations.com.

[14] Boole, G. (1854). *An Investigation of the Laws of Thought.* Palm Springs, CA: Wexford College Press, Accessed from Gutenberg Project online on 2010.

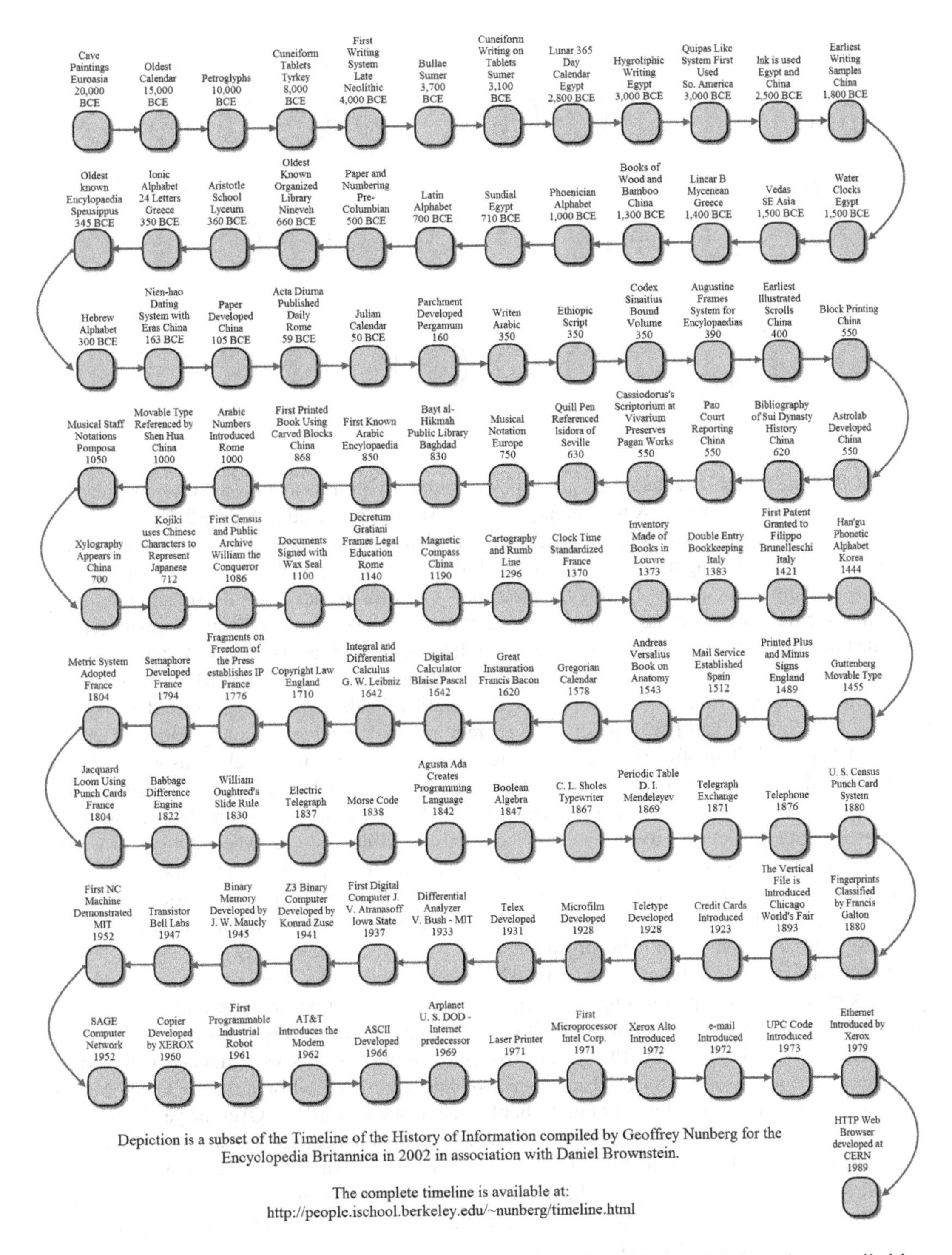

FIGURE 2.4 History information timeline depicting a subset of the history of information compiled by Geoffrey Nunberg for the *Encyclopedia Britannica* in 2002 in association with Daniel Brownstein. (The complete timeline is available at http://people.ischool.berkeley.edu/~nunberg/timeline.html.)

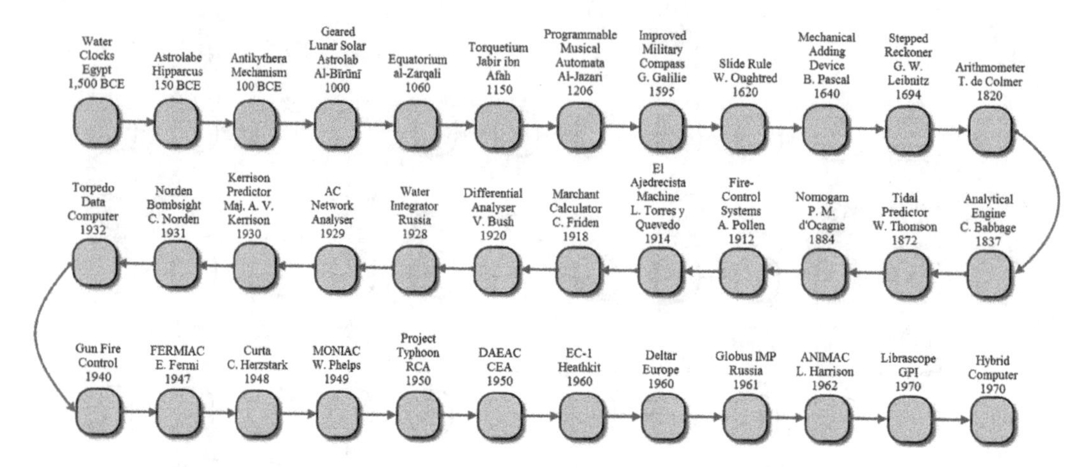

FIGURE 2.5 Analog device history.

100 BCE. Some of these included cryptography machines with information security counterparts existing in binary form today. A brief analog device history is provided in Figure 2.5.

The term "computer" was originally created to describe someone who does computation. From 1920 onward, "computing machine" was applied to machines that could make the same calculations as the human computers. During the electronic age, with the advent of the digital computing machine, the term was shortened to "computer." For the human computer, information management entailed detailed calculations and record keeping. Inventions such as the ACCO fastener,[15] three-ring binder, vertical and horizontal file, high-density compact mobile-shelving file systems, ledger sheets, and numerical filing systems using Rolodex rotating file devices were considered the highest form of GTS. Into the late 1980s, paper products required to document the configuration of a complicated deliverable item often weighed the same as or more than the item itself. This was certainly true for the delivery of the first commercial Titan launch vehicle by Martin Marietta in 1990.

A major component of GTS at this time included short-term and long-term data repositories. Short-term data repositories were managed locally using filing systems, whereas long-term data storage for data that was not accessed on a regular basis was managed using a warehouse solution. This gave rise to a new industry market sector dedicated to data storage and retrieval of paper records. Then, as now, the long-term data warehouse ran the risks of not having a backup for the data map and, occasionally, of data corruption.

Loss of the index of warehouse container content resulted in chaos. Data recovery was a daunting task of retrieval and performing a new inventory on every container's content to recreate the data map. This happens more often than industry would like to admit. It is due to a lack of centralized indexed records in the PD/PL system. Occasionally warehouses were damaged by natural and human-made causes. Historical Titan launch vehicle records were lost at the Lockheed-Martin Waterton facility due to a leaking roof in the bunker the data was stored in. Over the years, the carefully indexed documents were reduced to a 0.91 m deep pool of paper pulp that was only discovered when the watertight doors to the bunker were opened to retrieve Titan I archives pertaining to the U.S. government property. Today's archive systems are concerned with avoiding similar issues that result in what we can call *bit-pulp*, a condition that manifests when the server data map becomes corrupted and you have no backup.

The issue is ongoing. According to a 2018 article, "NASA has lost a wide array of historical spaceflight memorabilia such as an old lunar soil bag, former spaceflight hand controllers, and even a test lunar rover over the last few decades. That's according to a report from NASA's Office of

[15] A kind of two-prong fastener manufactured by ACCO company; see http://www.buyonlinenow.com/search.asp?1=1&keywords=PRONG+FASTENER&subcat=ULAA&manufact=1700005.

the Inspector General, which analyzed how the space agency oversees its historical assets. While procedures have improved at NASA, a few unique pieces of storied spaceflight property have either been misplaced or taken by ex-employees."[16]

With the ability to create and store records electronically, information security has taken on a new vibrancy. Not only is it possible to record and store data with metadata fields providing data mining capabilities only dreamed of a scant decade ago; but the influence of modern trends in how information is viewed from the corporate level is redefining what is acknowledged as information security. In many cases, it was found that maintenance of hard copy files is no longer seen as part of information security but rather is relegated to a sub-function of records management associated with governmental regulatory compliance. In some cases, it is performed by the data management function under a CM implementation and in others by a G&A function under the contracts department due to the regulatory aspect of many of the records so maintained.

Information technology (IT) is in the process of rebranding itself as the digital business group, integrated technology services, global technology solutions, and information services, and redefining itself as the study, design, development, application, implementation, and support or management of computer-based information systems. As the industry has yet to agree on a single name, we will use *global technology solutions* (GTS) throughout the rest of the book except in cases where IT is used in a quotation or a reference. In some cases, communications and video systems have been included in the definition, but these are being delegated to document operations and telecommunications in many market segments. Inherent in the definition provided above is maintaining security for those information systems and the content inside of them.

Modern technology generates new sets of terms. *Big data* is one of these; it refers to data collections so large that they must be managed on parallel servers instead of using single server and standard database management tools. GTS management must now include metadata trending in the CM implementation. According to International Business Machines, every day, 2.5 billion gigabytes of high-velocity data are created in a variety of forms, such as social media posts, information gathered in sensors and medical devices, videos, and transaction records.[17] Enterprise-driven data associated with any market segment is a subset of this dataset.

The IDC Digital Universe Study[18] sponsored by EMC, December 2012, projected that the digital universe in the United States alone would see growth from 898 exabytes to 6.6 zettabytes of data between 2012 and 2020. This is more than a 25% increase per year. This figure when combined with Western Europe will account for a scant 49% of the total data universe. Of the purely U.S. generated data, 25% is projected as requiring security, with 60% of that data or (15% of the total) having no protection at all. Additionally, 18% of the total U.S. dataset would be useful if properly tagged and analyzed, with an estimate that less than 0.5% is currently analyzed.

Conceivably, enterprise related data (although not evaluated for this book) would show a higher percentage of tagging and analysis due to its association with market segment dynamics and profitability. Yet, much information that is critical to other business aspects such as IP provenance remains only on paper or in pre-1990 record systems such as microfilm reels and microfiche sheets. A 30-year analysis of firms in the aerospace sector, when projected across industries supporting the firms used in the study, leads to the conclusion that data contained on drafting vellum sheets; microfilm; microfiche; 1.27 cm wide reels of nine-track magnetic tape; hard diskette platters; 20.32 cm, 13.335 cm, and 8.89 cm floppy disks; and Iomega zip diskettes in archived data may account for 30% of an enterprise's IP. Migration of this data to usable formats is difficult and expensive and rarely lends itself to automation.

---

[16] Grush, L. (2018, October 22) NASA lost a rover and other space artifacts due to sloppy management, report says. https://www.theverge.com/2018/10/22/18009414/nasa-lunar-rover-historical-artifacts-lost-space-shuttle,

[17] IBM. (2013, September 22). *What Is Big Data?—Bringing Big Data to the Enterprise.* http://www-01.ibm.com/software/au/data/bigdata/ retrieved 2014-08-12.

[18] EMC Corporation. (2012). *IDC Digital Universe Study: Big Data, Bigger Digital Shadows and Biggest Growth in the Far East.* EMC Corporation.

FIGURE 2.6 Rosetta Stone—British Museum.

This is borne out by failure to consistently upgrade facilities. Toward the end of the Space Transportation System (STS) era, the launch facilities at NASA's Kennedy Space Center in Florida were still using 086-based computers that had seen few upgrades since April 12, 1981, the launch of STS-1 (Columbia). Replacement parts were being sourced on eBay.

At the other extreme, New Horizons, a flyby mission to the Pluto/Charon system and the Kuiper belt, was the first in NASA's New Frontiers line of moderate-scale planetary missions and was the first mission to explore Pluto and its moons Charon, Nix, and Hydra. Launched on January 19, 2006, it flew by Pluto at 12,500 km above the surface on July 14, 2015.[19] This was followed by a fly-by of the Kuiper belt object Ultima Thule at 3,500 km from the surface, on January 1, 2019. Ultima Thule is 10.46 billion kilometers from Earth. It took six hours for data to be received on Earth. To avoid possible issues due to GTS upgrades and obsolete hardware, the entire flight operations system computers and software systems for the multispectral visible imaging camera and the linear etalon imaging spectral array were mothballed by the instrument provider shortly after launch.

The problem of data transferability is perhaps exemplified by the now famous Rosetta Stone (Figure 2.6) found July 15, 1799, by Napoleon Bonaparte's troops in the Nile delta. Created between 204 BCE and 181 BCE in the reign of King Ptolemy V, the stone contained 14 lines of hieroglyphic script, 32 lines of Demotic script, and 53 lines of Ancient Greek. It is often forgotten that the Ptolemys were Greek. Ptolemy I was one of Alexander the Great's generals. The Rosetta Stone chronicles Ptolemy V's victory over an Egyptian group that violently rebelled against Greek rule in what is known as the Great Revolt. Ptolemy V was about 14 at the time.

Despite its being viewed as clever propaganda to prop up the young king's authority, the triple inscription enabled the deciphering of Egyptian hieroglyphics. Jean-François Champollion (1790–1832) made a crucial step in understanding ancient Egyptian writing when he pieced together

[19] Reuter, D. (2008). Ralph: A visible/infrared imager for the New Horizons Pluto/Kuiper belt mission. *Space Science Review*, Vol. 140, No. 1–4, 129–154.

the alphabet of hieroglyphs that was used to write the names of non-Egyptian rulers. He announced his discovery in a paper at the Académie des Inscriptions et Belles-Lettres at Paris on Friday, September 27, 1822.

So, why do we include it in a book about CM? The answer lies not in the textual content of the Rosetta Stone, but in the fact that the knowledge contained on the stone was transferrable and decipherable some 2,000 years later. Many government programs are now lasting for 30–60 years after hardware delivery. Unless the historical digital records are upgraded to newer formats, all we will have to rely on are digitized copies of documents. This digital vs. digitized information conundrum rests squarely in the CM implementation.

Evaluating GTS storage needs is not an easy task. Often, metadata pulled from the CM implementation is mined to determine the historical storage size for data generated on a program-by-program basis. Once plotted, these trends can then be used as a predictor of future storage solutions. Although Moore's law provides some insight into data storage and retrieval solutions and needs by predicting a doubling in the number of transistors on integrated circuits (ICs) every 18 months, it does not consider rising sophistication in engineering software tools, additive manufacturing, smart 3D models in pdf format, or the potential impact of emerging nanotechnologies and heat dissipation improvements. Also missing from the mix is the move toward MBE.

Moore's law as a predictor of the number of transistors in ICs over time is always under scrutiny. It is based on single-layer IC chips. Samsung introduced a 128G V-NAND SSD 24-layer 3D flash memory chip in 2013, stepping around the 30 nm spacing issue required to eliminate transistor cross-talk. According to Moore's equation, this should not have happened until 2025 (Figure 2.7).

Eventually, Moore's law will be readjusted to account for the layered memory chip design. This may take the form of a simple step function, where the trend line simply ends for single-layer chip designs and continues at the same slope starting with the 128G V-NAND SSD 24-layer 3D flash memory chip, or it could change slope and Moore's law may be valid only for the single-layer chip density progression. In 2018, Intel demonstrated 3D packaging technology, called Foveros, which allows logic chips to be stacked atop one another with little loss of power. Adjustments to Moore's law will come about mostly due to clever physics and design.

The scaling of data storage needs is dependent on IC chip evolution. Needs must be evaluated based on adding additional capability due to demand. In this, it is much like evaluating manufacturing capability on the factory floor and adding additional production lines based on demand. In the case of storage, the needs evaluation slope is increasing in a non-linear way over time due to the rapid evolution in software used in product development as well as technological leaps forward. Åse Dragland stated in 2013 that a full 90% of all the data in the world had been generated over the previous two years.[20] A similar increase in data received and tracked by most GTS organizations has also been observed. This is due in no small part to how standard enterprise security information is gathered and processed. The similarities between infrastructure expansion to support manufacturing and GTS are shown in Figure 2.8.

Stanford University professor Nicholas Ouellette writes[21] in *Flowing Crowds* that a hydrodynamic model is the best one for studying crowd movement, whether it's a group of humans, a flock of birds, or a swarm of insects. The model can be used to better design pedestrian and vehicular traffic pathways. We believe that it can also be used for developing better GTS solutions. Leveraging the principles outlined in the article, GTS pressures may be better understood proactively as a function of employee density over time vs. reactively as individuals logged in. Peak employee densities in buildings map directly to the density of GST usage. Perhaps one day, bandwidth constrictions (slowing down of the Internet during peak hours) will be mitigated through advances by service provider based on peak

---

[20] Dragland, Å. (n.d.). *Big Data—For Better or Worse*. Retrieved September 28, 2013, from http://www.sintef.no/home/Press-Room/Research-News/Big-Data--for-better-or-worse/.

[21] Ouellette, N. (2019). Flowing Crowds, *Science*, 04 Jan 2019: Vol. 363, Issue 6422, pp. 27-28 DOI: 10.1126/science.aav9869

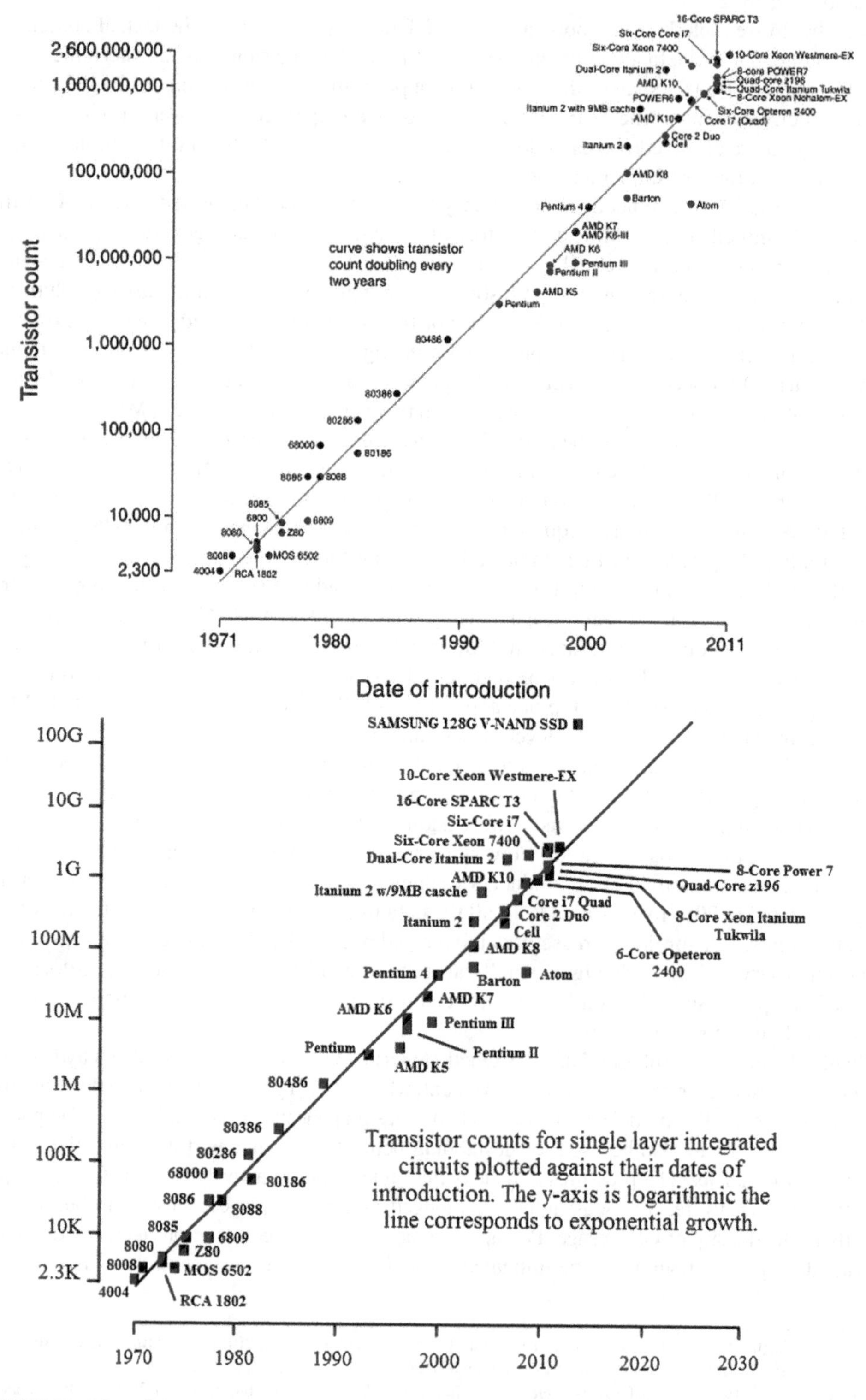

FIGURE 2.7 Moore's law.

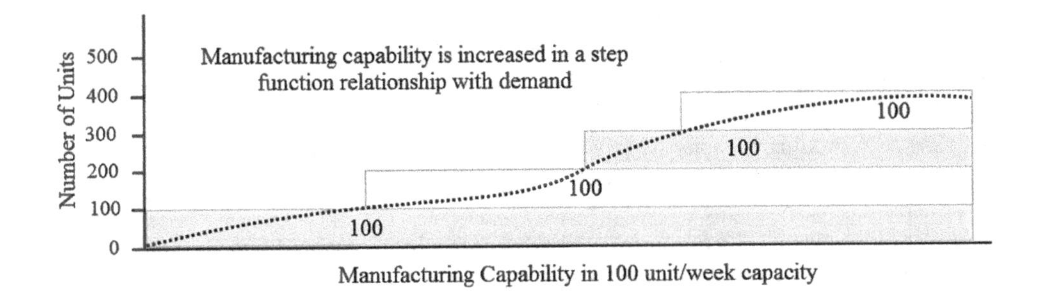

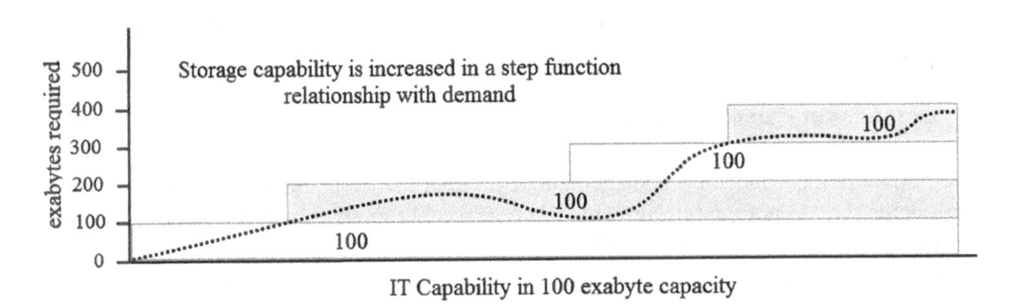

FIGURE 2.8 Manufacturing and GTS infrastructure block updates: (a) production needs increase based upon factory capabilities, and (b) GTS storage needs increase based upon server capabilities.

population densities. If not, we may experience rolling brownouts for Internet services just like we see in aging electrical grids during peak usage times. The collecting of data on peak population densities was already going on in 2018 by service providers analyzing cell phone locations.[22]

Prior to 1997, few firms implemented GTS solutions for monitoring employee movement between facilities, employee Internet and e-mail traffic, routine facility security checks, or e-mail attachment data content for political correctness, or other contents. Enterprise GTS solutions now include the collection and processing of such information. This adds to the non-linear growth of data storage and complicates the process of determining how to segregate data needed for daily transaction functions from that required in case of a physical security breach or security audit. In many respects, GTS of digital data lends itself to project management with strong CM implementation elements rather than to its early beginnings to manage the computing and server resources. To do this, many enterprises are moving to a form of cloud-based infrastructure like the Amazon Web Services platform introduced in 2002.

At this point, it is best if a few definitions are introduced.

- *Cloud computing:* Standardized elastic/scalable GTS-enabled services, software, and infrastructure components delivered through an Internet or intranet interface with built-in redundancy and failover.
    - In many respects, portions of the cloud are active, and use has been growing in every business model since the early 1950s. The 1950s were a time of large mainframe computers with dedicated terminals that were not only a large financial investment for firms and institutions, but also one that, when needed, used all its capabilities but, when not needed, sat idle for extended periods of time.
    - Recent introduction of IaaS/PaaS technologies to enable internal cloud computing.

[22] Kent, K., and Ng, V. (2018). *Smartphone data tracking is more than creepy – here's why you should be worried* The Conversation http://theconversation.com/smartphone-data-tracking-is-more-than-creepy-heres-why-you-should-be-worried-91110 retrieved February 28, 2019.

- *GTS resources:* The associated human resources, systems maintenance and upgrades, utilities, facilities, capital expenditures, physical and electronic security, disaster preparedness, and administration infrastructure.
- *Lead strategy:* Adding capacity to a GTS resource in anticipation of demand.[23]
- *Lag strategy:* Adding capacity when a GTS resource reaches its full capacity.
- *Match strategy:* Adding GTS resource capacity in small increments, as demand increases.
- *Cluster:* Independent GTS resources working as a single system based on reasonably similar hardware and operating systems (OSs).
- *Grids:* Organization of computational resources into logical pools that are loosely but collectively coordinated to constitute a virtual computer deployed on a middleware layer of resources.

2.3.1.1 Examples of Capability Demand

- Modification of product data management or product life cycle management (PLM) metadata fields that do not contain information constituting engineering data modifiable only by an engineering change order (ECO).
- Modification of established automated workflow to facilitate changes in policies, procedures, and practices (3P).
- Software updates across the infrastructure (e.g., Adobe Acrobat DC replacing v11.0, so that Word documents can be generated from other than a bitmap scanned pdf).
- A file server with restricted access.
- Rolling off one GTS backbone[24] infrastructure or server software suite to another to facilitate new software tools.

2.3.1.2 Examples of Capacity Demand

- A request to accommodate 3 EB of data generated from test data streaming via a frame capture utility or program.
- Increased computing or storage infrastructure due to a capability demand.

Capability and capacity demand are interlinked, with one often driving the other. Both are tied to the entirety of digital GTS (Figure 2.9). In any organization, CM of hardware and software will require both a physical GTS component and a digital GTS component. The digital GTS component is an element of OH. It continues to evolve as a profession, and in many companies, it has been allowed to set itself aside from EBIT considerations with a self-defined charter and mission statement. Although a level of independent self-determination is necessary in any organization, occasionally it can be to the detriment of the revenue-generating function of an enterprise if not done responsibly.

This is where CM implementation comes into play. Just as the survival of any species or product is dependent on its environment, survival of the enterprise is dependent on a digital GTS contingent, which enhances survivability. The GTS first line of support is the service desk or helpline. Other user interfaces may exist depending on the size of the organization supported. There are several basic GTS service desk infrastructure approaches: first in first out (FIFO), single-string approach based on criticality (SSC), and functional service request disposition (FSRD). This is shown in Figure 2.10.

---

[23] Erl, T., Puttini, R., and Mahmood, Z. (2013). *Cloud Computing: Concepts, Technology & Architecture.* Prentice Hall.

[24] * GTS backbone is the internally networked server structure that forms the backbone of the global technology services deployment within a company.

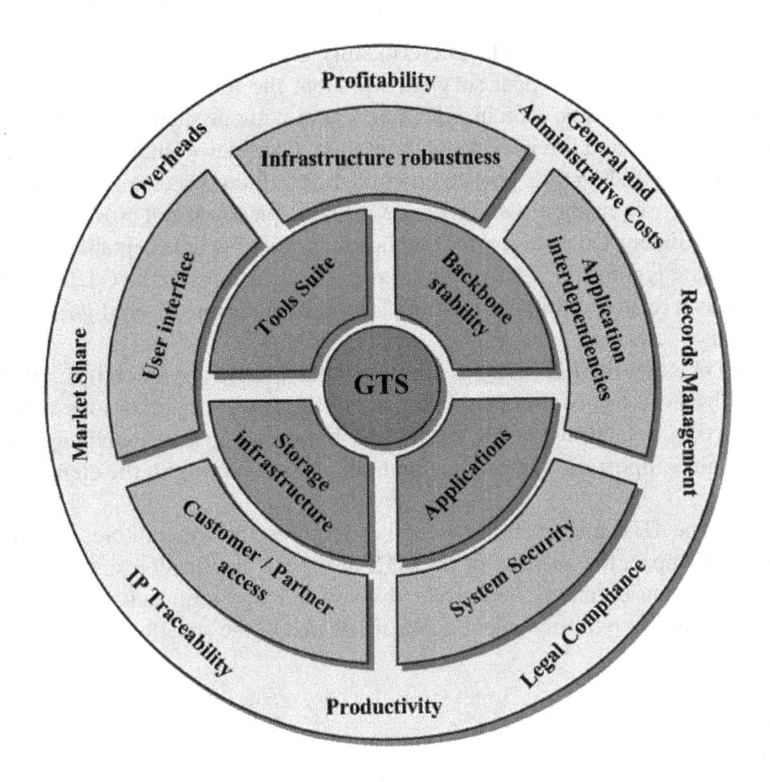

FIGURE 2.9 Scope of digital GTS.

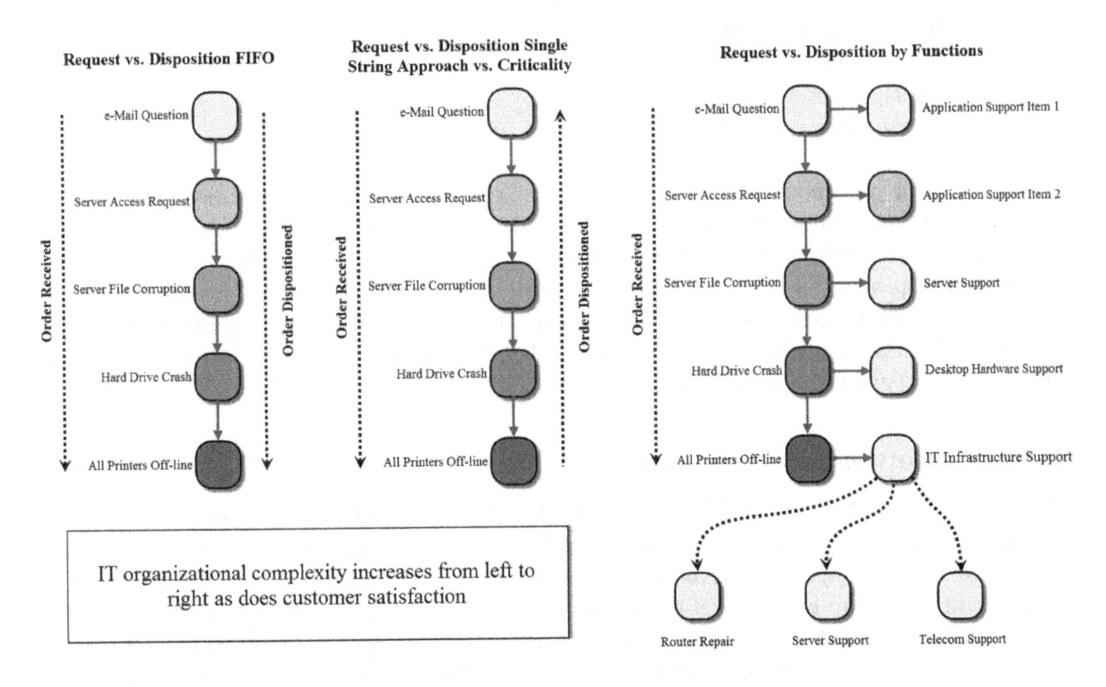

FIGURE 2.10 FIFO, SSC, and FSRD.

A FIFO approach does not consider the criticality of the request, and requests are processed in the order received. In a biological survival situation, the following survival guidelines apply. You can survive three minutes without air, three hours without shelter (except in extreme cold), three days without water, and three weeks without food. Consider what would happen if these were called in to the service desk in the opposite order of shelter, food, water, and air in a FIFO scenario. The internal customer requiring air would expire if that request was not moved to the top of the list and acted on immediately. Considering the implications, FIFO is unacceptable from a system GTS CM perspective, as it ignores the relative importance of other items in the GTS queue. We will discuss this further as it relates to the timing of implementation during the configuration control board (CCB) evaluations of changes.

SSC resolves some of the FIFO issues and may be a tenable approach in smaller enterprises. SSC has been observed to become increasingly ineffective in organizations with a GTS infrastructure supporting over 80 users due to the link between capability and capacity demand. A GTS infrastructure supporting 120 users soon finds itself adapting to the elements of the FSRD model.

Moving from one GTS architecture to another requires the expenditure of enterprise profits, which in turn may impact the stability of the GTS backbone, application suites, GTS system security, G&A costs, IP considerations, and productivity, which again impacts profitability. Utilization of a *best-of-breed* approach to implement applications tied to the specific needs of:

- Proposal management
- Program management
- Contracts management
- Systems engineering
- Assembly and test
- CM of hardware and software (includes firmware)
- Supply chain management
- Talent acquisition
- Data management
- Quality assurance

requires a larger support staff to manage the interfaces between the applications and may not provide an optimal solution. Choice of a single application solution that encompasses all the needs of the enterprise functional resources may not provide an optimal solution either.

The interdependencies between GTS, G&A, OH, IP, and ITAR necessitate a strategic view of the GTS system configuration, but focused management of GTS solution and planned phasing of updates to the technology involved to determine when block updates to the infrastructure will have the least impact and greatest payback.

2.3.2 Stability of Internal Infrastructure

Looking simply at the scope of GTS, as presented in Figure 2.9, it is hard to grasp the importance of the physical infrastructure required to support an intranet (internal network vs. Internet/external network or the impact failure in that infrastructure represents to the users of GTS services). The stability of end-to-end Internet paths is dependent on both the underlying telecommunication switching system and the higher-level software and hardware components specific to the Internet's packet-switched forwarding and routing architecture.[25] This applies to an intranet setting as well, as it is built on the same platform and utilizes transmission control protocol/Internet protocol,

[25] Labovitz, C., Ahuja, A., and Jahanian, F. (1998). *Experimental Study of Internet Stability and Wide-Area Backbone Failures*. University of Michigan, Ann Arbor, MI.

hypertext markup language, simple mail transfer protocol, and file transfer protocol and GTS infrastructure library (ITIL) methodologies.

2.3.2.1 GTS Backbone

Intranet stability is dependent on several factors that will be addressed briefly to provide a feel for the issues involved. This is intended not to make the reader an expert in these areas but to introduce the subject to make the subject and context familiar. This familiarization is necessary to facilitate a clearer understanding of the information provided in the chapters that follow.

2.3.2.1.1 Telecommunication Stability

Peter Racskó, in his article "Information Technology, Telecommunication and Economic Stability,"[26] points out that the increase of effectiveness of a system performance might lead to instability of operation. The conclusion is that we cannot exclude that unlimited development of GTS and telecommunication might induce structural instability.

Management of stability requires an understanding of what instability looks like, as well as the transitory states leading from a system's equilibrium toward the instability of the physical structure of the system itself. Stability concepts used in dynamic systems may require discrete iterative models or differential equations to demonstrate all stability phenomena in question.

Structural stability is a property of the system and not its equilibrium points. Dynamic similarity in this case means that the topological congruency is maintained between all systems and that fixed points in one system match fixed points in another system. Put simply, small perturbations do not change the quality of systems performance and are often not noticeable to the end user.

2.3.2.1.2 Autonomous System Routing Protocols

The routing information protocol (RIP) with a technology link to as early as 1957 is the oldest one still used today. RIP has lost popularity and has been supplanted by the open shortest path first (OSPF) and the intermediate system-to-intermediate system (IS-IS) routing protocols. These *interior routing protocols* are distinguished from *exterior routing protocols* connections between networks. Reactive protocols broadcast path request packets to find a route; proactive protocols maintain tables of possible routes and give them rankings. RIP, OSPF, and IS-IS are all types of proactive routing protocols. RIP is a *distance vector* protocol, counting the number of links to a destination and choosing the shortest. OSPF and IS-IS are *link state* protocols, monitoring performance of links in their libraries and adjusting their rankings accordingly.[27]

2.3.2.1.3 Router/Routing Stability

2.3.2.1.3.1 Router Configuration Errors April 25, 1997—A misconfigured router maintained by a small Virginia-based service provider injected an incorrect routing map into the global Internet. This map indicated that the Virginia company's network provided optimal connectivity to all Internet destinations. Internet providers that accepted this map automatically diverted all their traffic to the Virginia provider. The resulting network congestion, instability, and overload of Internet router table memory effectively shut down most of the major Internet backbones up to two hours. Incorrect published contact information for operations staff, and lack of procedures for inter-provider coordination, exacerbated the problem.[28]

---

[26] Racskó, P. (2010). Information technology, telecommunication and economic stability, cybernetics and information technologies. *Bulgarian Academy of Sciences*, Vol. 10, No 4 (Sofia). p. 5.

[27] Stein, A. (n.d.), *What Is the Purpose of a Routing Protocol?* http://www.ehow.com/facts_7260499_autonomous-system-routing-protocol.html#ixzz2iAnG97ad.

[28] Barrett, R. S., Haar, R., and Whitestone, R. (1997, April 25). Routing Snafu causes internet outage. *Interactive Week*.

2.3.2.1.3.2 Transient Physical[29] Problems with leased lines, router failures, or elevated levels of congestion lead to transient physical issues in intranet and Internet stability. Single-event faults, coupling faults, line degradation, and unconfirmed faults must also be expected and planned for. In some instances, hardware may enter what is deemed a petulant mode due to packet sequencing or other factors causing a reset simply as an undocumented feature of the hardware.

2.3.2.1.3.3 Data Link Problems August 14, 1998—A misconfigured critical Internet database server incorrectly referred all queries for Internet machine names ending in \.net "to the wrong secondary database server. As a result, most of connections to \.net" Internet web servers and other end stations failed for a period of several hours.[30]

2.3.2.1.3.4 Software Issues November 8, 1998—A malformed routing control message stemming from a software fault triggered an interoperability problem between core Internet backbone routers manufactured by different vendors. This problem led to persistent, pathological oscillation and failure in the communication between most Internet core backbone routers resulting in a widespread loss of network connectivity, and increased packet loss and latency. Backbone providers resolved the outage within several hours after adding filters, which removed the malformed control message.[31]

2.3.2.1.3.5 Human Error Misconfigurations caused by human error are often the reason for router instabilities that occur due to the intricacies of routing protocols. Studies show that almost 6% of bridging control protocol (BCP) updates are inconsistent and unable to show topological changes in the network. Seventy percent of the mis-advertised prefixes occur due to BCP misconfigurations causing routing anomalies such as invalid routes, persistent oscillations, routing loops, and link state advertisements (LSA) penalties.[32]

2.3.2.1.4 Internet Protocol Packets and Congestion

A large network running OSPF protocol may occasionally experience the simultaneous or near-simultaneous update of many LSAs. This is particularly true if an OSPF traffic engineering extension is used, which may significantly increase the number of LSAs in the network. We call this event an LSA storm and it may be initiated by an unscheduled failure or a scheduled maintenance event. The failure may bc hardware, software, or procedural in nature.

The LSA storm causes high central processing unit (CPU) and memory utilization at the router, causing incoming packets to be delayed or dropped. Acknowledgments beyond the retransmission timer value result in retransmissions, and delayed Hello packets (beyond the router-dead interval) result in neighbor adjacencies being declared down. The retransmissions and additional LSA originations result in further CPU and memory usage, essentially causing a positive feedback loop, which, in the extreme case, may drive the network to an unstable state.

The default value of the retransmission timer is five seconds and that of the router-dead interval is 40 seconds. However, recently there has been a lot of interest in significantly reducing OSPF convergence time. As part of that plan, much shorter (sub-second) Hello and

---

[29] Smith, M. A. Internet Routing Instability. www.cs.fsu.edu/~xyuan/.../michael_instability.ppt no date is on the PowerPoint presentation.

[30] North American Network Operators Group (NANOG) mailing list, http://www.merit.edu/mail.archives/html/nanog/msg03039.html.

[31] NANOG mailing list, Retrieved on September 10, 2014, from http://www.merit.edu/mail.archives/html/nanog/msg00569.html.

[32] Routing Anomalies – Their Origins and How Do They Affect End Users. Retrieved on September 20, 2013, from http://www.noction.com/blog/how_routing_anomalies_occur.

router-dead intervals have been proposed. In such a scenario, it will be more likely for Hello packets to be delayed beyond the router-dead interval during network congestion caused by an LSA storm.[33]

The network-working group recommends methods that can be used to improve the scalability and stability of large networks and should be adopted by smaller networks as a preferred practice. They are summarized below.

1. *Classify all OSPF packets in two classes:* A *high-priority* class comprising OSPF Hello packets and LSA packets, and a *low-priority* class comprising all other packets. The classification is accomplished by examining the OSPF packet header. While receiving a packet from a neighbor and while transmitting a packet to a neighbor, try to process a *high-priority* packet ahead of a *low-priority* packet.
2. *Inactivity timer:* Reset the inactivity timer for an adjacency whenever any OSPF unicast packet or any OSPF packet sent to AllSPFRouters over a point-to-point link is received over that adjacency instead of resetting the inactivity timer only on receipt of the Hello packet. Therefore, OSPF would declare the adjacency to be down only if no OSPF unicast packets or no OSPF packets sent to AllSPFRouters over a point-to-point link are received over that adjacency for a period equaling or exceeding the outerDeadInterval. The reason for not recommending this proposal in conjunction with the one above is to avoid potential undesirable side effects. One such effect is the delay in discovering the down status of an adjacency in a case where no high-priority Hello packets are being received, but the inactivity timer is being reset by other stale packets in the low-priority queue.
3. *Algorithms:* (Equation 2.5) Use an exponential back off algorithm for determining the value of the LSA retransmission interval (RxmtInterval). Let R(*i*) represent the RxmtInterval value used during the *i*th retransmission of an LSA. Use the following algorithm to compute R(*i*).

$$R(1) = Rmin$$

$$R(i+1) = Min\left[KR(i), Rmax\right] \quad \text{for } i \geq 1$$

where:

K, Rmin, and Rmax are constants
The function Min[.,.] represents the minimum value of its two arguments
Example values for K, Rmin, and Rmax may be 2, 5, and 40 seconds, respectively.

EQUATION 2.5 RxmtInterval.

4. *Implicit congestion detection and action based on that:* If there is control message congestion at a router, its neighbors do not know about that explicitly. However, they can implicitly detect it based on the number of unacknowledged LSAs to this router. If this number exceeds a predetermined *high-water mark*, then the rate at which LSAs are sent to this router should be reduced progressively using an exponential back off mechanism but not below a certain minimum rate. At a future time, if the number of unacknowledged LSAs to this router falls below a predetermined *low-water mark*, then the rate of sending LSAs to this router should be increased progressively, again using an exponential back off mechanism but not above a certain maximum rate.

[33] Choudhury, G. (2005). *Prioritized Treatment of Specific OSPF Version 2 Packets and Congestion Avoidance, Network Working Group, The Internet Society.* Retrieved July 24, 2014, from http://www.rfc-base.org/rfc-4222.html.

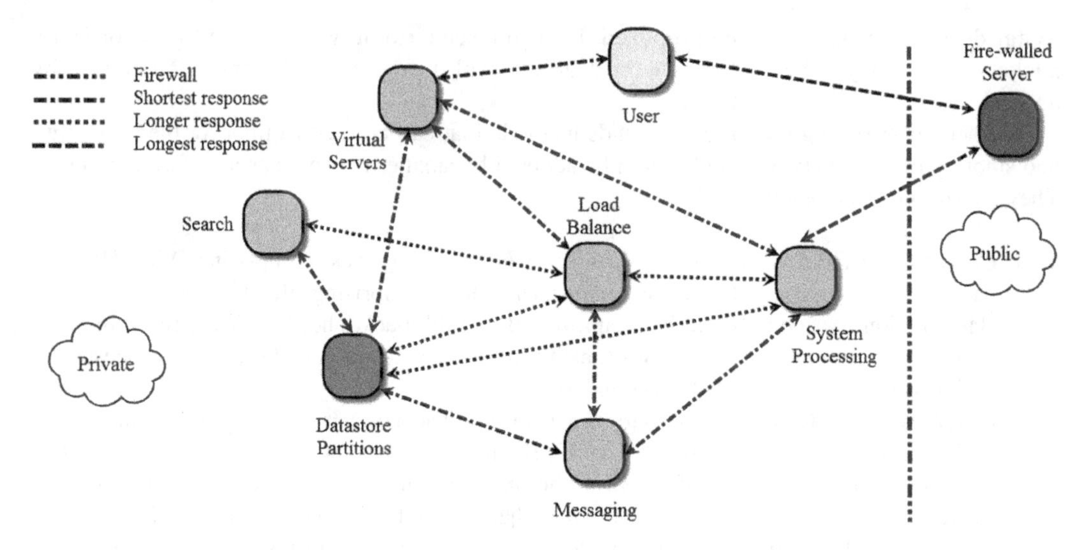

FIGURE 2.11 Prototypical GTS topology.

5. *Throttling adjacencies to be brought up simultaneously:* If a router tries to bring up many adjacencies to its neighbors simultaneously, then that may cause severe congestion due to database synchronization and LSA-flooding activities. It is recommended that during such a situation no more than n adjacencies should be brought up simultaneously. Once a subset of adjacencies has been brought up successfully, newer adjacencies may be brought up if the number of simultaneous adjacencies being brought up does not exceed n. The appropriate value of n would depend on the router processing power, total bandwidth available for control plane traffic, and propagation delay. The value of n should be configurable.

2.3.2.1.5 System Topology

Topology is a mathematical study pertaining to the connectedness, continuity, and boundary of shapes and spaces that began with geometric questions.[34] GTS system topology (Figure 2.11) is designed to support the needs of the enterprise and may be single-tier or multi-tier dependent on needs. Some of the considerations that must be addressed to provide enterprise event management are identified below.

Historical data: Provides the ability to capture and analyze event data from a set period or specific data span, thus facilitating the ability to spot long-term trending.

Immediate alerting: Ensures that real-time changes and code updates are identified, and impacts are tracked.

Integration flexibility: Provides the ability to utilize a library of adapters from vendor sources or developed in-house.

SaaS delivery: Removes the necessity for the added hardware and software upgrades, allowing a rapid response to get things up and running.

Scalability: Provides the capability to manage greater numbers of events per second and per day.

Real-time views: Allows real-time application chatter analysis to provide place in time understanding of the current application topology.

[34] Euler, L. (1735). *Seven Bridges of Königsberg*. St. Petersburg, Russia: Imperial Russian Academy of Sciences.

2.3.2.1.6 Border Gateway Protocol for Internet Connection

Border gateway protocol (BGP) concerns making critical core-routing decisions originally designed for transitioning to a decentralized system backbone and its associated networks. At the enterprise level, it uses several OSPF networks that are beyond the capability of OSPF to scale to the needed size. BGP also assists in assuring better redundancy to a single Internet service provider (ISP) or multiple ISPs for multi-homing (e.g., for computers and devices connected to more than one computer network.) BGP makes routing decisions based on path or configured network policies and rule sets. In some cases, BGP design will not allow delivery while routes are being updated. Intranet routing table growth may impact less capable routers unable to cope with the memory requirements. Routing table growth is often tied to load balancing or multi-homed networks. This can be mitigated using locator/identifier separation protocol gateways within the exchange points to avoid increases in the number of routes seen on the global BGP table.

Several variants of the multi-homing approach are used to eliminate network potential single point of failure.

Single link, multiple IP address: The host has multiple IP addresses but a single physical upstream link. When the single link fails, all connectivity is down.

Multiple interfaces, single IP address: The host has multiple network interface controllers and each interface has one or more IP addresses. When one link fails, the others may still be available.

Multiple links, single IP address: This is the accepted definition of multi-homing. With the use of a routing protocol such as BGP, when one link fails, the protocol stops traffic over the failing link and automatically reroutes it to remaining open links. It is not generally used for single hosts.

Multiple interfaces, multiple IP address: Utilization of a specialized link load balancer or wide area network (WAN) load balancer between the firewall and the link router providing the advantage of using all available links to increase network bandwidth.

2.3.2.2 Application Deployment

Application deployment is accomplished using what is known as a "circular strategy" with six key elements: plan, develop, integrate, deploy, monitor, and analyze. These are leveraged forward from the quality business process improvement cycle discussed in International Organization for Standards (ISO) and elsewhere. Damon Edwards,[35] in an interview posted in an HP Blog by Jim Gardner,[36] had an interesting observation directly applicable to both the GTS business improvement cycle and application deployment (Figure 2.12). Edwards implies that technology developers tend to view innovation as invention, whereas business management sees it as application of ideas to capture new customers and value for those customers. When both definitions exist in a single company, friction will occur. It is the challenge of every company to make it clear what innovation is and how it is to be used and measured.

In any form of communication, it is critical that all parties first agree on the terms to be used and strictly enforce those terms during conversations that pertain to all steps involved. Terms with precise meanings will do much to eliminate errors in interpretation and will reduce the overall timeframe that is involved.

[35] Damon Edwards, co-founder of IT consultancy DTO Solutions, has spent more than a dozen years working on web operations from both the IT and business angles.

[36] Gardner, J. (2012, December 19). *IT Stability and Business Innovation Are Not Enemies*. Retrieved December 20, 2012, from http://www.enterprisecioforum.com.

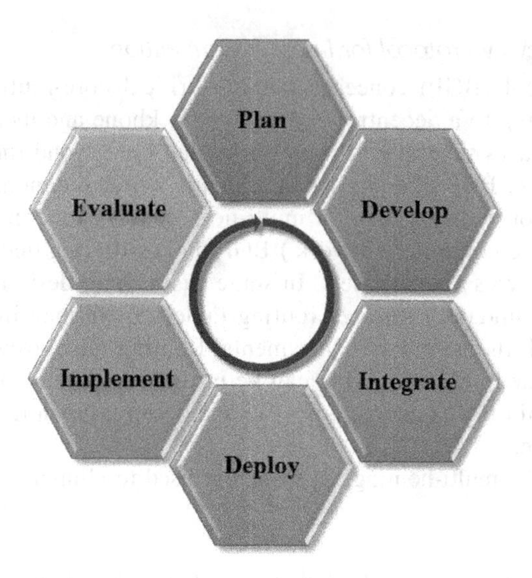

FIGURE 2.12 Business improvement cycle.

Change is constant, and change is not easy. Roger Hebden stated it in a simple formula in 1995[37] (Equation 2.6).

$$\text{Pain} + \text{Desire} = \text{Change}$$

EQUATION 2.6 Change is constant.

When the pain of using an existing application becomes so great that it impacts market survivability, the desire increases. When that desire is stronger than the perceived pain, change is the result.

For the deployment of a new application, planning is the first step in the cycle. For the deployment of an application upgrade or application replacement, this is the first step in the continuing cycle. Planning in this case is much more than the scheduling of events. It includes a complete blueprint of everything that is impacted by the application rollout. Planning starts with defining the nature of the problem and validating the need. This includes resolving incomplete, ambiguous, and contradictory requirements. Only then can you move on to:

- Examining trade studies to evaluate best value solutions for meeting the need;
- Evaluating in-house development or commercial off-the-shelf components (COTS) sourcing;
- Considering implementation strategies; identifying stakeholders, users, and cross functional impacts;
- Evaluating and mitigating known–unknown and unknown–unknown risks;
- Evaluating roll-out scenarios, training plans, material, timeline, new-user checkout and certification (if required), and success metrics/business intelligence (BI) metrics;
- Beta testing of the application in a sandbox/test bed area; and
- Considering data integration and possible platform upgrades or changes.

Once the planning is in place, software development takes one of two roads. Deployment of specialized software will involve development activities that consider the unique characteristics of the end user. In a PLM system, COTS development interface workflows and existing policies,

[37] Hebden, R. (1995). *Affecting Organizational Change.* New York: IBM.

procedures, practices, and handbooks are taken into consideration and terms unique to the deployment are identified and instantiated into the software prior to loading into the sandbox/test bed area for a test to break period. In this case, the resultant contract defines the scope, and many COTS providers will entertain a demonstration phase where some of the requirements are shown. Once the requirements are clearly defined, a contractual agreement is signed that specifies cost, schedule, dispute resolution procedures, and other terms and conditions required under the governing laws.

In an in-house development, a standardized version of the software development life cycle is followed. Several development models exist. Developers need to mix and match based on the scope of the project to meet end objectives and documentation requirements. Some are best suited for lab purposes, whereas others meet the full intent of and context of ISO 12207, quality management systems—guidelines for CM. The intent of what is presented below is to convey a sense of what is involved in each rather than an in-depth explanation of software development and software CM implications. In 2011, over 55 named software development methodologies were in use along with multiple hybrids.[38] Several of the more common models will be discussed.

2.3.2.2.1 Similarities and Differences Between Hardware and Software

Let us look at the similarities and differences between hardware and software before proceeding with the software development types. What follows is a paraphrase of information contained in "Product Data Management and Software Configuration Management—Similarities and Differences."[39]

2.3.2.2.1.1 Differences

- Hardware
  - Geared for part-centric production
  - Production is expensive
  - Designed to last
  - Managed complexity
  - Designed to make multiple copies
  - Deliverables stored on pallets or in bins
- Software
  - Geared to function
  - Production is cheap
  - Designed to change
  - Too much complexity to manage to the same level as hardware
  - Quantity is always 1
  - Deliverables are stored in a repository

2.3.2.2.1.2 Similarities Outweigh the Differences

- Both manage and track
  - engineering change requests.
  - requirements.
  - problem reports.
  - customer requests.
  - documents.

---

[38] Jones, C. (2011). *Evaluating Ten Software Development Methodologies*. Namcook Analytics LLC. http://namcookanalytics.com/evaluating-ten-software-development-methodologies/. Accessed on June 14, 2014.

[39] Dahlqvist, A. P. (2001). *CM i ett produktperspektiv—ställer hårdare krav på integration av SCM och PDM*. The Association of Swedish Engineering Industries.

  - projects/activities/tasks.
  - test cases.
  - built products.
  - test results.
- Traceability is still king.

2.3.2.2.1.3 Do Not Structure Software Like Hardware

- If you try to structure software like hardware, you'll get weak structure.
  - There's a clear functional mapping (allocation) between product function and design.
  - Both need to be mapped and present.
- Full traceability yields negative OH.
  - Good traceability and good traceability tools = lots of benefits.

2.3.2.2.1.4 Software and Hardware CM Differences

- Everything is built from within the repository.
  - Branching/merging is far more prevalent than in hardware CM.
  - Delta/difference reports and peer reviews are essential.
  - Nightly builds (or even continuous builds) are important.
    - Way too costly for most hardware—especially if retooling is required.
  - Product variants are easier to manage.
    - But only if they are handled correctly—wherever possible as run-time variants, not build-time!
  - Both functional and design product architectures.
  - Software repository should be all that is necessary for technology transfer to a new owner, or back to the original contractor.
    - There is so much more needed for hardware.

2.3.2.3 Software Methodologies

At present there are multiple software methodologies being used in industry. While this book is not software intensive, it is important to be familiar with the software development types.

2.3.2.3.1 Waterfall

The waterfall model (Figure 2.13) requires that strict application reviews take place at each stage prior to moving to the next. Revisiting prior decisions is discouraged and may be forbidden

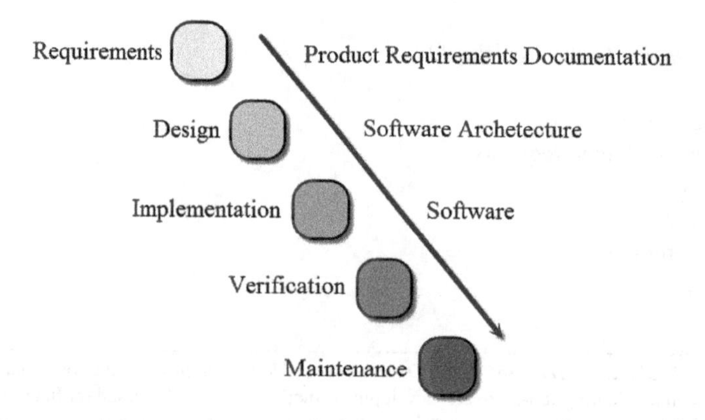

FIGURE 2.13 Waterfall model.

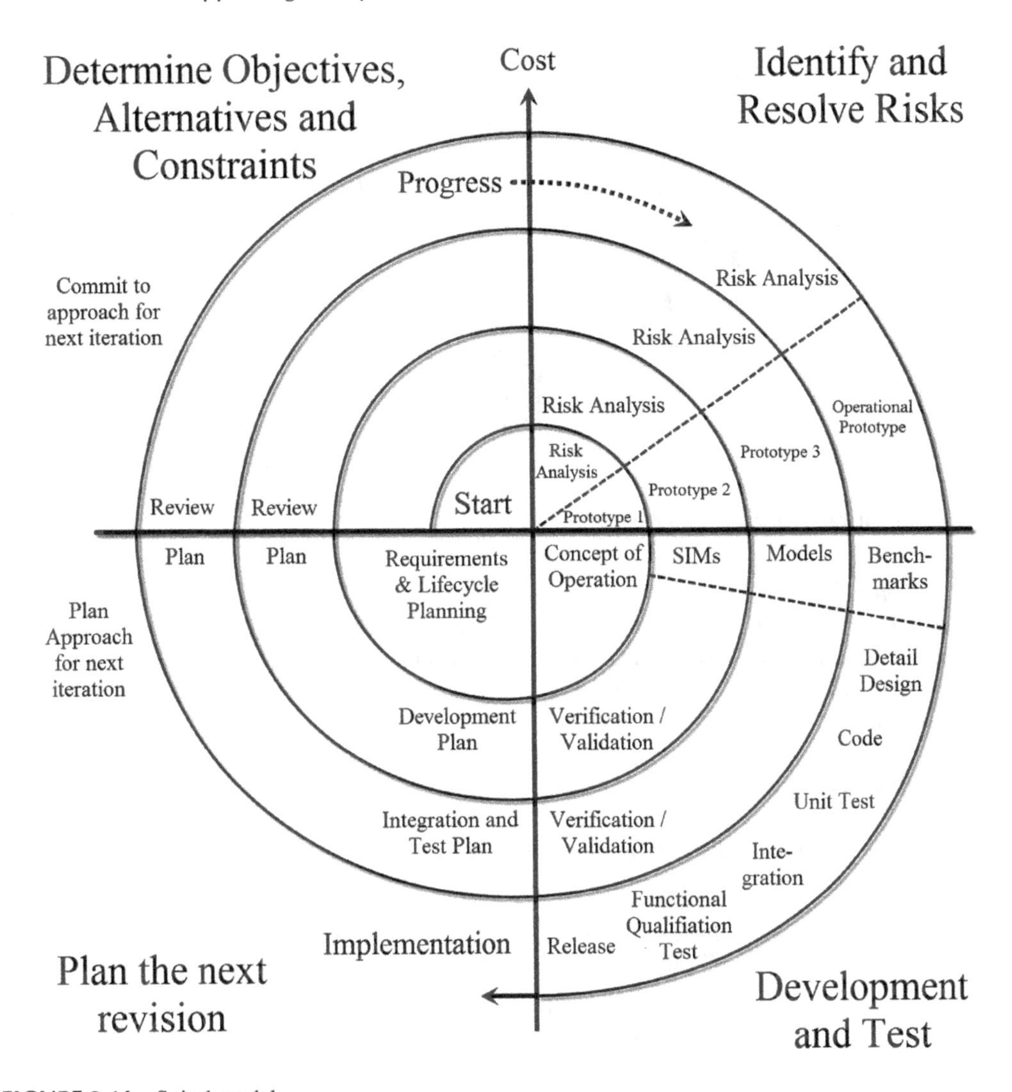

FIGURE 2.14 Spiral model.

depending on how the software development plan is written. The steps can be remembered using the mnemonic "A Dance in the Dark Every Monday" for *Analysis*, *Design*, *Implementation*, *Test*, *Documentation*, and *Execution*.

2.3.2.3.2 Spiral

The spiral model (Figure 2.14) builds in staged risk management during the development cycle, mitigating shortcomings in other models, and is well suited to complex systems.

In some cases, the cost versus risk avoidance posture of a project precludes using this model due to the impact on profitability.

2.3.2.3.3 Iterative and Incremental

Iterative and incremental refer to any combination of iterative design or iterative method (Figure 2.15) with the incremental build model.

Larger developmental efforts find it extremely useful as the number and nature of incremental builds may increase as needed. Such builds are scalable to the design effort. Elements of iterative and incremental methods are found in other methodologies.

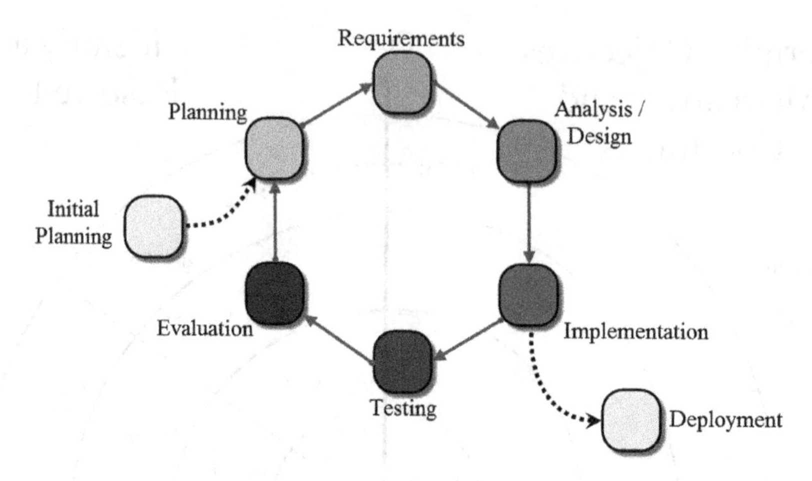

FIGURE 2.15 Iterative development model.

2.3.2.3.4 Agile Development

The Agile development (Figure 2.16) is derived from the Agile Manifesto which is based on 12 principles[40] that are used during the developmental cycle.

Extreme Programming[41] is a type of Agile development which can be visualized as 12 distinct steps, as given below:

1. Customer satisfaction by rapid delivery of useful software
2. Welcome changing requirements, even late in development
3. Working software is delivered frequently (weeks rather than months)

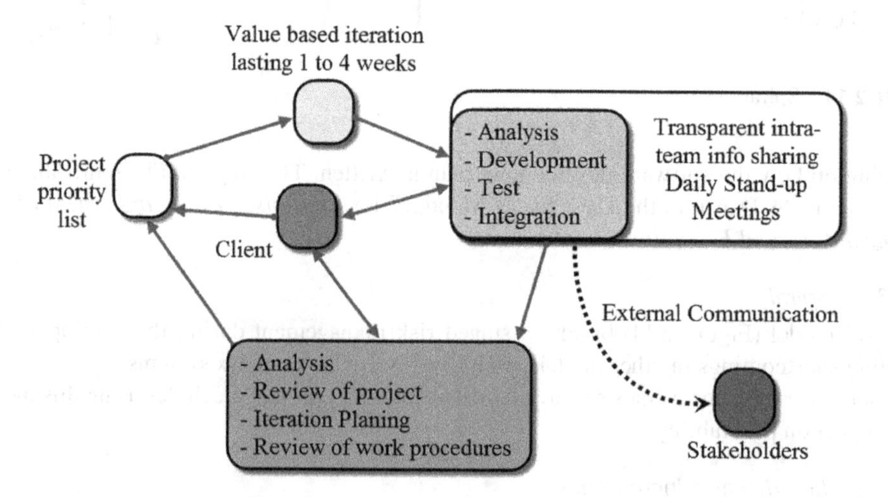

FIGURE 2.16 Agile development model.

---

[40] Beck, K. et al. (2001). *Principles behind the Agile Manifesto.* Agile Alliance.

[41] Copeland, L. (2001, December 3). *Extreme Programming.* Retrieved July 12, 2013, from http://www.computerworld.com/s/article/66192/Extreme_Programming.

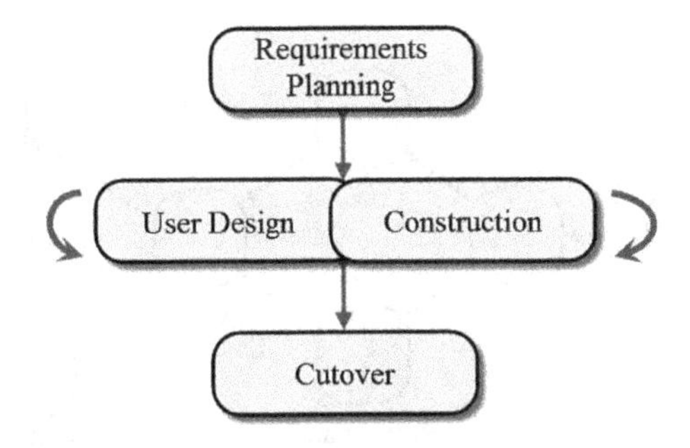

FIGURE 2.17 Rapid application development model.

4. Working software is the principal measure of progress
5. Sustainable development, able to maintain a constant pace
6. Close, daily cooperation between business people and developers
7. Face-to-face conversation is the best form of communication (co-location)
8. Projects are built around motivated individuals who should be trusted
9. Continuous attention to technical excellence and superior design
10. Simplicity—the art of maximizing the amount of work not done—is essential
11. Self-organizing teams
12. Regular adaptation to changing circumstances

2.3.2.3.5 Rapid Application

The rapid application model (Figure 2.17) is based on minimal planning in favor of rapid prototyping with planning interleaved with software coding. Structured techniques and prototyping are used to define requirements for end system design.

2.3.2.3.6 Code and Fix

Under code and fix methodology (Figure 2.18), actual coding starts with little design planning due to schedule pressures. The developer has control over the use of tools, language algorithms, and coding style. Often documentation is minimal or non-existent. During the test phase, programming issues surface and must be corrected prior to the deployment.

Testing and documenting the results is done with the goal of developing software that meets all the requirements/must-have items and as many of the desirements/would-be-nice-to-have items as possible. Once requirements have been met by the proposed software design, the

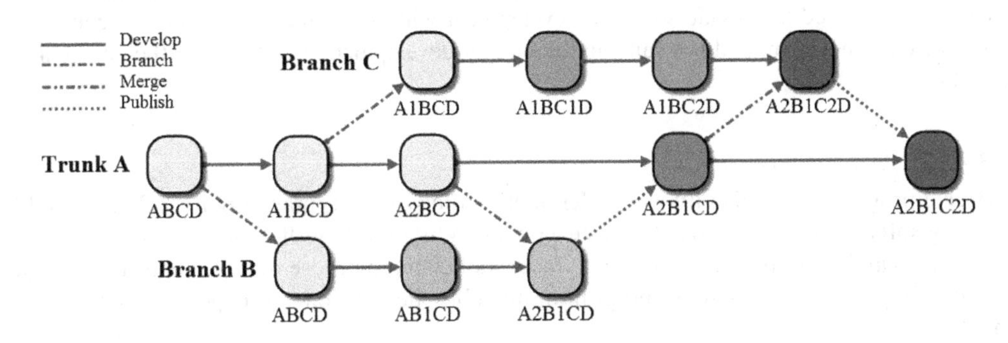

FIGURE 2.18 Code and fix methodology.

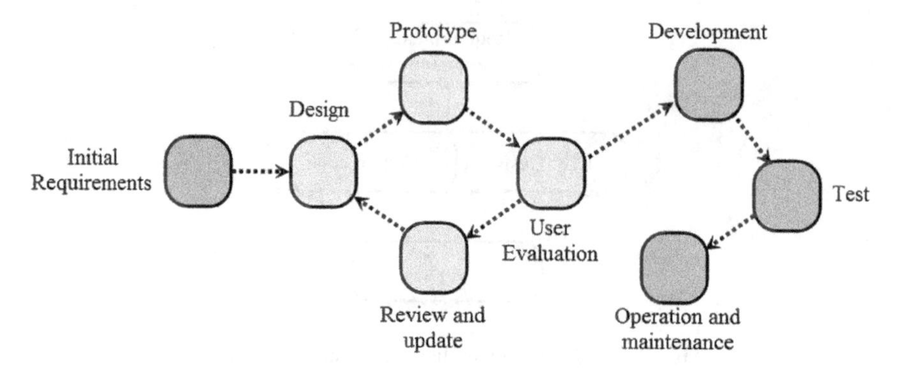

FIGURE 2.19 Prototype methodology.

desirements are incorporated using the concept of gathering the lowest hanging fruit first. This means that the easiest desirements are addressed first (low-hanging fruit) and the harder ones are evaluated using the formula, Pain + Desire = Change. The result of the development activity is the generation of a scope document that clearly defines the cost and limitations of the development.

At the end of any development, regardless of method used, success metrics/BI metrics are evaluated for positive or negative trending. Data gleaned from service-level automation and other sources is used to determine the level of acceptance and issues relating to the need for additional training or lack of a full understanding of the known and unknown-unknowns. During the evaluation period, everyone using the deployed application tests and provides evaluation and feedback in the form of service orders and recommended improvements. If the software deployment is for COTS software, this evaluation is centric to the user base across the industry.

2.3.2.3.7 Prototype

Prototyping (Figure 2.19) allows developers to demonstrate the functionality of various viable solutions before investing in end application development. The extensive involvement of the end user may result in scope creep and could result in extensive modification to the workflow due to the quantity of iterations.

2.3.2.3.8 Dynamic Systems Development

Dynamic systems development (Figure 2.20), originally based on the rapid application development methodology, is an iterative and incremental approach that emphasizes continuous user involvement. It was created to provide software development within specified time and cost constraints. It is costly to implement and not suitable for smaller development efforts where the cost makes it prohibitive.

2.3.2.3.9 Extreme Programming

Extreme programming, also known as XP methodology prototyping (Figure 2.21), is used for creating software in a very unstable environment. While it allows flexibility within the modeling process and lower requirements cost, the cost of changing those requirements as the project matures is prohibitive. Extreme programming relies heavily on the expertise of the people involved.

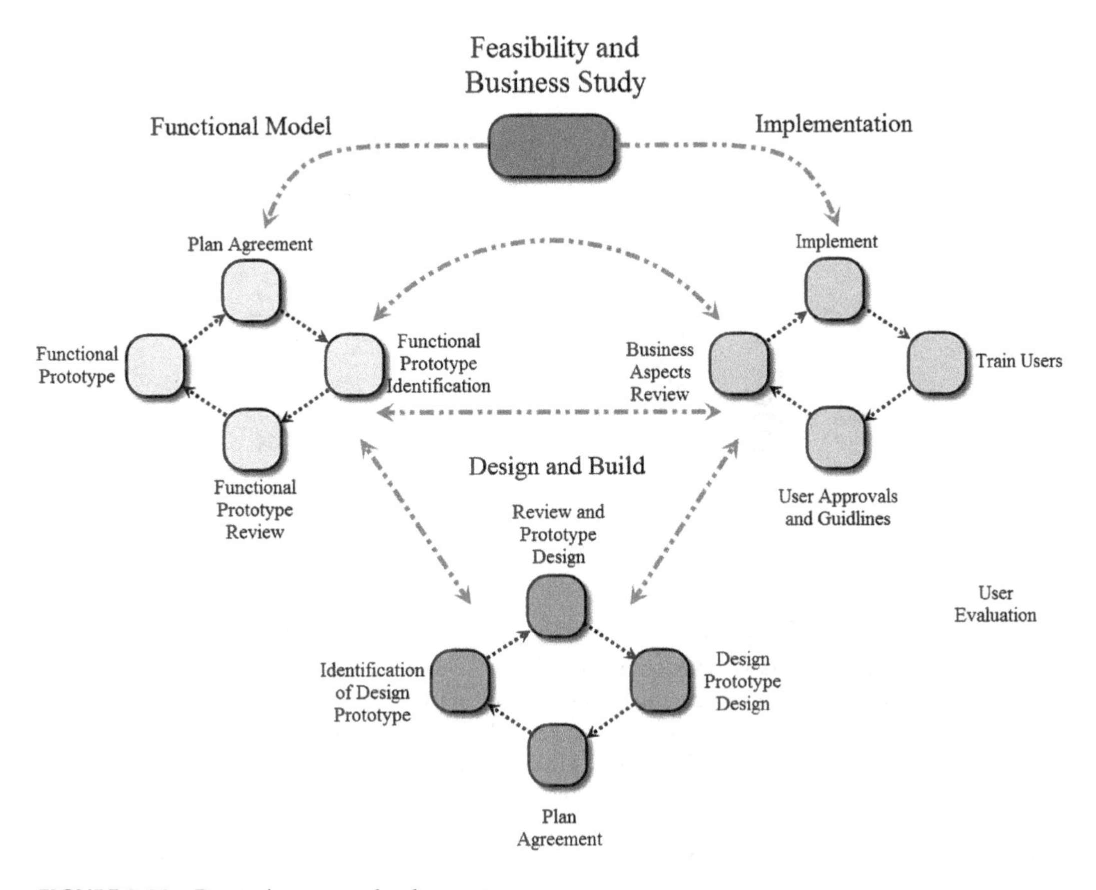

FIGURE 2.20 Dynamic systems development.

2.3.2.3.10 Feature Driven Requirements

The feature driven requirements methodology (Figure 2.22), also known as feature driven development (FDD), is used in an object-oriented environment. It is often the transition approach of choice for software teams moving from phase-based environments and requires a very large team to implement effectively. It also requires a strong lead versed in the roles of coordinator, designer, and mentor.

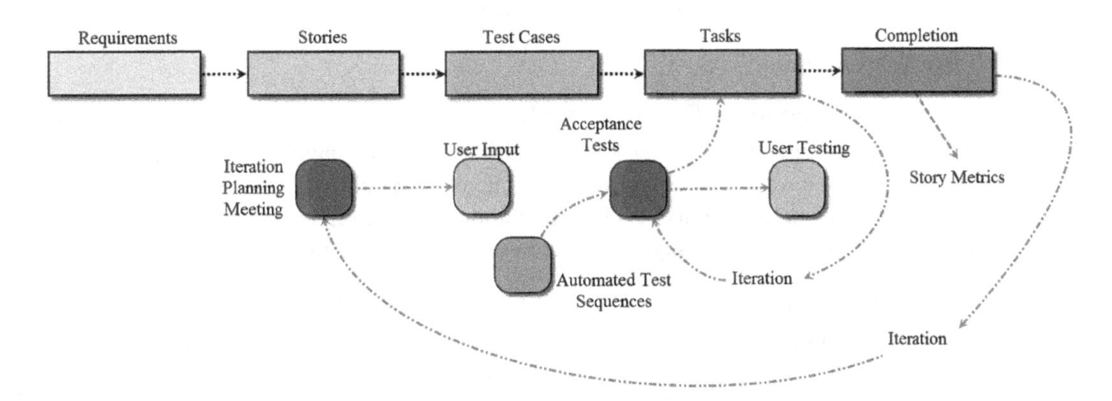

FIGURE 2.21 Extreme programming.

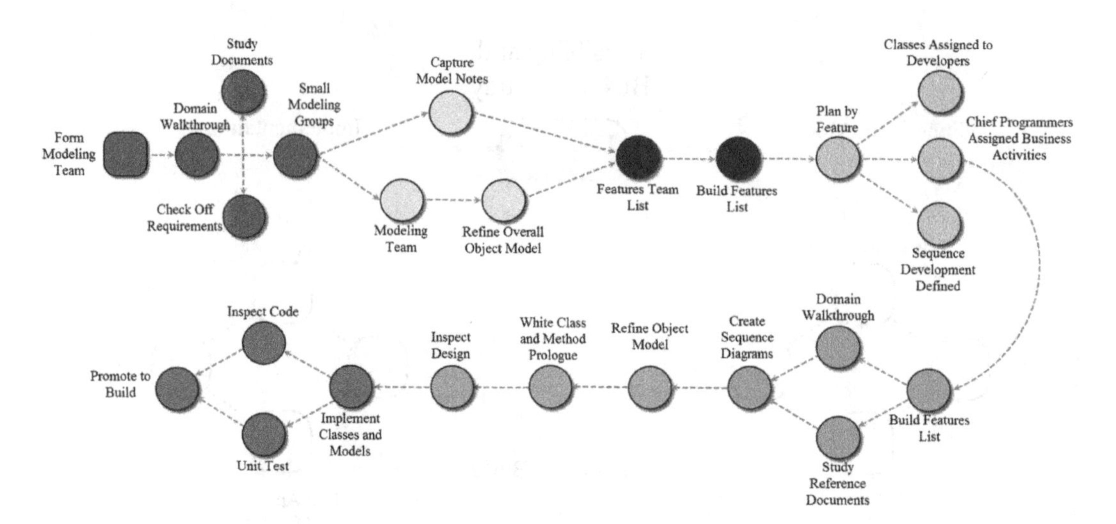

FIGURE 2.22 Feature driven requirements.

2.3.2.3.11 Joint Application Development

The aim of joint application development (JAD) (Figure 2.23) is focused on business issues rather than technical issues. It can generate quantities of quality information in a shorter timeframe with issue resolution in a near real-time setting through collaborative workshops. It requires a dedicated support team and is labor intensive. The joint application development model was leveraged from Das V-Modell, which will be discussed in Section 6.3.2.

2.3.2.3.12 Rational Unified Process

Rational unified process (Figure 2.24) is an object-oriented and web-based methodology that consists of four phases intersected by swim lanes with an emphasis on accurate documentation. The methodology is not organized along traditional software development trajectories and is very complex in its nature.

2.3.2.3.13 Scrum

Scrum (Figure 2.25) can be applied to most software projects that do not have a firm set of requirements. The rapidly changing requirements combined with new requirements require multiple software design iterations until requirements stabilize. It is unwieldy in large software development projects and requires the dedication of expert software developers.

2.3.2.4 Software Adaptation

2.3.2.4.1 Adaptation of Software Due to Adversity

Adaptive radiation, as discussed in Section 1.5.2.1, is driven by many environmental factors. The relatively recent generation of eclipse is a single example of adaptation of another kind. In most cases in nature, as well as in any product sector, adaptation of some kind is the rule. Evolution is the exception. Adaptation due to adversity, as shown in Figure 2.26, is rather common in software.

2.3.2.4.2 Trunks and Branching

Parallel development of an item is not a new concept. It exists whenever an allocation of requirements exists. In the case of software, it has come to be known under the general concept of branching and merging. Many believe it is centric to Agile development, but it has existed in multiple forms going back to programming using punch cards where subroutines were developed, tested, and then merged into the correct place in the card stack. Branching and merging are depicted in Figure 2.27.

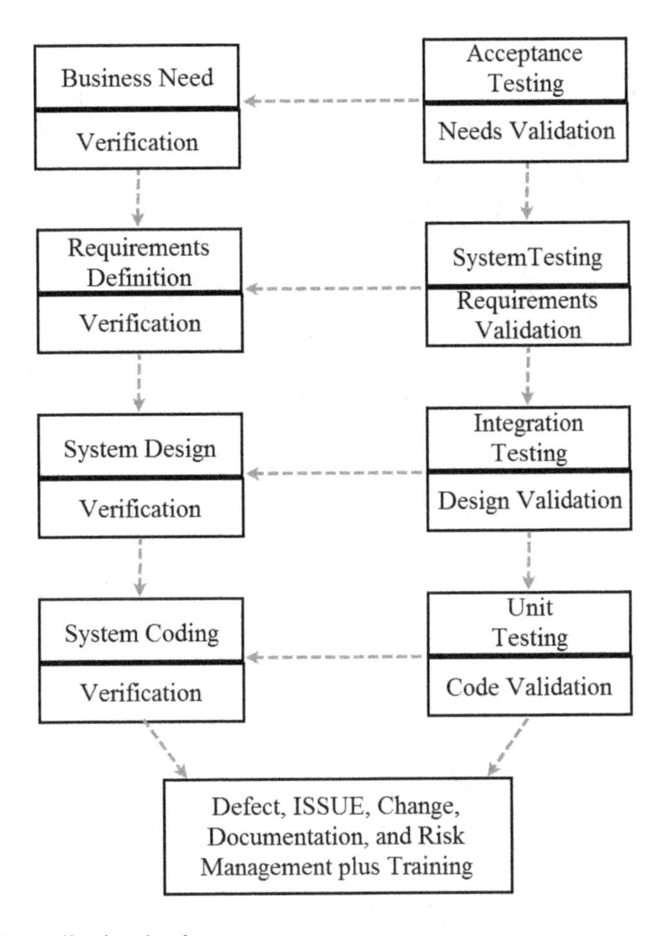

FIGURE 2.23 Joint application development.

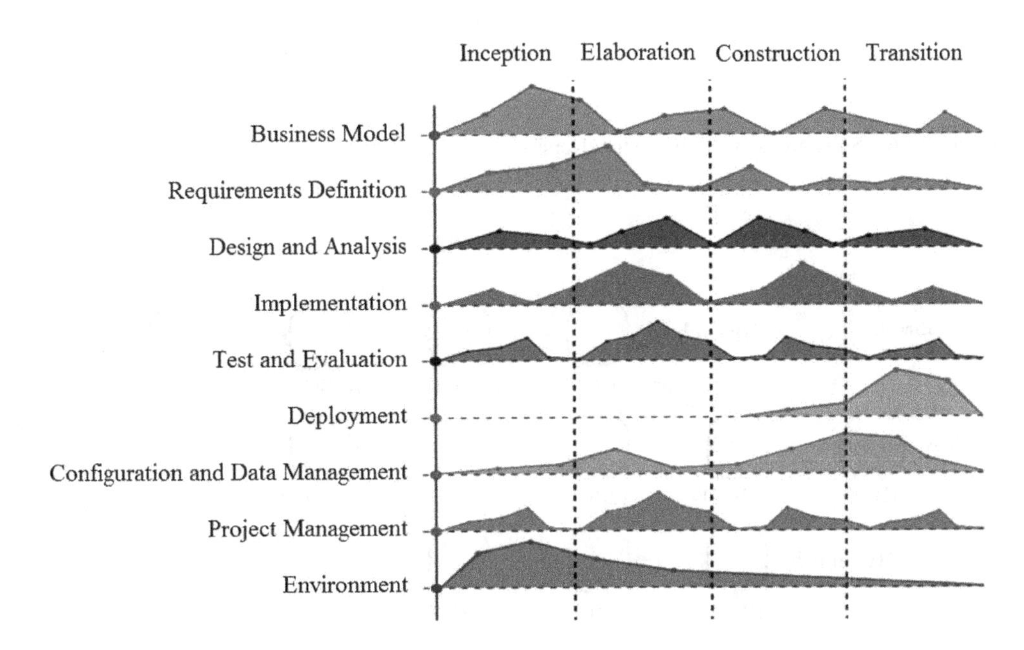

FIGURE 2.24 Rational unified process.

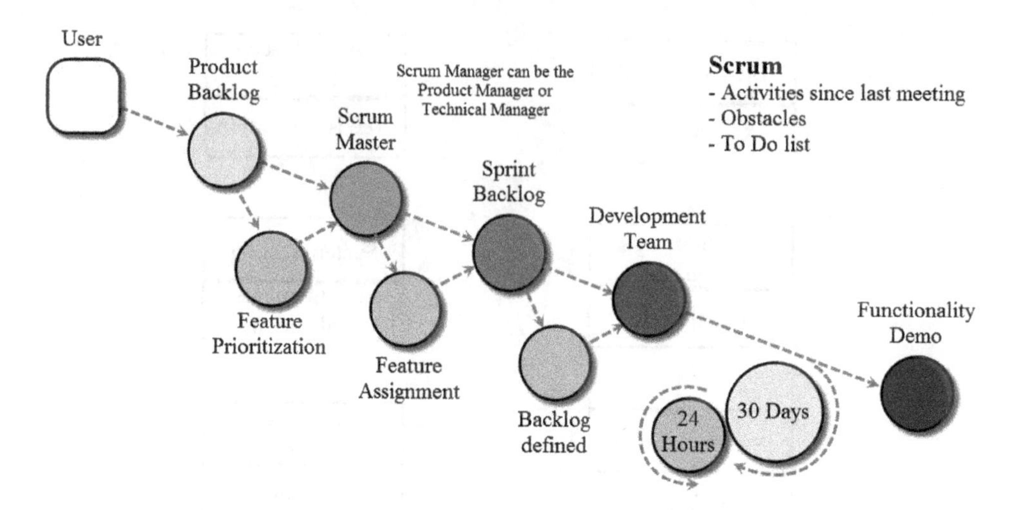

FIGURE 2.25 Scrum.

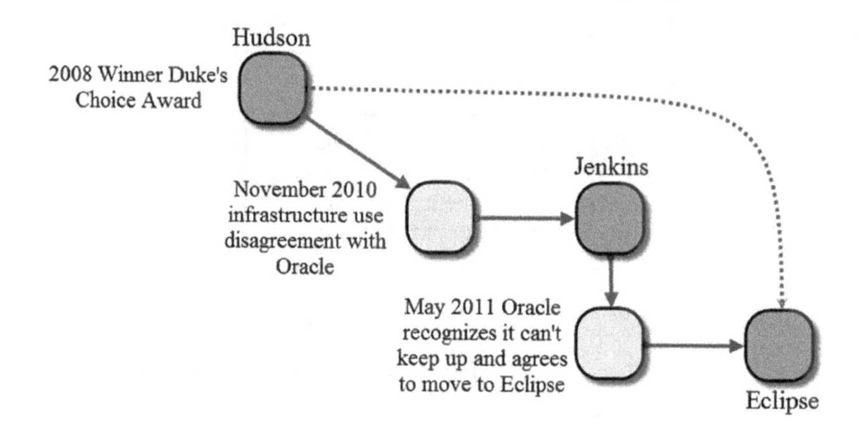

FIGURE 2.26 Software adaptation due to adversity.

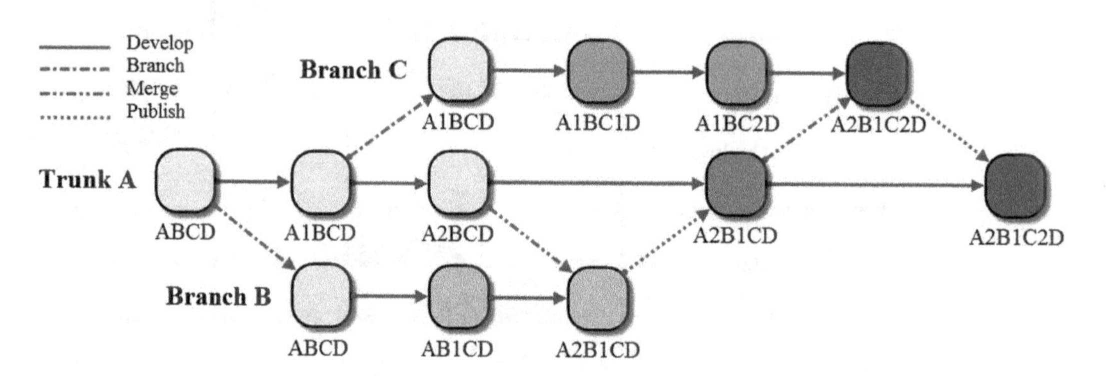

FIGURE 2.27 Branching and merging.

This method of development is frowned on by some who believe that maintaining multiple releases is a better method of managing the software configuration. They state with good reason that there is no 100% capability of merging with the current tool set until it is possible to merge at the semantic and syntax level. It is recommended that branching and merging be limited to merging blocks of text.

At present, the following issues exist even with merging of blocks of text:

- Failed publish
- Conflict clash and resolution (manual editing)
- Branch on conflict

An Agile Perspective on Branching and Merging is an excellent source of information.[42] Its authors point out that early and frequent feedback is critical in all Agile methods. This is aided by continuous or near continuous integration, which must be supplemented by the practice of versioning everything and versioning often.

2.3.2.4.3 Tool Upgrades

Tool upgrades are a specialized case of application deployment with the following considerations.

- Check the GTS system's configuration against the tool minimum criteria to assure system tool compatibility. In some cases, it may be necessary to update the GTS infrastructure prior to GTS tool upgrades to both enhance GTS system performance and avoid instability issues caused by tool deployment.
- Evaluate the total GTS system to detect stability issues and other problems and clean these up prior to the upgrade. These may include hardware and software issues, user account structures, sharing services, connections to peripheral devices and networks, firewalls, and other security measures and software configuration issues on the GTS infrastructure. Unless these issues with the GTS infrastructure backbone are resolved before the upgrade is implemented, existing software will not migrate with the tool upgrade.
- Make sure that all required basic and specialized maintenance routines are completed, resolving any server drive, partitioning, and other potential issues. Full general maintenance routines should also be run, ensuring that all cached data, temporary files, and any associated programmable random access memory data are reset throughout the system.
- Run updates on all COTS software and analyze the upgrades against the tool compatibility criteria. Install any required patches to assure compatibility prior to tool the upgrade.
- Run a full backup of the entire GTS server structure just prior to the tool upgrade. Do not limit it to the server hosting the tool as it may reach out to other servers in the GTS environment during and after the upgrade. Disconnect the backup from the GTS backbone.

At this point, proceed with the tool upgrade and run a full spectrum system check, assessing the health of the GTS infrastructure. If all is good, you can proceed to the implementation phase.

Tool upgrades include software versioning due to patch deployments. Looking at deployment of a single patch through the eyes of an environmentalist, imagine that the GTS/telecommunications implementation is like a high mountain meadow. As the morning light sweeps across the meadow (a new patch deployment), the first plants it illuminates wake, turn toward its rays, and start the day's work of changing carbon dioxide to oxygen and sugar (e.g., the patch deployed successfully). Eventually the patch is deployed across all user infrastructure. Some of the plants in the meadow only receive afternoon or evening light and are at a disadvantage. Companies that do not deploy the patch in a timely fashion run a higher risk of disruption.

---

[42] Berczuk, S., Appleton, B., and Cowham, R. (2005, November 30). *An Agile Perspective on Branching and Merging.* Retrieved November 22, 2013, from http://www.cmcrossroads.com/article/agile-perspective-branching-and-merging.

Often the software loading sequence on a new computer must be carefully managed due to incompatibilities between applications. Loading them in the wrong sequence reduces the stability of the computer and application performance. The same thing can happen with patches. Expanding the mountain meadow concept across the entire GTS/telecommunication system is analogous to spotlights (representing patches from multiple software vendors) illuminating the meadow instead of sunlight. Spotlight beams either enhance or subtract from each other, causing system disruption (just like out of sequence software loads) if patch deployment is not properly orchestrated.

2.3.2.5 GTS System Security

In January 2019, a company we work with sent out a test e-mail to see how many employees correctly followed the company policy on suspect e-mails. The results of the fake phishing e-mail campaign showed the following:

- 41% of employees reported the e-mail as suspicious by clicking on a "Report Phishing" button
- 23% of employees did nothing and deleted the e-mail
- 25% of employees opened the e-mail and clicked on the internal link; all were susceptible to the phishing simulation
  - 50% of these also forwarded the e-mail to other colleagues and asked questions about it (e.g., Have you seen this? or Is this real?)
- 11% of employees opened the e-mail but didn't click on the link, report it, or delete it

The risk from employees who unintentionally expose company data to malware by opening an e-mail attachment is not the only internal threat. Well intentioned workers attending trade fairs are especially vulnerable. It is easy to forget to scan a free memory stick containing marketing data before inserting it into a company universal serial bus (USB) port. It is equally easy to e-mail a document home, update it over the weekend, and e-mail it back with a virus attached because the home computer is infected. Employees are also far more likely to access and expose company information without evidence of intrusion simply because they are authenticated on the server domain. Employees are notorious for creating passwords that are easily deciphered and for leaving their portable devices on busses, trains, and airplanes, and in public places.

Drew Farnsworth[43] of Green Lane Design states, "External attacks, meanwhile, must exploit an outward facing connection, which often has much deeper security. The tools for purely external attacks such as structured query language (SQL) injection and distributed denial of service (DDoS) are limited in their scope. These attacks usually do not compromise all data on a network. Internal attackers can copy files without anyone having any knowledge of the source of the attacks."

Internal and external security mitigation must be considered one of the key GTS functionality elements. GTS breaches can take place before tools are deployed to protect valuable IP from theft or destruction. This is a failing in the operations and maintenance of the "As-maintained" configurations and in employee awareness. The Equifax breach is just such a case. We will address some of these failures in Chapter 11, "When Things Go Wrong." GTS security is not limited to the GTS backbone and applications; it also includes smart devices that interface with it.

Bruce Schneier's book *Click Here to Kill Everybody*[44] explores this issue in depth. Highlighted are car and truck vulnerabilities resident in vehicle digital versatile disc (DVD) players, OnStar navigation, and computers embedded in tires. Aircraft vulnerabilities exist in avionics via entertainment systems and ground-to-air communications. The CrashOverride incident at the Pyrkarpattyaoblenergo control center in Western Ukraine via a malware backdoor was only

[43] https://digitalguardian.com/blog/insider-outsider-data-security-threats, accessed January 9, 2019.

[44] Schneier, B. (2018): *Click Here to Kill Everybody*, September 4, 2018, W.W. Norton & Company, 15 Carlile Street, London W1D3BS ISBN 0393608883

thwarted because the station had a manual override allowing technicians to take the station offline. Vulnerabilities were exploited in 2017 when 150,000 printers were hacked. Once 3D technologies reach the point where bio-printing is common, a similar hack could cripple medical institutions' ability to assist the wounded. U.S. vice president Dick Cheney had his internal defibrillator replaced in 2007 with one that had the wireless option disabled due to fears that it could be hacked and cause his heart to stop.

The same issues potentially exist in any item containing a computer, from appliances to children's toys and baby monitors. Yet in 2018 we found little synergy in new product development between company GTS organizations and the product and product software design community. Without such synergy, it is certain that parts, mechanisms, assemblies, and subsystems with known vulnerabilities will be incorporated into new product design. In January 2019, the European Commission started funding 15 "Bug Bounties" totaling €851,000 in prizes to ethical hackers who actively search for security vulnerabilities in popular free and open source software. The U.S. GIDEP alert system and other systems like it are not robust enough to catch all of these.

Following standard security protocols leveraged forward into the GTS environment necessitates a highly defined degree of planning. Elements should draw from: TL 9000, *Quality management practice*; ISO/Electronic Industries Alliance (EIA) 9003:2004, *Software engineering—guidelines for the application of ISO 9001:2000 to computer software*; ISO/IEC 7064:2003, *IT—security techniques—check character systems*; ISO/EIA 9160:1998, *Information processing—data encipherment—physical layer interoperability requirements*; as well as other guidance sources such as ISO/EIA 14888-1:2008, *IT—security techniques—digital signatures*; and ISO/EIA 14888-2:2008, *IT—security techniques—digital signatures Part 2*. Even these do not give a clear picture of the CM impact as additional standards are under development addressing the necessity of establishing trust between a system and those needing to access it. The days of doing this with a simple electronic handshake are gone.

Two-factor authentication may not be enough as we enter or contemplate the impacts of wireless fifth generation cellular technologies (5G), and 6G and 7G telecommunication environments. Each G represents a 10-times increase in wireless capability: 4G = 100 MBPS, 5G = 1 GBPS, 6G = 10 GBPS, 7G = 100 GBPS, and 8G = 1 TBPS. London's National Cyber Security Centre (NCSC) Defence Condition DEFCON 658 (Cyber), Defence Standard DEFSTAN 05-138 (Cyber Security for Defence Suppliers), and similar requirements may not be adequate in the next decade simply due to wireless speed increases. In the end, it really comes down to managing the configuration of the GTS system and anything that interfaces with it holistically. This needs to center on proper planning, physical security, disaster recovery (DR), and an underlying familiarization with GTS infrastructure threats.

- DR planning for recovery from natural disasters and other unplanned events. Paul Kirvan provides an excellent summary and plan template.[45]
  - GTS DR plans (DRPs) provide step-by-step procedures for recovering disrupted systems and networks and helping them to resume normal operations. The goal of these processes is to minimize any negative impacts to company operations. The GTS DR process identifies critical GTS systems and networks; prioritizes their recovery time objective; and delineates the steps needed to restart, reconfigure, and recover them. A comprehensive GTS DRP also includes all the relevant supplier contacts, sources of expertise for recovering disrupted systems, and a logical sequence of action steps to take for a smooth recovery.

---

[45] Kirvan, P. (2009). *IT Disaster Recovery Plan Template v 1.0*. Retrieved October 27, 2013, from http://searchdisasterrecovery.techtarget.com/Risk-assessments-in-disaster-recovery-planning-A-free-IT-risk-assessment-template-and-guide reprinted with permission of TechTarget.com.

- Assuming you have completed a risk assessment, and have identified potential threats to your GTS infrastructure, the next step is to determine which infrastructure elements are most important to the performance of your company's business. Also, if all GTS systems and networks are performing normally, your firm ought to be fully viable, competitive, and financially solid. When an incident—internal or external—negatively affects the GTS infrastructure, the business could be compromised.
- Plan table of contents from GTS DRP Template v1.0
  - GTS statement of intent
  - Policy statement
  - Objectives
  - Key personnel contact info
  - Notification calling tree
  - External contacts
  - External contacts calling tree
  - Plan overview
    - Plan updating
    - Plan documentation storage
    - Backup strategy
    - Risk management
  - Emergency responses
    - Alert, escalation, and plan invocation
      - Plan triggering events
      - Assembly points
      - Activation of emergency response team
- DR team
  - Emergency alert, escalation, and DRP activation
- Emergency alert
- DR procedures for management
- Contact with employees
- Backup staff
- Recorded messages/updates
- Alternate recovery facilities/hot site
- Personnel and family notification
- Media
  - Media contact
  - Media strategies
  - Media team
  - Rules for dealing with media
- Insurance
- Financial and legal issues
  - Financial assessment
  - Financial requirements
  - Legal actions
- DRP exercising
- Appendix A—Technology DRP templates
  - DRP for <system one>
  - DRP for <system two> DRP for local area network (LAN)
  - DRP for wide area network (WAN)
  - DRP for remote connectivity
  - DRP for voice communications

  - Appendix B—Suggested forms
    - Damage assessment form
    - Management of DR activities form
    - DR event recording form
- Physical security to prevent unauthorized access to servers and DR archives
- Servers may be co-located for convenience or dispersed for added security
- Duplicate data recovery backups should be in several separate locations
- GTS infrastructure threats (leveraged from cloud computing protected[46])
  - Abuse and unauthorized use
    - Spam
    - Malicious code
  - Insecure application programming interfaces
    - Provisioning
    - Orchestration
    - Monitoring
    - Layered complexity
  - Malicious insiders
    - Integrated GTS management vulnerability
  - Shared technology vulnerabilities
    - Disk partitions, CPU cashes, GPUs compartmentalization is weak
    - Isolation properties are vulnerable
  - Data loss and leakage
    - Sharing of resources and access media with external customers and partners
    - Service integration may be compromised
  - Account, service, and traffic hijacking
    - Hijacking of user credentials results in
      - Communication interception
      - Wrongful transactions may be made
      - Authorized users may be blocked or simply monitored
      - Access can be a window for deeper penetration to controlled GTS systems and data content

Most enterprise-level GTS department provide some form of security architecture that describe how security countermeasures are positioned and how they interrelate with the overall GTS architecture. The website SecurityArchitecture.org defines the systems attributes as confidentiality, integrity, availability, accountability, and assurance services.[47] GTS insecurity consists of all threats such as vulnerability, exploitation, and payloads. GTS securities and insecurities are shown in Figure 2.28.

Security can take the form of hardware solutions, hardware-based solutions, and software solutions, or a combination of all three. Once a GTS system is compromised, it may be very difficult to remove the intruder without a complete system rebuild and sanitization of all affected data records.

[46] Rhoton, J., Clercq, J. D., and Graves, D. (2013). *Cloud Computing Protected: Security Assessment Handbook*. Recursive Limited.

[47] Open Security Architecture. (2006, January). *Definitions Overview*. Retrieved July 10, 2011, from http://www.opensecurityarchitecture.org/cms/definitions.

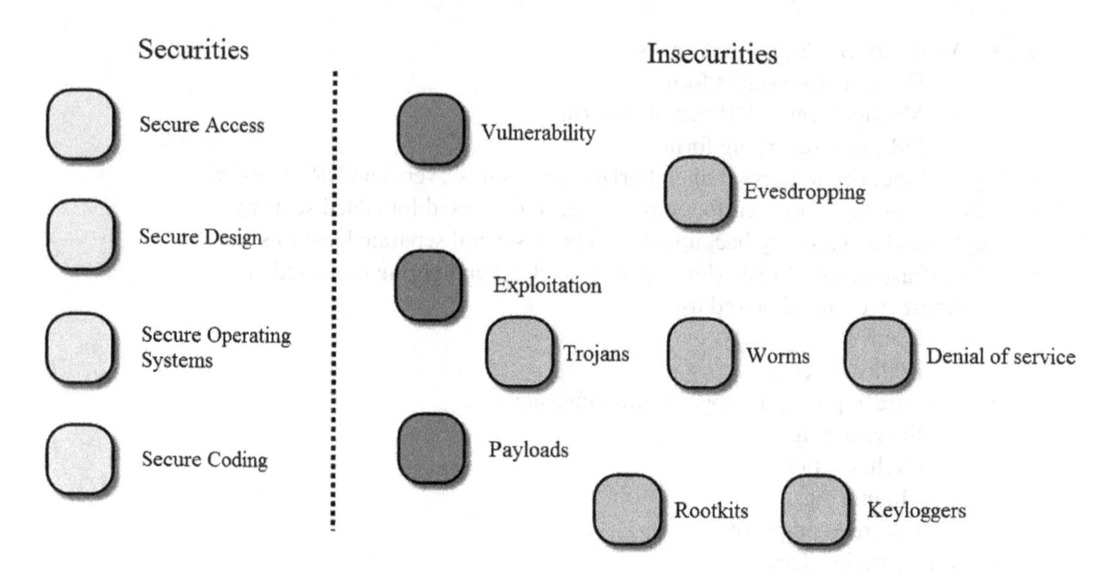

FIGURE 2.28 GTS securities and insecurities.

Secured by design[48] computer security principles applied to the GTS environment would include the following approaches:

- *Authentication:* Ensuring the individual signed in is who they say they are. Authentication methods should consider preventions of replay and man-in-the-middle or man-in-the-browser based attacks. It should also take into consideration loss prevention of tokens or code cards.
- *Least privilege:* Locks down portions of the GTS substructure to only those persons requiring access. This in effect compartmentalizes attacks to limited areas of the GTS infrastructure. Every user may have access to the basic functionality and, as security needs increase, privileges in the system are successively locked down in a multi-tiered approach. In a server access setting, this may mean that only the system administrator has access to everything on the server and no other single user would.
- *Fail secure rather than fail insecure defaults:* This is leveraged forward from safety engineering principles to prohibit making the systems insecure only in the case of a deliberate act made freely with knowledgeable forethought.
- *Unit testing and code reviews:* Assuring that modules are secured when formal correctness proofs are not feasible or possible.
- *Automated theorem proving or automated deduction:* Validating crucial software systems using mathematical cross-check functions to prove the installed algorithms and code meet specification.
- *Chain of trust techniques:* Ensuring loaded software is certified authentic by the system designers.
- *Microkernel implementation:* Underlies the OS to apply labeling restrictions to the OS itself.
- *Cryptography:* Defending interception of data between systems to reduce interception.
- *Firewalls:* Providing protection from online intrusion.
- *Data integrity comparisons:* Assuring data is the same between two sets of stored data and if different it was modified knowingly.
- *Endpoint security:* To prevent data theft and infectious abuse and unauthorized use.

[48] *Secured by Design* is the official UK police initiative supporting "designing out crime" owned by Association of Chief Police Officers.

Critical to any form of DR loss mitigation is assuring that the backup media is validated as good before it is put into storage. Automated backup is great, but the plan must call for verifying the media. In addition, the plan must call for offsite storage of regular backups, not just storage in another room in the same or a neighboring building. Natural disasters such as flood, fire, and wind can destroy multiple co-located buildings.

2.3.2.6 Platform Transitions

At times it may be necessary to transition from one GTS backbone structure to another (Microsoft to Linux) or from one integrated database to another using the existing GTS backbone. All transitions are difficult and the longer the transition time is, the more difficult it becomes.

An *everything-at-once* (EAO) approach or a *piece-by-piece* (PBP) migration can be used. Both offer advantages and disadvantages depending on the migration type.

The focus of this book is CM, so let us look at platform transitions and the related tool upgrades, considering data migration from a flat-file database structure used to record modifications to engineering artifacts to a multilayered database integrated PLM system. Such a transition must include management of all artifacts as discussed in ISO/EIA 10007:2003—Quality management systems—guidelines for CM, as it relates to ISO/EIA 20000-1:2011, IT—service management and ITIL (Figure 2.29 and Figure 2.30).

Perhaps the hardest aspect of all transitions is a lack of a systems engineering and CM perspective regarding the current state of the entire GTS infrastructure. Software-to-software interface control drawings (ICDs), system-to-system ICDs, and giver–receiver lists as well as specifications describing the GTS system and sub-system specifications describing each sub-element have been non-existent in companies reviewed prior to beginning this book.

2.3.2.6.1 Everything-at-Once Migration

Under the EAO migration concept, the customer would continue to utilize the flat-file database structure as well as other standalone software tools and their related data structures. Maintenance of the tools and old codebase during such a migration is critical. This migration type is best managed with a fixed timeline, as the longer it takes to accomplish, the more expensive it becomes. Data maps will need to be validated between systems in a duplicate copy of the associated databases before they go live date of the PLM system. Conversion of historical data to electronic format (known as digitization or digitized data) such as a portable data format (pdf) needs to be scheduled and funded

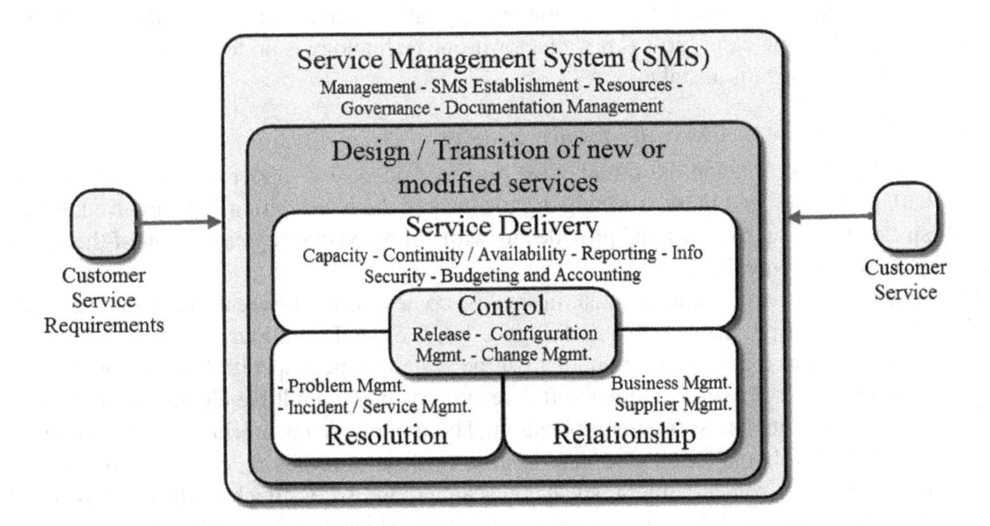

FIGURE 2.29 ISO 20000 and ITIL.

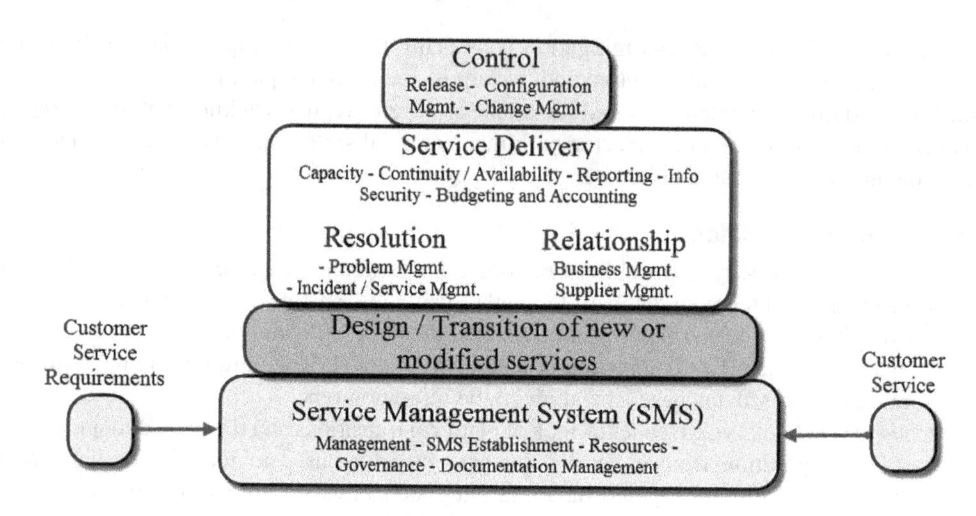

FIGURE 2.30 ISO 20000 and ITIL pyramid.

as part of the transition. Go-live activity will require that GTS services be made unavailable for a period, as the new PLM system is loaded to the active GTS infrastructure instead of being housed in a sandbox/test environment, data migrations take place, and system validation is done.

2.3.2.6.2 PBP Migration

The migration may be staged by function with an EAO migration plan for each. This concept allows for greater control over individual functional requirements as well as greater flexibility, as the migration can be spread over several months with some savings in overall cost. The same degree of rigor is needed as with the EAO migration plus the added need to control data mapping to shared portions of the code and functionality. This will drive costs and will necessitate management of multiple deployments along with the associated server resources. Before each migration's go-live activity, all interfaces between the existing live dataset and the *to be migrated* dataset will need to be verified for cross-functional incompatibilities.

This migration by data module is not trivial, and sound GTS and CM methodologies and controls must be in place long before each sub-migration takes place. It is wise to prioritize migration of features on an elimination of effort duplication priority rather than a customer needs basis if cost is a consideration. In some cases, cost is not a factor if one technology is no longer supported and the survival of the enterprise is at stake.

2.3.2.6.3 Importance of CM with Modular Systems

Central to the GTS function is sound CM practices identified under the control function. Figure 2.30, ISO 20000 and ITIL, is a mahjong tile view of the interrelationships involved. Rotated in a 3D space, it forms a pyramid and provides us with a more visual representation of the specific hierarchy of functions involved.

CM in modular systems, while different in specific context, is no different than CM of any other system when viewed from the systems level. Figure 1.2, U.S. DOD configuration management process, can be viewed as a conceptual flow that can be scaled to meet specifics addressing the need at hand. Just like other systems, modular software systems merge multiple elements (software and devices) to make a complete software application. These components are integrated into a larger application. A typical home-based computer system consists of monitors, computer, printer, keyboard, mouse, wireless router, camera, speakers, scanner, and so on. Each of these items may be designed and fabricated by different companies yet each is critical to the overall performance of the system. Each item is a component or module of the modular system, yet each is a system by itself.

Modular application development and integration into an enterprise application consists of COTS integrated together to form the entire system. Modular software development often proceeds from "best-of-breed" COTS software components connected through software and hardware interfaces.

Logical boundaries, components, and applications are established. Applications communicate through established data protocols, often set up by the OS developer for the software, that are compatible with that OS. COTS manufacturers follow CM methodologies throughout the product life cycle. Similar application of CM methodologies should be followed for the GTS landscape's evolving life cycle. The difference is that the GTS system life cycle is easier to understand from an ecological viewpoint, whereas the individual COTS components are viewed from a species viability viewpoint.

Impact printers (type box, wheel, and ball) succumbed to dot matrix printers that in turn succumbed to line printers (drum, chain or track, bar, and comb.) These succumbed to plotters, liquid ink (electrostatic), and laser printers. The GTS modular framework ecology did not change. A space still existed for a *printer* as part of the ecological mix. The specific COTS product occupying the ecological space was supplanted time after time by a product better able to survive in that space.

Until recently, modular system management was relatively straightforward in most enterprise settings. To facilitate security of the GTS backbone, many companies have instituted strict policies regarding the installation of user-purchased specialized software or freeware despite the point in time increases in individual user productivity. A 2003 article by Victor Luckerson[49] discusses the move in many market segments to allow employees to work using personal devices (laptops, tablets, phones, etc.). Bring your own (BYO) technology policies mean less expenditure on the part of the company, eliminate associated depreciation, and allow faster upgrade of the interconnected modularity of the system. Drawbacks include some loss of control with respect to managing company data and other standard GTS security issues. Depending on the industry, the costs savings versus risk posture is such that the convenience and cost aspects of BYO trump other concerns.

CM activities span the GTS electromechanical age, electronic age, and information age. GTS interfaces and mission needs statements often do not reflect this. Due to customer requirements, many CM data, records, and release activities require the use of paper and automated systems. GTS in the cleanest sense departs from the current definition. Today, many consider GTS as the management of computer-based information systems and all that entails. This brings forth images of paperless workplaces with data flowing seamlessly into and out of all aspects of the enterprise through the wonders of bluetooth connectivity and cloud-based technologies.

The paperless world has yet to become reality in many market sectors. It is also not yet fully embraced by many GTS organizations. Companies and customers still require documents signed in ink and either scanned and transmitted by e-mail or faxed transmission with the original sent via postal or other carrier service prior to assigning access codes. This was observed as recently as 2013 as customary practice between two Fortune 500 corporations. There are also CM implications presented by Deep Web technologies and successive encryption removal for uber-secure point-in-time transactions and information transmission. These will become increasingly important as we enter the world of 3D technologies. As onsite growth of medical transplants (keyed off the recipient's specific genetics) and the potential for other human capability upgrades come online, aspects of CM, security, and GTS protection of critical-to-life organ models will become critical.

Management of the configuration and its defining documentation necessitates that the practitioner work in both the old and new definitions of GTS. In a supposedly digital environment, there is a surprising resistance to utilizing software tools for CM of the GTS infrastructure. This mix of hard copy signatures and recorded test results with signatures and stamps on paper, technological edge applications used for performance monitoring, and relatively standard digital GTS is shown

[49] Luckerson, V. (2013, November 11). Bring your own tech: Why personal devices are the future of work. *TIME* (U.S. Edition).

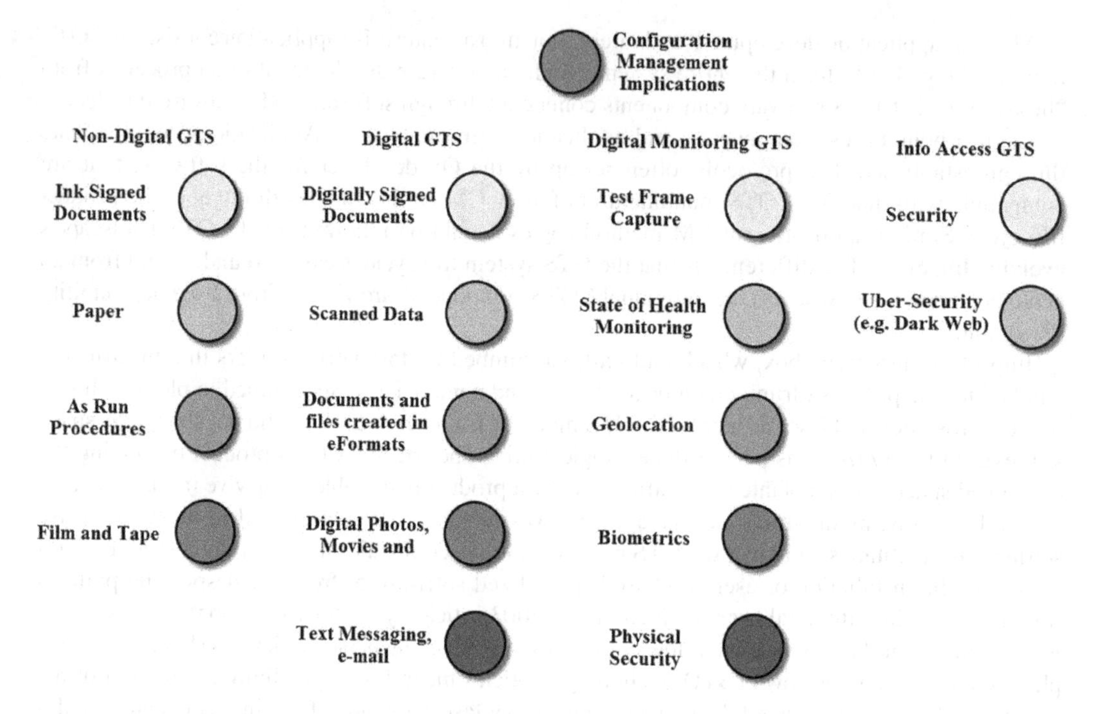

FIGURE 2.31 CM of the GTS environment.

in Figure 2.31. Although not all aspects are present at any single time, management and security implications are crucial and must be addressed. The more decentralized the organization and the wider the modularity, the more critical this becomes.

2.3.2.7 GTS Case Studies

Two case studies are presented to illustrate the complexities involved in management of configurations in GTS.

2.3.2.7.1 Case Study 1—Software Suite Decision

A provider of products to the commercial and government sectors was under contract to deliver associated data to its government programs in specific file formats and application software versions. All contract data requirements list items were subject to government acceptance or rejection under the contract terms. Penalties for not complying with the specified formats constituted a violation of the contract data deliverable requirements and contract termination. Termination included the contractor paying all re-procurement costs for opening a new competition and awarding the contract to another vendor.

The GTS organization was not tied into the revenue-generating element of the enterprise. It pushed out an upgraded software suite that allowed downward versions to be opened but did not allow saving a file in the previous format (e.g., opening *mydoc.doc* and only being able to save as *mydoc.docx*).

Because the extension and software version were contractual, this software suite upgrade resulted in issuance of a cure notice[50] by the customer for default due to non-compliance with the contract. The GTS action was reversed in time to avoid termination of the contract but not without financial penalty. The GTS professional responsible stated, "Software suite decisions are the purview of GTS

[50] Federal Government, U.S. (2013). *U.S. Federal Acquisition Regulations 52.249.X, Termination for Default sub-part (a)(1)ii and (a)(1)iii*. U.S. Government.

and should not be mandated by our customers." Mindsets like that expressed by the GTS engineer not only are uninformed but show a disregard for customers and stockholders. Contracts are bilateral agreements. If the contract states that files must be in defined formats and a contractor signs the contract, the contractor must comply.

The root cause of this incident was a lack of a sound requirements definition and CM practices in the GTS organization. Decisions to push out an upgraded software suite were based on a perceived cost savings if machine-based software were replaced with an intranet (internal network) solution with machine interfaces to the intranet (a precursor to cloud computing). Research into this case study shows no evidence that revenue-generating functions of the enterprise were consulted in the decision.

2.3.2.7.2 Case Study 2—PLM Decision

The GTS decision for the acquisition of a PD/LM system to replace its aging PD/LM system was based on the Figure 2.32 decision tree. The enterprise had recently purchased new financial system software. It used a different data backbone than the rest of the company and the chief financial officer (CFO) mandated that other systems infrastructure use the same backbone. This was the driving consideration and the winning PD/LM system was compatible with the financial reporting systems backbone. It also provided management reports based on metadata fields that the CFO had determined would be beneficial to the program managers based on metrics the CFO had used at a previous employer.

- Total number of drawings estimated
- Total number of drawings released to date versus the drawing releases planned
- Total number of drawings estimated versus total number of drawings used
- Average number of engineering changes per drawing

Several one-solution vendors were discussed and rejected without consideration, as their single-solution suites included a finance module. The winning PD/LM system had no ability to schedule multiple delivery reports (schedules, financial data, monthly reports, weekly reports, etc.), or to export documentation other than doing so document by document using manually entered requests. The top assembly item identification could not be entered, and all associated documents had to be output in a single command. Each document had to be deduced from an indentured bill of material report and output by creating individual user requests. Because the

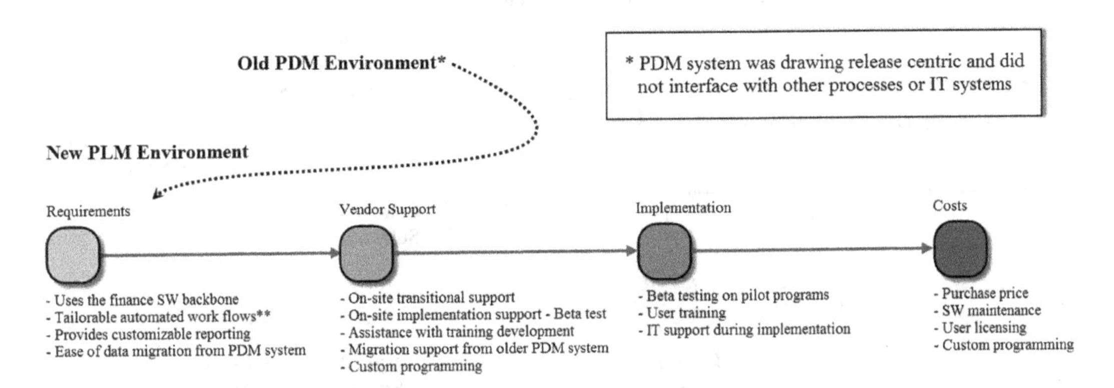

** It was determined that any new PLM environment must fit as closely as possible with established configuration management procedures to do this automated data flows needed to be tailorable.

Missing from this decision flow is any consideration with interfacing to supply chain management for materials procurement, interfacing with the manufacturing system, interfacing with the program scheduling system for deliveries of products and data associated with products, non-conforming product evaluation and disposition, or export of data from the system.

FIGURE 2.32 PLM procurement decision tree.

product data/life cycle management (PD/LM) system had been *drawing release* centric, other functionality was not considered. After implementation, this lack of broad reaching infrastructure functionality was addressed in what was deemed a best-of-breed approach versus a one-solution strategy (Figure 2.33).

As the enterprise matured, it found that the cost of best of breed was substantially greater than investing in a single solution would have been. Costs to maintain best-of-breed interfaces with the common data backbone increased over time as individual solution providers updated their

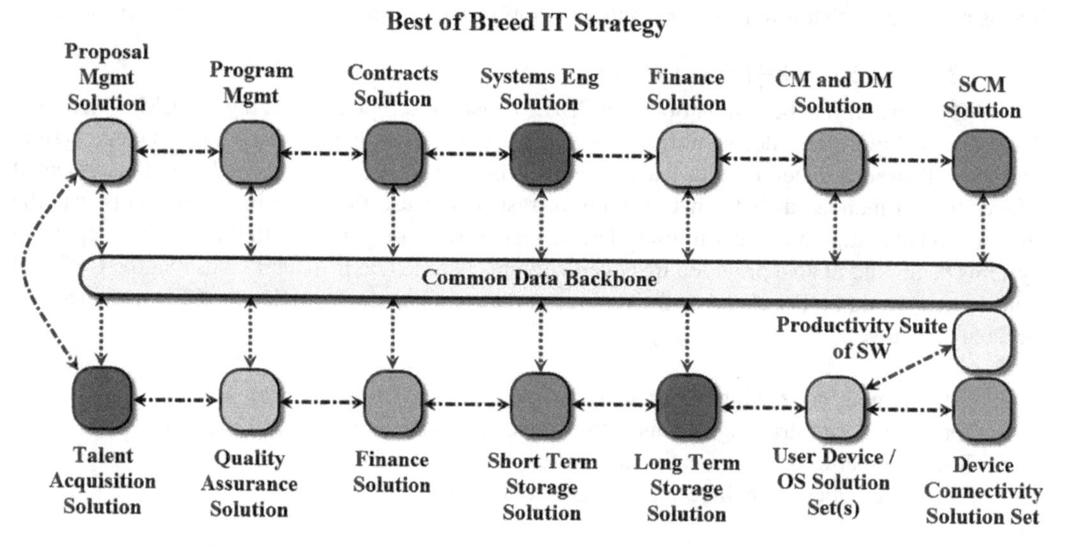

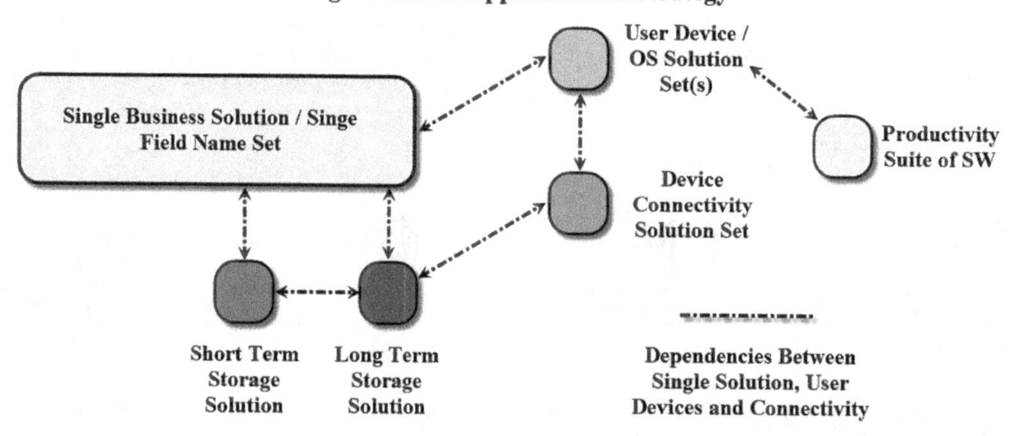

FIGURE 2.33 PLM implementation: (a) best-of-breed versus (b) single business application GTS strategies.

product, necessitating constant updating of the data field to backbone interface. Management of the entire GTS infrastructure also required expenditure of resources as individual solution providers also moved from machine-based to web-based software interfaces, at times reorganizing the software interfaces and workflows and interdependencies between the solution sets. A side effect of this was a noticeable increase in enterprise-wide cross-functional procedure updates and a growing frustration in the user community. This frustration was referred to as "noise in the system."

Throughout the implementation of the best-of-breed approach, GTS saw no need to generate a plan to manage the configuration of the entire infrastructure. User need requests were met through FIFO. Users were often unaware of the needs of other organizations. Without a centralized implementation strategy, the requests were found to be working at cross purposes. Requests acted on to help one area be more productive caused cost inefficiencies to appear everywhere else. A CM plan and a best-of-breed CCB could have easily halved implementation costs. Had the investment in the initial financial system been declared a mistake, an integrated one-solution software suite could have been implemented with an eventual savings of more than 30% long term just in software provider costs (Savings = Writing off the financial software + Purchase and maintenance of the one-solution suite).

The metrics determined by the CFO were not valid. The first was dependent on the computer-aided design (CAD) systems. The others could easily be manipulated by the design team.

- Total number of drawings was estimated.
  - Many firms use this metric despite increasing sophistication in design and modeling tools. It evolved out of the paper-drafting system.
  - Drafting boards allowed multiple details and assemblies to be shown on a single drawing. This facilitated the evaluation of the finished assembly on one set of drawings and associated lists.
  - As CAD systems evolved and CAD models became more sophisticated, the size of the file outgrew what could reasonably be uploaded and retrieved from the GTS systems. It became necessary to break the models into smaller sub-sets, thus generating more drawings to reflect the same number of details and assemblies shown on a single paper drawing set.
  - At the firm under discussion.
    - Drafting Board—30 details and/or assemblies = 1 drawing
    - AutoCAD 12—30 details and/or assemblies = 6 drawings
    - AutoCAD Inventor—30 details and/or assemblies = 9 drawings
    - Solid Works—30 details and/or assemblies = 21–30 drawings
  - Over time, the number of details and assemblies remained constant.
- Total number of drawings released to date versus the drawing releases planned.
  - Once engineers found out the metric was being used, drawings with minimal details were released (often with no content) to meet release schedules.
- Total number of drawings estimated versus total number of drawings used.
  - Once engineers found out the metric was being used, drawings with more details or assemblies than advisable were released to meet drawing count, often resulting in unstable CAD models.
- Average number of engineering changes per drawing.
  - Once engineers found out the metric was being used, changes were saved and packaged to show improvements, often to the detriment of parts procurement, manufacturing, and schedule needs.

A broader understanding of CM implementation results in developing sound metrics, but by establishing requirements to be utilized in the PD/LM choice that met engineering and CM implementation needs, CM considerations were excluded from much of the requirements development process and the resulting poor decisions proved to be expensive lessons learned. Chapter 11 will delve deeper into other examples of where a grasp of CM was not present when it could have added value to the decision-making process.

These case studies reflect the limitations of a supporter mindset in the GTS organizational structure as a close-knit organization that is secure from outside intervention and geared to respond to user needs based on their requests as described by Langer and York.[51] The disconnect between how the GTS organization sees itself and how others believe it can best serve the enterprise may be due in part to its developmental history.

Many software–user interfaces are moving to a cloud-based or cloud implementable–based user interface hosted on the enterprise GTS backbone regardless of the sophistication of the GTS infrastructure. However, grids may be deployed as part of a GTS cloud or part of non-cloud strategies for meeting GTS needs. The struggle is most complicated in finding the right mix of GTS resources to meet enterprise needs. No single resource strategy will meet all needs (lead strategy, lag strategy, or match strategy) unless a strong CM element is present. Demands for GTS resources can be classified for our purposes in the two broad categories of capability demand and capacity demand.

2.4 OVERHEAD, GENERAL AND ADMINISTRATIVE, AND OTHER COST ELEMENTS

This section introduces cost elements, techniques, and examples to further the understanding of later chapters. Its content draws on personal observations, from referenced materials, and from managerial accounting.[52]

Reviews of multiple business segments have revealed a growing lack of business acumen in those involved in CM implementation planning. This is not to be confused with the broader definition of managing the configuration. The lack of business acumen has resulted in companies introducing changes in their policies, procedures, and practices that work against sound business practices, against meeting financial responsibilities on contracted work, and against company stockholders.

Specifying the addition of a single mandatory signature for the release of an engineering drawing has far reaching profitability impacts. Time studies were performed to determine the average length of time required to log into a PD/LM system, evaluate an ECO and the associated drawing and part objects, and approve it electronically. It was found that for minor-to-medium-level complexity items, such a review could not be done in less than 30 seconds. Next, the total numbers of such ECOs for programs of diverse sizes were examined against total engineering objects at the same companies. The average number of ECOs per piece of engineering was found to be 6.5 with a range of 1.2–48. The average number of engineering objects (e.g., drawings, specifications, statements of work, test and other procedures, plans, etc.) per program was 2,600 with a range of between 300 and 29,000.

Using a sum of €174.67 for each direct hour worked (includes salary, benefits, and other associated indirect costs), we get Equations 2.7 through 2.12.

$$\left(\frac{\text{€174.67 per hour}}{\text{60 seconds per minute}}\right) = \text{€2.91 per minute}$$

EQUATION 2.7 Cost per minute.

[51] Langer, A. M., and Yorks, L. (2013). *Strategic IT: Best Practices for Managers and Executives (CIO)*. John Wiley & Sons.

[52] Maher, M. W., Stickney, C. P., and Weil, R. L. (2011). *Managerial Accounting: An Introduction to Concepts, Methods, and Uses*. Cengage Learning.

$$\left(\frac{€2.91 \text{ per minute}}{3}\right) = €0.97 \text{ per 20 seconds}$$

EQUATION 2.8 Cost of a 20-second review.

Leveraging this value as the cost per signature over the number of ECOs multiplied by the average number of engineering objects gives us the average impact to the average program.

$$€0.97 \text{ per signature} \times 6.5 \text{ ECOs per object} \times 2,600 \text{ objects} = €16,393 \text{ per program}$$

EQUATION 2.9 Impact per program.

The companies involved had a total of 1,496 programs impacted.

$$€16,939 \text{ per program} \times 1,496 \text{ programs} = €24,523,928$$

EQUATION 2.10 Program impact across all companies studied.

This €24,523,928 could have been used for risk reduction or budget or schedule maintenance, and be booked as profit if on a firm-fixed price contract.

Leveraging the information from the study further, the average time to locate an object on a screen and click a mouse button was found to be an average of five seconds. Using the same €2.91 for each minute of direct hour worked leads us to a cost per click of €0.2425.

$$\left(\frac{€2.91 \text{ per minute}}{60 \text{ seconds per minute}}\right) \times 5 \text{ seconds} = €0.2425 \text{ per click}$$

EQUATION 2.11 Cost per mouse click.

Keystroke data was reviewed over a five-day period for a single individual performing CM activities. The average number of times the mouse was clicked was observed to be 753 per eight-hour day. This equates to €182.60 or, rounded, €183 per day just to move the mouse and click.

$$€0.2425 \text{ per click} \times 753 \text{ clicks per day} = €182.6 \text{ per day}$$

EQUATION 2.12 Cost for clicking a mouse per day.

We assumed the individual studied is the norm and that great savings and efficiencies can be realized through establishing file structures and data management flows that allow access to data with a minimum of file nesting.

Diverting the stated purpose of a meeting to engage in a technical discussion rather than setting up a different meeting for technical resolution is another example of costs that should be applied more productively. Meeting invitees who do not need to be in the technical discussion still charge their time to the meeting while such discussions take place. Barco ClickShare has developed the online "Meeting Waste Calculator" resource to quantify the costs when this happens. It allows users to input the number of meeting attendees, average salary, minutes wasted per meeting, number of meetings per week, and number of years involved to calculate the projected loss.[53]

---

[53] https://clicksharedemo.barco.com/save-money.html?ccmpgn=T-00000769&utm_source=comptiasb&utm_medium=advertising&utm_c ampaign=mx-us-landgrab

These three simple examples show why some business acumen is required by those performing CM functions. Accounting is part of the information system of any company, and CM implementation either contributes to EBIT or detracts from it. CM practitioners should understand not only how their action or inaction impacts EBIT, but also how to read a financial statement. What follows is designed to impart a rudimentary grasp of the subject as it applies to this book. It is not an attempt to present the subject in depth.

Costs can be divided into two categories: direct and indirect costs. As with the section on GTS, some definitions for terms used in this book are required at this point to facilitate the discussion that follows.

- *Cost objective:* Any revenue-generating activity such as a product, service, program, or other activity.
- *Direct cost:* All costs directly associated with a single cost objective.
- *Indirect cost:* Fixed and variable costs not directly associated with a cost objective. Indirect costs include G&A, OH, and bid and proposal (B&P), independent research and development (IR&D) and authorization for expenditure (AFE). We use AFE as "authorization for expenditure" and not "average funds employed" throughout this book.

B&P and IR&D[54] are presented from a basic perspective and may be explored in more depth through other resources.

Paul Bauman Jr. states:[55]

> Overhead and G&A (General and Administrative) are two of the most loosely used terms in government contract accounting. Everyone uses them … few really know what they are.
>
> Let's start with what is the same. Both are indirect costs. That is, they are costs that you incur while running your company that you cannot easily (directly) charge to a contract. For example, rent, office supplies, accounting, management, etc.
>
> The difference is that G&A refers to that portion of your indirect costs that apply to your whole operation; whereas overhead applies to a portion of your operation. Examples of overhead are: engineering overhead, labor overhead or manufacturing overhead, material handling, sub contract management. All apply to a specific function or cost within the organization.
>
> Overhead pools are selected based on the nature of the operation. Generally, you select different overheads if the costs associated are different. For example, if your projects entail a combination of in-house labor and sub contract labor you might notice that your in-house labor entails indirect costs (benefits, payroll taxes, leave, etc.) that do not apply to your sub contracts. On the other hand, you might notice that there is a cost for management of the sub contracts that is not present for in-house labor. In this case you might want to set up in-house labor overhead and sub contract management overhead pools. Meanwhile you collect those costs that apply across the board (accounting, business development, etc.) into a common G&A pool.
>
> To get the cost of a job or a department in your company you take the sum of the direct costs (those you charge directly to the job) and a fair share allocation of indirect costs. Using our example above, your labor costs for a job would be the sum of direct labor (salaries you pay) plus a fair share of labor overhead plus a fair share of G&A. Meanwhile, your sub contract cost for a job would be the sum of the sub contract cost, plus a fair share of sub contract handling and G&A.

---

[54] Leveraged from information contained in United Technologies Corporation. (1998, November). *Independent Research & Development and Bid & Proposal Costs*. Retrieved November 3, 2013, from www.utc.com, www.utc.com/StaticFiles/UTC/StaticFiles/proposals_english.pdf.

[55] Baumann, P. (2007, April 24). *What's the Difference between Overhead and G&A?* Retrieved October 3, 2013, from http://www.theasbc.org/news/15315/The-ASBC-Community-News-Whats-the-difference-between-Overhead-and-GA.htm. Reprinted with the permission of the author.

For our purposes:

B&P: Part of the G&A rate. Those costs incurred in the preparation and submittal of solicited and unsolicited proposals to government and commercial customers.

IR&D: Part of the G&A rate. Basic research toward increased knowledge of a science. Applied research exploiting the scientific potential or improvements in technologies, techniques, materials, and methodologies/processes; development of scientific and technical application in product design, development, and test and evaluation of new or improved products and services; and systems and concept formulation studies to identify new or modified systems, equipment, and components.

The definitions that follow for OH and G&A are specific to a company. They are simply an example but not rules that always apply. It depends on the mission and expenses of the company. There is no rule that says a given cost is always G&A or always OH.

OH: OHs are an ongoing expense required for any business to function. As noted earlier, OH costs do not generate revenue. They can be fixed or indirect. Fixed OHs are thought of as static and do not change substantially based on activity. Indirect OHs vary based on activity.

- Insurance
- Supplies and manufacturing expendables
- Taxes
- Company (corporate) travel unrelated to a specific contract. In some cases, travel may be part of the G&A rate if associated with a G&A function such as accounting or contracts rather than an OH item.
- Depreciation
- In some cases, factory support labor
- Inventory management
  - Purchased materials and parts not worked on
  - Work in process
  - Finished goods inventory

It is a widespread practice to divide OHs into pools separating corporate from company and business unit. Overhead pools exist for department supervision, depreciation, training, and fringe benefits. These may be further subdivided based on organizational responsibilities under a business unit such as engineering, manufacturing, and material handling. OHs are allocated over the entire revenue producing labor (labor charged direct to a cost objective) hour base. In the case of engineering, the OH would be allocated over the total engineering direct labor hours.

G&A: Any expense incurred for managing and administering the whole business unit and allocated to final cost objectives. These expenses in the corporate ledger and financial report will generally appear under operating expenses. G&A may include the following:

- Rent
- Insurance
- Utilities
- Human resources
- Accounting
- Contracts
- Consulting expenses
- Public relations
- Legal

- Corporate management
- Equipment depreciation
- Outside audit fees
- Subscriptions
- IR&D
- B&P

Managing the configuration of a building or the design, development, integration, test, and installation of a new asset as a capital expenditure or updating manufacturing infrastructure will require the CM practitioner to size the strategic and tactical implementation to the management need.

Returning to the broader aspects of cost accounting, several sound business practices need to be observed to make direct and indirect cost data usable. The following is leveraged from paragraph 8-400 of the U.S. Defense Contract Audit Agency (DCAA) Contract Audit Manual.[56] Portions of the definitions are taken in their entirety from the DCAA Contract Audit Manual; others have been simplified to meet the scope of this text.

Consistency in estimating: Estimating, accumulating, and reporting of costs should be applied consistently.

Consistency in allocating: Costs incurred for the same purpose are allocated the same way to ensure that each type of cost is allocated only once and on only one basis to any contract or other cost objective. All costs incurred for the same purpose, in like circumstances, are either direct costs only or indirect costs only with respect to final cost objectives.

Allocating home–office expenses to segments: Expenses incurred for specific segments are to be allocated directly to those segments. Expenses not directly allocable, but possessing an objective measurable relationship to segments, should be grouped in logical and homogeneous expense pools and distributed on allocation bases reflecting the relationship of the expenses to the segments concerned.

Capitalization of tangible assets: Capitalize the acquisition cost of tangible assets in accordance with a written policy that is reasonable and consistently applied.

Unallowable costs: An expressly unallowable cost is that which is specifically named and stated to be unallowable by law, regulation, or contract. It must be separately accounted for and not allocated to a contract.

Cost accounting period: The cost accounting period used by a contractor must be either its fiscal year or a fixed annual period other than its fiscal year (e.g., a calendar year) across its entire business base.

Standard costs for direct material and labor: Such practices must be written and the established labor-rate standards and variances against the standard recorded and adjusted yearly (e.g., fabrication, minor assembly, final assembly, and test may have a standard rate). The functions performed within each department are similar, the employees involved are interchangeable, and the inputs of direct material are homogeneous. Each variance account is distributed annually based on the department's labor dollars. The practices are stated in writing and consistently followed, and the standard costs are documented in the books of account.

Cost accounting for compensated absence: Compensation paid by contractors to their employees for such benefits as vacation and other leave should be assigned to the cost accounting period or periods in which the entitlement was earned.

Depreciation of tangible assets: Estimated residual values must be determined for all tangible capital assets or groups of assets. The residual values must be deducted from the capitalized value.

[56] DCAA, U.S. (2013). *Cost Accounting Standards, Defense Contract Audit Agency (Chapter 8-400 Section 4)*. U.S. Government.

Allocation of business unit G&A expenses to final cost objective: G&A expenses must be grouped in a separate indirect cost pool and allocated only to final cost objectives. For an expense to be classified as G&A, it must be incurred for managing and administering the whole business unit.

Accounting for acquisition cost of material: The cost of material and allocation to cost objectives must be done according to written statements of accounting policies and practices. If the material is procured for a single cost objective, it should be allocated as direct, and if it is considered shop supplies, it can be an indirect cost allocation across the work center.

Composition and measurement of pension costs: There are basically two kinds of pension plans: defined contribution plans and defined benefit plans. A defined contribution plan provides benefits to retirees according to the amount of the fixed contribution to be made by a contractor. In a defined benefit plan, contributions made by the contractor are calculated actuarially to provide pre-established benefits. The cost of benefits under a *pay as you go* plan must be measured in the same manner as the costs under a defined benefit plan.

Adjustment and allocation of pension costs: Actuarial gains and losses represent differences between actuarial assumptions and need to be properly accounted for with the result that pension obligations can be met and clear and concise criteria for determining the funding status for pension plans exist and can be tracked as a financial liability.

Cost of money: There is a cost associated with borrowing money for capital improvements. This cost of obtaining facilities capital is a contract cost and commonly referred to as the "cost of money." Criteria for measuring and allocating an appropriate share of the cost of facilities capital identifiable with the facilities employed in a business should be established.

Deferred compensation: Deferred earnings should be paid out at the same rate that they were earned (e.g., if banked overtime was earned in 2013 and paid in 2018, the payout should reflect the salary scale at the time the hours were banked).

Insurance: Records to substantiate the amounts of premiums, refunds, dividends, losses, and self-insurance charges. They should also show the frequency, amount, and location of actual losses by major type of risk.

Cost of money as an element of the cost of acquisition: The cost of money applicable to the investment in tangible and intangible capital assets being constructed, fabricated, or developed for a contractor's own use shall be included in the capitalized acquisition cost of such assets.

Allocation of direct and indirect costs: A written statement of accounting policies and practices should exist for classifying costs as direct or indirect, which shall be consistently applied; indirect costs shall be accumulated in indirect cost pools, which are homogeneous; and pooled costs shall be allocated to cost objectives in reasonable proportion to the beneficial or causal relationships of the pooled cost-to-cost objectives.

Accounting for IR&D and B&P costs: Costs for IR&D and B&P activities are part of G&A and should be accumulated by project and allocated using the same procedure as G&A.

Another concept that configuration managers need to understand is that of cost estimating relationships (CERs). CERs are a method used to estimate future costs based on past performance. A CER is also referred to as a factor or a factor applied to a base. Across the companies studied, use of a CER for CM was found to be inconsistent. Those that used CERs accounted for every hour worked on a program regardless of the nature of the labor from an employee remuneration aspect (e.g., all hours, paid or unpaid, were recorded in the time keeping system against the cost objective worked on). Since all CM hours were accounted for, a valid cost estimating relationship could be generated.

CERs for related aspects of a program, product, or service are formed using the largest labor element of cost, from which the item factored costs are determined. The following example is provided as a means of explanation.

A contractor has been tracking all costs associated with programs over time. The total number of engineering hours per year was discovered to have a consistent relationship with the number of individual items (parts, assemblies) designed and produced. Non-engineering direct labor was then evaluated against the engineering labor and found to have a consistent relationship over time. Based on this evaluation, a CER was developed based on the number of individual items.

Each individual part = 280 hours of direct engineering labor

280 hours direct engineering labor = 9 hours of configuration management
5 hours of mission assurance
2 hours of reliability
0.5 hour of safety
1 hour of parts, materials, and processes

After validation, the company used a CER of 297.5 hours per part that included engineering, configuration management, mission assurance, reliability, and parts, materials, and processes (Equation 2.13). This CER was validated each year and adjustments made resulting from changes to business practices and efficiencies.

$$280+9+5+2+0.5+1=297.5\ \textit{hours}$$

EQUATION 2.13 CER calculation.

The largest drawback to developing any CER is not having the total cost basis of the item. One company studied was evaluating the implementation of a CER similar to the earlier example. Historical data was gathered for actuals, which resulted in a CER (adjusted to the earlier example) of 196.35 hours per part that included engineering, CM, mission assurance, reliability, and parts, materials, and processes. The accounting system did not account for every hour spent and those working on the program were donating 20 hours a week for every 40 charged. As a result, the CER was 33% lower than the actual number of hours spent. Had it been implemented, every program would be underestimated by one-third.

Occasionally, companies will attempt to estimate cost using parametric estimating methodologies to establish the cost of something they have never built before. In its simple form, a parametric estimate is based on the premise that if it looks like a box measuring 12 cm² × 12 cm² × 36 cm², and you just built a box measuring 12 cm² × 24 cm² × 36 cm, the new box should cost half as much as it is half the size. When applied to complicated and cutting-edge technologies, it can cause significant cost issues. During the 1990s, bids were placed by major corporations to repurpose aging U.S. government missiles. A proof of concept contract was awarded for conversion of two missiles to a company that bid using parametric estimating. The bid was evaluated by both internal and external sources for reasonableness prior to submittal. After the performance started, it was discovered that assumptions used to create and validate the parametric estimate were flawed. The company underestimated the effort by 70%.

2.5 INTELLECTUAL PROPERTY

IP is that broad category of tangible and intangible things that encompasses everything created as products of the human intellect. IP can be categorized as personal, familial, guild, tribal, religious, company, and, in some cases, national.

Personal: Inventions, arts, communications, photos, tweets, and anything posted on social media
Familial: Recipes, traditions, and unique-to-family ways of doing things
Guild: Inventions, trademarks, trade secrets, traditions, stories, and rituals
Tribal: Totems, beliefs, traditions, stories, and rituals
Religious: Icons, beliefs, books, codes, mysteries, and rituals
Company: Inventions, trademarks, and trade secrets
National: Identity and monopolies

Known as "traders in purple," Phoenicians first produced Tyrian purple from the spiny dye-murex, the banded dye-murex, and the rock-shell murex. The process was "Phoenician IP." They had a monopoly on purple dye and it became part of their national identity. For years it has been rumored that they dyed their sails purple. Despite much research, no evidence that Phoenician sails were ever dyed with Tyrian purple was found by the authors. As it was so valuable in trade, worth its weight in silver, it was probably not used for this purpose.

2.5.1 What Is IP?

IP brings up recollections of things such as copyright law, trademark law, patent law, trade secrets, *Sui generis*, and public domain. All these exist to protect the interests of one party from being encroached upon by another party. The understanding of each type of law and its application often involves a broad-based cultural mindset. In some countries, the application of a thing to a new purpose not intended by the original patent holder may be given a new patent (e.g., the application of a thing and not the thing itself is patentable). A different technical approach may also allow different people to be granted copyright for seemingly identical works. The diverse types of IP are shown in Figure 2.34.

In the United States, there are 45 categories or classes of trademark. Product (or goods) classes account for the first 34. Service classes account for the rest. Use of a trademark under license may be granted to one firm, whereas the true owner of the original trademark is verted in another. Licensing of a trademark originally held includes Class 12: vehicles and products for locomotion on land, water, and air; Class 14: jewelry products; Class 21: housewares and glass products; Class 25: clothing and apparel products; and Class 28: toys and sporting goods. We see this in "officially licensed" Jeep products.

2.5.2 What Drives IP?

Seeking and asserting IP rights provides the holder certain privileges to these rather intangible assets and forms of legal remedies for encroachment on them in a court of law. The application of markings notifying others of the rights held by the owner is critical. Enforcement relies on the reader of the markings to respect those rights. Enforcement of rights under cases of infringement is often difficult and costly.

2.5.2.1 Internal

IP rights are protected as they provide the company a competitive advantage because they compete within the ecological market niche they inhabit. In some cases, the advantage is perceived and in others it is factual, knowing internally that the copyright on the phrase "Now contains formula ZC241 to help reduce your salt intake" on a food container is perceived as a competitive edge by the consumer. Internally it is known that formula ZC241 is potassium chloride (also known as KCl). Holding dear the trade secret of how to create carbon nanotubes from lint would be a factual competitive advantage. Such factual competitive advantages may protect the company's products in the marketplace for several years.

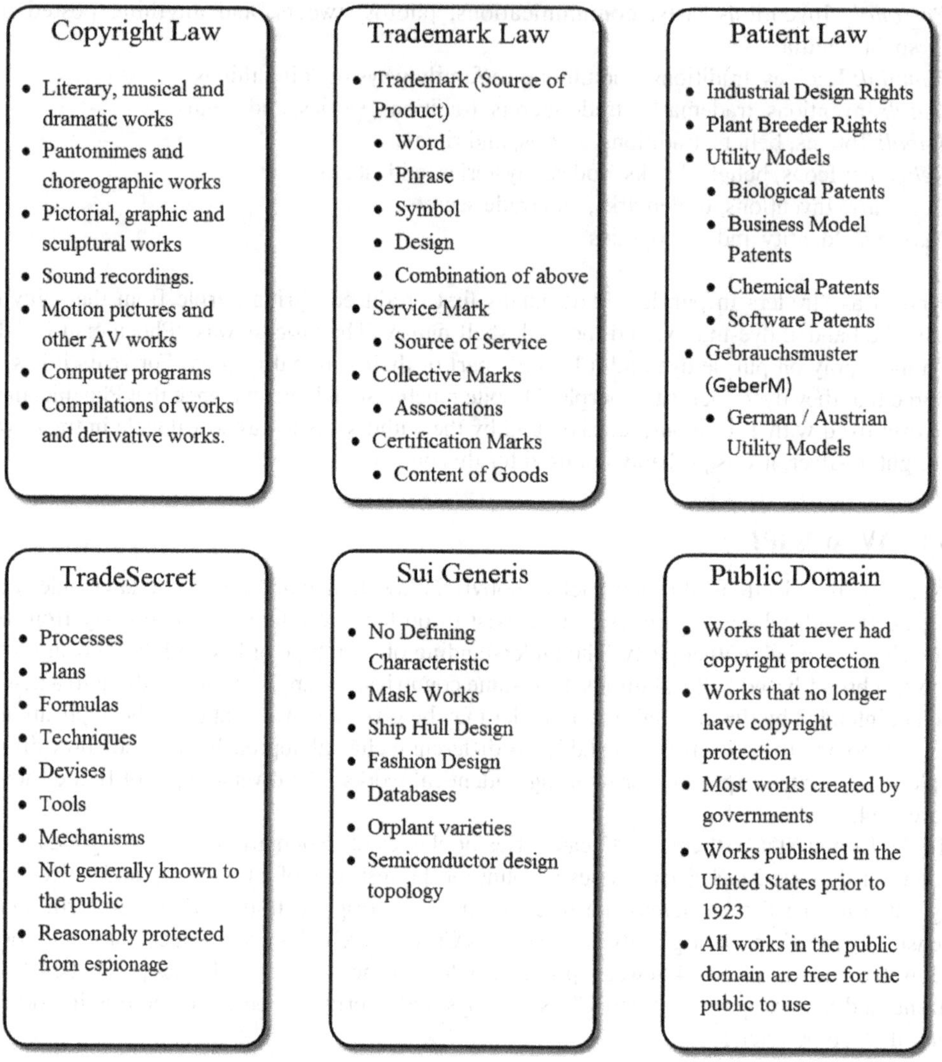

Copyright Law

- Literary, musical and dramatic works
- Pantomimes and choreographic works
- Pictorial, graphic and sculptural works
- Sound recordings.
- Motion pictures and other AV works
- Computer programs
- Compilations of works and derivative works.

Trademark Law

- Trademark (Source of Product)
  - Word
  - Phrase
  - Symbol
  - Design
  - Combination of above
- Service Mark
  - Source of Service
- Collective Marks
  - Associations
- Certification Marks
  - Content of Goods

Patient Law

- Industrial Design Rights
- Plant Breeder Rights
- Utility Models
  - Biological Patents
  - Business Model Patents
  - Chemical Patents
  - Software Patents
- Gebrauchsmuster (GeberM)
  - German / Austrian Utility Models

TradeSecret

- Processes
- Plans
- Formulas
- Techniques
- Devises
- Tools
- Mechanisms
- Not generally known to the public
- Reasonably protected from espionage

Sui Generis

- No Defining Characteristic
- Mask Works
- Ship Hull Design
- Fashion Design
- Databases
- Orplant varieties
- Semiconductor design topology

Public Domain

- Works that never had copyright protection
- Works that no longer have copyright protection
- Most works created by governments
- Works published in the United States prior to 1923
- All works in the public domain are free for the public to use

FIGURE 2.34 Types of intellectual property.

2.5.2.2 External

IP has existed external to the personal, familial, guild, tribal, religious, company, and national consideration since recorded time. For the purposes of the remaining discussion in this chapter, these groups will be referred to as "entities." IP exists anytime one entity has some knowledge it wishes to keep from another entity. Personally, it could be one's true name rather than public name or other attribute that protects one's identity or possessions. Between siblings, it may be the location of an item that if found could cause actual or perceived harm to them by a person of authority. Between guilds, it may be a process held only by masters of the trade. Between tribes, it may be the location of a relic or rites of passage from child to adult. From a religious aspect, it may be the location of or the ability to translate sacred writings or concepts or the location of revered objects. Nationally, it may be the ability to defend or attack.

Those nations with advanced technology desired to keep it away from others who did not possess it and took steps to do so. Communities that could make weapons of copper had an advantage over weapons made of stone. Communities with bronze had an advantage over copper and so on until the

FIGURE 2.35 Jian sword.

present. This weapons technology history is evident in the sword types in some forms of martial arts today. Fighting with cast iron swords or forged swords made with a single layer of metal required precise movements so as not to submit the weapon to impact forces that would break it. These older metallurgical methods gave way to the three-layer Jian sword (Figure 2.35). The technological breakthrough was to sandwich a layer of harder steel between two softer layers. It was not available to the Europeans until the first century.

2.5.3 How It Applies to CM

CM and data management activities are central to the management of documentation. In many enterprises, the configuration manager is the last set of eyes to look at a document prior to formal release. Working in close association with contracts and legal people (often IP legends and rules for their application are generated by the legal department), the configuration manager ensures that the proper legends are on the document. Should the item being produced be subject to governmental marking requirements where certain rights are passed to the government, company compliance with those marking requirements is also verified.

Typical items looked at prior to release include the following:

- Copyright marking relevant to unpublished works
- Proper use of trademarks
- Intellectual rights statements
- Citations pertaining to licensed technologies
- Technology transfer identifiers required by a technology transfer control plan (TTCP)

2.5.4 Why Is It Necessary?

The World Intellectual Property Organization Treaty[57] enacted by U.S. 17 U.S.C. §§ 512, 1201–1205, 1301–1332; 28 U.S.C. § 4001 defines IP generally as rights in relation to literary, artistic, and scientific works; the performance of performing artists, phonograms, and broadcasts; inventions in all fields of human endeavor; scientific discoveries; industrial designs; trademarks; geographical indication, service marks, and commercial names and designations; and all other rights resulting from intellectual activity in the industrial, scientific, literary, and artistic fields.

[57] World Intellectual Property Organization (WIPO). Retrieved September 4, 2014, from http://www.wipo.int/export/sites/www/freepublications/en/intproperty/450/wipo_pub_450.pdf.

As with *property* rights, IP rights are exclusionary—so that third parties are prohibited from their use and exploitation. With normal property rights, the nature of the object in which property rights subsist gives it a monopolistic character. For example, if the property is a bicycle, only one person can own or use it. If one person is using the bicycle, no one else can do so. However, IP rights are fundamentally different—concurrent use of inventions or copyright materials is possible. For example, one invention could be produced by several manufacturers and used by hundreds or thousands of people. A musical performance can be recorded and then broadcast and then replayed hundreds of times by hundreds of different people. Use and/or performance do not diminish value or cause the IP to perish, unlike the position with most normal property.

An uncomplicated way to distinguish the various categories of IP rights from each other is to recognize those that are *hard* and those that are *soft*. The World Intellectual Property Organization goes on to define copyright, database rights, unregistered design rights, and unregistered trademarks, trade secrets, and confidential information as soft rights. These require no registration, no fee, and no formalities. They arise automatically—you get *something for nothing*. It also defines hard rights as patents, registered designs, and registered trademarks.

In the United States, it is possible to register copyright to demonstrate ownership of a piece of work. One way of showing that you own copyright at a point in time is to make a copy of the work in question and send it through the post in a sealed envelope. The envelope should be date marked and preferably recorded/signed for delivery then left unopened. If required, you could dramatically prove as a matter of evidence by opening the envelope in a court of law that copyright existed on the marked date.

The growing international aspects involved require the differences in how IP is defined and marked be understood and applied when products are marketed in the international setting. Current laws and practices regarding IP are evolving. 3D printing—should it evolve into onsite growth of human tissue based on an individual's specific genetic makeup—will challenge long-cherished IP laws and justification, not to mention the management of the associated infrastructure and commodities used in the process.

2.5.5 IP and Legal Documentation (Incarnations or Permutations of the Design)

Most industries reserve first rights to the IP developed by an employee as one of the conditions of employment. First rights allow the employer to evaluate ideas and technologies based on its needs, and those that do not fit well with the focus of the enterprise may be given over to the individual who developed them.

The evolutionary aspects of products need to be explored. Just like the provenance of a work of art must be impeccably established, the provenance of a design must also be established before the intellectual rights of other designs leveraged from it can be claimed. In some cases, the concept of first reduced to practice may be all that can be relied on due to lack of formally controlled documentation.

U.S. government contracts generally are written to give the government unlimited rights to anything developed under a contract. Ownership resides with the contractor and the government can utilize the technology as it sees fit. Should the provenance of a design prove that it existed prior to the contract date, limited rights to the technology can be established that prohibit government use except for the specific purpose the information is provided for under the contract. These limitations are also extended to software items via an IP statement known as "restricted rights."

2.6 INTERNATIONAL TRAFFIC IN ARMS REGULATIONS

Restrictions in technology transfer have existed in one form or another, as have sanctions by one entity against another, since earliest times. The Arms Export Control Act (AECA), ITAR, and the U.S. Munitions List (USML) are a manifestation of this. The AECA was enacted in

1976 to restrict data concerning items with military application for countries making up what was then known as the Eastern Bloc (East Germany Soviet Union, Czechoslovakia, Bulgaria, Albania, Hungary, Romania, and Poland). These countries formed the Union of Soviet Socialist Republics[58] and were engaged in a sustained state of political and military tension with what was known as the Western Bloc [in 1949: Belgium, Canada, Denmark, France, Iceland, Italy, Luxembourg, the Netherlands, Norway, Portugal, the United Kingdom, and the United States; the other member countries were Greece and Turkey (1952) and Germany (1955)]. This period of tension was called the Cold War. Such tensions can also arise when import tariffs between countries are not negotiated and implemented, when trade agreements are renegotiated, or when international trading partners change.

The AECA is described in Title 22, Chapter I, Subchapter M, of the U.S. CFR. The USML identifies items that are controlled, and it is interpreted by the U.S. Department of State Directorate of Defense Trade and Controls (DDTC). Since its inception, items contained on the USML have changed. Information on the USML through the ITAR can only be shared with non-U.S. persons if authorized by the Department of State. Many items on the list are not intuitively arms related.

In 2013, the definition of a "U.S. Person" included U.S. citizens, persons granted political asylum, a part of the U.S. government, a business entity incorporated in the United States under the U.S. law, and permanent residents granted a U.S. right to work permit who do not work for a foreign company, government, or governmental organization. U.S. citizens who work for a foreign company, government, or governmental organization are not considered U.S. persons but representatives of that company or government.

Exports can take the form of foreign military sales, export licenses, warehouse and distribution agreements, technical assistance agreements (TAA), and manufacturing license agreements. Technology control plans (TCP) are defined under the National Industrial Security Program Operating Manual 10-509 and by the ITAR 22CFR § 126.13 (c). In addition, TTCP may specifically identify file formats and other restriction of data transferred. A company needs to register with DDTC and know what is required of them to be compliant with the ITAR and self-certify that they possess this knowledge and will implement it.

One critical CM aspect of the TAA is the requirement that each piece of data transferred be uniquely identified and tracked. This means that each revision or version of the data is uniquely identified. Generally, CM is not consulted when TAAs and TTCPs are started. This may result in overcomplicated requirements. One example would be if the customer is performing part of the design activity and requires the native files to be delivered while the TAA specifies that data will be delivered as pdf format only with the document number and revision acting as the unique identifier. This creates a dilemma as the pdf and native file cannot use the same document number and revision/version as the "unique identifier." In Figure 2.36, the pdf and native files are decoupled. Recipients of ITAR-controlled data are prohibited from transfer of the data to others unless this retransfer (also known as re-export) is specifically authorized in the applicable export authorization.

The TAA and TTCP may be so specific that discussions between the exporter and the foreign entity are restricted to specific locations or buildings. In such cases, if the named parties meet in a location not specified in the TAA (such as at an international conference), no discussions can take place. U.S. enforcement activities have risen since 1999 when the Department of State assumed control of export regulations for satellites, resulting in a decline of 23% from its high of 83% for the U.S. portion of the worldwide market segment. ITAR does not apply to items in the public domain or information related to the underlying principles of science, mathematics, and engineering.

One can imagine an early form of ITAR being practiced by the Vikings and what is now Iran in the period between 800 and 1000 with commerce along the Volga trade route. During this

[58] Сою́з Сове́тских Социалисти́ческих Респу́блик (СССР), also known as USSR.

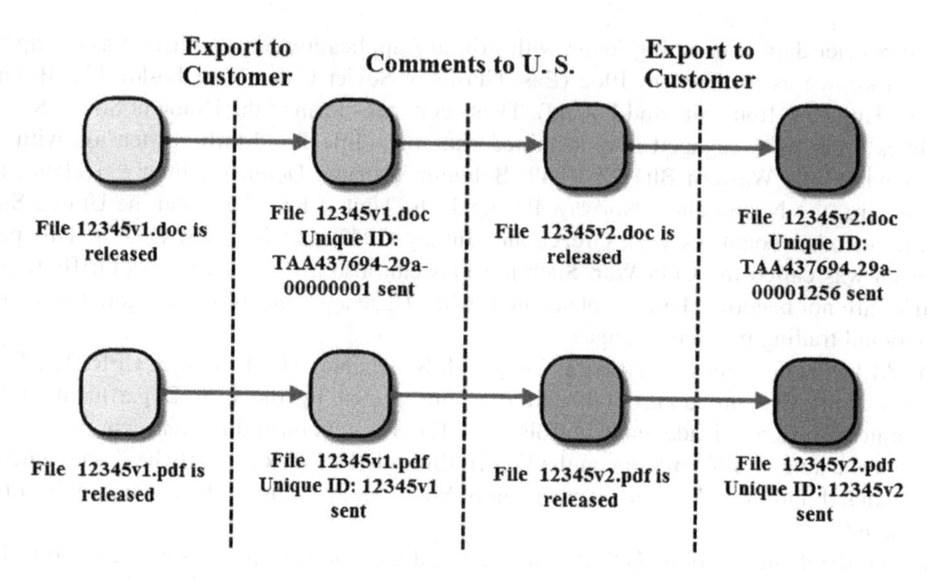

FIGURE 2.36 TAA unique identifiers.

period, very good carbon steel was found in both Damascus steel blades and the famed Viking Ulfberht swords. Conceivably Viking traders exchanged Nordic commodities along with those obtained along the trade route for ingots of this superior steel from which the swords were made. In the program "Secrets of the Viking Sword,"[59] narrator Jay O. Sanders makes the point that for half a century before the advent of the Ulfberht blade, warriors outside of Europe had been fighting with crucible steel weapons. Damascus steel blades were made from material similar in nature to the Ulfberht.

True "+ULFBERH+T" blade National Museum in Copenhagen, Denmark

Fake "+ULFBERHT+" blade Germanisches Nationalmuseum, Nuremberg

FIGURE 2.37 Ancient sword forgery.

[59] Yost, P. (Director). (2012). *Secrets of the Viking Sword—NOVA*. [Motion Picture] Boston, MA.

FIGURE 2.38 Forged Dead Sea Scrolls (digitally modified from an image contained in *The Yeshiva World* of October 23, 2018.)

2.6.1 Counterfeit Products Circa 800

So sought after were true Ulfberht swords with the name inlaid with soft iron into the blade and spelled "cross, u-l-f-b-e-r-h, cross, t" (+ULFBERH+T) that inferior blades made from steel, which was brittle due to the amount of inclusions contained in it, were also available with the name spelled "cross, u-l-f-b-e-r-h-t, cross" (+ULFBERHT+) (Figure 2.37).

This is the earliest record found while researching this book of a counterfeit product suitable to use as a figure. Historically, counterfeiting of coins and other items is recorded that predates the +ULFBERH+ sword.

2.6.2 Counterfeit Items Circa 2010s

The BBC stated on October 23, 2018, that there is a high probability that five of the 16 Dead Sea Scrolls that were housed in The Museum of the Bible in Washington D.C. are modern forgeries[60] (Figure 2.38). They have been removed from display. While forgeries of museum artifacts are well documented, we have also seen forgeries of documentation associated with titanium and other metals used in aerospace products.[61]

[60] BBC News (2018) Bible Museum says five of its Dead Sea Scrolls are fake. https://www.bbc.com/news/world-us-canada-45948986

[61] Flight Global (2014) Pratt & Whitney sues supplier of titanium used in F-35 engines https://www.flightglobal.com/news/articles/pratt-whitney-sues-supplier-of-titanium-used-in-f-403306/

3 Configuration Management and Product Management

3.1 QUESTIONS ANSWERED IN THIS CHAPTER

- How do changes in the market affect configuration management (CM) product development efforts?
- Describe how CM applies to marketing activities.
- Who performs the boundary spanning activity for the organization when it comes to the product definition and customer desires?
- What attributes of the software make identification much more difficult?
- What attributes of the embedded hardware make identification necessary?
- Is there, or should there be, a connection between marketing research and CM?
- How are marketing experiments influenced by CM?
- Theorize the connection between marketing message and the CM.

3.2 INTRODUCTION

Enterprise CM implementation plays a critical role in every aspect of company performance. This role is especially pertinent when it comes to product management. CM implementation influences can easily be observed in market selection, implementation strategies, market feedback, product development, and product sales and support. Maximizing market potential and bringing products to market is exciting and the results can surprise even seasoned professionals.

After listening to friends complain about problems they encountered caring for their pets, Cary Dahl decided to see if he could sell rocks to the public in April 1975. He marketed them as Pet Rocks starting in late June with a 32-page care and training manual. They were the perfect pet, requiring no feeding, watering, walking, bathing. or grooming. They never got sick. The fad lasted six months and saw 1.5 million sales at €3.54 each. His cost per unit was very low. Cleaned rocks cost pennies, the packing straw cost little more, and printing the booklet was tacked on to an existing contract, reducing its cost. After the fad ended, other innovators started Pet Rock Cemeteries, offering memorial services and video recordings of the event. Dahl's strategy took full advantage of innovation dependencies (Figure 3.1).

In a world where artificial intelligence development is outpacing safety research, it will become increasing important to step back and analyze innovation dependencies to determine just what CM implementation means for the company. We are just now seeing the fallout of algorithm data mining the infometrics associated with people's buying preferences, network groups, and political affiliations. Marketing firms blast their messages across our smart devices and robocalls invade our telephone services without regard to attempts to stop the invasion by placing our names on do-not-call lists. These trends suggest that artificial intelligence will allow marketing agendas to expand across all communication venues well before they contract.

3.3 MARKETING

Product management starts well before product development. Markets and sub-markets are targeted and surveyed to see if the company's "next great idea" is viable. If the answer is yes, a strategic approach is created drawing from the company's strengths, aversion to risk, and desired market

FIGURE 3.1 Innovation dependencies.

capture percentage and profitability. This strategy becomes an attribute of the company product mix and is used to achieve or develop market share and profit margins.

Recently, Jon M. Quigley had the occasion to purchase a car at a dealership. His General Motors Corporation (GMC) "Jimmy" had more than 402,336 km on it and the cost of operation and maintenance exceeded his cost–benefit analysis. While in the showroom, he viewed the GMC product line on a touch-screen monitor. His son noticed a button icon they hadn't yet explored, so he touched it. To their surprise, the software version of the touch screen product line marketing program was displayed (Figure 3.2).

According to the American Marketing Association, the definition of marketing is:

> Marketing is the activity, set of institutions, and processes for creating, communicating, delivering, and exchanging offerings that have value for customers, clients, partners, and society at large. (Approved July 2013)[1]

In Section 1.3.1, we brought up the idea that marketing relies on capturing the customer's attention using appeals to one or more of the seven deadly sins mentioned by Dante. To maximize exposure of the company's target market selection activities and organizational objectives, the legendary eight Os are employed[2]:

1. Occupants.
2. Objects.

FIGURE 3.2 Touchscreen identification information.

[1] American Marketing Association. (2004, October). *Definition of Marketing.* Retrieved March 14, 2014, from https://www.ama.org/AboutAMA/Pages/Definition-of-Marketing.aspx.

[2] Czinkota, M. R., and Ronkainen, I. A. (2003). *International Marketing.* Cincinnati, OH: South-Western College. Cincinnati, OH, p. 17.

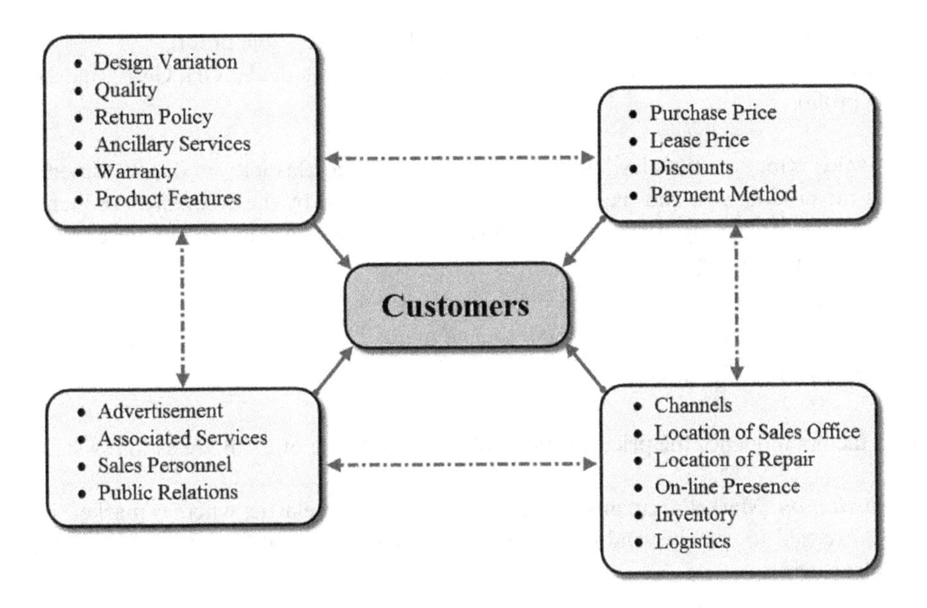

FIGURE 3.3 Marketing elements.

3. Occasions.
4. Objectives.
5. Outlets.
6. Organization.
7. Operations.
8. Opposition.

Marketing experts will manipulate the mix of variables to maximize the cost performance of the total product portfolio (Figure 3.3). The mix maximizes product, price, promotion, and place. These are referred to as the four Ps of marketing. Each will be discussed in turn in the sections that follow.

3.3.1 Product

In marketing speak, product means all the activities associated with the branding of the actual product or service. This also includes product images, logos, name, etc., that can be leased to outside firms. In the case of movies and television, the income derived from these ancillary sources provides a large revenue stream that continues to exist long after movie and television releases are past their prime. Even in product heavy companies, services associated with the product line continue to be purchased by the customer base. In the case of transportation, extended warranties and maintenance packages can be sold after vehicle purchase.

3.3.2 Price

The pricing policy is influenced by product demand, which is in turn influenced by the perception of the brand and competitive forces. Automobile manufacturers in the U.S. at one point maintained a five-tiered pricing approach. These were known as invoices.

1. Factory cost (the cost of manufacture including OH, G&A, and facility profit).
2. Factory invoice (the cost of manufacture plus the minimum company acceptable profit).
3. Dealer cost (factory invoice plus storage, sales OH, G&A, and factory profit).

4. Dealer invoice (dealer cost plus the minimum dealer acceptable profit).
5. Sticker price (dealer invoice plus transportation, handling, dealer OH, G&A, and salesperson profit).

Companies often test the elasticity of their pricing. Price elasticity is the measurement of the impact of the product cost and the change in product demand by the company. An increase in the sale price of the product by 10% may see a corresponding decrease in the purchase of the product. Equation 3.1 puts it simply:

$$\text{Elasticity of demand} = (\%\ \text{change in quantity}) / (\text{Change in price})$$

EQUATION 3.1 Pricing elasticity.

Many factors influence the price elasticity of demand. Some of them are as follows:

- *Substitution:* Market segments that are not saturated are elastic, whereas market segments that are tied to certain vendors are not elastic (e.g., personal hygiene items vs. hip joint replacement).
- *Luxury versus necessity:* Luxury items are elastic, whereas necessities are not (e.g., jewelry or a yacht vs. special medicines).
- *Income percentage:* Items that consume the greatest percentage of expendable income are elastic, whereas items that do not are inelastic (e.g., transportation or housing vs. salt).
- *Reaction time:* The longer the period the consumer has to react, the more elastic the demand (e.g., a three-hour sale generates less income than a two-week sale).

3.3.3 Promotion

In public relations, spin is a form of propaganda used to encourage public opinion in favor of a product. Promotion is a way of spinning a product in a way that drives demand. It is just as important to public relations as a world-class complaint resolution department. Promotions also rely on Dante's seven deadly sins. A combination of advertising, sales discounts, and publicity events are used to draw attention to the brand to increase profits.

Promotions follow a strategic plan and are accomplished using a variety of media forms:

- Stationary (billboards)
- Print (newspapers and magazines)
- Video and radio
- Digital (Internet, through social media such as Facebook, YouTube, etc.)

3.3.4 Place

Place refers to the distribution policy or variables in it. The two place elements are the logistics of getting the product to the desired destination and distribution channel management. These elements physically connect the customer to the product and ensure the product or service is available at the right place and at the right time. The goal is to ensure that the product is *on the shelf* for purchase before customers are looking for it and not lag behind the marketing campaign.

Distribution channel management entails the process of establishing and managing the intermediaries between the provider and the ultimate customer. The identification of distribution channels that improve product acceptance by the market is part of market planning.

3.3.5 Marketing and Market Research

Understanding the market is critical to the life of a product. It forms one of the cornerstones in the analysis of the competitive environment and is the basis for understanding and managing product perception. An understanding of the needs and views of the consumer base, combined with input from strategic and leveraged innovation business partners, is key in determining:

- Need versus availability.
- Market trending.
- Price analysis.
- Analysis of competition (domestic and international).
- Market segmentation (domestic and international).
- Brand strength (yours and your competition's).
- Risks (implied and actual).
- Media utilization based on targeted demographics.
- Marketing media mix.
- Test marketing.
- Perception management.

Market research and marketing research are closely related. The overlap is shown in Figure 3.4.

3.3.6 Customer Interface

The marketing department is typically the company's face to the customer base. The marketing department will perform market studies, market surveys, and demonstrations of different solutions to prospective customers to gather user feedback. They also explore different market segments to prioritize future developmental work. A company can then identify which of those segments optimize profitability and prioritize product development efforts on meeting those needs.

Marketing research quantifies the demands of the clientele and brings that information to the product development team. The product development team then segregates consumer requirements into needs, wants, and nice-to-have categories (desirements). This information establishes a new set of phased product requirements that are vetted with stakeholders using change control.

The marketing group sets about assessing areas of concern identified by prospective customers with the goal of quantifying the difference between an acceptable product or service and an excellent product or service.

3.3.7 Market Segmentation

During the days of Henry Ford's Model T, you could easily get the vehicle in black. Black paint dried the quickest, which maximized production rates and profitability. Providing multiple paint

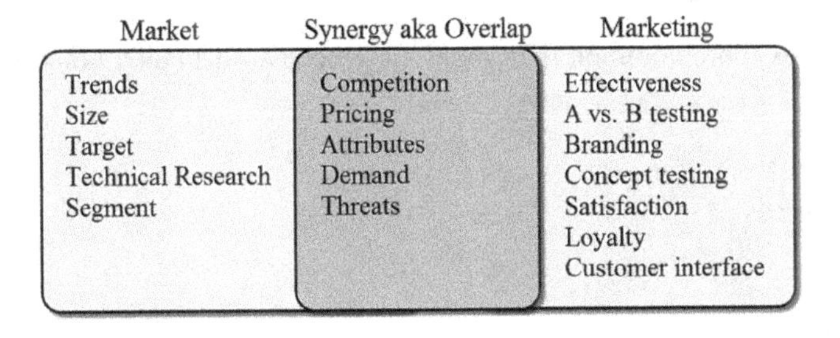

FIGURE 3.4 Market versus marketing.

FIGURE 3.5 Bespoke paint color choices of the Morgan Roadster.

colors slowed both. The automobile was a new commodity and the manufacturer could dictate the available features and colors. As competition increased, paint color became a customer-driven differentiator and companies adapted to meet the need or lost market share. Today, customers can choose the car type and color they want (Figure 3.5) within a wide range of possibilities based on the following multiple market drivers:

- Geographic
- Demographic
- Psychographic
- Behavioral

Ultimately, the marketing message is tailored to match the uniqueness of each of these drivers. Marketing may also tailor the price for the product, alter some of the feature content, and use other tactics to maximize marketing opportunities.

3.4 INCREASING THE COMPLEXITY OF THE PRODUCT

The advancement of the electronic age and processor improvements following Moore's law has driven complexity in the electrical aspects of products over the years. Moore's law is based on the observation that the transistor densities for microprocessors double every 18 months to two years. As these microcontroller improvements are made, product features increase with little added expense.

In an analogous way, increasing the regulatory pressure to achieve varying national or international priorities has driven how some products are made, what they can be made of, and how they are packaged, as well as how they comply with multiple environmental and safety rules. In the U.S., national priorities also drive the percentage of government set aside for small, small-disadvantaged, and minority business.

3.4.1 EMBEDDED SOFTWARE

Embedded software is more common than many may believe. It finds a home in many products such as:

- Microwaves
- Radios and stereos
- Automobiles
- Airplanes
- Medical equipment
- Consumer electronics and smart technologies

In embedded products, the development team has a sizable portion of the software effort located in the product itself. An embedded development project consists of software development wedded to a microcontroller or microprocessor-based hardware. The features of the product developed reside in the combination of hardware and software. The software provides the application and drives hardware function. The software can consist of an operating system for a complex item along with application code. Product-specific features reside in the application code. Even if the hardware can support a feature, it does not mean the software is available to capitalize on that capability. The hardware may have a universal asynchronous receiver/transmitter (UART) but that does not mean that the communication works on a product, as the feature may not be accessed or even enabled. These attributes are difficult to identify by visual inspection of physical properties. Embedded software allows additional functionality to be added as a maintenance activity or as a purchased upgrade. Embedded products are often subject to exploitable vulnerabilities.

3.4.1.1 Embedded Firmware

Some parts of software and the hardware differentiate poorly (hence the frequently used neologism "firmware"). Firmware weds hardware and software because hardware such as a programmable read-only memory (PROM) has no function without the software burned into it. Engineers work the hardware and the software development simultaneously to achieve a form of inorganic symbiosis; that is, the final product results from a dialectic exchange, a back and forth adaptation of each component to the other.

3.4.1.2 Physical Attributes of Software Not Visible

In the case of things like a UART, it becomes difficult to know the capability of the software after compilation. Software contents derive from a variety of sources. The digital thread starts at the product requirements and extends past the actual coding to the compilation of the software with the goal of making the software as error free as possible. This requires careful planning, requirement allocation, and control of functionality growth throughout the development cycle. The more distributed the product development (the longer the lines of communication), the more complex the digital thread becomes. A single developer would simply need to document all the functions put into the software. Software projects are often a team effort and do not rely on one person writing all the code. This exponentially compounds the degree of coordination required.

Comments in the software work only prior to compilation. This practice also makes extensibility a bit easier when the company grows or morphs functions after product launch. It is incumbent on those writing the code to keep a close connection between the requirements, source code, and compiled version. This is the only straightforward way to understand and manage the total software content.

3.4.1.3 Capabilities of Software Not Externally Obvious

Software, despite its uniqueness, often has capabilities that are not obvious to the end user. Software commands can be reliant on the content of tables containing parameters not resident in the software itself, but controlled and loaded into erasable PROM from a secondary data source. The software program may self-generate these data sets as it runs. This is often seen in the case of any built-in capability for learning based on previous inputs. Table updates may also be due to external analysis of system behavior. An example would be the modification of table data to increase the performance of a vehicle rather than increasing its efficiency. We see this in some vehicles giving the driver the option of engaging different driving modes.

3.4.2 Embedded Hardware

Hardware electronic memory components resident in the product subsystems are also called embedded. Compiled software is uploaded into the memory components. The hardware consists of the microcontroller or microprocessor core, along with the supporting circuitry to perform the functions specified. During development, the hardware features may see increases in functionality. This continues as prototypes iterate toward the final product. Hardware capabilities increase not only from prototype testing but also from building progressively more capable iterations. Functionality is often staged to get the most time possible with each product subsystem. For the UART example, the company may choose not to include transceiver interfaces in the first prototype if the functionality is not immediately required for testing.

3.4.2.1 Embedded Hardware Attributes Not Readily Visible

Like the software, the hardware feature content or capability may not be readily discernible. Many small parts are surface mounted. Some areas of circuit boards may be covered with conformal coating or other protectorate to meet environmental factors (e.g., protection from moisture, heat, cold, shock). Components like microcontrollers may have functions specified in the data sheet but not taken advantage of as utilized. Simple data communication applications may test performance during development but disconnect the interface in the final product.

3.4.2.2 Capabilities of Hardware Not Obvious Though Implied

Purchase of a vehicle implies that the vehicle is safe even though the built-in safety capabilities may not be obvious during the fabrication of the vehicle itself. Structural panels embedded into the doors and other frame members may act to divert the energy of a collision around the passenger compartment rather than through it. A spacecraft will endure vibration as well as acoustic shock on its way to orbit. Erosion due to atomic oxygen and ionic charging during its mission life or excessive torques if it has a single solar array is assumed to be mitigated in the design. These are examples of some capabilities that while not obvious are implied.

3.4.3 System Complexity

At one time, there were no *computers* or *microcontrollers* on a motorized vehicle. In the mid-1980s vehicle emissions were found to be hurting the environment and heavy commercial vehicles used a single device to meet vehicle emissions regulations. Modern vehicles have multiple controllers. The radio has morphed into an *infotainment system* capable of satellite navigation and communication with both the road you are traveling over and the vehicle manufacturer. Features and complexity are also increasing in small household appliances, televisions, cable boxes, microwaves, doorbell buttons with built-in cameras, and garage touchpads. Complex communication schemes are commonplace.

System complexity is not limited to the products produced. The system infrastructures used to produce these products have likewise increased in complexity as international development and production collaborations increase.

3.5 DISTRIBUTED PRODUCT DEVELOPMENT

The world is becoming ever more interconnected. This is true for almost everything, not just for embedded products but also for mechanical parts. Organizations acquire not only expertise but also efficiencies by looking globally to ensure that product safety, quality, and cost goals are met. A design may be developed in one location, or through a global collaboration. The sub-assemblies may originate from multiple factories in a variety of geographic regions. Products may be assembled worldwide to save on shipping costs.

3.5.1 Outsourced Coordinated Development

With the ease of access to communication tools, a global talent and manufacturing pool can be tapped for products. Global collaboration has some challenges despite the ability to share data. Managing the development direction and changes to the product not co-located can be a significant source of consternation. One organization studied has 10- to 12-week latency from the time of design change to the time the organization receives the first prototype parts for critique. During the design change to prototype availability time-lag, collaborative development efforts mandated a fresh design solution. The prototypes were scrapped, and the process started over.

It is not about just prototype part development, but also the nature of change and projects in general. As the saying goes, "The only thing constant in life is change."[3] Consider an entire system developed and assembled from parts that are sourced globally. This happens every day in automotive and in aerospace industries. Parts may be developed by one supplier and be sent to another supplier in another part of the world. The instrument cluster for an automobile developed in Germany ultimately contains heavy-tooled plastic parts from Asia and is manufactured and assembled in Mexico or Canada and then shipped to the United States and put onto the vehicle. This is only possible through implementation of a coordinated outsourcing work package.

3.6 HARDWARE HERITAGE

Adaptive radiation of hardware heritage is tied to both hardware qualification and intellectual property. Henry Ford could trace certain features of his designs from the Ford Quadracycle to the Model T to the Model A and firmly establish the design heritage and his rights to the intellectual property associated with it. Both are critical pieces of the proposal process today. Government customers are often risk averse and want to order systems with proven technologies. Ideally this means that the component on the new program is 100% the same as the component used on the last program. This is especially true with space hardware. It is also true that, to provide the expanded capabilities of the new systems, heritage designs need to be upgraded or used in some cases as a design concept. This places constraints on source selection teams as they weigh the competitive proposals. They want the best product for the money; but they do not want obsolete technology like microprocessors with known vulnerabilities that must never be used again. So, when is heritage hardware no longer heritage?

Our research was not able to answer this question. Even after studying Aerospace Report No. TOR-2013(3909)-1, Objective Reuse of Heritage Products,[4] we are unable to give a definitive answer. We can relay the following results:

- Most firms surveyed believe that if the design is 70% or greater in form, fit, function, and quality, they will consider claiming it as heritage.
- Source selection teams rarely give 100% of the available rating to any design having less than 90% the same form, fit, function, and quality.
- Only 65% of the firms surveyed can trace their independent research and development results (designs, processes, etc.) to their current designs. As a result, they cannot claim limited rights for hardware and restricted rights for software.

If a company's life blood is tied up in its intellectual property, there is severe intellectual property hemorrhaging going on in many industries. Part of the problem with not capturing the intellectual property source is observably a result of capture managers being overzealous in giving the customer whatever it asks for regardless of the downstream consequences. Non-disclosure agreements and competition sensitive agreements exist to protect a company's competitive edge. In some cases, unmarked intellectual property was freely distributed to all competitors simply because it had not been properly marked on all sheets of the document.

[3] François de la Rochefoucauld, 1613–1680 CE.

[4] Aerospace (2012): Objective Reuse of Heritage Products, 2012-04-30 prepared for the National Reconnaissance Office, 14675 Lee Rd., Chantilly, VA 20151-1715

4 A Configuration Item and What It Implies

4.1 QUESTIONS ANSWERED IN THIS CHAPTER

- What is a configuration item (CI)?
- What criteria need to be considered before designating something as a CI?
- What is involved in the configuration identification process?
- Does configuration management (CM) implementation apply if something is not designated a CI?
- What is serialization and why is it important?
- Is CM concerned with all phases of the product life cycle, or only with management of the allocated and product baseline up to the point of sale?
- Does CM only apply to documentation?
- What CM implications are associated with 3D printing?
- What is meant by *traceability*, and why is it important?
- What do the terms *allocation*, *verification*, and *validation* mean?
- What are the outputs, activities, inputs, constraints, and enablers associated with meeting functional requirements?
- What CM elements are associated with management of intellectual property (IP)?
- What is the CM life cycle process?
- How do the U.S. government mission stages relate to baseline management?
- Is the developmental configuration a baseline?
- What is an aligned product development model?
- What is a technical data package?
- What is involved with implementing change control?
- What is the purpose of status and accounting and what does it imply?
- Why are audits necessary?

4.2 INTRODUCTION

We have discussed many factors that influence configuration management (CM) implementation and how it relates to the market segment and enterprise resources over the product life cycle. Those who work in a CM environment are often so close to the day-to-day implementation that they lose track of the philosophy behind why CM exists and the intent of the terms that are related to it.

CM standards have been leveraged forward from decade to decade, with little broadening of the applicability of the terms and the underlying concepts (see Section 6.3.1). Methods, technologies, markets, viability of market solutions, and tools used are constantly evolving. The viewpoint that the customer dictates the level of control required to manage the configuration of a product, while true in the case of contractual obligations, often falls short of the entire scope of CM implementation involved.

This chapter will focus on the concept of designating something a configuration item (CI) or configured item, what that implies in the form of information, and how that information is managed. Throughout the book, *CI* will be used as the acronym for both configuration item and configured item. Despite the controversy across some market sectors and between practitioners; the *regulatory* use of the term *CI* only applies to government procurements. Steven Easterbrook of CMPIC has traced the evolution of the use of the term *CI*. His analysis shows that CI application to things other than hardware and software (later firmware) resulted from out-of-context interpretations of the underlying U.S. Department of Defense CI criteria not unlike the childhood game of Telephone. CM-specific terms are evolving, making the task of speaking a common CM acronym-laden language with universally accepted definitions harder despite the core concepts behind them remaining constant. Steven's discussion can be found at https://cmpic.com/PDFs/CMTrends_Issue29_2017_01.pdf. We found that this may be true in the U.S. environment, but elsewhere the acronym *CI* was used to denote a configured item.

Some companies we studied tended to overuse the designation, assigning CIs at every level of assembly (COOLLAR program). Others are doing their best to reduce the numbers of CIs (historically, the Minuteman ICBM suite had over 6,000 CIs; now, 2,600 are active).[1] Deciding how many CIs are good or bad really necessitates the involvement of the customer. This will become clear as you read on.

The remainder of this book will also introduce many discussions reflecting current thought on the implementation of CM. In *CM in the Twenty-First Century and Beyond*, Jack Wasson[2] states,

> At one time, Configuration Management was only about establishing proven configurations for delivered products and controlling the changes to them; now it has to accommodate changes and manage the outcome in the best possible way. It is about maintaining accurate and valid data to retain corporate knowledge and history (i.e., lessons learned) so we don't have to keep paying for it over and over.

During November 2013, the LinkedIn North Atlantic Treaty Organization (NATO) CM Symposium Group[3] came up with the following definition of a CI:

> A configuration (configured) item is defined as any product that requires formal release and control of supporting information prior to acceptance by the user; where the supporting information is detailed with sufficient precision for all internal and external stakeholders to understand.
>
> A configuration (configured) item shall be assigned a unique configuration (configured) item identifier (CII) that when associated with the place of origin assures the item's provenance and content are irrefutable.
>
> Lower-level elements of a product may themselves be designated as configuration (configured) items where that level of control is identified as being required; for example, because of:
>
> - Being produced by an entity external to the enterprise;
> - Regulatory factors; or
> - Other factors (e.g., safety, integral to product functionality, unique requirements, etc.)

As designating a product as a configuration (configured) item imposes additional information constraints on the product that increase over the product life cycle; it is advisable to perform a cost/risk vs. derived benefits analysis prior to designating a product as a configuration (configured) item.

Supporting information includes but is not limited to all enterprise- and vendor-level information regarding formulation, safety, handling, test, operation and installation, environmental and biological impacts due to use, other known hazards, and similar data associated with the product, items it comprises, and processes used during product creation.

---

[1] Lindsey, B. (2017). *Intercontinental Ballistic Missiles (ICBM) 50 Years of Sustainment.* 2017-08-27 presentation at CMPIC Seminars, Workshops, and Training event Rosen Centre Hotel. 9840 International Drive, Orlando, Florida 32819.

[2] Wasson, J. (2008). Configuration Management for the 21st Century. Retrieved November 23, 2013, from https://www.cmpic.com/whitepapers/whitepapercm21.pdf. Reprinted with the permission of the author.

[3] Wessel, D., Pep, G., Sounanef, Pickering, C., Robertson, K., Watson, M., DeAleida, R., Hartley, D., and LinkedIn NATO CM Symposium Group. LinkedIn discussion "What is a configuration item and which consequences does the selection of CIs have to your organization," moderated by Dirk Wessel, started on July 1, 2013. Reprinted with the permission of the author.

This definition is much broader than the definition provided in the following sources despite its closeness to the aliases, related terms, and examples in SAE/EIA 640-C. CM activities continue beyond product acceptance through product maintenance to end of life.

- ISO 10007:2003, Quality management systems—Guidelines for configuration management
  Configuration Item: entity within a configuration that satisfies an end use function.
- SAE/EIA 649-C, *Configuration Management Standard*, Table 2
  A product, allocated components of a product, or both, that satisfies an end use function, has distinct requirements, functionality and/or product relationships, and is designated for distinct control.
  Aliases, Related Terms and Examples
  - Computer software configuration item (CSCI)
  - Configuration object
  - Configured product
  - Designated item
  - Product
  - Functionally significant item

Remember that CM applies to everything and not just to CIs. SAE/EIA 649-C states that a CI is designated for distinct control. This means that the items were selected because the customer will have greater insight into the product's development through reviews, audits, and other means. It does not mean that only CIs are configuration managed.

- MIL-STD-490, Specification Practices
  - 1.4.2 Configuration item. Hardware or software, or an aggregation of both, which is designated by the contracting agency for configuration management.
  - 1.4.3 Hardware Configuration Item (HWCI). See Configuration item.
  - 1.4.4 Computer Software Configuration Item (CSCI). See Configuration item.
- MIL-STD-973, Configuration Management
  - 3.13 Computer Software Configuration Item (CSCI). A configuration item that is computer software.
  - 3.23 Configuration Item (CI). A configuration item is an aggregation of hardware or software that satisfies an end function and is designated by the government for separate configuration management.
- MIL-HDBK-61A, Military Handbook Configuration Management—Guidance, Section 1.3.

A configuration item (CI) may be an individual item, or may be a significant part of a system or of a higher-level CI. It is designated at an appropriate level for documenting performance attributes and managing changes to those attributes. The CI concept has confused some people into thinking that the level at which CIs are designated is the point where configuration management stops. In reality, the CI level is where configuration management really begins; the process encompasses, to some degree, every item of hardware and software down to the lowest bolt, nut, and screw, or lowest software unit. This does not mean that the acquiring activity, the prime contractor, or even subcontractors have visibility or configuration control authority over every part. Rather it means that some organization within either the supply chain or the standardization process has configuration documentation and change control responsibility for each part. The attributes of configuration items are defined in configuration documentation. Configuration baselines are established to identify the current approved documents. Configuration items are uniquely identified. They are verified to make sure they conform to, and perform as defined in, the configuration documentation.

The NATO CM Symposium Group definition of a CI introduced the concepts of provenance and content to the definition of a CI. Provenance—from the French word "provenir," which means "to come from"—covers an object's complete documented history. It was chosen for use in the definition as it includes documentation throughout the product's life cycle and not simply to the point of creation. Content was similarly chosen as content may change over time.

MIL-HDBK-61A definition of a CI makes the proviso that the "CI level is where configuration management really begins."

Configuration management begins at the point that the decision is made to enter any market segment and continues throughout the product's development, introduction, growth, maturity, and decline. It is not necessary to designate a product as CI for CM to take place. In many cases, when performing on a contract to produce items for a specific customer, it is inadvisable to designate something as a CI until decisions have been made regarding its use and maintenance. ISO 7001 identifies a set of items that should be considered when selecting a CI.

- ISO 10007:2003, Quality management systems—guidelines for configuration management, Selection criteria should consider:
  - Statutory and regulatory requirements
  - Criticality in terms of risks and safety
  - New or modified technology, design, or development
  - Interfaces with other configuration items
  - Procurement conditions
  - Support and service

The standard goes on to caution:

> The number of configuration items selected should optimize the ability to control the product. The selection of configuration items should be initiated as early as possible in the product life cycle. The configuration items should be reviewed as the product evolves.

The list of things to be considered in ISO 1007 is incomplete and, to be meaningful, should also consider the following:

- Cost versus benefits
- Quality
- Reliability
- Estimated life

Under the area of support and service, the following points should be added:

- Reparability
- Location of repair
- Difficulty of repair

CIs should be designated at the highest level possible, and too few CIs rather than too many CIs in the final product are advised. This is at odds with the view taken by many companies involved in providing items to governments. It must be remembered that CIs can exist at any level of the supply chain. The fact that you are unaware of CIs at the lower levels does not negate their existence (Figure 4.1). This blindness could leave the prime contractor open to contract non-compliance issues if end item data package requirements and data marking requirements are not properly flowed down to all levels.

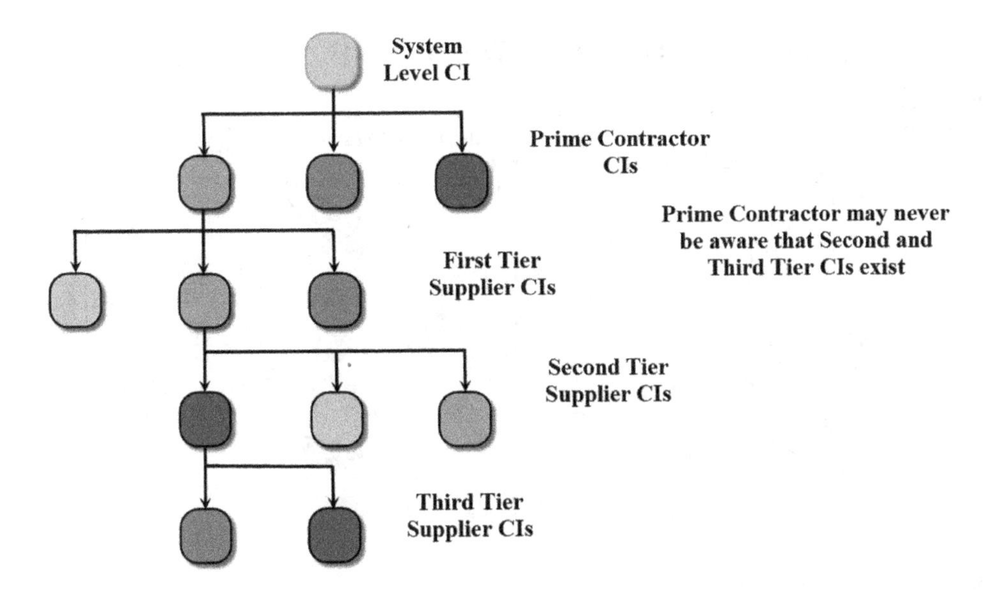

FIGURE 4.1 Configuration items tree.

Designation of a product as a CI implies that there is an identifier associated with it (CII). This CII is often overcomplicated and, in some cases, misunderstood. A CII can be anything that specifically distinguishes a CI from the other CIs and, when paired with its provenance (pedigree), is part of the product's identity.

In the general context, some who do not agree that CIs only apply to government contracts believe a person's national identity code (social security number (SSN), in the United States) is their personal CII number. They contend that all medical records, employment history, education, and such are tracked against it. The country of residence benefits, as it can generate revenue from that CII; trace the person's actions and medical history; and access credit history, current accounts, and any other desired information about each CII (person) in the system. A person's national identity code may be coupled to their passport (if they are not the same) and movements abroad may be monitored as well for government use.

U.S. currency is another government use of the concept of what, in a very broad sense, is a CII. U.S. currency in use in 2013 had a series date, the Federal Reserve district number (also appears as the prefix of the serial number), a note number position, a plate serial number, and a serial number associated with its denomination. The provenance of a U.S. currency note can be established irrefutably to its place of origin and position on the sheet of notes printed (Figure 4.2).

While series numbers change with each new Secretary of the Treasury and part number changes as the design of the note denomination changes, the serial number is unique to the note in that series and denomination and serves as a general representation of the relationship between a CII and serialization. A single dollar note has the same buying power against other currencies on the day it is spent. This would lead one to expect the one dollar note to be the same in form and format. This is far from the truth. The U.S. legal tender one dollar note from 1861 to 1864 featured the face of Salmon P. Chase, U.S. Secretary of the Treasury. The more familiar image of George Washington did not appear until 1869. At various times, the one dollar note has gone by many names that reflected its worth. Some one dollar silver certificates in 1891 featured the image of Martha Washington. Those issued during 1896 featured both George and Martha Washington. Those issued in 1899 featured the American bald eagle.

Reserve Bank	Letter	Designation
Boston	A	1
New York	B	2
Philadelphia	C	3
Cleveland	D	4
Richmond	E	5
Atlanta	F	6
Chicago	G	7
St. Louis	H	8
Minneapolis	I	9
Kansas City	J	10
Dallas	K	11
San Francisco	L	12

FIGURE 4.2 U.S. dollar provenance.

Every major form of currency has relatable information sets. In the case of the United States one dollar note, some make the case that the series date plus the plate serial number constitute the CII for the note being printed. CII numbering schemes are as varied as the market segments they exist in. A sample CII numbering schema is shown in Figure 4.3.

First Order	Second Order	Third Order	Forth Order	Example
Sequential Number	Variant			Sequence 10426 Variant A = 10426A
Product Type	Sequential Number	Variant		Special Test Equipment Sequence 0157 Variant A = STE0157A
Motor Type	Place of Manufacture	Subtype	Variant	F Type Motor Japan Short Wheelbase Removable hard top = FJ40V

FIGURE 4.3 CII numbering schema.

4.3 SERIALIZATION

The decision to assign a serial number (also known as the manufacturer's serial number, product key, or other unique code) to differentiate one unit from another unit in the same production run is not a trivial one. "Serial number" is a misleading description, as it can be composed of any string of alphanumeric characters or symbols applied to a unit in a sequence that can later be decoded. Some industrial sectors prefer to serialize and then add what are known as "nominal characters" after the serial number to denote changes in the item status during the product life cycle as a quickly recognizable item genealogy.

XZY2460001	Assigned serial number
XZY2460001T	Serial number after test
XZY2460001TV	Serial number after test and validation
XZY2460001TVA	Serial number after test, validation, and acceptance
XZY2460001TVAM1	Serial number after test, validation, acceptance, and mod kit 1
XZY2460001TVAM1M2	Serial number after test, validation, acceptance, mod kit 1, and mod kit 2

This is not unlike the tradition of keeping family genealogy by adding the mother's maiden name to the middle name of each child in some cultures.

Peter Merrill VahDonk
Sam Merrill Ensign VahDonk
Joshua Merrill Ensign Fitzroy VahDonk
Joseph Merrill Ensign Fitzroy Sharma VahDonk

Serialization exists on almost every item we use in our daily lives and offers many advantages to the producer as well as to the consumer. It can be applied or associated with tangible as well as intangible goods. Serial numbers may also be necessary to validate certificates and certificate authorities through the application of mathematically rigorous serial numbers and serial number arithmetic not identifying a single instance of the content being protected. In 2010, the U.S. Food and Drug Administration (FDA) published final guidance for pharmaceuticals serialization on prescription drug packages to assist in the prevention of counterfeit medicines in the marketplace.[4] It follows standard global trade item number protocols used worldwide by 23 industrial sectors, including healthcare, and adopted by 65 countries to uniquely identify pharmaceutical products. Serial numbers are also utilized in GTS network protocols and generally enforced using special rules, including sophisticated lollipop-shaped sequence number spaces proposed by Dr. Radia Perlman.[5]

A serial number distinguishes multiple copies of any item under configuration management from one another as an aid to traceability. There is no set formula to serial number application or creation. It transcends market segments and market niche. If futurists are correct, the following from the movie *Blade Runner*[6] may be only decades away in one form or another.

Cambodian Lady: "Finest quality. Superior workmanship. There is a maker's serial number 9906947-XB71 …. Snake scale."

There is no distinct key that can be used as guidance for the creation of a serial number. General direction is that it is unique to each copy of the product. In some market sectors, there

[4] FDA, U.S. (2010). Guidance for Industry Standards for Securing the Drug Supply Chain—Standardized Numerical Identification for Prescription Drug Packages: Final Guidance, Silver Spring, MD: U.S. Department of Health and Human Services Food and Drug Administration. U.S. FDA.

[5] Perlman, R. (1983). Fault-tolerant broadcasting of routing information. *Computer Networks*, Vol. 7, 395–405.

[6] Scott, R. (Director). (1982). *Blade Runner*. [Motion Picture]. Warner Brothers Studios, Burbank, CA.

Lotus 1996 and 1997 CE Model Year

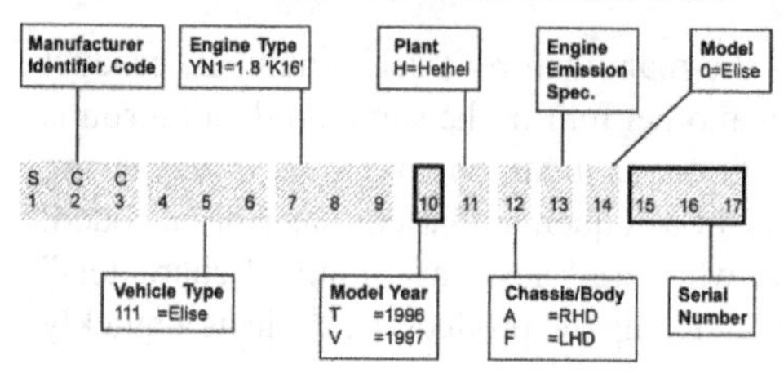

KHS Bicycles

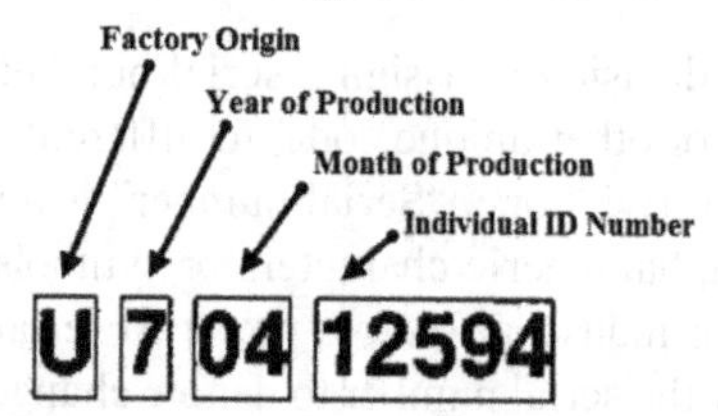

U. S. FDA

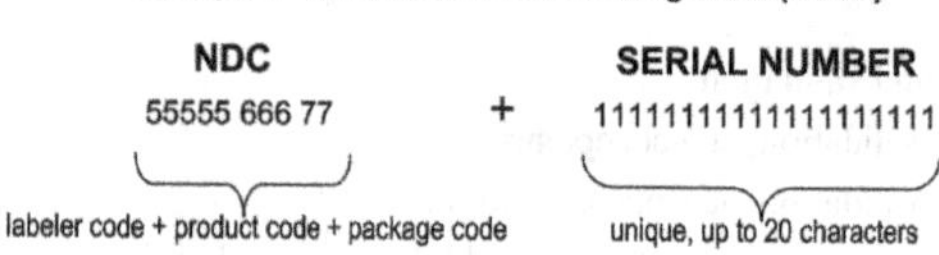

Case Steam Tractor

Case Steam Production Data

Year	Starting Serial No.	Ending Serial No.	Total	HP Sizes
1876	1	75	75	10, 8
1877	76	184	109	10, 8
1878	185	421	237	10, 8
1879	422	665	244	8, 10, 15
1880	666	975	310	8, 10, 12, 15, 20
1881	976	1386	411	8, 10, 12, 15, 16, 20
1882	1387	1892	506	8, 10, 12, 16, 20, 30
1883	1893	2484	592	8, 10, 12, 16, 20, 25, 30
1884	2485	2786	302	10, 12, 14, 16, 20, 25, 30
1885	2787	2981	195	8, 10, 12, 14
1886	2982	3163	182	8, 10, 12, 30
1887	3164	3399	236	8, 10, 12, 16, 30
1888	3400	3679	280	6, 8, 10, 12, 16, 20, 30
1889	3680	3976	297	6, 8, 10, 12, 14, 16, 20, 25, 30

FIGURE 4.4 Serialization approaches.

is an attempt to tie the serial number to the drawing associated with the top assembly of the CII. This is not always the best approach to serialization of an item. How serialization will manifest itself as a best practice in the future is unknown. Once the capabilities of 3D printing evolve, it may result in identification coded into the product itself, as portrayed in *Blade Runner*; combined with the evolution in nanotechnologies, this is entirely possibly. Currently, configuration data for hip and other biomechanical replacements are serialized and tracked to the patient. It is inconceivable that similar tracking of 3D-printed internal organs would not be implemented. Serialization approaches vary, and some of these approaches from different industries are shown in Figure 4.4.

The following rules for serialization are used in many market segments.

- Once a serial number is assigned, it can never be changed.
- If a serialized product is installed into a higher-level product, the higher-level product is said to have consumed the lower-level one.
- A single instance of a design item is typically used in association with repair and warranty repair/replacement traceability.
- Candidates for serialization include the following:
    - The product, if designated as a CI
    - Lower-level products incorporated into the top-level product
    - Any item that can be replaced in any logistics activity
    - Spare lower-level products available for sale as a standalone item

Going back to the earlier representation of a person as a CI, this might include any 3D organ grown inside or outside of the body and replacement parts such as knees and hip joint components. This is one possibility envisioned by Hod Lipson and Melba Jurman.[7] We will discuss 3D printing further in Section 5.4.1.

[7] Lipson, H., and Kurman, M. (2013). *Fabricated: The New World of 3D Printing*. Hoboken, NJ: John Wiley & Sons.

4.4 TRACEABILITY

Understanding the concept of traceability may be made easier by referring to the websites www.wheresgeorge.com and www.trackdollar.com. These websites allow you to enter the denomination and serial numbers of a piece of U.S. paper currency and, should the same denomination and serial number have been entered previously, information regarding everywhere the note has been is displayed. This is like the way international movement of individuals can be traced through their passport number. As with many aspects of life, the day-to-day exigencies of work at times prevent aspects of CM from being obvious. Reasons why traceability is important appear almost weekly in the news and social media. In 2012, record numbers of automobiles,[8] children's toys,[9] food products,[10] and other items were recalled due to inadequate management of product configurations and associated processes.

The U.S. FDA, suppliers, and consumers may soon have standard traceability mechanisms based on commonly collected data facilitating risk reduction, supplier control, and traceability, and minimizing recall damage. These are critical components of any food safety plan. Regardless of a company's position in the supply chain, dependence on those companies above and below you in the supply chain creates liability and exposure. In agriculture, traceability is critical through long processes and zones (farm, blend operations, distribution, picking operations, and retail). This issue was the subject of Food Seminars International's 2013 seminar, "Using Traceability and Controls to Reduce Risk and Recall: Integrating Food Safety Systems."

Traceability goes both ways when applied to products and services. It is not simply the tracing of a requirement to the system or subsystem. Traceability is flowed down throughout the product life cycle. The form, fit, function, and quality of the item specified in the engineering is verified against any item produced or service performed. Validation assures that product or service requirements are met through test, analysis, inspection, or demonstration. Understanding the heritage of the design and how it applies to the whole and also the individual elements that make up the finished product is critical. This may be a new way of looking at traceability in CM implementation to many. Figures 4.5, 4.6, and 4.7 provide a pictorial look at three forms of traceability needed for product management.

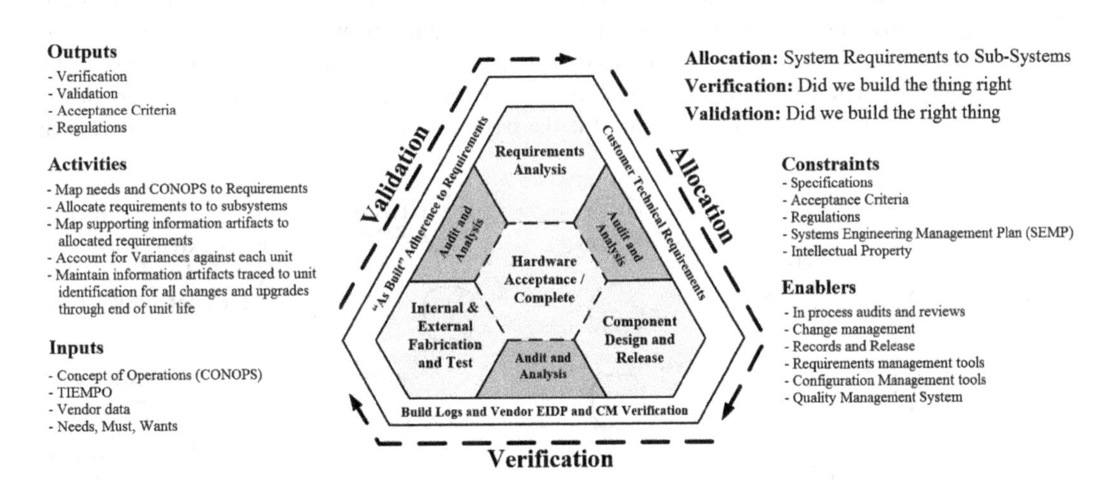

FIGURE 4.5 Functional requirements.

[8] Gorzelany, J. (2012, December 29). Biggest Auto Recalls of 2012 (And Why They Haven't Affected New Car Sales). Retrieved November 26, 2013, from http://www.forbes.com/sites/jimgorzelany/2012/12/29/biggest-auto-recalls-of-2012/.

[9] Sole-Smith, V. (2013). How to Protect Kids from Lead Toys, http://www.goodhousekeeping.com/health/womens-health/poisonous-lead-toys-0907. Retrieved December 26, 2014.

[10] Bottemiller, H. (2012, October 8). 2.5 Million Pounds of Recalled Canadian Beef Entered U.S. Retrieved November 26, 2013, from http://www.foodsafetynews.com/2012/10/2-5-million-pounds-of-recalled-canadian-beef-entered-u-s/#.U9FDCPldV8E.

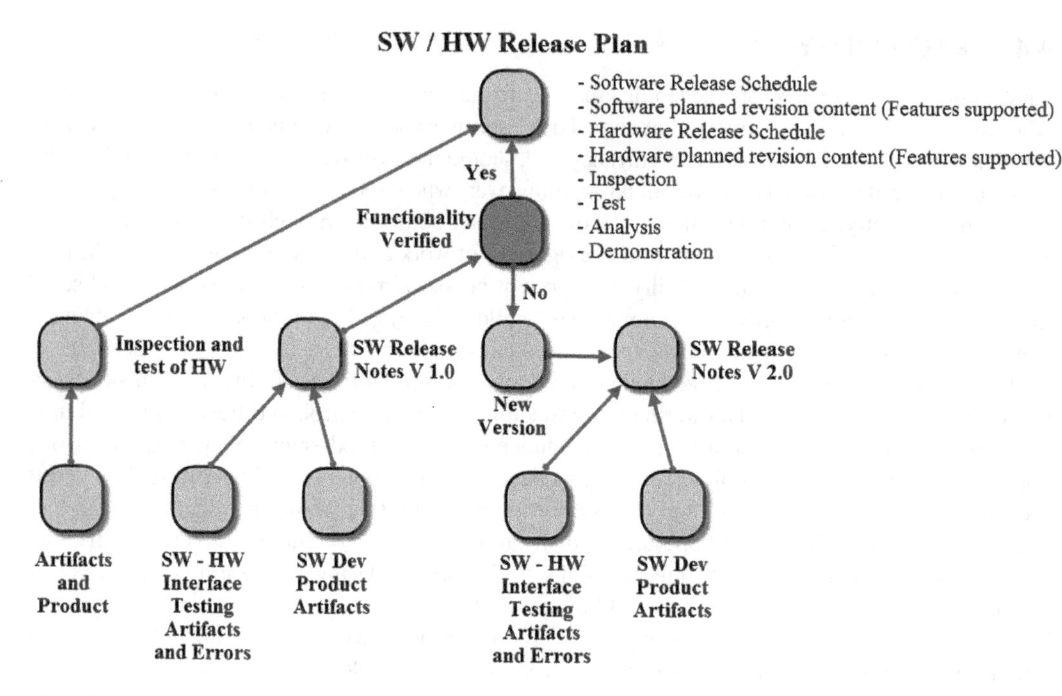

FIGURE 4.6 TIEMPO.

Functional requirements traceability (Figure 4.5) is but one element of traceability. It is also what is generally thought of as traceability by most people. As soon as all allocations have been established, every supporting information artifact must be gathered and mapped (traced) back to requirements. A single artifact may fill multiple requirements and cross-traceability must be established (e.g., the artifact must be mapped to more than one requirement in the digital fabric). Lower-level elements of a product may be produced within the company or they may be produced by another company. The associated make-or-buy analysis or trade study also becomes part of the traceability of the requirements allocation. It is the basis for the generation of a vendor statement

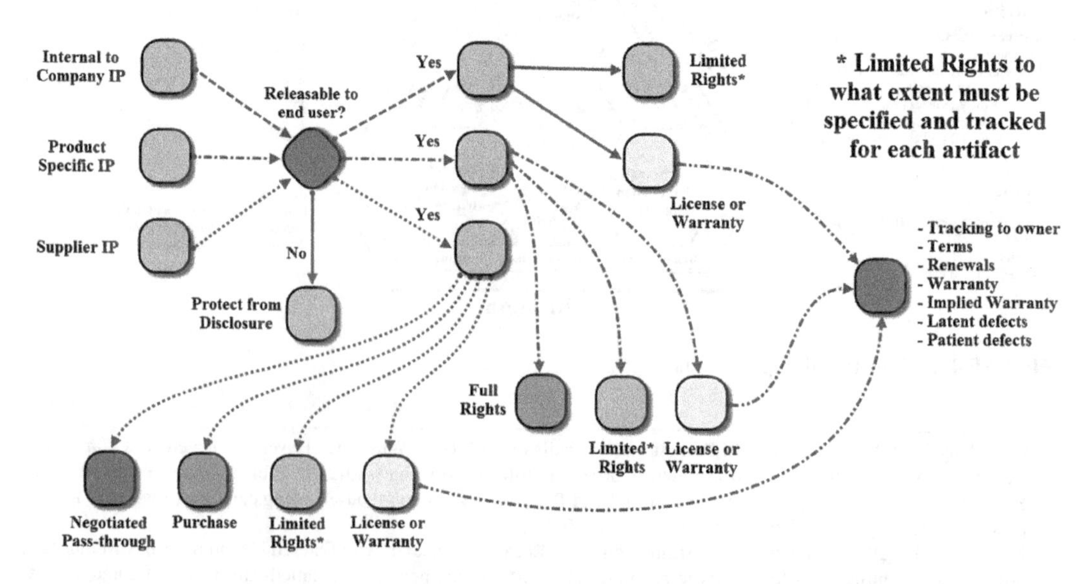

FIGURE 4.7 Intellectual property.

of work, specifications, interface controls, and negotiation of vendor intellectual property (IP) considerations. The supporting information includes, but is not limited to, all enterprise-level and vendor-level information regarding formulation, safety, handling, test, operation and installation, environmental and biological impacts due to use, other known hazards, and similar data associated with the product, items it comprises, and processes used during product creation.

Test Inspection and Evaluation Master Plan Organized (TIEMPO) (Figure 4.6) is a useful tool for the collection of the necessary artifacts and leverages forward aspects of the U.S. Department of Defense's Test and Evaluation Master Plan (TEMP),[11] adding in inspection as well as analysis and demonstration. These align with Section 3.6.3 in the California Department of Transportation systems engineering guidebook (verification requirements) of the U.S. Federal Highway Administration and California Department of Transportation systems engineering guidebook. So, TEMP becomes TIEMPO, Spanish for "time" or "weather" depending on the context used. TIEMPO as a tool simply means that if you do not take the time, you will hit some very bad weather on your project. TIEMPO is discussed further in Chapter 12.

One key aspect of TIEMPO is the mapping of the product delivery or product growth through the development process. In the plan, the content in each individually developed product iteration since inception is described. These development snapshots are recorded to document the test, evaluation, inspection, or demonstration, verifying that specified requirements are being met. This provides a comprehensive view of the product development and level of quality obtained at each iteration.

The IP aspects of each product require a level of traceability that reaches far beyond requirements for product verification and validation (Figure 4.7). IP may include implied and actual warranties, software leases, and remedies for latent and patent defects.

Supporting information may need to be gathered and associated with the product, subassemblies, and parts it comprises. This information may also be part of the documentation suite needed to meet the internal process improvement and regulatory needs. Traceability does not start with the designation of an item as a CI; neither does it end with delivery. It may continue indefinitely if it pertains to manufacturing process, plant operations, geopolitical, or environmental issues.

4.5 HIGHER LEVEL OF MANAGEMENT CONTROLS

IEEE Standard 828-2012[12] describes seven primary lower-level processes and two special instances of applying the lower-level processes.

- The primary lower-level processes are
  - Planning
  - Management
  - Configuration identification
  - Configuration change control
  - Configuration status accounting
  - Configuration auditing
  - Configuration release management
- The special instances are
  - Interface control
  - Supplier CI control

---

[11] DOD, U.S. (2011). *MIL-HDBK-520a, Department of Defense Handbook, Systems Requirements Document Guidance.* Washington, DC: U.S. Department of Defense. U.S. Government, Pentagon.

[12] IEEE. (2012). IEEE-828-2012—IEEE Standard for Configuration Management in Systems and Software Engineering. New York, NY: IEEE.

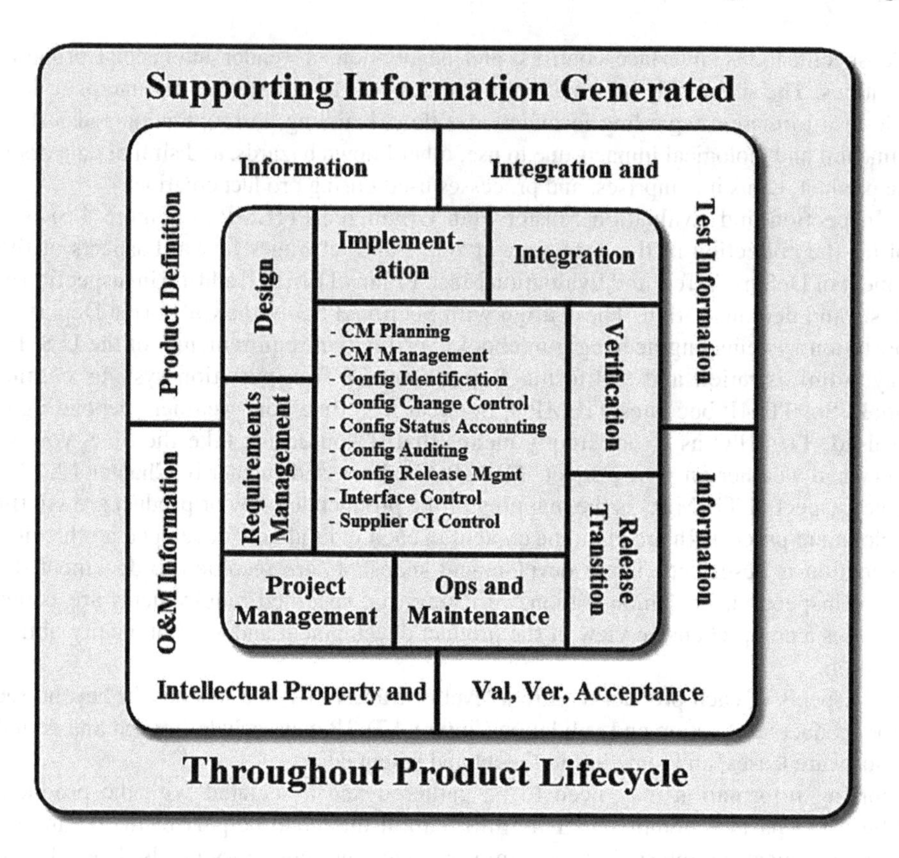

FIGURE 4.8 CM life cycle processes.

CM life cycle processes are shown in Figure 4.8.

4.5.1 CM Planning

SAE/EIA 649-C[13] provides the following guidance in Section 5.1, "Configuration Management Planning and Management."

CM planning and management over the life cycle of a product are essential to achieving effective, predictable, and repeatable CM processes. CM processes shape the application of solid, practical procedures that result in cost avoidance and enhanced configuration quality and stability.

Comprehensive CM planning and management includes:

- Applying the appropriate level of CM functions throughout the product's life cycle.
- Implementing policies and procedures, resulting in effective configuration management.
- Assigning CM functional responsibilities to appropriate organizational elements.
- Training of CM personnel and any others who have CM responsibilities.
- Determining and applying adequate resources to implement the planned CM system.
- Establishing CM performance measures to serve as a basis for continuous improvement.
- Ensuring appropriate performance of configuration management by the supply chain through insight and oversight.
- Integrating the organization's product configuration information processes

[13] SAE. (2019). SAE/EIA 649-C-2019-02, Configuration Management Standard. SAE International. Warrendale, Pennsylvania.

Arrival at this definition was not a trivial process and it should not be taken lightly or out of context. Performing CM planning activities is more than completing a boilerplate CM plan based on a set of criteria established using customer, company, or content suggested by IEEE Standard 828, *IEEE Standard for Configuration Management in Systems and Software Engineering*; ISO 10007-2003, *Quality Management Systems—Guidelines for Configuration Management*; or SAE/EIA 649-C, *Configuration Management Standard.*

Earlier, we presented the cost and influence of critical design decisions plotted over time (Figure 1.3). One of those design decisions involves CM planning. Planning the management of the configuration requires a knowledge of not only CM principles but also their application. The application must be tailored to the market sector where the product resides. This requires a knowledge of the product itself, company functional resources, consumer needs, and expectations, as well as the IP and regulatory facets involved. It is not as complicated as it sounds. Planning an excursion to a location you have never visited requires research as well as a deeper understanding of multiple parameters. Considerations for a visit to a hydrothermal vent in the abyssal zone on the Comfortless Cove Hydrothermal Field (4° 48′ South 12° 22′ West, elevation –2996 m) would be quite different from a visit to HaLong Bay or the Bay of Descending Dragons, Vietnam (20° 53′ 60″ North 107° 6′ 60″ East).

Each of the elements cited from SAE/EIA 649-C requires the same level of comprehension. There is no one-size-fits-all solution for CM implementation. Similarities will exist between companies and programs, but the solution you arrive at will be unique to the product and its place in the product life cycle. Larry Bowen, in his presentation on advanced configuration management,[14] observed that in the GTS area, the following considerations should be made regarding the planning, CM, and cost assumptions for implementation.

WALK, THEN RUN!
THE PLAN IS NOTHING. PLANNING IS EVERYTHING!
CLAIM EARNED VALUE AS YOU IMPLEMENT!

- Work from a plan
- Assume a phased implementation
    - Involve end users in developing the technical plan, implementation strategy, and training plan
- Assume that legacy data, systems and procedures will be incorporated
- Satisfy, don't optimize
- Use pilot projects to test assumptions
- Gather user feedback and incorporate it
- Collect and analyze metrics to measure results
- Remain flexible; adjust or abandon unsuccessful efforts early
- Form a customer/contractor team to:
    - Analyze requirements
    - Maintain all information electronically
    - Manage real-time upgrades and installation of enhancements
- Implement a constrained system
- Plan rollout to Production
- Establish a configuration control board (CCB) to baseline the system and control enhancements/upgrades
- System engineering, information technology, data management, engineering, manufacturing, procurement, logistics, and configuration management must work as a team controlling the development to a requirements document that is used to implement user requirements

[14] Bowen, L .R. (2007). *Advanced Configuration Management.* Boulder, CO: Larry R. Bowen.

- Select a standard set of tools to minimize as much as possible customized code/applications to make the PDM system work
- Hit the long poles in the tent first—use metrics to drive the implementation
- Burden fund the system—make its use free to programs and projects. Data entry should be paid for by the project/program. The PD/LM system and its maintenance are paid for by the company

A good rule of thumb after the initial implementation of a new system is that the deployment will cost approximately €524,007 for maintenance and €69,867.60 for continued development until the desired system functionality is achieved. These costs are based on an enterprise involving 8,000 employees with no more than 400 users accessing the system at the same time. A users group as one of the stakeholders is critical to evaluating system upgrades that meet both internal and external requirements. Bowen cautioned that companies should not underestimate the cost of migrating legacy data into the system. We found that legacy data transfer was often scoped at about 60% of the actual costs involved.

4.5.2 Configuration Identification

Configuration identification goes beyond specifying what CIs are going to be assigned.

IEEE Standard 828 states,

> 8.1 Purpose
> The purpose of configuration identification is to determine naming schemes for configuration items (CIs), identify the items that require control as CIs, and apply appropriate names to them. Additionally, the physical and functional characteristics of the CIs are identified.
>
> The scope of configuration identification includes:
>
> a. determining the CIs that are to be managed and determining what documentation information is to be used for describing the physical and functional characteristics of each CI
> b. planning for the collection, storage, retrieval, and change control of baselined versions of the items and their descriptive documentation information
> c. establishing and maintaining associations between versions of each item and its descriptive information
> d. establishing versioned assemblies or collections of CI versions that satisfy the totality of end use functions; establishing and maintaining associations between versions of such assemblies or collections and the descriptive documentation information of the physical and functional characteristics of the assembly or collection, and
> e. describing the product structure through the selection of CIs and identification of their interrelationships.
>
> 8.2.2 Identify configuration items – Note 3
> NOTE 3 – Items that are likely to need controlling include, but are not limited to, baselined requirements specifications, interface specifications, designs, code, builds, build data, database-related items such as triggers, schema and SQL scripts, unit and coverage tests, and the standards that were used to create such items. In addition, the following are typically included: design drawings, parts lists, reference models, baselined models or prototypes, and maintenance and operating manuals. The determining factor is whether an item or information will be needed if the project needs to reinstate a previous baselined position in the life cycle to once again move forward on the build cycle.

9.2.4.2 Control changes to baselines
For controlling baselined software or systems as CIs, the following attributes shall be available in the Configuration Management Data Base (CMDB):

a. **Attribution:** Attribution shall be provided to the CCB that is accountable for the affected baseline.
b. **Impact analysis:** A documented impact analysis reviewed by all the CCB members shall be recorded. Impact analysis shall consider severity of any problem to be corrected by the change as well as how frequently or how widely such problem will affect the product's users. Impacts to the product development project shall also be considered.
c. **Rationale and approval:** The CCB shall approve/reject the change using its established decision process and shall provide the reason for their decision.
d. **Notification:** Notification to affected parties shall be part of the formal process.
e. **Reversibility:** Baseline change control shall be supported by versioning systems that allow individual reversibility of constituent configuration items. This, aided by the set of changes requirement, assures full reversibility.
f. **Set of changes:** The affected baseline and the impacted artifacts (from the impact analysis) form the set of changes to apply to the baseline and shall be documented.
g. **Audit trail:** A record of the change request forms containing the process evidence shall be kept for auditability purposes.

MIL-HDBK-61A, Military Handbook Configuration Management Guidance gives what many believe is the typical definition of activities that take place during configuration identification.

4.5.2.1 Configuration Identification Activity

Configuration identification incrementally establishes and maintains the definitive current basis for control and status accounting of a system and its configuration items (CIs) throughout their life cycle (development, production, deployment, and operational support, until demilitarization and disposal). The configuration identification process ensures that all acquisition and sustainment management disciplines have common sets of documentation as the basis for developing a new system, modifying an existing component, buying a product for operational use, and providing support for the system and its components. The configuration identification process also includes identifiers that are shorthand references to items and their documentation. Good configuration control procedures ensure the continuous integrity of the configuration identification. The configuration identification process includes:

- Selecting configuration items at appropriate levels of the product structure to facilitate the documentation, control, and support of the items and their documentation
- Determining the types of configuration documentation required for each CI to define its performance and functional and physical attributes, including internal and external interfaces. Configuration documentation provides the basis to develop and procure software/parts/material, fabricate and assemble parts, inspect and test items, and maintain systems
- Determining the appropriate configuration control authority for each configuration document consistent with logistic support planning for the associated CI
- Issuing identifiers for the CIs and the configuration documentation
- Maintaining the configuration identification of CIs to facilitate effective logistics support of items in service
- Releasing configuration documentation
- Establishing configuration baselines for the configuration control of CIs

Effective configuration identification is a prerequisite for the other configuration management activities (configuration control, status accounting, audit), which all use the products of configuration identification. If CIs and their associated configuration documentation are not properly identified, it is impossible to control the changes to the items' configuration, to establish accurate records and reports, or to validate the configuration through audit. Inaccurate or incomplete configuration documentation may result in defective products, schedule delays, and higher maintenance costs after delivery.

Jack Wasson goes on to give the following definition of configuration identification:

> This process involves identification of documents comprising the configuration baselines for the system and lower-level items (including logistics support elements) and identification of those items and documents. When an item is identified, it is known as a configuration item (CI). Configuration identification determines the makeup of any and all products along with their associated documentation. It defines performance, interface, and other attributes for configuration items; provides unique identity (i.e., drawing, document, or ID numbers) to products, components, and documentation; specifies identification markings (if required); modifies product and document identifiers to reflect major changes; maintains release control and baseline definition; provides reference for changes and corrective actions; and correlates document revision level to product configuration, which enables users to distinguish between product versions, allows people to correlate a product to the appropriate instructions, and correlates items to service life. What it boils down to is that configuration identification determines how document control numbers and version numbering are applied and used so that everything is labeled correctly and understandably as prescribed by the configuration item manager of that product.

The concept of baseline is mentioned in several definitions. Many familiar with CM as something done after receipt of a contract will be familiar with the concept of milestone events and the baseline gate normally associated with them. The mission stage timeline shown subsequently is roughly relatable to the product phases of development, introduction, growth, maturity, and decline discussed in Chapter 1. For many, their exposure is limited to contracted work in a governmental setting, and little consideration is given to the product phases, as they relate to the market segment in which the product is being offered. One misconception is that a CI can only be in a single baseline state when, in actual practice, parts and sub-assemblies that make up the CI may be in any phase (e.g., the CI top assembly may be in the allocated baseline phase, yet individual sub-assemblies and parts may have already passed some form of incremental functional and physical configuration audit).

Common design items incorporated into the CI may be in the product baseline stage. Some subcontracted items may be in the development configuration phase nearing the product baseline. Internal to the company, CIs may be in the developmental configuration phase just entering the first of many major reviews. Figure 4.9 shows the relationship between baseline and mission stage.

The term *unified change control* may not be familiar to many involved with CM implementation. It was chosen as it is better understood globally and equates to integrated change control. "Unified" implies a larger scope and requires that all aspects of a change be evaluated not only against a product procurement but also against every other product in the entire system prior to being authorized. It is recommended that an aligned product development model be used to assure a full understanding of the scope of the items whose configurations are being managed. Figure 4.10 illustrates a prototypical aligned hardware–software model using generally recognizable milestones and baselines.

A change to the focal plane in a remote sensing device that increases image quality across the visible spectrum while degrading it in the infrared and ultraviolet may appear a simple enhancement. This enhancement impacts the ability to use algorithms developed for the evaluation of the degraded frequencies and results in the product not being suitable for its intended use.

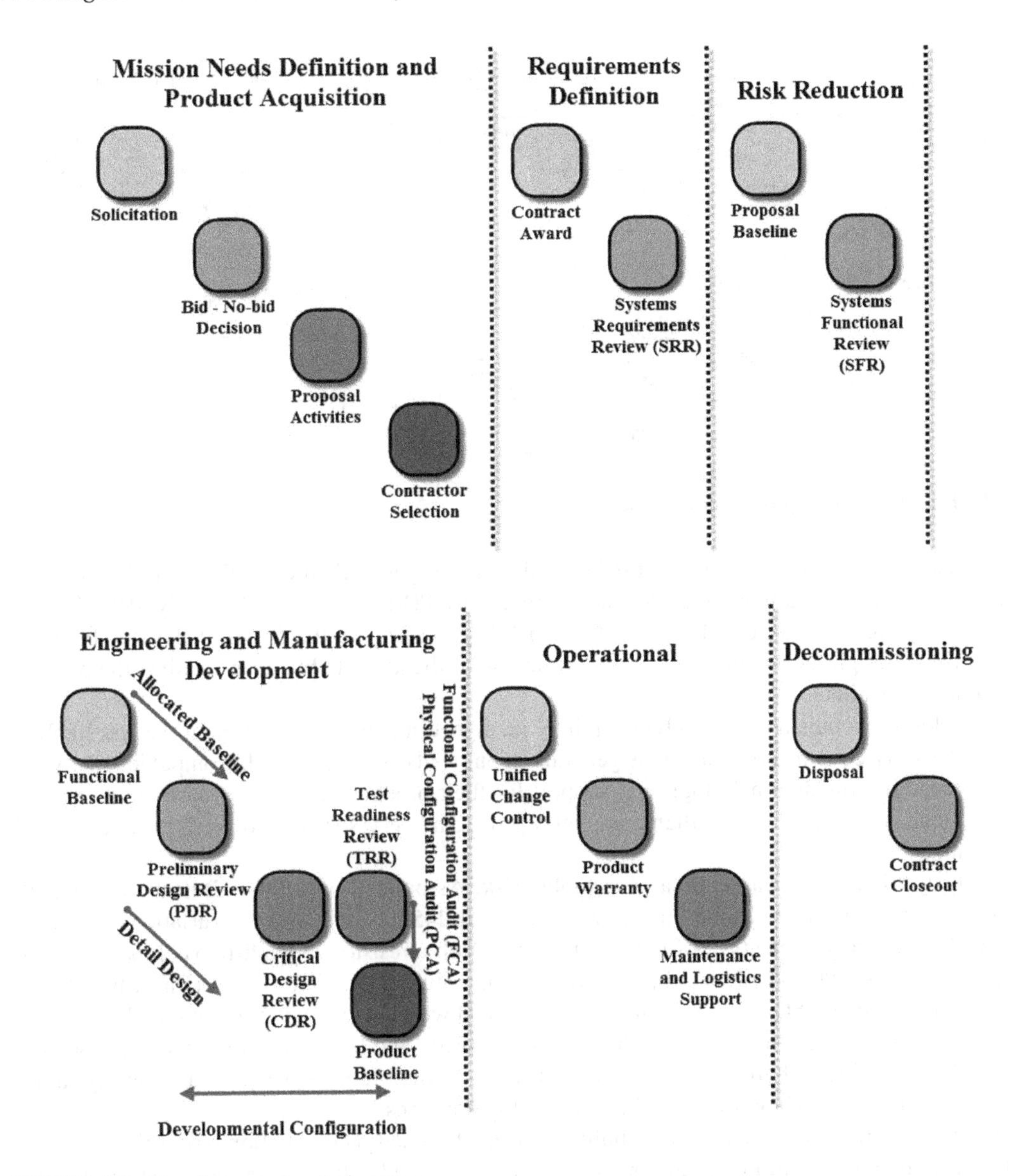

FIGURE 4.9 Baseline versus mission stage.

During development of a now decommissioned U.S. ballistic missile, the missile stage contractors produced stage diameters that were not all the same. The lack of controlled management of the configuration interfaces rippled throughout the launch system, impacting the thickness of pads used to support the stages in the launch tube, the ground support equipment, stage-to-stage interfaces, and missile assembly procedures, as well as launch system deployment and logistics support of deployed missiles.

This identification of incompatibilities across all systems through the proper evaluation of impacts prior to implementing a change is the realization of CM in its purest sense. The IEEE Standard 828 citation identifies a key and often overlooked element of the baseline process. "Reversibility: Baseline change control shall be supported by versioning systems that allow individual reversibility of constituent configuration items." Reversibility in this context means that the CM system can identify the release status of all parts in any CI on the date the baseline was established, as well as for each serial number of the CI on the date it was accepted for shipment or sale. Of the companies

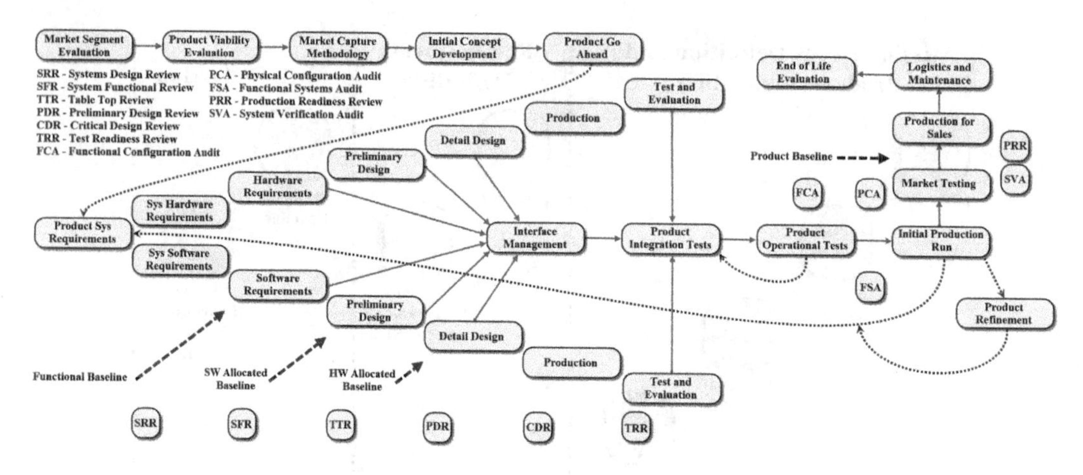

FIGURE 4.10 Aligned product development model.

evaluated for this book, reversibility of software design was generally present. Reversibility of hardware design was generally not present. Hardware-centric PD/LM systems often only identified the current approved change level for parts. No capability to compare the design difference between preliminary design review and critical design review or the allocated baseline against the product baseline was available.

The closer a company was to relying entirely on a government revenue stream, the less likely it was to have baseline reversibility. This generalization needs to be qualified. Companies that were experienced in providing field logistical support for their products understood baseline reversibility and managed it much better than companies producing very specialized products in a one-off environment.

At first, this may appear to be an acceptable risk. As pointed out in the CMstat August 2018 blog,[15] a printed wiring assembly may have as many as 145 different as-built variations that meet the as-designed requirements if a single 1/4th watt 100-ohm resistor with a fifth band resistor value tolerance + or – 5.00% (Gold band), sixth band temperature coefficient of 250 (Black band), and cleaned for use in an ISO class-4 clean room is replaced with either a better quality or cleaner part. Extrapolate this to encompass all the electronics, software, and hardware in your deliverable, and suddenly producing as-built versus as-designed reconciliation reports becomes a necessity rather than something nice to have due to safety and quality concerns.

Companies did often produce an as-built versus as-designed reconciliation report that included all critical, major, and minor variances and authorized part substitutions. In most cases it was not searchable and not associated in the CM system with the specific serial number. Part substitutions and variances were not tracked in the CM system due to incompatibilities between quality management, build, and design databases. We also found that records associated with delivered products received back for upgrade were not available. The lack of data availability in an automated system for each serial number is disastrous to the maintenance of anomaly resolution of complex systems. It makes recall of sold products due to latent defects or poor design nearly impossible. In one case, design development (proof of concept evaluation) units built with unqualified parts were installed in an operational satellite. The issue was not discovered until final assembly and test.

No CM system or method of configuration identification is perfect; errors will occur, and some of these are documented in Chapter 11. The key to reducing critical issues is the proper sizing of the CM application to the product. This is done using standard methodologies from all functional

[15] CMStat (2018): Hardware Configuration Items in As-X Configuration Management by Aerospace & Defense Contractors (Part 2 of 2), August 2018, CMstat, 3960 Howard Hughes Parkway, Suite 500, Las Vegas, NV 89169, https://cmstat.com/hardware-configuration-items-in-as-x-configuration-management-by-aerospace-defense-contractors-part-2-of-2/

resources and tying together metadata generated by each (e.g., the digital threads) in such a way that configuration identification is adequately established, and the provenance of each serial number is irrefutable.

Product provenance and logistical implications of product maintenance are critical to the longevity of the company. Four conditions normally exist after sales in the life of a product.

1. The original equipment manufacturer (OEM) is still producing the product and is providing repair/maintenance services.
2. The OEM is still producing the product and sells repair elements for third-party installation/repair/maintenance.
3. The OEM is still producing the product and is providing repair/maintenance services and is in competition with other vendors providing repair/maintenance services using non-OEM elements.
4. The company is no longer producing the product and a third party is producing and selling repair elements for third-party repair/maintenance.

Put into an easier to grasp context:

1. You can purchase a vehicle for personal transportation and have the vehicle dealer repair/maintain it.
2. You can have repair/maintenance services done by someone other than the dealer or do them yourself using OEM parts.
3. You can have repair/maintenance services done by someone other than the dealer or do them yourself using non-OEM parts.
4. Your vehicle is out of production and restoration/repair/maintenance services can only be done using non-OEM parts by someone other than the dealer, including yourself.

Those providing products designed specifically for government use will be requested to provide a technical data package in accordance with a standard format and content structure (e.g., MIL-STD-31000[16]) to facilitate re-procurement of products or re-competition on a build-to-print basis. MIL-STD-31000 was updated in 2018 to bring it up to current data standards (Figure 4.11).

Offloading of production is a common activity in commercial markets. TDP packages for government and commercial environments are similar in nature and contain all information required for someone else to produce the item without reference to internal documents or company-protected IP. In some cases, dies, jigs, and fixtures are provided along with drawings, test procedures, inspection points and inspection procedures, and final acceptance criteria.

4.5.3 Change Control

Change is a fundamental component of our universe. Change goes on around us regardless of personal desires to hold elements in our lives, products, and businesses static. The best we can hope to accomplish is to manage change instead of allowing it to manage us. So, what is meant by change management as it applies to a company's products, services, and internal operations? Figure 4.12 shows that it is not enough to react to change. Change management is not complete until the total implementation of the change has been verified. This includes modifying units that may be in process or in stock waiting for delivery. It also includes all recall and rework activities associated with product delivered to a sales organization and rework of all sold items. The reality of change management encompasses much more than is typically included with many product data/life cycle

---

[16] DOD, U.S. (2009). MIL-STD-31000B, U.S. Department of Defense Standard Practice: Technical Data Packages. Washington, DC: U.S. Government, Pentagon.

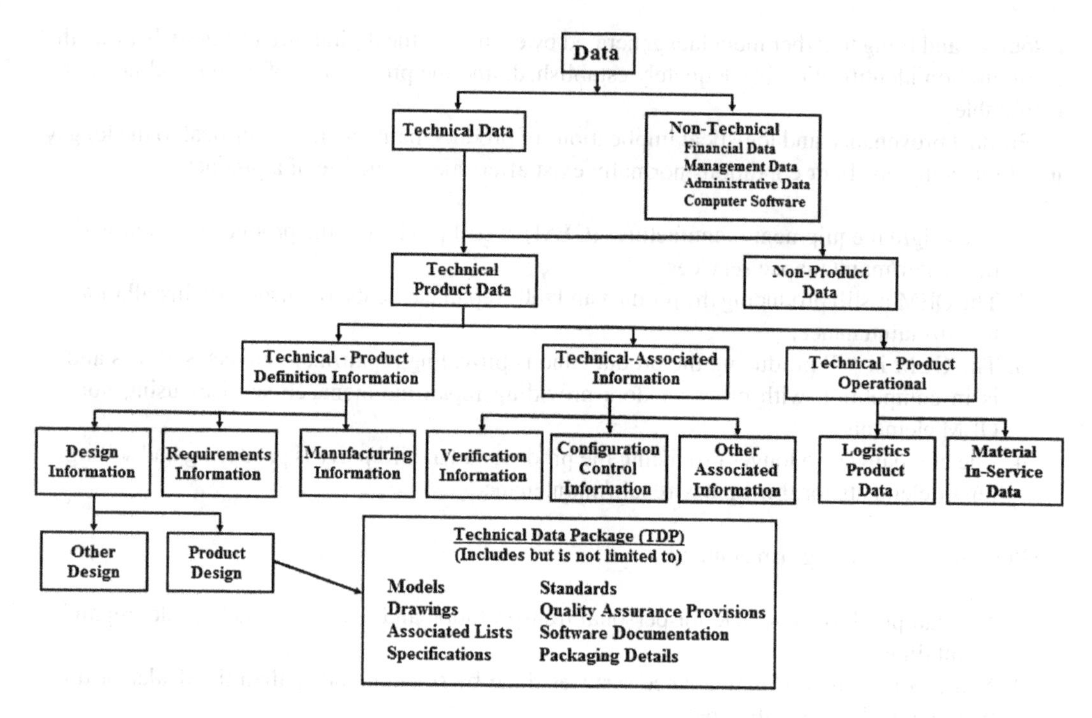

FIGURE 4.11 Technical data package (TDP) relationships (redrawn from MIL-STD-31000 B Figure 1).

management (PD/LM) systems. Change management is the process of utilizing change control capabilities from the identification of a need to change through implementation and verification that the change is implemented.

The prerequisite to change control is understanding what is being developed, how sub-systems interact, what information should be developed, how that information will be used, what the product structure and data structure look like, and how they are interrelated. Change control must be an end-to-end approach that considers not only the what, why, where, and when of change implementation, but the impact to cost and schedule of the product. It also needs to account for test temporary configurations; critical, major, and minor variances; and what are known as "block updates" to the design that establish, upgrade, and repair liens against all delivered units. Block updates to the

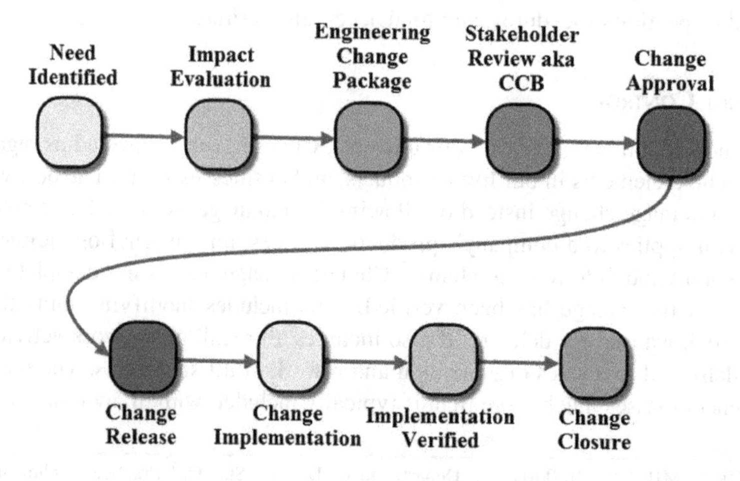

FIGURE 4.12 Change management.

design are an everyday occurrence in consumer software and software-inclusive products such as Blu-ray players and telecommunication devices. In the automotive sector, they form the basis for recall of units sold and are rolled into products under production or available for sale from new vehicle sales locations. Non-critical block updates of automotive parts may be tied to changing manufacturing capabilities as well as the standardization of product components and are tracked against parts lists for vehicles sold on a component cross-reference basis.

A recent test of this capability involved the purchase of a new ignition switch for a 1971 Ford 350 Econoline van. The existing switch was removed and the Ford dealer pulled a new unit from stock. It came with an adapter ring to allow its use as a replacement for the original part. Form, fit, function, and quality of the new unit were superior to the old unit and CM had been maintained up to the point of interchangeability as a logistics support activity. This showed a clear understanding and implementation of configuration control throughout the product life cycle of the 1971 van as well as an impressive wide-reaching enterprise-level CM implementation that allowed for real-time GTS searches for replacement and repair activities.

The engineering change request (ECR) is one element of change control and change management that is often misunderstood and misused. The ECR is often the precursor to a block update of the design. ECRs define what needs to be changed; the complete rationale for the change; known system, cost, and schedule impacts; and a recommended point of implementation (lot, serial number, date, etc.). Once the ECR is approved, it exists in the PD/LM system as a lien against the engineering, and a change can be evaluated by all stakeholders prior to implementation or rejection. ECRs can be as simple as a lien against design due to sub-element obsolescences and replacement or as complete as a product phase-out due to lack of sales or fabrication capability. A simplified change flow is shown in Figure 4.13.

Mentally merging the simplified change flow and the aligned product development model provides some idea of the complexities involved with managing a configuration. This is essentially the same change flow in place during development of the U.S. Saturn V rocket used to launch Skylab and Apollo astronauts. At the time, Saturn V was the most complex electromechanical device conceived and fabricated. The entire program was managed using paper drawings with no computer-based PD/LM system. The single most significant difference between the CM implementation on Saturn V and similar activities going on at companies such as SpaceX today is the thousand-fold increase in the amount of data that must be managed today and the PD/LM software that allows us to manage it.

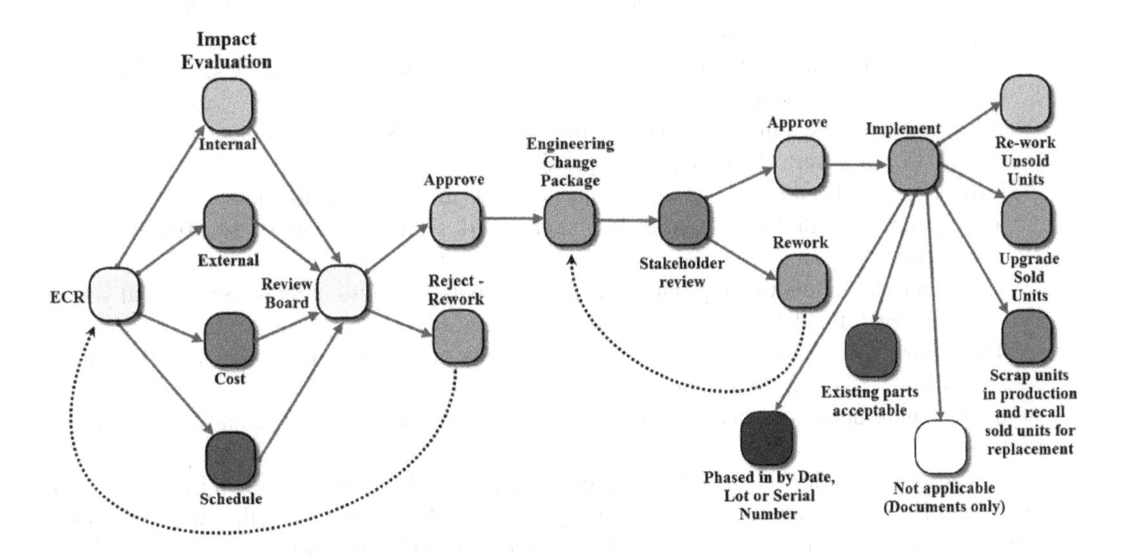

FIGURE 4.13 Simplified change flow.

The contextual relevance of the data associated with the change must be maintained. If it is not, the evolution of the change will not be complete and that will influence the integrity of change evaluation and implementation. Because of this, many companies are starting to insist that a business case be made at a different level of granularity than a bottom-up impact analysis of the change itself can provide. If the business case cannot be made to support the change, it proceeds no further. Change risk assessment must consider the impact if the change is not approved.

4.5.3.1 Change Effectivity

We found that change effectivity is a misunderstood aspect of change control in management and engineering circles. Change effectivity impacts the way you need to plan changes to the configuration. Silos of expertise must work together to understand what can be done and what cannot be done to the existing configurations that are in production, that are complete, or that have been delivered and are being maintained. The entire project organization must be involved in vetting any proposed change to make sure that implementation is well thought out and planned. This effort prevents overlapping changes and out-of-sequence change approvals. This book partially addresses the issue of out-of-sequence change approval in Section 7.11.2.10.1, but effectivity goes much deeper. At its simplest, change effectivity determines how you are going to implement the change for the units being made. Effectivity in PD/LM systems consists of the following cases.

1. Document changes; the effectivity is "Not Applicable."
   a. Documents do not fabricate parts.
2. The change adds substitute parts; the effectivity is "Use" or "Use as is."
   a. Engineering does not care at what point the change is incorporated.
   b. Used when the change is to improve manufacturing operations such as a change in the processes used.
   c. Change is compatible across all platforms that use the item being changed.
3. No deliveries have been made and all impacted units are still in production.
   a. If you rework the units in production, the effectivity is "Rework."
      i. The use of existing parts without first modifying them is prohibited.
      ii. Change is backward compatible across all platforms that use the item being changed.
         1. If you cannot retrofit the changed unit into previously delivered units (e.g., the asset is on orbit), you need to re-identify the unit.
         2. You cannot define the part being changed as having been disposed of if delivered units are still in use.
   b. If you scrap the units in production, the effectivity is "Scrap."
      i. Use of or rework of the parts is prohibited.
4. Deliveries have been made and some impacted units are still in production.
   a. If you rework the units in production and rework all delivered units, the effectivity is "Rework."
   b. If you scrap the units in production, the effectivity for those units is "Scrap" and you must recall the delivered units.
5. The change will be implemented when existing stock has been depleted or at a set point in time; the effectivity is "Phase in."
   a. You must define the date or lot or serial number when the change becomes mandatory.

Some companies we looked at failed to implement approved changes across all impacted serial numbers, citing that the units not modified were spares and if they were needed they would be modified at that time. Not doing full effectivity implementation is both costly and risky, as NASA found out with the Apollo 13 oxygen tanks discussed in Section 11.6.1.

One aspect of effectivity process that is often overlooked is that scrap and rework dispositions also need to include a schedule for completion, a verification method, and a definition of what organization is responsible. The use of family identification numbers (FDN) as described in *Fundamentals of Configuration Management*[17] has been found useful in many companies. A suffix is often added to the part number to indicate if an IC is backward compatible. (For example, if the last digit is a 9, the chip is backward compatible with the last eight versions. If the last digit is a 2, it is only backward compatible with the last version.)

4.5.3.2 Re-identification

We would like to start with a quote from Martijn Dullaart of ASML, "What is often forgotten is that while form, fit, and function (FFF) still applies, there are changes to the way service has to deal with the part. This is something that is not seen if you let the design engineer make re-identification decisions where manufacturing engineers or service engineers have a better view on these types of things. Next to that traceability is often overlooked in re-identification decisions. For instance, the need to re-identify a part because you need traceability that otherwise cannot be ensured." This is often tied to phase-based item re-identification, which will be discussed in Section 9.5.1.

Frank B. Watts[18] states, "When in doubt change the part number." Frank goes on to say that if the change is required to meet specification requirements, the part is not interchangeable and must be re-identified up to the level that interchangeability is re-established. If a part is an improvement over the specified requirements, then the part is interchangeable and does not have to be re-identified. What this means is that the next higher-level consuming assembly must be re-identified as well and so on up the product structure until interchangeability is re-established. We found that there is a proliferation of the use of phantom BOMs by engineering-centric companies that are light on their CM implementation. Phantoms rarely get re-identified, so traceability and the level of interchangeability are lost. This trend toward gamesmanship to avoid re-identification while saving little on the front end has costly ramifications when it comes to change implementation.

Regardless of how items are re-identified, you cannot have two or more parts in use that are identical in every way but have different part numbers. The corollary to this is that you cannot have two or more parts in use that are not identical in every way but have the same part number. If this happens, the results can be catastrophic, as in the General Motors Corporation switch recall discussed in Section 11.7.1.

4.5.4 Configuration Status Accounting

SAE/EIA 649-C, *Configuration Management Standard*, Table 2 defines configuration status accounting as:

> The CM function that formalizes the recording and reporting of the established product configuration information, the status of requested changes, and the implementation of approved changes including changes occurring to product units during operation and maintenance.
>
> Aliases, Related Terms and Examples
>
> - Configuration database
> - Configuration records
> - Status accounting

[17] Samaras, T. T., and Czerwinski, F. L. (1971), *Fundamentals of Configuration Management*. Wiley-Interscience, a division of John Wiley & Sons, Inc. New York.

[18] Watts, F. B. (2000), *Engineering Documentation Control Handbook*. Noyes Publications Park Ridge, New Jersey, U.S.A.

MIL-HDBK-61A, Military Handbook Configuration Management Guidance gives a unique perspective of what configuration status accounting entails.

4.2.1 (d) Configuration Status Accounting (CSA)—All of the other CM activities provide information to the status accounting database as a by-product of transactions that take place as the functions are performed. Limited or constrained only by contractual provisions and aided or facilitated by the documented CM process and open communications, this activity provides the visibility into status and configuration information concerning the product and its documentation.

The CSA information is maintained in a CM database that may include such information as the as-designed, as-built, as-delivered, or as-modified configuration of any serial-numbered unit of the product as well as of any replaceable component within the product. Other information—such as the status of any change, the history of any change, and the schedules for and status of configuration audits (including the status of resultant action items)—can also be accessed in the database.

Metrics (performance measurements) on CM activities are generated from the information in the CSA database and provided to the management and planning function for use in monitoring the process and in developing continuous improvements. To the extent that contractor and government databases and processes are integrated, the government CM manager may also be able to monitor contractor performance trends.

In general terms, CSA is the activity involved in the collection of, management of, accounting for, and reporting on all the information that is generated relative to a product regardless of source, as well as data mining of that information to be used to:

- Identify where in the life cycle every element of a product is.
- Identify what design, production, test, and evaluation operations have been performed on it.
- Identify the direct and indirect cost allocations associated with it.
- Identify all pending and approved variances, ECRs, and changes associated with any established baseline.
- Identify transition gates between baselines and reversibility of all elements to any previous baseline.
- Collect metrics associated with the product and any or all of its elements.

CSA is done as part of the management of a configuration if all accounting and transactional systems are tightly integrated, or it may be done by different functional elements. Configuration and data management activities are far reaching. CSA requires that all functional resources work collaboratively to implement preplanned data capture, integration, and application of consistent metadata nomenclature and integrated intra-database functionality, so that information contained in separate databases can be utilized across all management functions.

4.5.5 Configuration Audit

The most common reviews and audits are identified in Figure 4.9. They are the standard kinds of reviews that many CM implementations are familiar with. These audits can be considered snapshots at predetermined intervals during the product life cycle. They are often mandated by government requirements or take the form of *in process* audits performed on each change as it moves through initiation, review, approval, implementation, and verification.

Some events people refer to as audits are really reviews:

- System requirements review (SRR)
- System functional review (SFR)

- Tabletop review (TTR)
- Preliminary design review (PDR)
- Critical design review (CDR)
- Test readiness review (TRR)
- Production readiness review (PRR)

Customer mandated audits are generally:

- Functional configuration audit (FCA)
- Physical configuration audit (PCA)
- Functional system audit (FSA), previously called formal qualification review (FQR)
- System verification audit (SVA)

Each review or audit answers specific questions.

- SRR: Have all system requirements been identified, known unknowns mitigated, and unknown unknowns identified?
- SFR: Have all interfaces between system elements been documented and intersystem dependencies allocated?
- TTR: Does the design approach meet the needs and have all trade studies been done against the design to ensure it meets requirements and maximizes EBIT?
- PDR: Does the design approach satisfy the requirement?
- CDR: Does the design approach satisfy the requirement?
- TRR: Is the item ready for testing, have test parameters been defined, and have any test temporary configuration departures been identified?
- PRR: Are all outstanding changes dispositioned, are all other reviews and audits completed, and is production able to start with no liens against the product or the production and test facility infrastructure?
- FCA: Have all functional requirements specified in the baseline specifications been demonstrated?
- PCA: Is the hardware and software built in conformity with the engineering documentation?
- FSA: Have the system functional requirements that were not evaluated at the CI level been demonstrated?
- SVA: Do all systems associated with a product conform to total system requirements?

SVAs are not being performed if a single company is responsible for the production of all the system components. SVAs generally exist in multinational development and in systems developed and deployed by a government.

Although audits are applicable on all cases, the same cannot be said for reviews. Conducting reviews and audits to the levels specified by the U.S. government must be tempered by an evaluation of not only the product being produced, but also the market segment for which the product is produced. In a non-government sales environment, following the government mandate criteria may be overkill.

Audits were established to give government customers assurance that the contractor was performing to the contract and that the product met government needs. Some form of physical and functional assessment is required prior to any product being authorized for production. The FCA and PCA events were specifically created to establish a product (e.g., production) baseline to which all succeeding delivered products to that government would comply. Stringent and inflexible government procurement rules often require the same degree of adherence to FCA and PCA format and content regardless of product ordered or the government's intent to order more than a single unit. These audits are contractual and curtail the ability to take full advantage of lean methodologies.

Six Sigma and lean methodologies can be implemented for processes regardless of audit criteria. Six Sigma and lean methodologies were not created to be broadly applicable in a one-off manufacturing environment. Some manufacturing processes in a government–customer environment will benefit from Six Sigma and lean application; however, total production runs in the tens, hundreds, and occasionally thousands significantly reduce the benefits possible in a commercial environment.

Refined review and audit methodologies exist in the commercial market. Throughout the process, in-process reviews and audits happen on continual bases, allowing rapid realignment of all product parameters with significant cost savings. Continuous review and audit are tied to the same survival criteria that allowed the product to enter and either maintain or increase its market share. Management of the configuration in the commercial environment is done in an empowered environment, where each person involved in new product development is a member of a product development team and all teams are in synergy.

4.5.6 Metrics

Inherent in any integrated process is the need to understand how well that process is working. Development of metrics and reporting against them is one very helpful management tool. Many times metrics are tied to key performance indicators (KPIs). A list of program and enterprise level KPIs will be provided as an appendix.

So, what constitutes a good metric? From those firms studied, four fundamental metric criteria emerged:

1. It must be meaningful.
2. It must be measurable.
3. It must be something that cannot be manipulated.
4. It must be timely.

Anne Mette Jonassen Hass[19] provides the following guidance:

> Data may be objective with no personnel involved. These are often enumeration, and tools may be used to collect measurements. Conversely, data may be subjective, involving an element of evaluation. This entails some uncertainty, but the data is often cheap to collect, and its usefulness should not be underestimated. There is nothing wrong with a metric for which the measurement is a verbal answer of yes or no. This could be, for example, whether the convention is used or not.

Three types of data are employed in metrics.

1. Raw data is collected for things like direct and indirect charges, frequency drift, power outages, etc.
2. Direct measuring extracted from raw data such as frequency of occurrence over a set period.
3. Indirect measuring calculated from direct measuring such as recurring trends.

Metrics that have been pre-established in out-of-the-box PD/LM system reports we looked at have been tied to things that the programmers thought were important, rather than things that were important to CM implementation at either the enterprise or program level. Looking at the health of

[19] Hass, A. M. J. (2003), *Configuration Management Principles and Practice*, Addison-Wesley Pearson Education, Inc. 75 Arlington Street, Suite 300, Boston, MA 02216.

the program and the health of the organization relative to CM implementation means that metrics really must be created for both. Metrics on how a company is doing in meeting its Six Sigma goals relative to CM processes/improvements are much different than metrics measuring the health of the program relative to program-specific key performance indicators (KPIs). CM metrics for processes/improvements are focused on risks and opportunities that impact the enterprise. CM metrics related to program KPIs are focused on how the program is doing, and KPIs can vary from program to program based on the program risk profile, profit incentives, and other factors.

4.5.6.1 The Number of Engineering Drawings on a Program—Poor Metric

This metric fails all of the four fundamentals.

1. Work is constantly scheduled and rescheduled, and may even be eliminated during the performance of a program due to changes, false starts, manufacturability issues, and test results. The belief at the start of the program may be that 1,600 drawings are required, but the true number may increase or decrease as the program moves along. Entire assemblies originally planned for an in-house design and build may be outsourced. Outsourcing can reduce the internal drawing count by as much as 15%.
2. In many cases, drawing numbers are not reserved until concept development discussions are completed. During concept development discussions, the drawing may exist in conceptual form (in computer-aided design, software tools, etc.), but the information does not reside in the PD/LM system. Data mining of the PD/LM system result in reports blind to documents that haven't been entered into it.
3. Our studies show that, once it is understood that the number of drawings is being measured as a metric, design engineering will increase or decrease the number of details and assemblies defined on any single drawing or software version description document. This negates the metric. (For example, if more details and assemblies are being put on one drawing, the total number of drawings required goes down. If drawings with multiple details are changed so each detail is now on its own drawing, then the total number of drawings goes up.)
4. Data pulled from the PD/LM system may be required to support weekly or monthly reviews. As a result, the data is needed for the report much earlier than the meeting date. This means that old information is being presented.

4.5.6.2 The Number of Engineering Orders on a Program—Poor Metric

This metric fails three of the four fundamentals.

1. Measuring the number of engineering orders against a drawing as an indicator of program performance increases program costs and does nothing to help manage program performance. Change is part of the design, build, test, and redesign process. Changes are a good thing. As designs mature over time, changes decline. Engineering change to design is driven by not only the stability of the functional requirements, but the complexity and stability of the development itself. As the technological boundaries are pushed from known solutions to state-of-the-art development, the percent of engineering orders rises steeply.
2. It is measurable.
3. We found that, once it is understood that the number of engineering orders per drawing is being measured, design engineering will package updates. They hold off making small engineering modifications until they have reached a state where production is impacted. Only then do they put them all in single change. This lowers the total number of changes.

4. Changes to the engineering design need to be understood by the entire team and not just by the design team. The faster an engineering change is coordinated through the change review process, the more efficiently the entire team operates and the greater the overall cost savings.

4.5.6.3 The Number of CM and Software CM Requirements in the Contract—Good Metric

This metric meets all of the four fundamentals.

1. An in-depth review of contract requirements is critical to CM planning and implementation.
2. As requirements are identified, they can be measured.
3. The contract at award established the functional baseline, and each change to the contract modifies those requirements. We will explore this further in Figure 9.2, "Requirements drift versus baseline."
4. The review is performed at contract award and against each contractual change implemented. This is flowed to the entire program team, giving adequate time to adjust how it is managed. It also allows the program to perform incremental closeout of those requirements as the program progresses through the product life cycle.

4.5.6.4 How Long Engineering Orders Take to Be Approved and Released—Good Metric

This metric meets all of the four fundamentals.

1. Critical to program success is the timely release and incorporation of engineering orders into a product. This pinpoints hotspots and is used as an indicator of lack of responsiveness. The information allows root cause analysis to be done to see if it is due to poor change requirements, lack of staffing, or other reasons.
2. Most PD/LM systems allow for tracking of change approvals by individual or department. Such reports can be run on a prescheduled or ad hoc basis.
3. Standard review times should be established during the CM planning phase. A 48-hour rule was found to be the most common. Any change approvals taking longer than 48 hours failed the metric.

4.5.6.4.1 Other Sound Metrics

- The number of audit and/or corrective action findings or observations
  - The number is more than a pre-established "norm."
- The number of days required to close audit and/or corrective action findings or observations
- The number of months since a review of company Level I, II, III, and IV data for obsolete references to other documents has been made
  - Eighteen months is a good standard to measure against.
  - This metric, while a good one, can be eliminated if referenced documents are removed from the document and put into a document bill of material in the PD/LM system.
- Length of e-mail strings on technical topics
  - More than three generally indicates the need for a meeting.
- Percent of design restarts due to the original design solution being too costly, not adequate, or ill conceived

- Percent of lessons learned analysis input into a relational database
    - Lessons learned not in a relational database are simply lessons documented.
- Percent of leveraged forward designs that have metadata with upward and downward cross-references in the PD/LM system
    - Older design should cross-reference to newer leveraged design, and newer design should cross-reference to older design
    - This is the only way in a PD/LM system that you can track the design heritage.
- The time it takes an approved change to be incorporated into all units (in-production, completed but not yet delivered, and delivered)

5 Data Definition, Data Types, and Control Requirements

5.1 QUESTIONS ANSWERED IN THIS CHAPTER

- Why are there different control requirements for each configuration level?
- Do the configuration levels facilitate mass customization?
  - If so, how?
- What function does version control fulfill?
- Describe the connection between version control and product verification.
- What is the role of configuration management (CM) in managing enterprise-level data associated with the business but not the actual product designated configuration item or configured item (CI)?

5.2 INTRODUCTION

Data and management of data is the lifeblood of any organization. Key to management of data is an understanding of not only the level of control that is needed, but what that level of control means in the context of the CM implementation. Data consists of all data—not just data associated with a single item such as a configured item (CI). Most firms evaluated were found to have limited control over the interrelated enterprise data structure. This finding was disturbing because of the criticality and the need for data identification and change control on all but the most insignificant forms of data.

5.3 DATA DEFINITION

The availability of information of all sorts has never been greater than it is at present. Tool refinements allow us to measure, analyze, report, discuss, form opinions, and interconnect in a way that is beyond the conception of those who developed the original standards for CM. Attempts to sort through what information has meaning at any point in time are often clouded by the rising glut of information (also called noise) generated in documenting and sharing everyday life. The 35 mm slide shows of an assembly process created in the 1950s gave way to videos of similar events in the 2000s which have again given way to 24/7/365 news feeds of such events which will possible give way to total immersion 3D experiences in the future.

To put data in perspective, it is necessary to discuss information in general and how a piece of information relates to CM in specific terms.

Configuration data: Information relative to the design, development, and delivery of a product (or service).

Business data: The subset of all information relative to the business itself. Although not data specifically related to acceptance of CI, business data is controlled data.

Metadata: Data fields that do not form part of either configuration data or business data are not controlled data and help us to either sort the data or gather information about it for reporting purposes.

Controlled data: Data under configuration control using versions or revisions (letter, date, time stamp, etc.). Controlled data also includes event data such as part numbers, which do not require revision of version control.

Information: Anything that can be perceived, deduced, or formulated based on the intellectual curiosity of the species. This includes any written or unwritten information received through the senses as well as those things that can be conceived of based on personal knowledge or experience. All data is information.

Intelligence: Meaningfulness of the information, for example, the ability to derive intelligence pertinent to the information being examined at a point in time.

GEIA-859a[1] defines data relative to three areas: product, business, and operational:

> Data is information (e.g., concepts, thoughts, and opinions) that have been recorded in a form that is convenient to move or process. Data may represent tables of values of several types (numbers, characters, and so on). Data can also take more complex forms such as engineering drawings and other documents, software, pictures, maps, sound, and animation.

MIL-HDBK-61[2] echoes this definition in Section 9.1:

> In this age of rapidly developing information technology, data management and particularly the management of digital data is an essential prerequisite to the performance of configuration management. Digital data is information prepared by electronic means and made available to users by electronic data access, interchange, transfer, or on electronic/magnetic media. There is virtually no data today, short of handwritten notes that do not fall into this category. Configuration management of data is therefore part of data management activity; and management of the configuration of a product configuration cannot be accomplished without it.

Yet these definitions are insufficient from an enterprise CM perspective. Data comes in many forms. Graphical and photographic, sound recordings, receipts, letters, memos, faxes, internal web pages, engineering notebooks, databases, and others … all are important. Someone working on a single program or a part of a program often finds most of their activity artificially limited relative to the larger scope of the business, but this does not negate their responsibility to think and act with the larger scope in mind. CM planning must account for this larger picture.

Consider the following case study:

> The Savings and Loan Crisis of the late 1970s was one of the largest financial scandals in U.S. history. It reached the international stage in the 1980s and ended in the early 1990s. Savings and loan deposits were insured by the U.S. Federal Savings and Loan Insurance Corporation (FSLIC) and depositors continued to put money into these institutions despite the indication that all was not well. Complex factors eventually resulted in the insolvency of the FSLIC. As a result, the U.S. government bailed out the savings and loan institutions, spending some €108,290,000,000 in taxpayer dollars. The U.S. government's Resolution Trust Corporation (RTC) liquidated 747 insolvent institutions. Toward the end of the RTC's efforts, government officials asked, "How much had been spent on contracts to women owned, small, or small and disadvantaged owned businesses?"
>
> The Denver RTC contracts area was not able to provide a complete answer simply because that information had not been deemed significant enough to track as part of its data management activities despite its being a requirement in every contract let by the RTC Contractor Selection and Engagement Contracts Branch.
>
> One individual interviewed about the RTC case study stated, "No one told us until the last two months of the RTC charter that the information was going to be required so we didn't make any provisions for capturing it in our database. The RTC contracting function was not done using a paperless system. It would have taken the entire Denver office staff at least six months to properly redesign the database, open the paper files on every contract award, and enter the women owned, small, or small and disadvantaged owned information to support the kind of data mining being requested."

---

[1] TechAmerica. (2012). *GEIA-859a, TechAmerica Standard: Data Management.* Washington, DC: TechAmerica Standards & Technology Department.

[2] DOD, U.S. (2001). *MIL-HDBK-61a, Military Handbook: Configuration Management Guidance.* Washington, DC: U.S. Government Pentagon.

From the mid- to late 1990s, data management and data managers reported to the CM function in most companies doing business with the U.S. Department of Defense and, by association, with other U.S. government agencies. The CM department controlled all incoming and outgoing communication between the contractor and the supply chain (including the customer), controlled all internal memos, assisted with a functional decomposition of requirements, maintained *official* copies of those documents in annotated form (updated to reflect the latest contract changes), prepared proposed specification change notices and proposed engineering change notices, and in many cases maintained a conformed contract for the legal or contracts department. They verified the content of every piece of deliverable data and all programmatic information prior to delivery. CM implementation included establishing and maintaining a program document library containing the official version of all the requirements documents associated with a program.

By 2010, data management had morphed into the distinct subareas of knowledge management, information management, and data management. Much of what had been done manually started to involve data capture and analysis using sophisticated databases. These were often unique to each organization and not tied into the product data/life cycle management (PD/LM) systems. Data quality assurance aspects performed under the older paper data management paradigm were mislaid. What we found in discussions in LinkedIn and other groups relative to "traditional" CM and the "new" CM is that some see them as akin to the movement from the Gothic to the Renaissance in the fourteenth through the seventeenth century. Nothing could be further from the truth.

From before the time of Charlemagne (768), art was used as the major form of communication to a lay public. Events from scripture or centric to the life of the Christos were depicted in stained glass, sculpture, tapestries, frescoes, paintings, and drawings. These visuals contained a biblical and an extra-biblical narration[3] of the events leading to the resurrection in a form understandable to the unlettered populace. They portrayed a loving and empathetic deity until the time of the Black Death, ca. 1348–1350, after which the Christos was portrayed as the ultimate law giver with graphic portrayals of penalties for not following the exactness of church law. Decades after the Black Plague came a period of Renaissance (literally, *rebirth*), merging the traditions of the Roman Catholic Church with aspects of the re-emerging beliefs and science of ancient Greece. This is exemplified by events such as Cosimo de' Medici, in the fourteenth century, purchasing scrolls of Greek texts and having them translated into Latin. He also paid Marsilio Ficino to translate the complete works of Plato and collected a vast library of other Greek and Roman works. Copies of these were used to establish two libraries that influenced Renaissance intellectual life.

In a comparable way, proponents of *traditional* CM are viewed as holding to old ways, based on *paper-world* definitions and methodologies. This is because the current trend is to distribute many "traditional" CM data and status accounting roles to other organizations and functions, such as systems engineering, knowledge management, and GTS. Those who believe in a parallel between traditional CM and whatever will emerge as the Renaissance of CM are often frustrated by what they see as clinging to the past despite the changing data-intensive world of the present. Yet proponents of the *new* CM appear to be just as reliant on International Organization for Standards (ISO) and American Society of Mechanical Engineers (ASME) standards. Many are still being crafted by the same bodies that cling to the "traditional" CM practices and much is lost along the way. We highly recommend proponents of *traditional* CM, *new* CM, and what can be referred to as *digital fabric* CM be proactive and join committees like SAE G-33 to help craft requirements necessary to manage configurations.

The ASME Y14.35M standard also follows *traditional* methods by excluding the letters I, O, Q, S, X, and Z from the list of approved revisions due to their resemblance to the numbers 1, 0, 0, 5, 8, and 2.

---

[3] Voragine, de J. (2013). *The Golden Legend*, Translated by W. Caxton. Seattle, WA: CreateSpace Independent Publishing Platform.

TABLE 5.1
Line Item Number Formats

Accounting Classification Reference Number (ACRN	0001 through 9999
Contract Line Item Number (CLIN)	0001 through 9999
Sub-CLIN Numbering	0001AA through 9999ZZ
Info-CLIN Numbering	00101 through 00199
Exhibit Line Item Number (ELIN)	Generally 001 through 009, then 00A through 00Z, 010 through 019, then 010A through 010Z ... 090 through 099, then 090A through 090Z ... 990 through 999, then 900A through 900Z
From PGI 204.71 – Uniform Contract Line Item Numbering System DFARS 204	
Provisioning Line Item Number (PLIN)	AAAA through 9999
From DED 309, Provisioning Line Item Number (PLIN)	

This prohibition was established in a time when drawing copies were stored in filing cabinets rather than in electronic form and runs contrary to similar numbering schema used elsewhere in U.S. contract documents where contract line item numbers (CLINs), exhibit line item numbers (ELINs), and provisioning list line item numbers (PLINs) generally are alpha or alphanumeric in nature and exclude only the use of alphas I and O. Hopefully these restrictions will eventually be eliminated as the capabilities of automated databases are embraced by CM practitioners and customers alike. Table 5.1 shows line item number formats currently in use by the U.S. government buying offices.

In a world that some feel is on the verge of techno-meltdown brought on by data overload, it may be hard to determine what is important to pay attention to in a company setting. The CM implementation needs to make provisions for this as DM goes beyond the capture of informational data needed to allow reproducibility of a vetted design. Thomas Redman's[4] analysis of data as the most important business asset within corporations points out what in most CM implementations has been forgotten: Data is a very valuable asset that needs to be as professionally and aggressively managed as other assets. Data must be put to productive use and must also be easily accessible and of high quality, so that sound business decisions can be made. His research indicates that the cost of poor data can result in a loss of 10–20% of revenue. His other findings suggest:

- Knowledge workers spend 30% of each day looking for data with a 50% success rate.
- Data, if found, contains between 10% and 25% data inaccuracies.
- There is a lack of data cross-references and coordination.
- There is inadequate data definition, resulting in the data being misinterpreted.
- There are inadequate data safeguards.
- There are multiple sources of truth for the same data.
- There is too much data and not enough intelligence.

[4] Redman, T. C. (2008). *Data Driven: Profiting from Your Most Important Business Asset.* Boston, MA: Harvard Business Review Press.

5.3.1 Knowledge Workers Spend 30% of Each Day Looking for Data with a 50% Success Rate

This 15% loss in productivity adds up over a year. In many U.S. firms, there are 2,080 work hours in a non-leap year. Adjusting this down for national holidays (10 days) and personal holidays (21 days) and sick days (10 days) leaves gives us 1,752 billable hours a year (Equation 5.1):

$$2{,}080 \text{ hours} - \left(\left[10 \text{ days} + 21 \text{ days} + 10 \text{ days}\right] \times 8 \text{ hours per day}\right) = 1{,}752 \text{ hours}$$

EQUATION 5.1 Billable hours a year adjusted for holidays, vacations, and sick leave.

If we adjust this downward again by the 15% loss in productivity, we have a loss of 262.2 hours/year/employee simply looking for and not finding data (Equation 5.2).

$$1{,}752 \text{ hours } \times 0.15 = 262.5 \text{ hours}$$

EQUATION 5.2 Billable hours a year with 15% productivity loss.

Using the previously cited engineering rate of €174.67 an hour, we have a €45,798 loss in productivity per person (Equation 5.3).

$$262.5 \text{ hours} \times €174.67 \text{ per hour} = €45{,}850.875$$

EQUATION 5.3 Average cost of a 15% loss in productivity.

An additional 15% loss in productivity also occurs in non-engineering support hours across the entire company. Assuming a medium-size program utilizing 100 engineers, we end up with a yearly loss in productivity of €4,585,087.5 (Equation 5.4).

$$€45{,}850.875 \text{ per employee} \times 100 \text{ employees } = €4{,}585{,}087.5$$

EQUATION 5.4 Average cost of a 15% loss per 100 employees.

The €4,585,087.5 loss is only for the engineering portion of the project. This loss implies that if a company hired 12 additional engineers to manage data, it would cost the company no more than the loss it is seeing in just the engineering department on the program (Equation 5.5).

$$\frac{€4{,}585{,}087.5}{2{,}080 \times €174.67} = 12.62$$

EQUATION 5.5 15% loss in productivity equals 12 additional engineers.

This is approximately the number of data managers on a pre-1990 government development program worth €436,670,000 (at that time, there was one data manager for every eight configuration managers).

Adding 12 additional data managers today may be excessive simply due to advances in technologies. Adding two data managers properly trained in creating the infrastructure, data links, quality control, and data cross-linking per 100 engineers would result in substantial benefits and a reduction in time wasted in fruitless searches. Simply including cross-references in metadata fields between superseded and superseding documentation and for design heritage alone would provide untold benefits. These two steps would allow a better understanding of the current design and technical baselines, and also enhance tracing the evolution of intellectual property (IP) by establishing its provenance. In a governmental setting, such knowledge would preclude providing unlimited rights to data and software when limited or restricted rights were justifiable.

5.3.2 Data, if Found, Contains between 10% and 25% Data Inaccuracies

If 10% of your data contains errors, how reliable are decisions being made when relying on it? Gentry Lee describes a case where an investigator stated that, "The atmosphere follows a Gaussian (normal bell curve) distribution ...."[5] The investigator had not considered other distribution types such as normal, bimodal, or lognormal, or if the curve distribution could be mesokurtic, leptokurdic, or platykurtic in nature. Lee corrected the investigator and the scope of the scientific investigation was dramatically changed. Data errors observed in the research leading to the writing of this section were not as critical. The general observations fell into:

- Reliance on old data for a new analysis rather than using the latest data.
- Analysis using incorrect formulas or assumptions.
- Typographical errors.
- The wrong document attached to the document object (e.g., PL/DM object 1234567 Revision A had the image for 1324567 Rev A attached to it).
- Transposition of numbers in equations and results.
- Data used to test database functionality not being cleared out after the test.
- Incorrect work authorization numbers in PD/LM metadata fields.
- Incorrect program associations in PD/LM metadata fields.
- Corrupted PD/LM document object attachments.
- Attachments with incorrect references.
- Inconsistencies in people's names in metadata fields resulting in null sets on data searches.

These may sound relatively harmless, but even typos are costly. During the development of the handling equipment for a now decommissioned intercontinental ballistic missile in the United States, a requirement was placed in the functional specification to filter the air used by the motors on diesel semi-truck tractors at the test site to remove blowing sand particles. Beach sand is approximately 100–2000 μm in size. The requirement was written as 1.00E0 2 μm and larger. The document copier used by the lead designer to copy the master specification had a white-out streak on the glass. It produced a document that unfortunately read 1.00E0-2 or 0.01 μm (the size of tobacco smoke). Without questioning the typo, the engineer designed a truck engine air filter that was larger than the Subaru 360, shown in Figure 1.11.

IP marking such as limited rights, restricted rights, and commercial rights is unique to each contract. Two contracts with the same provision specifying the legend to be used could not have identical legends on documents because the contract number is part of the legend text that must be included. To use released data on multiple contracts, one company removed all IP markings from document formats and a "keep-out" space was established for use by the PD/LM system report writer. Program unique IP legends were stored in the PD/LM system as Text Source objects. The legends could be selected using a pull-down menu and applied during the export of released engineering in portable data format (pdf). The concept was beta tested using test Text Source objects created by the developers that ranged from *test* to paragraph-long descriptions of different cuts of meat.

The company was happy with the beta testing and the improvement was rolled out. Document objects were exported from the PD/LM system and delivered to customers. It was used for five years before anyone at the company became aware that the test legends had not been replaced with the contractual ones. Since the test legends were used for five years, one customer ruled it was clearly the intent of the contractor not to comply with the contract data marking requirements. As

[5] Lee, G. (2007). *So You Want to Be a Systems Engineer?* video presentation. Pasadena, CA: Jet Propulsion Laboratory.

a punishment, the customer removed the test legends and freely distributed the contractor's documents to other competitors, as was the customer's right under the contract provisions. The resultant loss of IP could have been avoided, as could the resultant loss of customer respect for the company.

The more distributed the responsibility for the overall configuration of databases becomes, the greater opportunity there is for data errors. CM implementation concern had been central to the beta test. The CM implementation team assumed that the developers would be replacing the test legends with the real ones. The developers assumed that the users would be replacing the test legends with the real ones because they had no idea how the legends in the Text Source objects should read. The lack of a comprehensive implementation plan with defined roles and responsibilities was not deemed critical to the implementation. Besides the loss of critical intellectual property, the company was forced to implement processes mandating that all outputs from the PD/LM system be opened and read to ensure that the proper legends were on the pdf. Data managers had never been required to open the files they were delivering prior to this change.

5.3.3 Lack of Data Context

The term "digital thread" started to gain popularity toward the end of 2018 as a means for referring to some aspects of CM. The concept behind it suggests that there is a linked set of data objects that wind their way from the inception of a program to the finished product. The search for the oldest use of "digital thread" leads us to the U.S. Department of Commerce, National Institute of Standards and Technology, on December 4, 2014[6]. The phrase has rapidly gained popularity among PD/LM developers. Unfortunately, the concept behind the "thread" is too simplistic in most descriptions of everything it entails.

It is easy to tease out a single thread for a test report back to the as-run test procedure, to the test procedure, to the test plan, to the set-up, to the liquid nitrogen purchase order, to the liquid nitrogen, to the liquid nitrogen material data sheet, and finally to the liquid nitrogen vendor. Unfortunately, this thread doesn't give insight into how the information in the thread relates to the product. The thread makes the information interesting but not particularly useful in the long run.

The multiplicity of digital threads combines to make a digital fabric. The weave is dependent on the CM implementation approach. The CM implementation must be right sized to meet the needs of the program. Thus, the complexity of the digital threads woven into the digital fabric (what was once referred to as the integrated paper trail) grows or shrinks across a company based on program specific needs. Anything from a digital cheesecloth to a robust and enduring digital carpet can be created if needed. The digital thread complexity needs to be properly documented in the CM plan.

Lack of data context occurs between documents, design teams, departments, business units, and partners. In its simplest form, lack of context occurs when there is no cross-reference between the original data source and items using it (e.g., "This analysis is leveraged from report XYZ" or "Policy 1234567 is superseded by 2345678"). Other examples are as follows:

- One team designing a part that interfaces with another without the knowledge that the interface has changed.
- One organization embarked on a two-year effort designing a GTS upgrade based on existing technologies. They were sequestered off campus and never knew a replacement system went live months before their project was completed.
- Corporate yearly profitability projections were presented to stockholders. They were based on month-old data that did not reflect cancellation of major programs due to lack of funding.
- The lack of any relationship in the PD/LM system between the items tested, the test procedure, the as-run test procedure, and the test report resulted in functional configuration audit findings.

---

[6] https://www.nist.gov/el/systems-integration-division-73400/enabling-digital-thread-smart-manufacturing

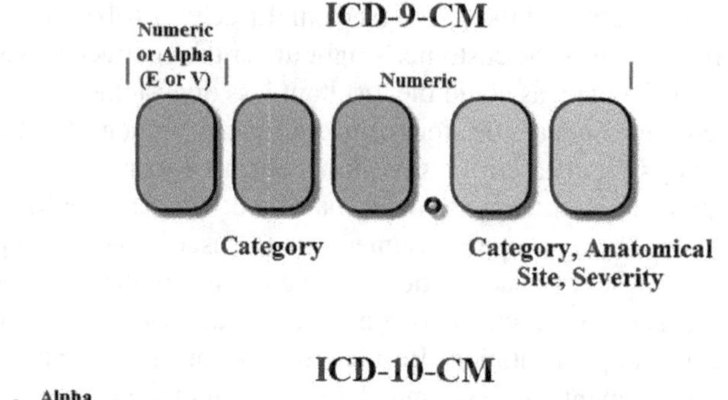

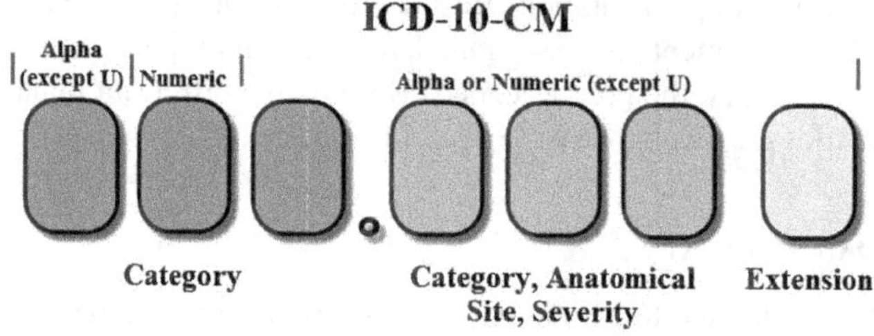

FIGURE 5.1 ICD-9-CM versus ICD-10-CM code schema.

The library science concept of cross-indexing data on multiple levels is just as applicable to industry as it is to the Bibliothèque nationale de France. We see this increased need for specificity and cross-referencing in most service industries, for online marketing smart merchant agents, and in the medical profession. Yet it is woefully missing in other product sectors.

The implementation of International Classification of Diseases (ICD)-10-CM[7] for medical coding is an example of this. ICD-9-CM utilized the same code for similar injuries on opposite limbs. ICD-10-CM provides much greater specificity with approximately 68,000 different codes while allowing for the addition of new codes, compared to 13,000 codes of ICD-9-CM (Figure 5.1).

This increase in specificity is like the change from a bar code to quick response (QR) codes to radio frequency identification (RFID) (Figure 5.2).

Like many things, much is gained, and much is lost with every new capability. ICD-10 implementation refined information pertaining to patient diagnosis. This allowed a more complete medical history to be available with a corresponding increase in coding capability at the expense of database complexity. Evolving technologies present challenges for sound configuration data management in other ways. Increased granularity in information pertaining to container content using technologies such as RFID facilitates inventory management. The downside is that such improvement can be exploited as the data can be targeted by those wishing to divert the items in the container for other uses. This kind of theft is known as either *scan, grab, and run* or *five-finger discount.*

FIGURE 5.2 Barcode, QR code, and RFID.

[7] Centers for Disease Control and Prevention. (2012). *International Classification of Diseases—Tenth Revision-Clinical Modification (ICD-10-CM)*. Atlanta, GA: U.S. Government.

TABLE 5.2
ICD-9 vs. ICD 10

ICD-9-CM Coding and Diagnosis	ICD-10-CM Coding and Diagnosis
453.41 Venous embolism and thrombosis of deep vessels of proximal lower extremity	I82.411 Embolism and thrombosis of right femoral vein

5.3.4 Inadequate Data Definition Resulting in the Data Being Misinterpreted

ICD-10-CM changes the data definitions from what was contained in ICD-9-CM to eliminate issues resulting from data being misinterpreted. This is accomplished in two ways. Procedural terms have been modified and diagnostic codes now reflect laterality and detail regarding medical conditions, as shown in Table 5.2.

Data definition errors have led to some rather spectacular failures in the engineering world as well. English versus metric conversion (or lack of it) resulted in the loss of Mars Climate Orbiter in 1999. The software interface specification required data to be provided in metric units, but the code had been written in Imperial units (also known as English units) by the contractor, Lockheed Martin. Jet Propulsion Laboratory (JPL) used the specified metric units. Instead of impulses needed for maneuvers being transmitted to JPL in Newton-seconds, they were provided in pound-seconds or a 4.45 error factor, which over time resulted in the spacecraft being approximately 47 km above the surface rather than the planned 150–170 km.[8]

In some cases, data definitions refer to data table locations (Table 5.3) that have a different name than the one intended. This may result in the data being misinterpreted. Should *source* be confused with *result*, software-dependent systems could easily be over- or undertaxed, with a resultant systems failure.

5.3.5 Inadequate Data Safeguards

Every piece of information that is generated by a company is subject to loss of one kind or another.

- File media become obsolete and the means of retrieval is not maintained.
  - Data records for a standard production item were converted to flat-film reels as technology improved aperture cards (kinds of microfilm media).
  - Due to the growing lack of availability of readers and printers, combined with concerns about the environmental hazards associated with the print media utilized, a ban on the use of aperture cards and other kinds of microfilm was put into effect within a green company.
    - Readers and files in the reader area were destroyed.
    - Destruction represented a loss of 25 years of engineering heritage.

TABLE 5.3
Data Table Locations

Line	Input	Source	Result
1	A1	1.8	.6
2	A48	2.7	.8
3	A109	3.1	.9

[8] Harland, D. M., and Lorenz, R. D. (2005). *Space Systems Failures: Disasters and Rescues of Satellites, Rockets and Space Probes*. London: Praxis Publishing Ltd.

    - A duplicate set of cards and tapes stored in a location other than the readers for disaster recovery was located.
        - One CM department member knew of their existence. It was not documented.
        - A non-disclosure agreement was put into place with a third party to convert the images to sequenced pdfs of the correct document size (e.g., all sheets of a single document were collected in a single pdf).
    - Data was eventually loaded into the PD/LM system.
- File media is assumed to be of one kind but is of another kind.
    - Data management procedures required that an 8.255 cm disk containing a native and pdf file of all engineering be collected along with the paper copy representation.
        - 1987: The CM department used Macintosh products.
        - 1995: Engineering mandated a move to personal computers (PCs).
        - 1998: A decision was made to move to a PD/LM system.
        - 1999: 8.255 cm disks were created on a Macintosh computer; were unreadable on a PC and assumed to be corrupted.
            - 90% of the disks were reformatted for PC use before it was discovered that the data was still readable with a Macintosh.
                - One Mac FX had been archived specifically for that purpose and the data could have easily been saved to a server for import into the PD/LM system.
    - The company moved to a PD/LM system and flat-file vellums and paper copies were scanned to pdf at 80 dpi (if they could be found) with an estimated loss of 15% of the existing product line engineering heritage and an 80% loss in readability compared to 300 dpi scans.
- No consistently funded and applied data upgrade functionality.
    - The Space Transportation System launch of STS-135 in 2013 was made using a suite of mission control ground equipment based on the Intel 086 chip architecture. Replacement parts were being sourced on eBay.
    - In November 2018, the U.S. Navy ordered more than €3,493,000 Xilinx and Intel-Altera FPGAs to keep F-35 and other military aircraft flying. Intel closed its acquisition of FPGA designer Altera Corp. in December 2015.[9]
- Standard file retention and destruct policies are applied to non-standard programs.
    - Mission life for an Earth observatory was extended 14 years well beyond its planned three-year mission life.
    - Record retention files were scheduled for destruction at launch plus 10 years.
    - Data management had the foresight to mandate that records be retained for a minimum of 20 years and that they be contacted prior to any data destruction.

5.3.6 Multiple Sources of Truth for the Same Data

The worst possible position that a company can be placed in is one of "data sprawl," which can be defined as storage of data in so many different GST service locations that it is impossible to determine the location and authenticity of the information you need. In one case, we found 15 different files with the same name and revision level on a single file server and none of them matched the released revision in the PD/LM system. Yet, 85 people on the program were found to be working to at least one of those 15 iterations and only one individual surveyed was working to

[9] Keller, J. (2018) Navy orders diminishing manufacturing sources (DMS) FPGAs to keep F-35s and other military aircraft flying military and aerospace electronics. https://www.militaryaerospace.com/articles/2018/11/diminishing-manufacturing-sources-dms-fpgas-military-aircraft.html?cmpid=enl_mae_weekly_2018-11-21&pwhid=4255a1f5be8f3980ba2f34063c0cb109b8319d5b33b4ad8f1ac4c718d2e3fed61b376465447d983e7f0e5204c306467c323993fa2563ea30d31f3da86303db7f&eid=417926188&bid=2306576

the released version. It made for interesting discussions and disagreements in technical meetings until it was straightened out.

In another instance, management of the customer's file server was assigned to the contractor's configuration manager. Access to the server was limited to the customer, the contractor's configuration manager, the contractor's systems engineers, and the contractor's product leads. Strict rules were agreed to between the customer and the contractor regarding data transmission to the customer's site. It was found that engineers at both the customer and the contractor's house disregarded the rules requiring use of controlled files. Just prior to acceptance, the customer's GTS department ran a file redundancy review on the server and 30 different copies of the same document with different file saved dates were discovered. All documents had the same file name and only one of them was the released version and under change control. The impact was compounded by the fact that 7,040 files had been delivered as CDRLs and an average of 19 unreleased versions of these existed on the customer's GTS system. As Equation 5.6 shows, this resulted in a 5% average data accuracy rate.

$$\frac{1\text{CDRL copy with provanance}}{19\text{ unreleased copies}} = 0.52631\text{ accuracy rate}$$

EQUATION 5.6 Average data accuracy rate.

The data listing was provided to the contractor's configuration manager for analysis. A task order for 120 hours was negotiated and implemented for the contractor configuration manager to purge the customer's file system of all unofficial copies so that a single point of truth could be re-established.

5.3.7 Too Much Data and Not Enough Intelligence

One of the great challenges of all data users is to put that data into context, so that it can be used. This means understanding the data source and its relationship to other data, as well as its place of origin and its frequency of use. Things like the speed of light are static data (299,337.984 km/hour). There is little likelihood that it will change, and it is hard for those who use it to misinterpret.

Dates are static but contextual data. The date 12/11/06 could mean any of the following if the contextual date schema is not known, documented, and disseminated. This could be considered interface critical in international collaborations and needs to be defined as part of configuration identification in the CM plan.

- December 11, 2006
- 12 November 2006
- 2012, November 06

Dates may also lead researchers astray. In preparing this book and delving into the evolutionary history of technologies, the location and date of various events was a key consideration. What we know as the Gregorian or western calendar was adopted to bring the date for the celebration of Easter to the time of year that the First Council of Nicaea had agreed upon in 325. Mind, the date in 325 for Easter is based on the equinox and not a specific and static calendar date. Easter is symbolic, allowing the celebration day to fluctuate from year to year.

Our research pointed to two reasons why this could have been done. The vernal equinox was the time that older forms of deity worship held fertility, regeneration, and rebirth observances. They gave thanks for the fecundity aspects of life, honoring Aphrodite in Cyprus, Oestre (or Oeastre) in Northern Europe, Hathor in Egypt, and Ostara in Scandinavia. At the First Council of Nicaea, Christians agreed to celebrate the passion, crucifixion, and resurrection of the Christos by timing Easter with the traditional renewal associated with spring to make it acceptable by and relatable to the older religious celebrations. We still employ the dichotomy between the old religions and the new with ancient images (flowers; youthful innocence; painted eggs; baby chickens, ducks, and rabbits) combined with new images of someone's torture, sacrifice, and death as a door to the afterlife. It was only natural that Sunday was appropriated by the Christian church for Easter. The other six days people had to work. If you are of the right mindset, you can squint sidewise at the past and see very similar incongruities throughout history.

Under the Julian calendar, the date of Easter continued to drift farther forward in time from the spring equinox due to the lack of a leap year to account for the difference between lunar cycles and the calendar. The Gregorian modification added a 29th day in February to every year exactly divisible by 4 except in years exactly divisible by 100 unless the year could be exactly divisible by 100 and by 4. Pope Gregory's bull in 1582 implemented the Gregorian calendar and, for that year the date March 10, 1582, was followed by March 21, 1582. The skipped 10 days were necessary to realign March 21 with the vernal equinox.

Unfortunately, the Gregorian calendar was not accepted everywhere at the same time and the bull had no influence in non-Catholic countries. The 13 colonies that would eventually join to become the United States of America adopted the change on September 2, 1752, with an 11-day jump. The adoption decree stated that the day following September 2 would be September 14. Alaska adopted the Gregorian calendar in 1867 along with a change in the International Date Line, which moved from its eastern to its western boundary, causing Friday, October 6, 1867, to be followed again by Friday, October 18, 1867. Due to differences between the Greek Orthodox Church and the Roman Catholic Church, the Gregorian calendar was not adopted by Greece until 1923.

Historical dates are further complicated by the fact that January 1 was not always considered the start of the New Year. An event that was reported to have occurred on New Year's Day left us having to do additional research, as stating it was New Year's Day was meaningless. During the middle ages, many Christian countries moved the New Year from January 1 to other important festival dates such as March 1, March 25, Easter, September 1, or December 25. As a result, dates can easily be interpreted out of context and they must either be normalized or otherwise explained. Dates referenced in this book have not been normalized unless done by the authors of the source material. The simple declaration that September 2 would be September 14 and the new year would start on January 1 in the American colonies meant that George Washington did not have a birthday in 1731. He did become 11 days older and one year younger[10] due to the change.

Funny old world, isn't it? Despite the centuries of disagreement (sometimes violent ones) between eastern and western Christian religious sects, they mostly agree that the date for observing Easter or Pascha is on the first Sunday after the first full moon after the spring equinox, but always after Passover. They still can't all agree on what calendar to use and can't all agree what to call it; but what the heck, the calculation is *traditional* CM, had change control, and was released, and we can't muck with the calculation. It's not quite on par with Planck's constant ($\hbar$), but still under configuration management. If they ever write a CM plan for Easter, they can sort out that niggling calendar thing in the Item Identification section.

5.4 A STRICT DATA DIET

Many adherents of *traditional* and *new* CM were found to hold to the belief that all configuration data is controlled data and that configuration data falls into the two broad categories of (1) data pertaining to the configuration required to assure product conformance, and (2) data specified as deliverable, which places this data in a distinct subset of all data relative to the product. They were also convinced that this encompassed the entirety of the controlled data that fell under the purview of their profession. Only proponents of *digital fabric* CM embraced a wider view of what data falls under configuration management.

[10] Culkin, J. M. (1991). *The Year Washington Missed a Birthday*, New York Times March 1, 1991 p. 26.

DM is part of CM. DM has a much broader reach than simply the management of information associated with a configuration-controlled product or a CII. Data management applies to all data regardless of form or format that is received or generated by a company in its daily business activities.

This idea of a strict data diet limited to managing only data associated with a deliverable item or associated to a data requirement listing (DRL) begs a few questions concerning where both *traditional* and *new* CM appear to be heading. MIL-HDBK-130, ISO, and EIA requirements (which many now cite as their CM guidance) are clear that CM must be sized to the environment to which it is applied. GEIA-HB-649[11] goes further in declaring that, on commercial programs, the line between reviews and audit is blurred and to some extent non-existent, whereas in a government-based market, the difference is very distinct. This places a greater reliance on the CM implementation to be based not only on the specific guidance documents but also on understanding the environment in which CM is being implemented.

The days of data being limited to reports, drawings, analysis, and version description documents for executable software are long past. CM implementation must also contend with DRLs consisting of articulated models, STEP, intelligent pdfs, video, data streams, apps, MathCAD set-ups, photos, and other file types. On the horizon is an environment that many implementing CM solutions were found to be reluctant to plan for. This includes the following:

- 3D printing technologies where the idea of a paper representation of a manufactured item is as relevant as using a wire recorder to hold music or dictation (Figure 5.3).
- The implantation of quantum computing.
- Database transfers.
- The expanding roles of social media, real-time video capabilities, and text messaging in a business environment.

5.4.1 3D Printing

Our ability to create a model of something in a design program and print it in a variety of materials is radically modifying how data is managed and how designs are conceived. This is not some futuristic vision: It is reality at Boeing, Airbus, Rolls-Royce, Northrop Grumman, SpaceX, NASA,

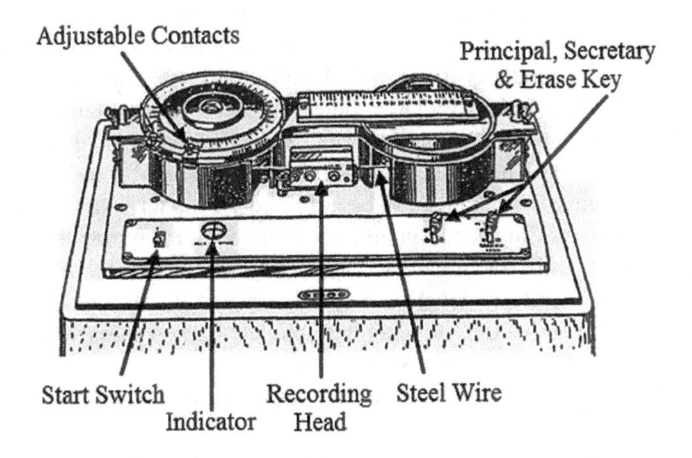

FIGURE 5.3 Dailygraph wire recorder, 1930s.

[11] SAE. (2005). *GEIA-HB-649, SAE Bulletin: Implementation Guide for Configuration Management.* Warrendale, Pennsylvania.

DOD, and commercial firms worldwide. Suddenly the idea of having a product baseline of anything simply to assure producibility of a vetted design has become way more complicated.

3D technologies have transformed how engineers approach design. They are no longer constrained to using traditional mills, lathes, forgings, and electric discharge machining. Sophisticated design iteration tools have enabled designs to become more fluid while maximizing distribution of static and dynamic loads. Using additive manufacturing, the need for tooling and jigs to support complicated shielded metal arc welding, gas tungsten arc welding, tungsten inert gas, metallic inert gas, and laser weldments is eliminated. The 3D process builds the support as the object is printed. NASA estimates the fabrication time for making one of its 3D-printed propellant distribution and engine nozzles fell from 18 weeks to two weeks. The changeover also equated to savings on material. They observed a reduction in material costs of 60% as there was no scrap.

Some 3D designs cannot be depicted on paper from a computer-aided design (CAD) output, unless you use slice views such as we see in magnetic resonance imaging (MRI). What we give to a customer as a DRL deliverable needs to be managed more like a cross between a thermal model and a software executable. This is due to the iterative nature of the 3D design. Portions of the design that meet requirements become static, like portions of the software code, as the rest of the design matures. Integration and test (I&T) activities can be run incrementally using 3D parts made with materials that can simulate the heat, stress, and wear you expect, to pinpoint areas of design concern before investing in a qualification unit. This prototyping reduces the need for multiple qualification units as design issues can be addressed before you move to formal mechanism, subsystem, or system qualification.

Interfaces between parts can be evaluated using the 3D prototype parts. Items where the design is finalized are manufactured using production materials. Items where the design is still in flux can be printed in lower cost polymers. The parts are then assembled for testing purposes. As testing results are obtained, the design gradually evolves until the design of all parts are finalized.

Configuration implementation will address IP and data marking differently. It is conceivable that spare parts will evolve to a catalogue of 3D print models, and model use rights are leased for manufacturing at the need location. Built-in 3D model use counters to limit the number of copies printed before the license must be extended will evolve. 3D fabrication complicates how distribution statements and unlimited and limited or restrictive rights are evaluated. Some we talked to believe they will become convoluted or meaningless.

Determination of latent defects and patent defects is also compromised. Consider a 3D model developer who specifies "This 3D model must be printed on a printer X or better which has been calibrated with each new material cartridge load." If a user prints the item at point of need, with a printer that is not the same model or that is out of calibration, what happens if the item contains defects? The customer rightfully makes a latent defect or patent defect claim; yet how would they evaluate if the defect is truly latent or patent if the defect was caused by an out-of-calibration printer or one that is different from the unit specified? This and other legal implications have yet to be studied in any depth. Safeguards would have to be built into the model, so a third party is not able to modify the model and pass it off as just as good as original equipment manufacturer (OEM). If 3D printing evolves to the point of tissue, organ, and joint replacement, what are the implications? One may be that a company such as Merck develops the gold standard 3D heart model, but your company insurance, Medicare, or nationalized health care has mandated that you must pay extra for anything better than the generic equivalent from 3D Body Parts-R-Us.

Questions that all CM professionals should be pondering are as follows:

- Should the 3D construct in the modeling/design environment be controlled the same as hardware, software, firmware, photos, and thermal models, or as a new, completely different form of configured object?
- What does on-demand manufacture at the user's site mean to acceptance and product warranties?

- How is IP monitored and controlled and what safeguards need to be put in place to protect the product design from being pirated or compromised (e.g., how do we protect the design from theft or alteration by a hostile third party)?
- What safeguards are needed to protect your company's IP and prohibit customers from sharing the 3D model with others?
- What happens to product liability if the model is printed by a third party using inferior materials or on an uncontrolled and out of spec printer?
- Does on-demand printing require on-site contractor personnel to be present, with printing limited to contractor-owned and contractor-maintained printers?
- How are the configuration and legal considerations of a 3D model impacted by leasing the model to a third party for production?
- How are data retention timelines impacted if 3D models eliminate design obsolescence? (For example, if I can always make another exact copy of the product, does the design ever go *out of production*?)
- Are parts printed from 3D models considered to be OEM parts if printed by a third party?

Let us explore this further. When you think of product design and development (PD&D), what first comes to mind? Is it the understanding of company business objectives (scope) followed by functional decomposition of requirements and their allocation to systems and subsystems to achieve that objective? Is it design to manufacture with designers, facility, and work center collaboration to assure cost savings through coordinated and producible designs? Or is it early and iterative design validation prior to manufacture? However you view it, additive technologies can play a significant role in not only better designs but also less expensive manufacturing.

> I believe we're on the verge of a major breakthrough in design for manufacturing in being able to take something from the concept of something from your mind and translate that into a 3D object and really intuitively on the computer and then take that virtual 3D object and to be able to make it real just by printing it. It's going to revolutionize design for manufacturing in the 21st century.
>
> **Elon Musk**[12]

The idea of being able to rapidly generate a prototype to validate solution through a construct of the proposed functionality, interfaces, and best-case design solutions was unheard of a decade ago. It is becoming the norm, along with highly accelerated life testing using shortened test parameters. The parameters are based on wear parameters for differences on material properties (e.g., making a prototype of one material to validate testing parameters for another before final production and test). Additionally, high-quality prototypes in recyclable wax deposited in half-thousandths layers using printers such as the Solidscape's R66 Plus can be made for visual evaluation and improvement prior to lost-wax casting or manufacture of a production mold.

Once reserved to the fringes of aerospace or small development labs, the results of 3D prototyping are translating into the printing of 3D parts in aerospace, jewelry, and automotive applications. We have also found companies developing additive manufacturing capabilities for building construction and composites. December 2013 saw the introduction of 3D-printed components created on a British Aerospace (BAE) Tornado fighter, the announcement that GE Aviation's LEAP engine will use 3D parts, and the announcement that NASA successfully developed a 3D-printed rocket engine component. Rolls-Royce, Pratt & Whitney, and GKN Aerospace are also investing in additive manufacturing of finished products. An early 3D-printed part is shown in Figure 5.4.

Several obvious advantages of the 3D capability soon become apparent. With additive processes, costly material such as titanium is utilized only in required quantities (adjusted for finish).

---

[12] Brewster, S. (2013). "Elon Musk shapes a 3D virtual rocket part with his hands." https://gigaom.com/2013/09/05/elon-musk-shapes-a-3d-virtual-rocket-part-with-his-hands/. Accessed April 22, 2014.

FIGURE 5.4 3D prototype and final part.

In a production environment, prototyping of designs that would not be possible, that consume many hours, or that would be cost prohibitive using standard manufacturing methodologies can be explored. In the field, product improvements can be scanned, prototyped, and more rigorously tested prior to being incorporated into the product line. One of the problems with earlier prototypes is their ability to stand up to the sort of testing that the final product would experience. For example, nobody would place a stereo lithography part on a vehicle and expect it to survive vehicle durability and reliability testing. Those concerns are reduced through 3D printing.

Rapid prototyping in 2012 was estimated as a €1,940,000,000 industry. On October 14, 2013, Chris Fox of Pddnet.com stated a business insider quoted Goldman Sachs as saying 3D printing is "one of eight technologies that are going to creatively destroy how we do business." On December 14, 2018, the 3D Printing Media Network reported that the 2018 additive manufacturing market was worth €8,110,000,000. This is a 480.41% growth in six years.

Assume that you have just received a contract to produce replacement mechanical piece parts for a product with faulty or non-existent design data packets. The customer can provide a copy of each part from inventory but, due to concern about material fatigue, wishes your company to reevaluate the part and beef up the design to eliminate the problem. Under normal PD&D operations, critical parts data is obtained in your dimensional lab and provided to the designer. Once the design is captured, improvements can be explored that mitigate the customer's concern before production and test can start.

Recent technologies allow 3D scanners to be used to establish a cloud mapping of the part that is then ingested into a CAD package. In CAD, part irregularities are corrected and material fatigue concerns mitigated by implementing part improvements resulting from fatigue actuals. The configuration-controlled model is then converted to the file type utilized in additive manufacturing. This enables a print-on-demand capability at a substantial savings over the cost of reverse engineering each part. The cloud map, original CAD ingestation, and all versions are under configuration control.

5.4.2 Quantum Computing

The development of quantum computing and the introduction of products such as the D-Wave q-2000[13] may radically change the way software and computer data are managed. D-Wave operates its processor at 0.038889 Kelvin or, if you prefer, −273.111111 degrees Celsius or −459.6 degrees Fahrenheit. As quantum computing evolves, configuration and data management methodologies associated with traditional base-2 systems will need to morph to encompass data that can exist in multiple states at the same time instead of a polarized true or false (1 or 0) state. Originally postulated by Richard Feynman, such digital data is represented in quantum bits (qbits) rather than bits (Figure 5.5). Quantum mechanics allows a qbit (also known as qubit) to be 1, 0, or both at the same time. This is known as superposition.

[13] D-Wave Systems Incorporated, Vancouver, British Columbia, Canada.

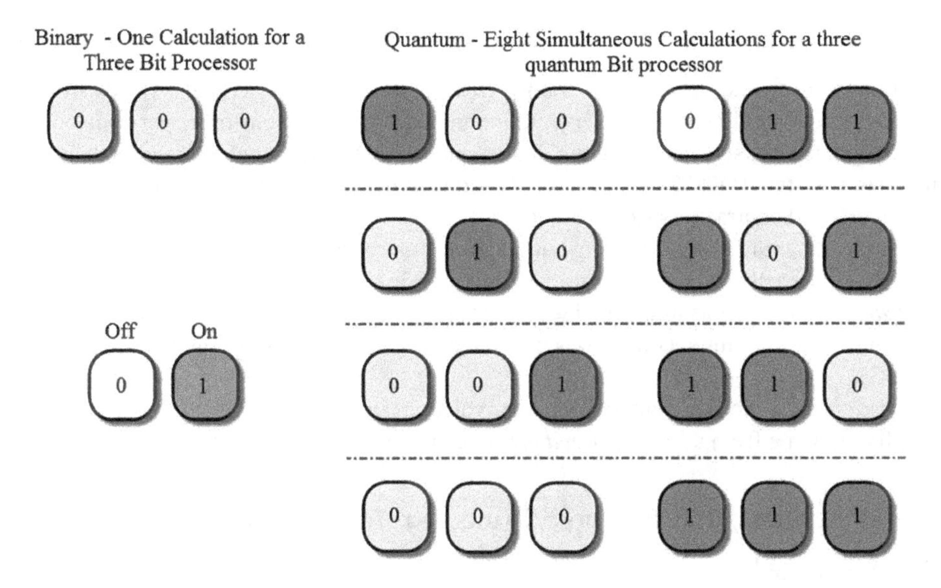

FIGURE 5.5 Binary versus quantum computing.

Products like the D-Wave Two and 2000Q (with a reported 2,000+ qbits) use an approach known as quantum annealing. Other firms are pursuing a topological quantum computing architecture. There is ongoing controversy over D-Wave's qbit claims. In 2018, Google, Intel, and IBM were fighting for quantum supremacy at the 50 qbit mark, with the goal of reaching 1,000 qbits by 2025. It appears that 50 qbits is the minimum number of qbits needed to surpass the capabilities of the fastest supercomputers of 2017. IBM announced in early 2019 the introduction of its commercially available 20 qbit IBM Q System One. It is advertised as the first quantum computer designed for businesses via the IBM Cloud. Scheduled to open in late 2019 is the IBM Q Quantum Computation Center for commercial customers in Poughkeepsie, New York.

Quantum computing capabilities being pioneered at D-Wave Systems, IBM, Microsoft, Intel, KPN, MagiQ Technology Inc, 1Qbit, Accenture, Google, NASA at QuAIL, RIKEN, and others will see the same kind of developmental continuum seen at the start of the PC, resulting in the Power Mac G4, which first offered more than 1B floating point operations per second. It appears that the quantum computing industry has yet to agree on a qbit benchmarking standard for machine capabilities. Until that happens, comparisons between quantum computers offered by different competitors may be considered nothing more than marketing hype. We all know "My dog's better than your dog. My dog's better than yours. My dog's better than yours because he eats Kennel Ration. My dog's better than yours!"[14] Quick, which of the marketing deadly sins from Section 1.4.1 does that represent? If you said Superbia (pride), you are correct!

Given a similar trajectory, it is not inconceivable that everything we know and manage in a CM implementation will be unrecognizable in two decades. Quantum qbit processors are generally following Moore's law for computer processor development, with a doubling in the number qbits every 18 months. Additional performance gains may be possible if superconductive materials, found in some meteorites, allow the qbit processors to operate at higher temperatures.[15] It is incumbent on the CM community at large to start addressing these developments as they appear.

---

[14] Actually, Rufus and Gris eat Old Roy.

[15] Conover, E. (2018) https://www.sciencenews.org/article/some-meteorites-contain-superconducting-bits. Retrieved March 7, 2018.

5.4.3 Database Transfers

Requirements are tracked and incrementally closed out as the development moves from a functional to a product baseline. CM implementation includes ingesting customer performance requirements into a database, such as Siemens's Teamcenter or IBM's Rational Dynamic Object Oriented Requirements System (DOORS). This activity requires a great expenditure of human resources on the customer's and contractor's sides (Figure 5.6).

Some companies are now partnering with their customers in a model-based engineering environment (MBE) to transfer requirements to the contractor in database form. This corrects many of the *lost-in-translation* moments associated with the traditional use of statements of work, contract data requirements lists, systems specifications, interface control specifications, and drawings/documents (Figure 5.7).

This is seen as an exciting potential for cost avoidance in cost-constrained procurements. The net effect is like moving from a *best-of-breed* approach to a totally integrated solution.

5.4.4 Social Media, Real-Time Video/Images, and Text Messaging

5.4.4.1 Social Media

Social media such as Facebook is credited with changing up events worldwide, such as flash mobs, partnering between like-minded people, tipping the scales in favor of one political candidate over another, and other events generally external to the standard paradigm associated with configuration-controlled items. Social media is also playing a much greater role in professional networking and is used exclusively to gather information. LinkedIn was used in the writing of this book to gather information and discuss certain concepts with professionals to whom it would be hard to reach out by other means. Facebook, Flickr, FeedBurner, SurveyMonkey, YouTube, and Twitter have found a place on the home pages of most corporations as a form of social engagement. It is only a matter of

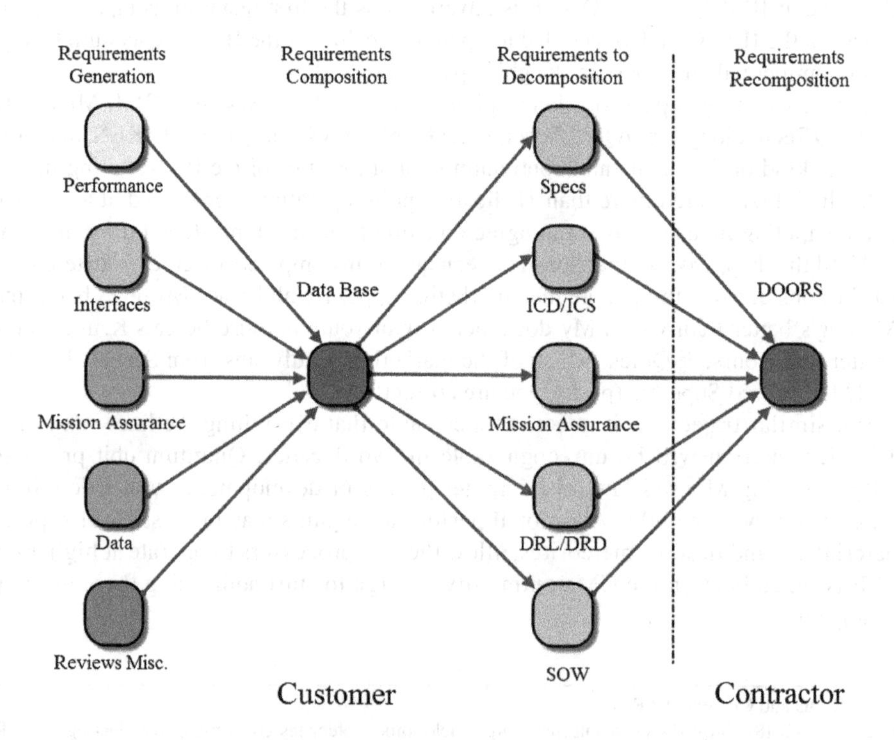

FIGURE 5.6 Traditional contracted requirements versus database transfer.

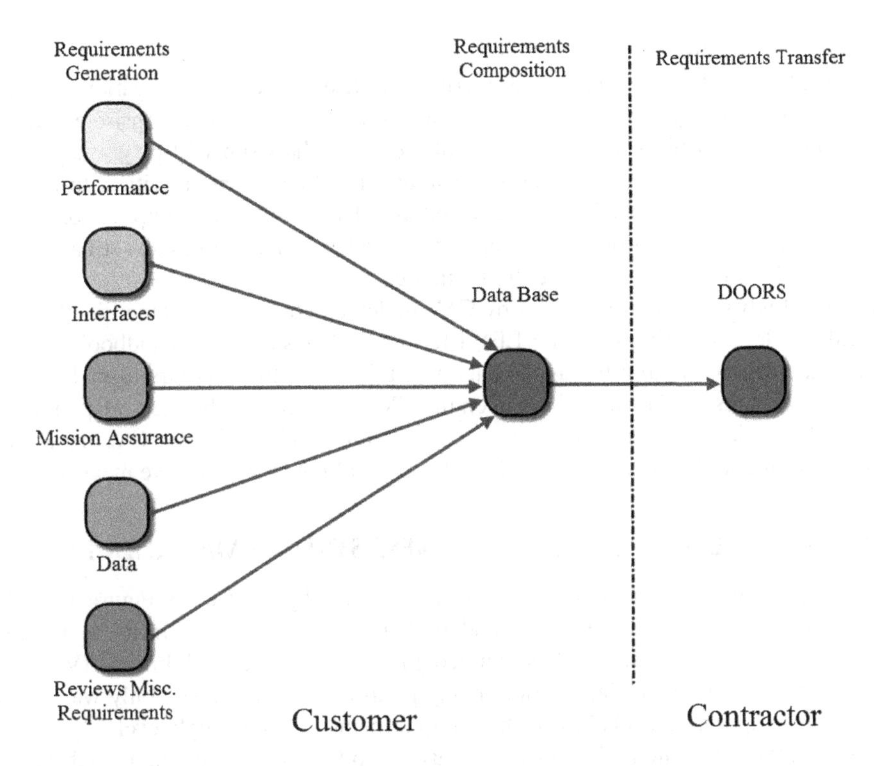

FIGURE 5.7 Database transfer with no decomposition.

time before further acceptance of these and other communication types is applied to the management of product development and results in new ways of viewing data management.

5.4.4.2 Real-Time Video and Other Images

This capability exists on most personal communications devices as well as dedicated digital image-capturing devices. Devices such as the GoPro, JVC Adixxion, Emerson Go Action Cam, Spypoint, Vivitar, Sony, BlackView, Sunco Dream, Veho, and others have met the need for personal image capture. We are also aware of their use in cleanrooms assisting with hardware transfers in and out of vertical thermal chambers. The inherent ruggedness of the designs makes them candidates for other applications, where eyes are needed to manage critical operations previously unobservable except by mirror.

Video and still image naming conventions generally were not found to be addressed in configuration and data management plans, despite the pictorial record (including video capture being critical to meeting requirements by inspection). When naming conventions are specified, they are often addressed by many companies using the *American Standard Code for Information Interchange* convention such as YYYY-MM-YY_0001.ext (e.g., 2016-02-28_0001.mov or equivalent) with no other tie to the operation, item, or content. The data forms a part of the acceptance data package or variance approval package and must be tied to it in some retrievable way in the PD/LM system.

One solution to this could be capturing the images specific to the observation in a report or other document released in the PD/LM system. Another solution could be releasing each file individually in the PD/LM system and placing them in the bill of documents of the part drawing, test report, or variance. Regardless of methodology, the items were rarely managed, often due to their volume, file size, and lack of staffing.

5.4.4.3 Text Messaging

WhatsApp is a Facebook-owned unlimited international text messaging application. Text messaging as a means of gathering quick answers to questions in a fast-moving enterprise environment is gaining acceptance. Capabilities such as those offered by WhatsApp will play a key role in the international business environment for communications that do not contain critical-to-the-business information. This technology needs to be followed and data management implications studied as the tools see wider implementation. Text string capture and export as well as text stream encryption appear to be needed to weave a complete digital fabric.

Technological advances will be lagged by CM implementation as envisioned in ISO, Institute of Electrical & Electronic Engineers (IEEE), EIA, and other standards, handbooks, and guidelines. The organizations writing the documents may take three to five years to make changes to the documents. As a result, it is incumbent upon the CM practitioner to become adept at evaluating innovative technologies and to informally agree on a forum for addressing the CM impacts of these technologies. We found that SAE G-33 and CMPIC were taking such proactive measures.

5.5 CM AND LEVELS OF QUALITY BUSINESS SYSTEMS MANAGEMENT

GEIA-859 contains two views of a four-level document hierarchy (GEIA-859 Figure 4 and Table 4) that should be expanded to include metadata and guidelines. Levels increase from the general to the specific as the level number increases. One implied goal is to create levels I through IV documentation that does not have to be deviated from on any program or project. The only way this can be accomplished is to approach levels I through IV requirements at a very high level.

A revised six-level data hierarchy much more useful to CM of products in a database environment is shown in Figure 5.8.

CM of six levels of plans, policy, process, procedures, metadata management, and guidelines was found to vary within the companies studied. In many cases, configuration control was assumed by the organization responsible for mission assurance, and the documents associated with levels I through IV were released in a PD/LM system. Stakeholder reviews were not inclusive, and many impacts only surfaced after document release. This in and of itself was not surprising, as CM responsibilities are distributive across the entire company. Unfortunately, we found few companies

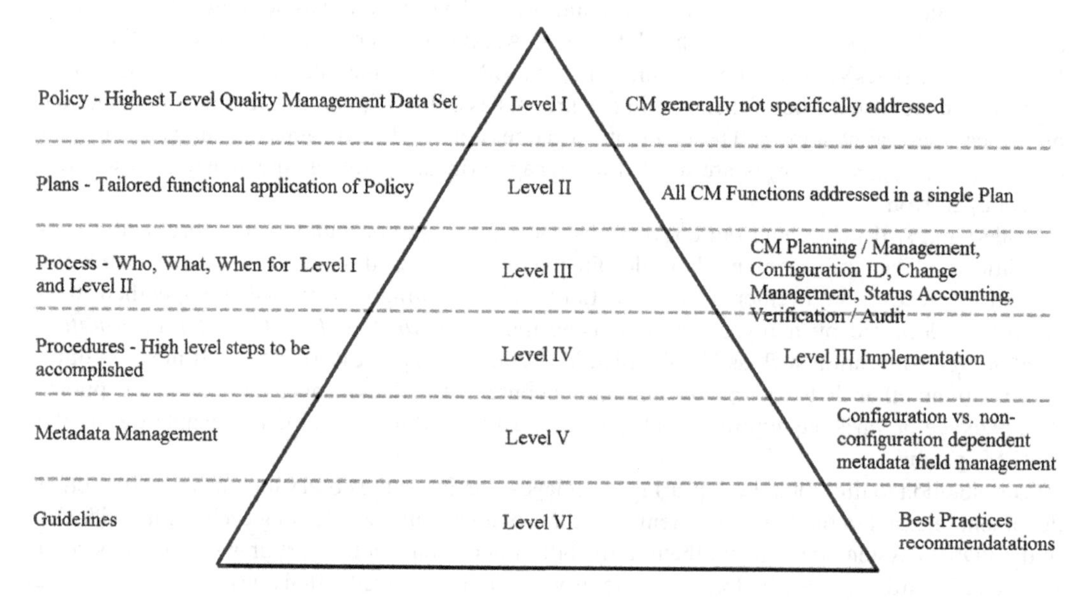

FIGURE 5.8 Six-level data structure.

with a company-level CM plan allocating CM activity to functional organizations and the relationships and interfaces between them. What follows is a summary of the consensus view of levels of control for the various quality management documents in a company without regard to allocation of CM implementation responsibilities.

Another goal of the data structure is to avoid having higher-level documentation place cost-driving requirements on lower-level documentation. A requirement levied in a policy to always use the latest template for a report, drawing format, etc., is innocent on the surface, but it sets the company up for findings of non-compliance when the system is audited by external organizations. A clear distinction between auditable requirements and unauditable requirements must be established (e.g., any documentation below level IV is unauditable). Flawed arguments are made that company mandated templates for things such as variances need to be auditable by outside organizations. Use of a template for a variance may well be a requirement, but templates may be specified or modified for a program based on contractual or other requirements and should not be subject to audit except by the customer or program that mandated the use. Arguments that all audit *findings* need to be disclosed to a customer are also flawed if those *findings* are against requirements below level IV. Any non-compliance discovered on items below level IV should be observations.

5.5.1 Level I, Policy Management

CM activities are not generally defined as a separate policy; however, all policies are subject to standard vetting with stakeholders, approvals, and release via change control, and top-level application of CM is inherent in these documents. The change control function may be delegated to the quality business system functionally but that transfer needs to be documented in a company-level CM plan. Additionally, all data releases should be consistent with established change control requirements and should utilize the same PD/LM system used for management of other data. In many ways, policy represents the mission-level requirements performed by different organizational units in the company.

5.5.2 Level II, Plans Management

If level I documentation is the mission statement, it follows that level II plans take on the role of mission sub-element descriptions. CM activities are generally defined at this level in a single plan subject to standard vetting with stakeholders, approvals, and release via change control. The change control function may be delegated to the quality business system functionally, but that transfer needs to be documented in a company-level CM plan. Additionally, all data releases should be consistent with established change control requirements and should utilize the same PD/LM system used for management of other data. The plan should not identify level III or lower-level documentation. In rare cases, CM plans may exist for management of all other plans and level II through level IV documentation. While not a requirement, there are distinct advantages in having an enterprise-wide configuration and data management plan defining the roles of each organization relative to CM implementation (e.g., allocation of CM activities to GTS, quality business systems management, program management, product management, design libraries, design media, systems engineering, training, certification, finance, contracts, facilities, etc.). The more vertical a company's organizational structure, the more critical it is for CM implementation to integrate all functional areas. The role of stakeholder becomes increasingly important in level III and level IV documentation.

5.5.3 Level III, Process Management

If level II contains mission sub-element descriptions, it follows that level III processes take on the role of mission sub-element systems specifications. Level III processes should not refer to the specific level IV documentation that implements them, just as subsystem specifications do not call out the specific documentation required to comply with them. Level III documents provide very high-level descriptions of who, what, when, where, and why activities are accomplished.

Product CM:

- Configuration planning and management
- Product identification
- Supplier management
- Interface management
- Configuration identification
- Configuration change management
- Approval of design and changes
- Configuration status accounting
- Configuration verification and configuration audit

Program interfaces with:

- Project management
- Engineering
- Production
- Finance
- Legal
- Contracts
- Supply chain management
- Facilities

Tools management:

- PD/LM
- Report generation

Almost all the companies were found to be silent regarding CM program interfaces and tool management in their efforts to reach ISO certification. Unfortunately, they relied on the guidelines identified in ISO 10007[16] in their quest to become ISO certified. ISO 10007 is product centric and ties back to ISO 9001.[17] Equal attention needs to be paid to ISO 10006[18] regarding program interfaces when crafting a CM implementation.

5.5.4 Level IV, Procedures Management

If level III processes take on the role of mission sub-element systems specifications, it follows that level IV procedures take on the role of subsystem requirements implementation. Level IV procedures should not refer to any level V or level VI documentation. Procedures should be extremely well vetted with all stakeholder organizations and should not assign responsibilities to other organizational elements (e.g., a production procedure should not assign a task to mission assurance, contracts, or engineering).

If such an interrelationship exists, then these organizations should be stakeholders. Each organization would then have an analogous level IV document supporting the requirements with the requirement defined in the organization's level III policy. It was generally found that CM procedures were overly complex and at too low a level to facilitate the flexibility needed. Companies surveyed often fell into the trap of having level IV documents written with the mantra *process is everything.*

[16] ISO. (2003). *ISO 10007:2003, International Standard—Quality Management System—Guidelines for Configuration Management.* Geneva, Switzerland: ISO.

[17] ISO. (2008). *ISO 9001:2008, Quality Management Systems—Requirements.* Geneva, Switzerland: ISO.

[18] ISO. (2003). *ISO 1006:2003, Quality Management Systems—Guidelines for Quality Management in Projects.* Geneva, Switzerland: ISO.

While this precept of ISO/IEEE 12207[19] should be understood, it should be viewed considering systems and software engineering and software life cycles and not as a blueprint for the entire quality business system approach.

Avoid procedural statements such as "The first release of a document or part is considered a revision and must be processed using an engineering change order (ECO). ECOs shall be created by the document originator with all affected items being identified on the ECO affected items. The ECO proceeds through one of the PL/DM system workflows shown below." A much better line is, "The PD/LM system shall be used to capture all revisions and versions of documents and their associated elements." Detail regarding specifics of PD/LM use should be reserved for incorporation into a PD/LM user's handbook released in the PD/LM system.

5.5.5 Level V, Metadata Management

Metadata falls into either configuration-dependent metadata field management or non-configuration-dependent metadata field management. Metadata specific to a program needs to be defined in the CM plan.

5.5.5.1 Configuration-Dependent Metadata Field Management

This metadata type is controlled by the release documentation [e.g., ECO, engineering change request (ECR), etc.] and consists of the following data fields:

- Object number (e.g., document or part/assembly identifier)
  - It is recommended that a revision level never show up with an associated part or assembly.
  - This is generally a misunderstanding of requirements resulting from drafting board practices of including the product structure on the face of the drawing instead of using a separate parts list.
  - PD/LM systems decouple these elements and the disposition of related parts is enough if the quality system is properly implemented.
- Object revision
- Object title/description (This should match the title of the object. If a drawing with title is "instrument mechanical interface control drawing (ICD/MICD)," the object title/description should be "instrument mechanical ICD," not "instrument MICD.")
  - Program name should not be in the title of any document as it is a metadata field and prohibits design reuse without expenditure of additional dollars to change.
- Change number (e.g., ECO and ECR)
- Responsible engineer (Staffing does change during the life of many products and the last revision of the object should always reflect the current responsible engineer.)
- Revision level (e.g., A, A1, A2, … B or 1.0, 1.1, 1.2 … 2.0 …)
  - It is recommended that this data field not be subjected to ASME restrictions unless contractual.
  - The prohibition against the use of revisions O, Q, S, X, and Z because they look like the numbers 0, 0, 5, 8, and 2 is a holdover from the days of drafting boards, flat files used to hold vellums, and filing cabinets.
- Revision state (e.g., preliminary, review, approval, etc.)
- Record creation date (e.g., for any record or change type)
- Stakeholder comments (e.g., reviewer comments)
- Stakeholder approve or reject date

[19] ISO. (2008). *ISO/IEEE 12207:2008, Systems and Software Engineering—Software Life Cycles. Geneva, Switzerland*: ISO.

5.5.5.2 Non-Configuration-Dependent Metadata Field Management

This metadata type is not dependent on the release documentation (e.g., ECO, ECR, etc.). It is managed as an administrative function and consists of the following data fields:

- Program work authorization number (e.g., project number)
  - As a design proceeds through the product life cycle, the project charge number may change from funded research and development to design and to production and then to disposal.
  - Each phase may have a different number.
  - Many companies also issue different numbers as the product moves from one baseline to the next.
- Program name (program names change as a product goes through its life cycle)
  - It may be called one thing during the bid and proposal or concept development phase, something else during design, and something else again after delivery.
- Contract number (may change several times over the life of a product)
- IP legend application
  - Generally, the best solution is to apply after the document release rather than embedded in the document to facilitate design reuse.
- Distribution statements (generally a U.S. Defense department requirement and subject to change as the program moves from development to production)
- Life cycle phase
- Change analyst (staffing changes during the life of many products and the latest revision should always reflect the current analyst)

The distinction between the two kinds of metadata is critical to program performance and EBIT. By decoupling the specific metadata field from the change agent (ECO, ECR, etc.), the analyst can rapidly access the state of the uncoupled data fields and manipulate them through upload and replace methodologies in the database itself, saving thousands of hours over the 10- to 40-year life of many hardware items.

Such an administrative change could be the reclassification of every program document from an allocated baseline to a production baseline. This capability necessitates a firm grasp of database management and presupposes that capability resides within the company to adequately manage PD/LM metadata (Figure 5.9). It also presupposes that the PD/LM implementation contains the capability to segregate metadata for managing each change agent and each object independently by latest release (or pending) change level. This capability ensures that the administrative change is only applicable to data at a single level of release status and is not propagated backward to historical releases (e.g., if Revisions A, B, and C already exist, the change is made to Revision C metadata only).

5.5.6 Level VI, Guidelines Management

Guidelines are generally regarded as a documentation of best practices and may contain information specific to a tool or process implementation. Management of guidelines in the larger context requires change control in the classical sense with review by the applicable stakeholders and subject to revision or version control. In some cases, guidelines may take the form of handbooks. It is critical that performance to guidelines not be subjected to external audit, and there needs to be a clear statement that this is the case in the level I through level IV quality documents. Guidelines and handbooks should be purged of all "shall" statements. While employee annual reviews may include adherence to such things as a CM handbook or PD/LM best practices, ISO and customer audits should never audit to that level. In many companies, level VI data is never acknowledged as existing in level I through level IV documentation for just this reason. Those companies refer to this level VI documentation as information that is ancillary to company operations.

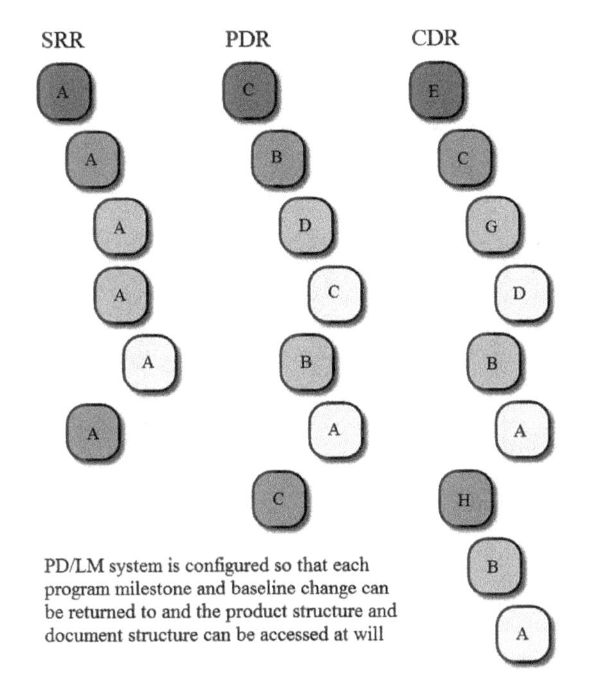

Systems Requirements Review (SRR)
Rudimentary design elements established and released
Design done on Charge Number 234521A
Change Analyst: Yen, Samantha
Contract Number: F342521-A234
Distribution Statement Attachment B: 231
All designs: Limited Rights Data

Preliminary Design Review (PDR)
Preliminary design elements established and released
Design done on Charge Number: 234521B
Change Analyst: Sabaria, Peter
Contract Number F342521-A234
Distribution Statement B: 235
All SRR designs: Limited Rights Data
All Post-SRR designs: Unlimited Rights Data

Critical Design Review (CDR)
Mature design elements established and released
Design done on Charge Number: 534211
Change Analyst: Sambuko, Asok
Contract Number F342521-B321
Distribution Statement B: 355
All SRR designs: Limited Rights Data
All Post-SRR designs: Unlimited Rights Data
Product baseline may be established at or near this event

FIGURE 5.9 Metadata management.

5.5.7 Level I, II, III, and IV Requirements Management and CM Planning

Our research indicates that, despite a sophisticated management system across all market segments, there is very little implementation of any form of requirements management system associated with level I, II, III, and IV data. Companies that were well versed with managing technical requirements in their products and on their contracts did not have any similar CM implementation for company-level imposed requirements.

The inability to quantify the interrelationships and dependencies between policies, plans, processes, and procedures made conflict resolution a slow and costly process. It also put these companies at risk of violation of their ISO and other certifications simply because of conflicting requirements in released level I, II, III, and IV data. CM must be implemented at the enterprise level and not simply implemented at a program or contract level if it is to be an effective management tool.

5.5.8 The Use of *Shall, Will,* and *Should* Is Changing

The use of *shall*, *will*, and *should* in requirements writing and contractual language is changing. There is a movement in the United States and other nations to replace *shall* with *must* in government regulations and contracts. *Must* is now being used in the legislation of Australia and some Canadian provinces (British Columbia, Alberta, and Manitoba) instead of "shall." Requirements interpretation acts have been modified to say that *must* is to be interpreted as imperative. The authors believe that the paradigm shift will be complete by the end of the next decade.[20]

[20] A summary of the U.S. government legal arguments for using *must* vs. *shall* can be found at: http://www.plainlanguage.gov/howto/wordsuggestions/shallmust.cfm.

6 Configuration Management

6.1 QUESTIONS ANSWERED IN THIS CHAPTER

- What tools and techniques can we use to identify configuration items (CIs)?
- Explain how configuration management (CM) impacts product performance.
- What is Das V-Modell and how does this German tool translate to contemporary views of CM?
- Explain and defend the impacts of poor or *ad hoc* configuration and change management on organizational exposure to risk.
- How does CM control drawing changes?
- What sort of software and hardware items should be put under CM control?
- Should an organization's processes fall under configuration and change management control?
- What does it mean when we say, "Trace requirements through to test cases"?
- What happens when we fail to identify requirements?

6.2 INTRODUCTION

We thought it would be fitting to start this chapter with the following quotation from Jack Wasson.[1]

> The lack of modern Enterprise CM processes in large organizations is rapidly becoming a major barrier to the deployment of reliable, secure, and correct products and systems. Those who know CM, know that it is everywhere. If you understand it, you see it daily, and even hourly. Just listen to the news or read the newspaper and observe how many issues are faced by organizations that lead to extreme costs, deficient performance and undelivered products. These are nothing more than a failure to apply the fundamental principles of Configuration Management. Many in business or the government are unaware of CM or may not see any reason for CM at all.
>
> Are they correct? Or are they limiting the practice of CM, ignoring the obvious, or even damaging it—or their careers and ours? We can either embrace it as control function, "a series of 'checks and balances' throughout the life cycle of a system or systems," or we can ignore it at a costly price (i.e., invalid data, uncontrollable or excessive costs, failed programs and inferior products).
>
> Configuration Management (CM) is a common yet usually misunderstood concept. It is the most important and often neglected business process of any organization. CM allows managers to better identify potential problems, manage changes, track progress and performance of any product during its lifecycle. It keeps them compliant with laws through recordkeeping, performance metrics, and recovery needs (knowing what you have, why it is, what changes have been applied and why). So, what is CM anyway? In layman's terms, it is the plan (or blueprint) for a product, a process or a document before, during its lifecycle and beyond. It seems obvious, yet many people have no idea what Configuration Management is, how it is performed, or even how to begin doing it. Some are overwhelmed by it, believing CM to be too difficult or too costly. Then, they tend to discount it or just assume someone else is performing it as part of some support function. Many organizations continue to apply conventional configuration management methods, then blame this core discipline for failures, or abandon CM altogether out of frustration. Practitioners of CM often face constraining resources in their organizations with widespread unawareness of CM and lack the expertise in organizational dynamics (politics) to sell or enforce their CM concepts—outside of their own group. In the past, budgets were often trimmed of CM-related costs even before the program or project is funded in a misguided attempt to save money or to even use it as a sort of financial cushion for contingencies. In today's environment, this practice is totally unacceptable. CM has a key role to play in the planning for, procurement and lifecycle of any system or product as another related function, Project Management.

[1] Wasson, J. (2008). *Configuration Management for the 21st Century.* Retrieved November 23, 2013, from https://www.cmpic.com, https://www.cmpic.com/whitepapers/whitepapercm21.pdf. Reprinted with permission of the author.

The standard definitions of CM have been changing and diverging over time and is often appropriated by one market segment and then further refined within that market segment. This results in narrowing the understanding of CM philosophy and definitions until the reasoning behind the need to manage configurations has been lost. CM is not market sector specific. CM implementation reaches across every market sector and is known by many different names. Section 4.4 referenced the November 26, 2013, Food Seminars International seminar, "Using Traceability and Controls to Reduce Risk and Recall: Integrating Food Safety Systems." The announcement for the seminar stated:

> This webinar will cover the concepts needed to enable the use of commonly collected data into a meaningful system designed to meet FDA, supply chain, and company requirements. A complete integrated food safety system design will be reviewed. An understanding of how to integrate data that food supply chain members already possess will enable players to protect their companies, the food they ship and the customers they serve.
>
> Risk reduction, supplier control, traceability and minimizing recall damage are all critical components of any food safety plan. Regardless of a company's position in the food supply chain, dependence on those companies one up and one down the ladder creates liability and exposure no one wants to think about.
>
> Areas covered in this webinar are as follows:
>
> - Legal Issues: Food Safety Modernization Act (FSMA), FSIS, Current Good Manufacturing Practices (CGMP)
> - Recall
> - Traceability Through Long Processes and Zones: (Farm, Blend Operations, Distribution, Picking Operations, Retail)
> - Electronic Real-Time Traceability Systems
> - Integrated Food Safety System Design and Functionality

Relating the subject of the Food Seminars International sponsored seminar to the definition of CM provided in the Australian, Technical Regulation Army Material Manual, we find that these are all issues related to the management of the configuration of an item as it progresses through the product life cycle.

> "CM is the most important discipline for the establishment and maintenance of technical integrity of material" AND "With the increasing sophistication, complexity and cost of modern systems, Defense is demanding the more effective use of Technical Data to improve operational capability and readiness at a lower life cycle cost." In the past fifty (50) years technology has evolved from resisters and switches to vacuum tubes to transistors to integrated circuits, etc. and the requirements for effective design and management have increased a thousand-fold demanding the need for more sophisticated management of the assets and documentation.

CM is often expected to help answer most continuing management questions:

- What is the requirement?
- How is it designed?
- Can we claim intellectual property rights?
- How can we prove we developed it?
- Why is it designed the way that it is?
- Are we building the right thing?
- Are we building it correctly?
- Does it interface with anything?
- Were there design changes?

- Why were the changes made?
- Where is it now?
- What state is it in?
- What is it doing?
- Is it deployed correctly?
- What does it have to do with my problems?
- What if we change it?
- Do we need more/less of it?
- Can we get rid of it?
- What is affected by changing or removing it?[2]
- Who is paying for it?
- Who made that decision?

6.3 OVERVIEW OF CM

CM is very much a product of the ability of *Homo sapiens* to produce more than one nearly identical copy of something. This capability to produce tangible and intangible goods has evolved over time and the concept of "nearly identical" has come to mean different degrees of precision at different points in history. A variance in size of the arrowheads associated with the terracotta warriors of less than 1 cm down to less than 0.22 mm in 210 BCE at one time was the accepted definition of identical parts. The same can be said of interchangeable Carthaginian ship keel components that can be traced back as far as the Punic Wars (264 to 146 BCE) through archaeological remains of boats in Museo Archeologico Baglio Anselmi, Marsala TP, Italy (Figure 6.1). They also met the interchangeability criteria for their time.

Production of exact copies of executable software from giants such as Microsoft Corporation with a variance dependent on only the physical makeup of the media on which it is housed is commonplace today. Yet even such precision as is available through digital video discs (DVDs) and Blu-ray discs is not without CM issues. Many titles on Blu-ray discs (*Iron Man*, *Saving Private Ryan*, *Star Trek: The Next Generation* boxed sets, Madonna's *MDNA World Tour*, and others) have been recalled due to configuration issues.

FIGURE 6.1 Carthaginian ship in Museo Archeologico Baglio.

[2] Australian, Technical Regulation Army Material Manual, Department of Defence, Canberra, ACT, Australia.

CM is so intertwined with safety, quality, and reliability that it is sometimes hard for the novice to grasp the extent of the interrelationships. CM is implemented across all industries and products—from the food we eat, to the things we buy, the entertainment we enjoy, and the systems we use. Keith Mobley contributing editor at *Plant Services*, states:

> American industry spends more than a trillion dollars annually maintaining its plants and facilities. At least 50%, €436,670,000,000, of this expense is caused by breakdowns, partial loss of function, frequent rebuilds and other reliability-related problems. Statistically speaking, at least 85% of reliability problems, asset use and escalating life cycle cost problems are directly attributable to deficiencies in, or total lack of, enforced configuration management.[3]

The University of Glasgow School of Computing Science—in association with the Advanced Space Operations School (Colorado Springs), 50th Space Wing AFSC (Schriever AFB, Colorado), ESR Technology Ltd. (Glasgow), and Air Force Safety Center (Kirtland AFB, New Mexico)—published an article on CM stating:

> Configuration management is a vital part of the safety of any space system, and a staple to those in the safety business. What is configuration management? It's simply knowing exactly what the piece of equipment is and what it does at any point in its life, cradle to grave. This includes knowing which software version is being used and how changes might affect the desired effect.[4]

6.3.1 History of CM

How CM is viewed in the U.S. government, commercial, and international communities is changing due to technological advances, global competition, and market segment. The essence of CM remains the same despite the methodologies, such as how it is done in one company, market subsegment, and market segment or what department is tasked with doing CM functions. Everyone performing a CM function must understand how it is evolving to maximize the benefits derived from it in the form of higher quality, lower cost, and ability to locate critical data quickly. Changes due to technological factors and their impact on business are major factors that should be considered. If the technology is not understood and as a result is not fully utilized to achieve the desired cost control, it will weigh heavily on those performing CM functions.

The specific names and the associated definitions in accordance with these documents may be so different as to be unrecognizable by many in a strictly U.S. government sales environment. Please do not be overly concerned about the acronyms used. The contextual construct of CM implementation is the important thing. The great reduction in government specifications and standards and the acceptance in industries not funded by government contracts require that those performing CM have a global perspective. Not all definitions have transferred to the commercial sector, and individuals performing CM in a PL/DM system are variously known as change managers, database administrators, document control specialists, software control specialists, product definition professionals, infrastructure coordinators, or, in some cases, data muckers. If you do happen to be called a data mucker, wear the name proudly. The profession has been around since our earliest ancestors could count higher than 20.

Words such as "data" should be interpreted to mean *information*, as "data" has specific connotations and "information" is more widely understood. Words such as "standard" and "standardization"

---

[3] Mobley, K. (2003), Major asset life cycles can be managed effectively with configuration management, *Plant Services*, https://www.plantservices.com/articles/2003/182/

[4] Fletcher, L., Kaiser, J., Johnson, C., Shea, C., and Cole, B. (2009) Configuration Management, http://www.dcs.gla.ac.uk/~johnson/papers/Wingman/Config_Management.pdf.

need to be put into context and tempered with an understanding of what standardization is and that standards are not static but driven by change. Casey C. Grant describes it in these words,[5]

Ancient World

One of the earliest examples of standardization is the creation of a calendar. Ancient civilizations relied upon the apparent motion of the sun, moon and stars through the sky to determine the appropriate time to plant and harvest crops, to celebrate holidays and to record important events.

Over 20,000 years ago, our Ice Age ancestors in Europe made the first rudimentary attempts to keep track of days by scratching lines in caves and gouging holes in sticks and bones. Later, as civilizations developed agriculture and began to farm their lands, they needed more precise ways to predict seasonal changes.

The Sumerians in the Tigris/Euphrates valley devised a calendar very similar to the one we use today. 5,000 years ago, the Sumerian farmer used a calendar that divided the year into 20 or 30-day months. Each day was divided into 12 hours and each hour into 30 minutes. An extra month was added periodically to keep the 354 day lunar year in step with the solar year of 365.242 days.

The Egyptians were the first to develop the 365-day calendar and can be credited with logging 4236 BCE as the first year in recorded history. They based the year's measurement on the rising of the "Dog Star" or Sirius every 365 days. This was an important event as it coincided with the annual inundation of the Nile, a yearly occurrence that enriched the soil used to plant the kingdom's crops.

Industrial Revolution

With the advent of the Industrial Revolution in the 19th century, the increased demand to transport goods from place to place led to advanced modes of transportation. The invention of the Railroad was a fast, economical and effective means of sending products cross-country. This feat was made possible by the standardization of the railroad gauge, which established the uniform distance between two rails on a track. Imagine the chaos and wasted time if a train starting out in New York had to be unloaded in St. Louis because the railroad tracks did not line up with the train's wheels. Early train travel in America was hampered by this phenomenon.

During the Civil War the U.S. government recognized the military and economic advantages to having a standardized track gauge. The government worked with the railroads to promote use of the most common railroad gauge in the U.S. at the time which measured 1.435 m, a track size that originated in England. This gauge was mandated for use in the Transcontinental Railroad in 1864 and by 1886 had become the U.S. standard.

20th Century

Cities experienced tremendous growth in the 20th century, bringing increased prosperity to America and attracting more and more people to urban centers. As cities became more sophisticated and their infrastructures more complex, it became apparent that a unique set of national standards would be necessary to ensure the safety of city dwellers.

In 1904, a fire broke out in the basement of the John E. Hurst & Company Building in Baltimore. After taking hold of the entire structure, it leaped from building to building until it engulfed an 80-block area of the city. To help combat the flames, reinforcements from New York, Philadelphia and Washington, DC, immediately responded—but to no avail. Their fire hoses could not connect to the fire hydrants in Baltimore because they did not fit the hydrants in Baltimore. Forced to watch helplessly as the flames spread, the fire destroyed approximately 2,500 buildings and burned for more than 30 hours.

[5] Grant, C. C. (2007). *A Look From Yesterday to Tomorrow on the Building of Our Safety Infrastructure, Official Contribution of the National Institute of Standards and Technology; Not Subject to Copyright in the United States.* Quotation is from the *NIST Centennial Standards Symposium*, NIST, Boulder, CO. March 7, 2001, from https://www.govinfo.gov/content/pkg/GOVPUB-C13-992cf96b57768b2553927fc04850fd99/pdf/GOVPUB-C13-992cf96b57768b2553927fc04850fd99.pdf.

It was evident that a new national standard had to be developed to prevent a similar occurrence in the future. Up until that time, each municipality had its own unique set of standards for firefighting equipment. As a result, research was conducted of over 600 fire hose couplings from around the country and one year later a national standard was created to ensure uniform fire safety equipment and the safety of Americans nationwide.

Standardization is an ongoing CM activity. The October/December 2011 *Defense Standardization Program Journal*[6] reported,

> Standardization also facilitates radical changes. For example, Turkey, which entered NATO together with Greece in 1953, changed its 630-year-old military map symbols and the colors used to denote friendly and opposing forces through the implementation of NATO Standardization Agreements (STANAGs). Indeed, many nations use only NATO operational STANAGs and no longer produce their own. Examples of the contribution of standardization to military operations are innumerable. In fact, without standardization, multinational interoperability could not be achieved, and NATO operations would not be possible.

The publication also contained articles entitled "First-Ever Transition of a NATO Standard to a Civilian Standards Development Organization," "NATO Revises CM Guidelines," and "NATO Adopts ISO/IEC 15288."

The History

- Army, Navy, Air Force (ANA) Bulletin 29 (1953)
  - Established the engineering change proposal (ECP)
- ANA Bulletin 391 (1956)
  - Defined ECP priorities
  - Expanded application of configuration beyond the airplane business sector.
- MIL-D-70327 (1959) engineering drawing requirements (forerunner of MIL-D-1000 and to some extent MIL-STD-100)
  - Established two drawing classes
  - Referenced and pulled together some 30 related military specifications
  - Established the rules for part numbering, serialization, and interchangeability
- Air Force Systems Command Manual (AFSCM) 375-1, configuration management during the definition and acquisition phases
  - Seen as the first CM "Bible."
  - Pulled together the now "classical" elements of configuration management (identification, control, status accounting, and audits)
  - Established a government CM office
  - Established a "uniform specification program"
  - Instituted "first article configuration inspection," the predecessor of the physical configuration audit (PCA)
  - Introduced the idea of a "compatibility" ECP
- AFSCM 375-7, configuration management for systems, equipment, munitions, and computer programs
- NPC-500-1 (NASA) *Apollo* configuration management manual—almost identical to AFSCM 375-1
- NSA/CSS 80-14 (NSA) configuration management[7]

---

[6] DOD, U.S. (2013). *The Defense Standardization Program Journal*. Retrieved December 26, 2013, from http://www.dsp.dla.mil/, http://www.dsp.dla.mil/APP_UIL/displayPage.aspx?action=content&accounttype=displayHTML&contentid=75.

[7] Items marked with * are contextually the same as they comprise the joint regulation on configuration management of the U.S. Department of Defense (DOD).

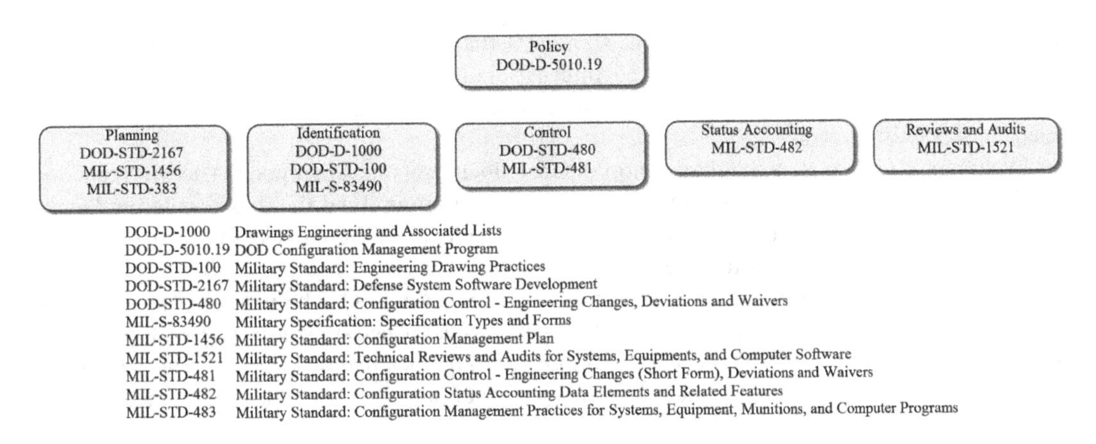

FIGURE 6.2 U.S. DOD requirements 1985.

- AMCR-11-26 (USA) configuration management*
- AR 70-39 (U.S. Army) configuration management*
- AFR 65-3 (U.S. Air Force) configuration management*
- MCO 4130.1 (U.S. Marine Corp.) configuration management*
- AMC Regulation AMCR 11-26 (Army) Army's equivalent of AFSCM 375-1
- NAVMATINST 4130.1 (U.S. Navy) Configuration Management—A Policy and Guidance Manual

Starting in the 1960s, the U.S. government attempted to amalgamate and standardize its CM requirements with the result that, by the early 1980s, the requirements landscape looked much different (Figure 6.2).

This resulted in the areas of responsibilities shown in Figure 6.3 for CM implementation and can, for our purposes, be considered as the CM fundamentals from which all aspects of CM seen across product lines in every market sector evolved.

Please do not be confused by the previous statement, as it is purposely a generality. Certainly, CM existed before the U.S. military. Many areas, such as manufacturers of medicines and

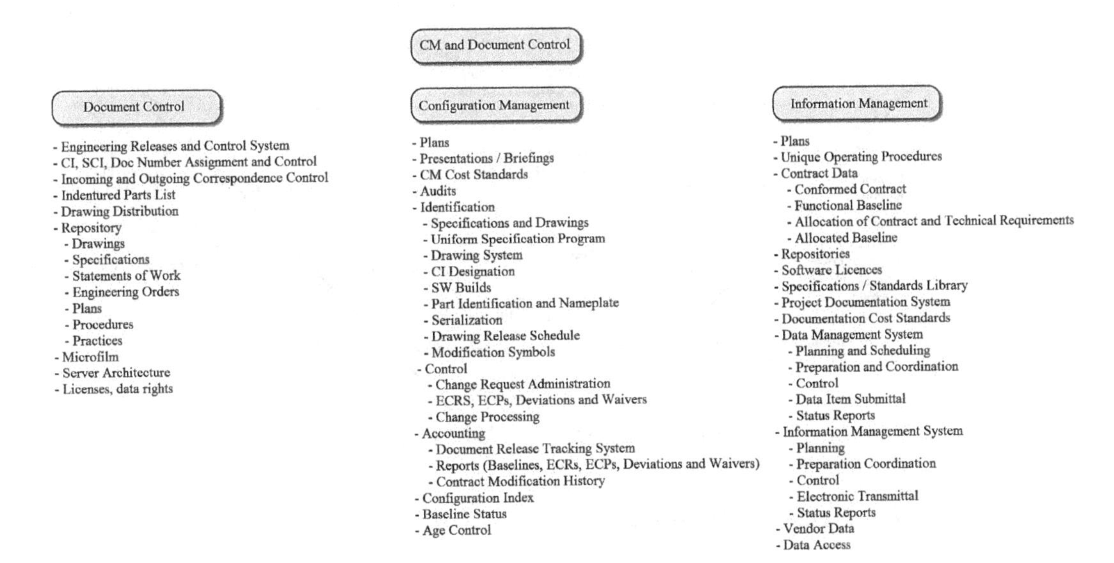

FIGURE 6.3 CM implementation as seen by U.S. DOD.

medical equipment, have such stringent standards that those CM activities imposed on contractors by the U.S. government and the International Organization for Standards (ISO) can be viewed as the minimum necessary to ensure production of duplicate copies of products that meet acceptability norms.

Evolution of CM-related standards and guidance documents never stopped. The divergence in product and product sector unique requirements is a continuing trend that has resulted in heated disagreement over the meaning of basic terminology simply due to refinement of practices specific to a single industry-specific instance of CM.

The best source for current CM thinking is found in EIA 649-C. The work by the SAE G-33 standards committee on Revision C of SAE/EIA 649 was published in February 2019. It brings the requirements up to current practice across all industries, defines five internationally accepted CM functions, and treats the 40 CM principles as statements of fact. The principles are defined in SAE/EIA 649-C Table A1.

SAE/EIA 649-1 (*CM Standard for Military*) and SAE/EIA-649-2 (*CM Standard for Non-Military Aerospace*) will be made mandatory once they are updated to reflect SAE/EIA-649C. Neither SAE/EIA 649-1 nor SAE/EIA 649-2 can be made contractual without tailoring to the specific program.

Hardware-centric configuration managers argue with software configuration managers over concepts such as branches and twigs being a software product structure. Tool-specific requirements also generate their own unique-to-tool CM language, which will in turn fall by the wayside as new tools appear, leaving only a basic understanding of how CM activities evolved. This Tower of Babel effect will be discussed further in Section 10.4.

A list of current CM-related standards is provided in the appendix.

6.3.2 Das V-Modell

The German V-Model "Das V-Modell" (Figure 6.4) is the official project management methodology of the German government. It is directly intertwined with CM activities. There is a lot of confusion when discussing V-Models, as the concept has been found to be applicable to multiple disciplines and subdisciplines, most notably in software development. It is used heavily by systems engineering in complex systems.

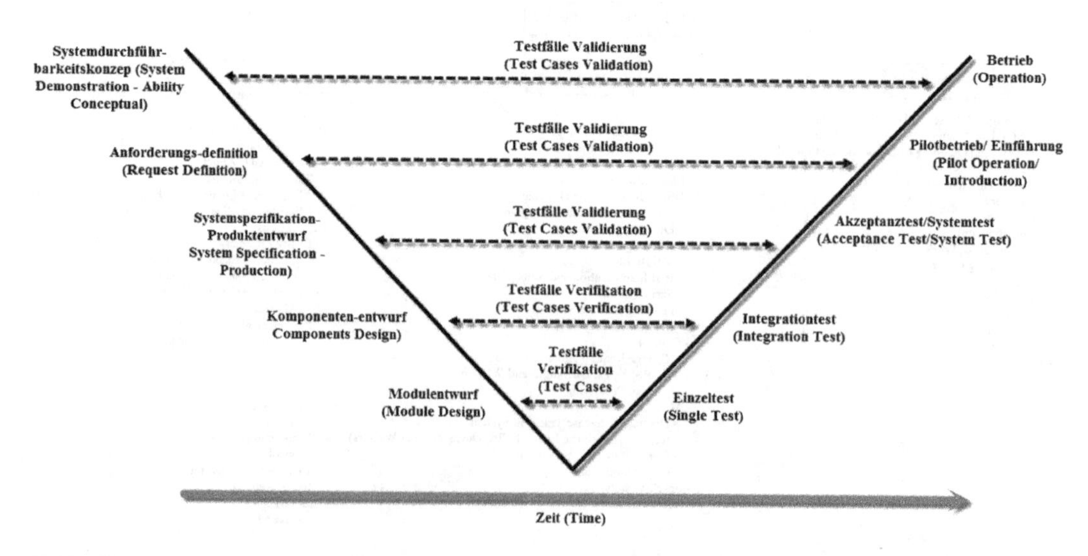

FIGURE 6.4 Das V-Modell.

As with the history of many things, the V-Model concept evolved independently in Germany and the United States in the 1980s. The German version was developed by Industrieanlagen-Betriebsgesellschaft mbH (IABG) in Ottobrunn, near Munich, in cooperation with the Federal Office for Defense Technology and Procurement in Koblenz, for the Federal Ministry of Defense. The U.S. version was developed for satellite systems involving hardware, software, and human interaction. The first U.S. V-Model appeared in a proposal from Hughes Aircraft in 1982 for the U.S. Federal Aviation Administration's advanced automation system program. From the V-Model came our current conceptions of allocation, validation, and verification.

In instances where companies are moving toward model-based engineering (MBE), the V-Model is mirrored and offset along the timeline to assure that as the model is created it reflects the functionality of the item being modelled. It will be interesting to watch over the next few years the convergence and intersection of MBE with CM. As viewed now, there is an almost fractal aspect to model development, with a combination of recurring waterfall method data collection requirements verification and validation combined with the development of the hardware item(s) being modeled. At the forefront of the MBE initiative is the proposed standard for sharing modeling and simulation information in a collaborative systems engineering context (MoSSEC) built on existing ISO standards (see *ISO 10303 (STEP) MoSSEC edition 1 dated 30.Sept.2016*). MoSSEC is based on establishing contextual data between the customer, the contractor, and the supply chain through:

1. Distributed infrastructure with secured communications for locations, organizations, software platforms.
2. Distributed processes supporting a multitude of modeling and simulation tools and simulation driven changes under PD/LM control.
3. Distributive data that includes modeling and simulation data, V-Model cycle metadata, with efficient sharing, synchronization, and integration.

Conceptually, MBE CM is much like the formidable and persistent issues being faced with intelligent transportation systems described by Sumit Ghosh and Tony S. Lee.[8] A similar holistic approach will be required for CM in an MBE environment to avoid becoming hopelessly entangled in the lower-level complexities of implementation. As with intelligent transport, the ideal system should be conceptualized and modeled first, and implementation of its attributes be added as the lower-level interfaces develop. The correlation between MBE requirements, allocation, and verification (traffic patterns) and the implementation of CM core principles (traffic loads) need to be carefully orchestrated. Jay Galbraith[9] and Robert Kazanjian speak to this. From their writings we have come up with five principles for CM in an MBE environment.

1. It is necessary to specify the required behaviors in advance of model execution, or the volume of information from points of action will overload the model hierarchy. If this is not done, the information context is lost.
2. The model needs to move from functional design to distributed design so the functionality within the model includes all resources (inputs, influencers, and metrics) to perform a task.
3. When the number of exceptions or unanticipated events becomes substantial, the model must be refined. Since the model includes multiple processes formally subject to continuous process improvement, metrics on the model behavior now need to be part of the model.

[8] Ghosh, S., and Lee, T. S. (2010). *Intelligent Transportation Systems*. CRC Press/Taylor & Francis Group, Boca Raton, FL.

[9] Galbraith, J. R., and Kazanjian, R. K. (1986). *Strategy Implementation Structure, Systems and Process*. West Publishing Company, 50 West Kellogg Boulevard, P.O. Box 64526, St. Paul, MN.

4. The model must selectively employ lateral decision processes that cut across lines of authority, moving decisions downward to where the level of information exists. This decentralizes decisions without creating self-contained silos of information.
5. The model needs to constantly evolve to encompass changes to the tool suite and other environmental constraints without compromising its core integrity. This goes back to the first principle. CM in an MBE environment consists of two models. The first is the specified required behaviors (end state) and the second is the actual enterprise-wide implementation of how that end state is met. The end state must remain constant while the implementation evolves over time.

What exactly is deliverable in the CM MBE environment as part of the technical data package is currently unknown. This is partially because major reviews in the MBE environment will take place using the model itself, and partially because access to the piece of the model needed to perform work is group limited. Only the piece of the model needed by groups such as manufacturing and test will be accessible based on job function. This adds another level of verification and validation to ensure that the output from the model reflects the model.

When MBE is combined with additive manufacturing, where the model representation in STEP or parasolid is optimized to use less material while maintaining strength, interfaces, and other criteria, is that optimization considered a design change? If the answer is yes, we need to consider what that means if design optimization results in a part with a completely different topology over time. Are they form, fit, and function interchangeable? The authors tend to believe that they are and that parts manufactured using additive or subtractive manufacturing methods should be considered to have the same form if the quality, fit, interface, and function remain constant. See Figure 6.5 for an example of product evolution.

Three looming questions have been identified relative to CM in a model centric environment:

1. How do you define the line between CM and ERP in the model?
   a. This is complicated because with development and operations (DevOps) a single line of code can deploy changes to the model across all impacted environments.
2. How do you control changes in a machine learning environment?
   a. Machine learning will define, analyze and allocate requirements.

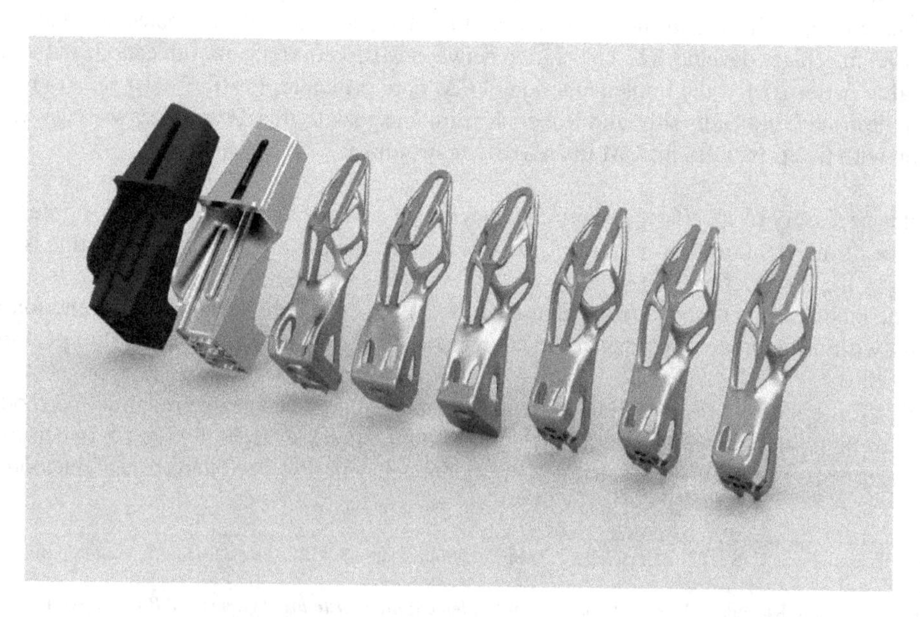

FIGURE 6.5 2018 BMW i8 Roadster 3D printed bracket evolution (courtesy of BMW).

3. How is CM of additive manufactured items done?
   a. If machine learning converts 300 parts to a single 3D printed unit, it has a great influence on the scope of requirements and how the requirements are verified.

This is not a blue-sky picture of the future. In 2018, Amazon had 18,000,000 updates released. That is over 49,000 changes a day. This DevOps approach is what is being pushed by systems engineering organizations and CM needs to find a way to accommodate it. One method gaining traction is to release the model and drawing together until the level of comfort is established, and then to transition to only model releases. Design reviews in real-time 3D is going on in many companies today, with CM as the backbone of the trust engine.

6.3.3 What Does CM Apply To?

In the narrow sense, as defined by the regulations cited in Section 6.3.1, CM applies to software, firmware, hardware, and drawings among other things. In a much broader context, CM applies to everything in some regard or another. This includes any item whose provenance is necessary to know and, in a different context, how something is hooked together with other items. Under this broader interpretation and somewhat outside the context of the current suite of standards, this includes art; music; literature; companies; organizations; people; architecture; community, town, city, state, and national infrastructure; scientific research; and all tangible and intangible goods. Understanding the configuration of anything requires the application of CM to establish that elusive thing known as "provenance." Many practitioners of CM will disagree with this statement, yet it is an irrefutable fact that unless a configuration is known our species is at a disadvantage. To put this in context, we need to go back and explore how we arrived at our current needs to understand the world around us and how we arrived at where we are today.

Our ancestors understood where edible fruits and vegetables grew, the movements of the animals they used as a food source, and the vegetation those animals consumed throughout the seasons. They knew the configurations of the animals. They understood how to harvest each type of creature in the best season and how each portion of the animal could best be used. The fact that a mammal's brain matter—when mashed in warm water along with its liver and the urine from its bladder—can be used to cure its skin to make leather for clothing was a process passed on from generation to generation.[10] They knew that deer harvested in the fall were the fattest and had the best skin for making buckskin leather. They knew the configuration of materials such as the types of wood and rocks for clubs, knapped tools for chopping, hand axes, knives, and spear heads. They knew how to utilize every bit of bone, meat, skin, feathers, plant fibers, fruits, and seeds to the best advantage. They understood that plants with four-petal flowers produced seeds and fruits generally harmful to our species and those with five-petal flowers were generally edible. Hunter-gathers knew these things and they carried only essential items with them as they could make anything else they required. The loss of an arrow or spear did not mean the loss of the hunt or endanger the family group's survival. More of anything could be easily made. This at-place-of-need manufacture (the 3D printing of the day) did not negate that they knew and practiced the management of configurations. In many cases, our predecessors outlived species they were contemporary with, a testament not only to their adaptability but also to their understanding of the changing configuration of the environmental niche they inhabited.

Boucher de Perthes (1788–1868) and his contemporaries had shown, as early as 1850–1860, that humans had lived at the same time as animal species that had disappeared at a very early period. The find made by Larte and Christy in the shelter of Madeleine (in the Dordogne) of a drawing of a mammoth on a fragment of mammoth tusk may be said to have given a double proof that humans were contemporary with the mammoth and documented the relationship.[11]

---

[10] Recipe from Jaeger, E. (1999). *Wildwood Wisdom*. Shelter Publications, Inc, Bolinas, CA.
[11] Bordes, F. (1968). *The Old Stone Age*, Translated from the French by J. E. Anderson. Weidenfeld & Nicolson, London.

Our nomadic ancestors eventually formed groups, tribes, seasonal communities, towns, cities, states, and nations. As their societies grew, the complexities of survival necessitated skill diversification. Community living led to the specialization and refinement of manufacture and commerce. Individual knowledge became group knowledge and group knowledge became guild knowledge. This resulted in a dispersed and shared capability for gathering the things needed to survive. People could make arrows and trade them for shoes made by someone else, trade livestock for grain and grain for metal goods, and pay a portion to a group to hire and pay someone to protect themselves and their fields from the folks over the hill. Suddenly the configuration and content of your flock, your fields, your household, your larder, and your purse became more important than where the nearest source of flint was or when the edible berries were going to be ripe. That was the worry of the green grocer of the day, as you were providing wool to the weavers' guild. You could pay the grocer for your fruit; but he had better "do his bit."

The rudiments of trade developed, including intellectual property, legal concepts, management of data in the form of yield, exchange rates (e.g., 1 lamb = 1 pair of shoes, 3 fish = 1 measure of grain), and ways to keep track of it (bullae, knots on a string, cuneiform, ruins, etc.). The local hard men you paid to protect you got organized and demanded standardization of the weapons they used. Someone else wanted a set of crockery identical to that used by the head man. Everyone wanted better control over the exchange rate. Every measure of grain needed to be equal in weight. Exchange rates changed: one weight of fish now equaled one weight of grain. Suddenly quality mattered. Needing to know the configurations of things around them became necessary.

Every guild, individual, and community had a way to do things. The standardization of the exchange rate for gems, where one carob seed equaled a carob weight stone (now called a "carat weight"), became more refined as carob seeds varied too much in size. Other standardizations took place as well. A town's official length measurement was often defined by metal gauges imbedded in stone near the town hall or on other public buildings. Dealers and craftsmen calibrated their own measuring tools to comply with the official town length when selling goods (Figure 6.6).

FIGURE 6.6 Official measurements: Altes Rathaus Regensburg, Bavaria, Germany.

As you travel through Great Britain and Europe you will see many of these. There is one showing the official measure for Prussian Ell and Prussian Feet at the town hall of Bad Langensalza, Thuringia, Germany. There are also the Trafalgar Square Standards installed in 1876 showing the official lengths of a poll (also known as a perch), Imperial foot, Imperial yard, one link and 40 links at 62 degrees Fahrenheit.

Accounting methods, units of measure, accounting practices, quality (purity, fabrication, etc.), performance, and all things measurable went through the process. This includes the evolution of the periodic table of the elements, which celebrated its 150th anniversary in March 2019. Russian chemist Dmitri Mendeleev created the first periodic table of the elements based on familial relationships among the known chemical elements that enabled prediction of elements that had not yet been discovered. The periodic table revolutionized chemistry and is an example of DM data analytics at its finest.

Management of configurations, processes, weights and measures, time, ocean currents, and information about those configurations evolved in every market segment in every country. CM, data management, and status accounting exist universally. Knowing that you have an extra five-pound note hidden in the boot of your car or one more Royal Coachman fly in your fly box is as much DM as the management of a multinational project such as the International Space Station.

Generically, knowing the configuration of something is having enough information about it that you know where and how it came to be as it is, as well as where it is and how it differs over time until that information is no longer relevant.

CM, as practiced in accordance with the standards, specifications, handbooks, and other source material shown in the appendix, is a very limited subset of what CM implementation entails. This documentation often pertains to specific industries, customers, and products and was written to mitigate dramatic failures. Yet, in the broader context of CM, issues involving our food supply[12] or clinical data trials are just as important. Organizations once considered to be outside of CM—such as the Association for Clinical Data Management (ACDM)[13]—are finding their activities encompassed by mainstream CM principles.

What follows are a series of definitions relating to software, firmware, and hardware from U.S. MIL-STD-483A, *Configuration Management Practices for Systems, Equipment, Munitions, and Computer Software*. This document was chosen as it represents what many firms selling to the U.S. government feel is the quintessential source for what they consider CM.

> MIL-STD-483A: 1.3 Application. Configuration management requirements established by this standard apply during the applicable system life cycle phases of configuration items whether part of a system or an independent configuration item. Contracts invoking this standard will specifically identify the appropriate applicable paragraphs and appendixes or portions thereof as defined in the contract work statement depending upon the scope of the program, other contractual provisions, and the complexity of the configuration item being procured. The contractor shall ensure that all software, hardware, firmware and documentation procured from subcontractors is generated according to the requirements of this standard.

6.3.3.1 Software Items

MIL-STD-483A: 5.1(e) Computer Software logically distinct part of a Computer Software Configuration Item (CSCI). Computer Software Components may be top-level, or lower-level.

MIL-STD-483A: 3.4.7 Computer Software Configuration Identification. CSCI specifications, design documents, and listings shall define software requirements and design details for a single CSCI. The subparagraphs below identify the specifications and design documents of the CSCI.

---

12 Warriner, K. (2013, December 19). *Food Seminars International, Food Safety Culture: When Food Safety Systems Are Not Enough*. Retrieved December 19, 2013, from http://foodseminars.net/, http://foodseminars.net/product.sc;jsessionid=9A15A4DB9773831A41A5B6D46E006C1F.m1plqscsfapp03?productId=163&categoryId=1.

13 Association for Clinical Data Management (ACDM) established 1988, https://www.acdmglobal.org/.

6.3.3.2 Firmware Items

MIL-STD-483A: 5.1(n) Firmware. The combination of a hardware device and computer instructions or computer data that resides as read-only software on the hardware device. The software cannot be readily modified under program control. The definition also applies to read-only digital data that may be used by electronic devices other than digital computers.

6.3.3.3 Hardware Items

MIL-STD-483A: 5.1(f) Configuration item. Hardware or software, or an aggregation of both, which is designated by the contracting agency for configuration management.
MIL-STD-483A: 5.1(q) Hardware Configuration Item (HWCI). See configuration item.

6.3.3.4 Drawings

We looked at drawings and drawing requirements briefly in Chapter 1. Drawings are included in this chapter as they relate to design engineers' daily life. Yet, as with many terms used in CM, "drawing" may be understood to mean many different things when viewed from the international and market segment perspective. In the broadest sense, a drawing may be defined to mean: "Any documented record from which an end item may be produced. Including but not limited to the formulation of chemical compounds (solids, liquids, gasses); hardware, firmware; software; patents; and licensable processes."

The U.S. American Society of Mechanical Engineers (ASME) Y14 series of documents identify specifics for "drawing" formats and types. It must be remembered that this series if written from a mechanical engineering perspective only. While it serves a CM implementation well in most cases, it is not as far reaching as will be required when the broader definition of a configuration item (CI) is introduced. It raises a good question.

If a drawing creates item identification and no drawing is produced from the drawing model, is the model the drawing, or is a drawing only something that can be shown in an old-style drafting table representation (sometimes called *paper-space*)? This is an interesting question. The model can be printed in 3D or sent in the form of a STEP[14] file to be produced on computer numerical control (CNC) machines or, after conversion, to an STL, OBJ, X3G, Collada or VRML97/2 file for common 3D printers. If the model isn't the drawing, then what is it? If the model is the drawing, is our concept of item identification outmoded?

Strictly speaking, the answer is a resounding no. The concept that a drawing is the only form of documentation creating item identification is evolving. Therefore, we introduced you to the use of 3di pdfs as the drawing deliverable in Section 1.5.2.3.2. That concept is only the first hurdle in a paradigm shift as model-based engineering evolves.

As always, the need to weigh the technical correctness of the documentation against the usefulness of the data contained in the documents is paramount. Document formats can no longer be bound to requirements necessary to operate in a paper centric CM implementation. If fields on the document other than the document identifier and revision are contained in the PD/LM metadata, do they provide useful information? If so, then we must ask the question, "To whom is it useful?"

Perhaps the clearest conceptualization of what constitutes a document is provided in SAE/EIA 649-C, Table 2.

- Records
- Specifications

[14] ISO. (2002). *ISO 10303-21:2002, Industrial Automation Systems and Integration—Product Data Representation and Exchange—Part 21: Implementation Methods: Clear Text Encoding of the Exchange Structure.* ISO, Geneva, Switzerland.

- Engineering drawings
- Drawing lists
- Pamphlets
- Reports
- Standards

For a meaningful definition of drawings, we have started with MIL-STD-100G, Engineering Drawing Practices, as our source material for the ASME standards. The ASME standards establish many drawing types and classifications:

- ASME Y14.100, Engineering Drawing Practices
- ASME Y14.24, Types and Applications of Engineering Drawings
- ASME Y14.34, Associated Lists
- ASME Y14.35, Revision of Engineering Drawings and Associated Documents

ASME organizes drawings into broad classifications, some of which are said to "create item identification." This is another way of saying there is enough information on them to make something. The term "does not establish item identification" simply means that you have insight or information, but it is not enough to make something.

Drawing types that establish item identification per ASME Y14.24:

- Detail drawing (mono detail and multi-detail)
- Assembly drawing (assembly and inseparable assembly)
- Hardware installation drawing (only if creating a work package or kit)
- Altered item drawing
- Selected item drawing
- Modification drawing (when required for control purposes)
- Wiring harness drawing
- Cable assembly drawing
- Printed board and discrete wiring board drawing sets (printed wiring assembly and printed wiring board [PWB])
- Microcircuit drawing set
- Kit drawing
- Tube bend drawing
- Matched set drawing

Drawing types that do not establish item identification per ASME Y14.24/MIL-STD-100F:

- Layout drawing
- Arrangement drawing
- Interface drawing
- Identification cross-reference drawing
- Mechanical schematic
- Electrical diagrams
- Functional block diagram
- Single-line schematic diagram
- Schematic diagram
- Connection diagram
- Interconnection diagram
- Logic circuit diagram
- Envelope drawing

- Undimensioned drawing
- Contour drawing
- Software installation drawing
- Alternate parts drawing
- Drawing tree

Design information drawings (do not establish item identification) per ASME Y14.24:

- Layout drawing
- Contour drawing
- Installation drawing
- Interface drawing
- Mechanical schematic diagram
- Single line diagram
- Schematic diagram or circuit diagram
- Connection diagram
- Interconnection diagram
- Wiring list
- Logic circuit diagram
- Drawing tree

Control drawings (establish item identification) per ASME Y14.24:

- Procurement control drawing
- Vendor item drawing
- Source control drawing
- Envelope drawing
- Identification cross-reference drawing
- Microcircuit drawing set

Lists (do not establish item identification) per ASME Y14.34:

- Artwork list (AL)
- Data list (DL)
- Index list (IL)
- Parts list (PL)
- Wire list (WL)

List prefix identifiers were originated to keep all items associated with a root drawing number (e.g., 1234567) together in a filing cabinet:

- 1234566
- Artwork list AL1234567
- Data list DL1234567
- Index list IL1234567
- Parts list PL1234567
- Wire list WL1234567
- 1234568

Prefix usage is gradually being phased out as companies move to PD/LM systems because of sorting issues (e.g., due to the prefix, documents sort by prefix vs. root number).

FIGURE 6.7 *Study of a Tuscan Landscape.*

CM as defined by these standards is not necessarily concerned with all documentation associated with the management of the configuration. It is concerned primarily with the documentation that establishes a CI's provenance. The concern is that CM implementation includes management of much more than a CI. The CI designation simply denotes that the government is involved in the design effort. The provenance of all the non-CIs is just as important.

A drawing such as *Study of a Tuscan Landscape* (Figure 6.7)[15] is an expression of communication in visual form. It contains details that would enable a knowledgeable person to construct town defenses depicted based on its content. It is conceivable that, as the cross-market segments and international aspects of CM are understood, the term "information" will suffice regardless of format with the single criterion that from it a product can be made. Word pictures painted by a story teller left in the minds of the listeners all information necessary for knowledgeable craftsmen to fabricate something. The premise is borne out in the explorations of Thor Heyerdahl's *Kon Tiki*[16] balsa raft and *Ra*[17] papyrus reed boat and by Tim Severin, who painstakingly researched and built a boat identical to the leather curragh[18] believed to have carried Brendan on his epic voyage to the North American coast. The meticulous trial-and-error watercraft construction techniques leading to these voyages by both Heyerdahl and Severin would have been common knowledge at the time the stories were first told. They could have been built by any *competent person*.

Undoubtedly, the configurations of items whose pictures were drawn in oral traditions were managed. They were reproducible. They were as nearly identical as necessary for their intended use. Oral traditions painting similar pictures exist today in sagas such as the Finnish epic known as the *Kalevala*.[19] The British folksong *John Barleycorn* is a similar form of oral tradition pertaining to the process for making the beverages such as whiskey and beer with broadside (also known as a

---

[15] da Vinci, L. *Verrocchio's Workshop, 1466–1476 (CE). (Study of a Tuscan Landscape* was drawn when Leonardo da Vinci worked in Andrea del Verrocchio's workshop for 10 years, thought to be during 1466 to 1476.)

[16] Heyerdahl, T. (1950). *The Kon-Tiki Expedition: By Raft across the South Seas*. Rand McNally & Comp, Skokie, IL.

[17] Heyerdahl, T. (1971). *The Ra Expeditions*. Doubleday, New York, NY.

[18] Severin, T. (2005). *The Brendan Voyage*. Gill & Macmillan Ltd, Dublin, Ireland.

[19] Lönnrot, E. (2008). *The Kalevala: An Epic Poem after Oral Tradition*. Oxford World's Classics, New York, NY.

"broadsheet") copies dating back to the 16th century. Scottish songs with a similar theme exist, such as the *Quhy Sowld Nocht Allane Honorit Be* from as early as 1586.[20]

A modern incarnation of the idea that a sound, practical understanding of the craft is all that is required for a competent individual to produce something is seen in the relatively recent change from detailed written directions and installation drawings to simpler arrangement drawings. Arrangement drawings are now used when connecting modern electronic devices using wireless technologies. Michael Jakubowski, the founder of Open Source Ecology, has leveraged this idea of the *competent person* forward in the Global Village Construction Set (GVCS).[21] GVCS is a modular, do-it-yourself (DIY), low-cost, high-performance platform that gives a competent person enough information for easy fabrication of the 50 different industrial machines required to build a small, sustainable civilization with modern comforts. It is also present in the instructions provided by companies such as Lego©, Meccano©, and Erector Set©.

The idea of what constitutes a drawing is morphing from that of a 2D representation to a 3D representation and eventually to the model itself. The model itself is considered a data set that removes all experience-based interpretation issues both within an organization and in cross-cultural multinational collaborations. Many we talked to are struggling with this due to the ubiquity of the 2D representation in our lives. ASME Y14.47 will address both model-based definitions and the model-based enterprise when it is released. One major hurdle associated with using the model as the data set is that of design representation sustainability. Archives of drawings on vellum lasted hundreds of years but digital data archives do not. To proactively manage obsolescence, the data formats need to be defined before any contract is put out for bid. It is highly recommended that the digital requirements for each model type be incorporated into contracts as a mandatory flow down throughout the supply chain. The definitions need to address the maturity state, geometry state and annotation, and attribution states of the model. The organizational framework should be incorporated into the identification section of the configuration management plan. It consists of naming conventions, associated groups, presentation states, product definition elements, and metadata.

6.3.3.5 Specifications

The word "specification" is of Roman origin and comes from the word "species." Under Roman law, the term "specification" refers to the acquisition of a new (nova) species arising from a change to an existing species. The concept of species identification/classification underlies much of engineering development, specification management, and management of configurations. The task of species identification is not static. The Entomological Society of America recently updated its master list of insect names to reflect decades of genetic and other evidence that termites belong in the cockroach order, *Blattodea*. On February 15, 2018, Mike Merchant (chair of the Common Names Committee at Texas A&M University in College Station), announced, "It's official that termites no longer have their own order."[22]

Ownership of a non-natural species (generally, as we discuss it here, a manufactured article) was defined by Procukians under Roman law, which held that the manufactured article should be understood to belong to the creator of the article. In a conflicting view, Sabinus/Cassus held that the manufactured article should belong to the owner of the material substance.[23]

In many ways, both arguments are with us today and define how new products are viewed, not only from the standpoint of laws governing sales to the government and in the commercial marketplace. These concepts also flow back into the degree of risk a contractor assumes when involved with

[20] Bannatyne, G. (1896). *The Bannatyne Manuscript. Retrieved on May 20, 2014, from* https://openlibrary.org/books/OL7034966M/The_Bannatyne_manuscript.

[21] Open Source Ecology. (2014). *Open Source Blueprints for Civilization Build Yourself.* Retrieved July 25, 2014, from http://opensourceecology.org/#sthash.BtrJ7jWL.dpuf.

[22] Miles, S. (2018). It's official: Termites are just cockroaches with a fancy social life, *Science News*, Vol. 193, No. 6, March 31, 2018, p. 7, accessed at https://www.sciencenews.org/article/itsofficial-termites-are-just-cockroaches?utm_source=editorspicks030418&utm_medium=email&utm_campaign=Editors_Picks.

[23] Zulueta, F. D. (1946). *The Institutes of Gaius*. Clarendon Press, Gloucestershire.

government contracts. Under government contract provisions, goods and services may be obtained under commercial terms and conditions (such as FAR Part 12), where the contractor assumes all risk and the product and all residual materials belong to the contractor (just as they would in the commercial marketplace until point of sale). Goods and services may also be obtained under government terms and conditions (reference FAR Part 15, "Contracting by Negotiation"), where everything produced under the contract or ordered under the contract resides with the government. In 2013, this was increasingly true in the United States even with firm-fixed price contracts under FAR Part 15.

The Roman idea of species and subspecies is well known in the natural sciences as the *scientific name*. It involves a two-part taxonomy called *binomial nomenclature* allowing scientists, regardless of nationality, to specifically identify every species past or present. A similar taxonomy is used in language defining the relationships between words. This concept finds its way into CM methodologies such as those defined in the ANSI/EIA 836-B-2010, *Configuration Management Data Exchange and Interoperability*, a canonical model of data transfer using a data dictionary.

Additionally, the idea of specifying something that exists has found its way into a methodology for identifying something that does not exist. It is in this context that most CM implementation recognizes it. Specifications *defining a valid set of requirements* used to form the functional baseline, have long recorded history that includes the Egyptians, Greeks, Aztecs, Romans, Norsemen, Huns, Saxons, Germans, French, Russians, and Chinese. A fair summation of the history of specifications is that every advanced civilization has evidence of their existence. The first requirement for specification preparation in the United States is MIL-S-6644, *Instruction for Preparation of Specifications*, April 13, 1953. These U.S. specification requirements continue to evolve. The current requirements are defined in MIL-STD-961, *Specification Preparation*.

Traditionally, the adage "If you can't draw it you can't spec it" was challenged early on and has been gradually replaced in some engineering circles with "If you can't imagine it you can't spec it." Many countries utilize a six-part engineering specification.

1. Scope
2. Applicable documents
3. Requirements
4. Mission assurance provisions
5. Preparation for delivery
6. Notes
7. Addenda

Descriptions in specifications are reduced to their technical essence and the specification identifies what is required, not how to do it. Ronald Schumann[24] states:

> A formal specification is intended to be a document that clearly and accurately describes the essential technical requirements for items, packing, materials, or services, and that includes the procedures by which a determination can be made that the requirements have been met.

Currently three major divisions of engineering specifications are used.

1. Performance specification
   a. A specification that states requirements in terms of the required result with criteria for verifying compliance, but without stating the methods for achieving the required results. A performance specification defines the functional requirements for the item, the environment in which it must operate, and the interface and interchangeability characteristics.

[24] Schumann, R. G. (1980). *Specifications, Issue 13, Government Contracts Monograph Government Contracts Program*. Reprinted with permission of the George Washington University, Washington, DC.

2. Detailed specification
    a. A specification that specifies design requirements, such as materials to be used, how a requirement is to be achieved, or how an item is to be fabricated or constructed.
3. General specification
    a. A general specification is prepared in a six-section format and covers requirements and test procedures that are common to a group of parts, materials, or equipment.

The evolution of specifications in the U.S. progressed from the general to the specific, leading to an almost crippling case of procurement micromanagement through specifications. At one time, in the U.S., there were 24 data item descriptions related to specification types.

Recently there has been a change of focus from micromanagement to macromanagement, and the trend that started in the 1990s is to provide only the level of detail needed to adequately specify an item.

When writing a performance specification, it is important to consider the following questions:

- Is the requirement stated real?
- Is there any way it could be misconstrued?
- What am I asking the designer to do?
- What am I doing to the company or contractor?
- How will the customer react to this specification?
- Have all the necessary requirements been stated?
- Will the customer know exactly what we are providing?
- Are there any *gold-plated* items that increase cost?
- How will someone prove the requirements are satisfied?
- Are the requirements bounded or open ended?
- Is there enough information to assure that each requirement can be satisfied by an artifact (test, analysis, demonstration, and inspection)?
- Are references tailored to bound additional cost drivers?
- Are the designer's prerogatives restricted or is there enough flexibility to facilitate a cost-effective design solution?
- Have I fallen into the trap of stating requirements are as implemented in another document or is the specification self-contained?
- Have I specified as requirements compliance with other documents that are not required or contradictory?
- Have I eliminated all "to be determined," "to be resolved," "to be negotiated," and "to be specified" statements?
- Have I progressed from general to specific requirements in each section?
- Have I used the word "shall" to mean a provision that is binding on the contractor?
- Have I used the words "should" or "may" to mean a provision that is not binding on the contractor?
- Have I used the word "will" to mean a provision that is binding on the customer?
- Have I included a clear order of precedence regarding the hierarchy of referenced documentation (e.g., the requirements in specification 1234567 Section 3.0 take precedence over documents referenced in Section 2.2.1 of the specification)?

It is critical that all requirements be clear, concise, and valid in terms of CM implementation. In addition, they should be uniquely identified to create a traceability of the requirement to the end configuration. The requirements need to flow from the highest level to the lowest level in order to tie the documentation artifacts from test, analysis, inspection, and demonstration to the product configuration.

Ronald Schumann goes on to state:

> Probably the most fundamental cause of undue restrictiveness in specifications is the writing of specifications to describe features of a particular item believed to be capable of satisfying the customer need, rather than to describe the need itself and those details of design or performance which are in some way necessary to satisfy the need.

Part of the specification writing task in this setting is to assure that a clear trace of requirements is established from the contract downward. One essential element of this is the generation of a contract specification tree as part of the overall product documentation tree. Think of these trees as a document bill of material (BOM).

6.3.3.6 Material Lists

Lists of material in the sense of the existing standards associated with DOD, American National Standards Institute, ASME, ISO, TechAmerica, and others consist only of the items that are incorporated into or used to produce the item. They do not include hardware and software items, or configurations used in the manufacturing process. The determination of these items is generally associated with the procedures and facilities in which the effort is performed. Cleaning is an example of this. A drawing may state clean in accordance with cleaning standard C-12. This cleaning standard discusses the steps to use, the facility it takes place in, the materials required (gloves, chemicals, disposable items, etc.), drying times between steps, and bagging of the item to ensure it does not get contaminated. With the development of electronic tools to facilitate the linking of data, the current practice involves many companies identifying the expendable materials in the item product structure with the documentation pertaining to it as well as its use in the BOMs for the expendable item.

Often product data management (PDM) and product life cycle management (PLM) tools are not properly configured during the PD/LM implementation phase. This results in few safeguards against circular references in the associated databases. Critical to avoiding these issues is understanding the PD/LM database structure and how it is populated by the PDM or PLM tool, as well as having a copy of the database that can be used as part of the CM implementation to beta test modifications to their material list strategies. Material lists should be thought of as items associated with the item being produced and those items that lose their individual identity (e.g., items consumed) to make the item being produced.

6.3.3.7 Document Material List (Document BOM)

Every document can be said to have a BOM. It consists of documents referenced in the document itself. Where companies traditionally go wrong when portraying this in the product data/life cycle management (PD/LM) database structure is simply in not paying attention to the order of precedence of the documentation. The order of precedence, once established and followed, will eliminate circular references (Figure 6.8).

A good example of this is in contracted items. The simple order of precedence is

- The subcontract
  - The subcontract calls out the statement of work (SOW) and the specification or control drawing
- The SOW
  - States that it takes precedence over the specification or the control drawing
  - States what items are being provided by the contractor to the subcontractor
  - States what items are being provided by the subcontractor to the contractor
- The specification or control drawing
  - Interface control drawing
  - Mission assurance requirements (MAR)
  - Any lower-level specifications

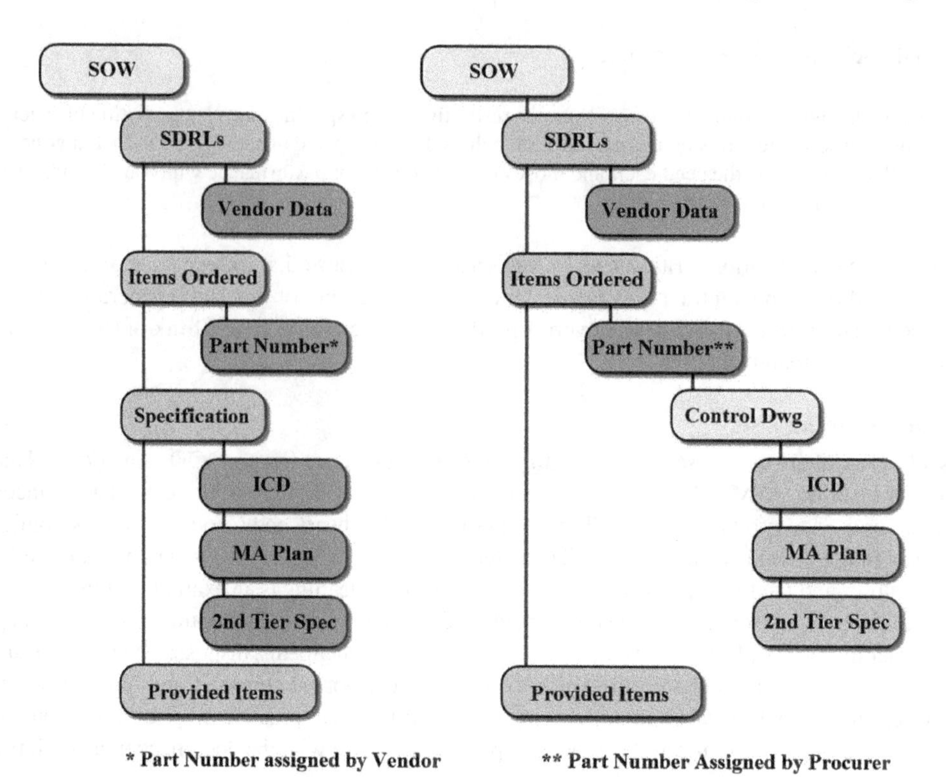

FIGURE 6.8 Document BOM.

Under this order of precedence, lower-tier documents do not call out higher-level documents. Figure 6.8 shows the document BOM structure in cases where:

1. The product is being procured using a specification and the item number is assigned by the supplier.
2. The product is being procured using a control drawing (no specification) and the item number is assigned by the procurer.

Document BOMs tied to metadata fields allow automated generation of various kinds of status and configuration accounting information, such as drawing trees with the associated revision status. The document structure creates a BOM of its own called the "table of contents." In a comparable way, creating contracts through carefully managed BOM configurations is very efficient as much of the contractual text is repetitive between one contract and the next, between one specification and the next, or between one SOW and the next.

6.3.3.8 Produced Item Material List (Item BOM)

Item BOMs are rather straightforward but still have circular reference issues. These again are tied to order of precedence. The item produced contains the document that defines it, and not the other way around (Figure 6.9).

6.3.3.9 Manufacturing Lines

Management of manufacturing lines from a CM perspective may appear alien if you have only worked with set definitions for a single market segment such as government sales. From the government sales perspective, very strict interpretation of not only the definition of CM but also interpretations of the scope of CM itself exist. Unfortunately, in many companies, this has resulted in

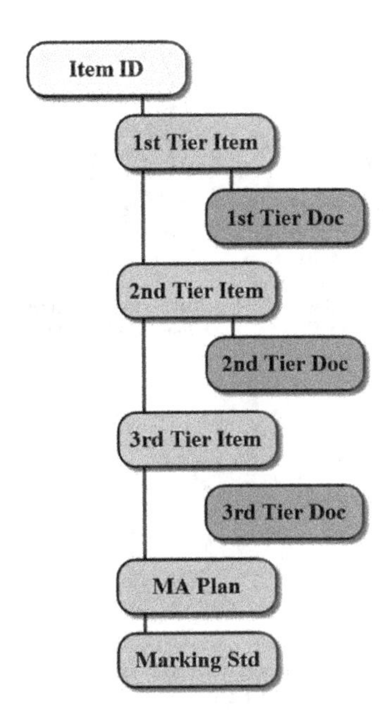

FIGURE 6.9 Produced item BOM.

a very limited understanding of the distributive nature of CM responsibility and accountability. It was noticed that incomplete understanding of contractual limitations combined with a lack of understanding of basic business functions resulted in non-contractual scope creep. This reduced the profitability by as much as 5%. Training costs to assure a broader understanding of contracts, program management, finance, and supply chain more than offset the additional staffing required to adequately manage the programs.

Increased profitability, lower price points, and increased quality and product safety are critical to a company's survival. The firms that have done best in each of these areas view CM from a company-wide perspective and it has become part of their cultural business ethic, engrained in their approach to quality business practices.

6.3.3.9.1 CM and Performance Optimization

Performance optimization and CM implementation are viewed differently by each of the functional resources. Each group has a specific idea of how CM methodologies help them accomplish the specific role they play in the product life cycle. Each is part of the entire scope of the application of CM methodologies in any market segment. In this case, the sum of the parts is not greater than the whole. Any business methodology cannot be applied to a single function, by a single department, for a single market segment, or applied to one regional or state production facility. It must be embraced by an enterprise as a way of doing business; not as something that can be partially applied or only applied for a specific period or specific instance. ISO, Electronic Industries Alliance (EIA), and SAE International standards are based on this crucial concept and CM methodologies are found throughout the documentation suites.

CM binds Kaizen, total quality management, just-in-time manufacturing, lean process, and information data management systems together. Figure 1.9 provided a view of market performance interdependencies. Performance optimization relies on the exploitation of these interdependencies in a way that capitalizes on lessons implemented as opposed to lessons documented. Too often companies add additional process to shore up what is perceived as a process failure due to lessons learned activities without driving deep enough to find the true root cause; in 85% of the 300 cases investigated, addressing the root cause would have reduced process instead of increasing it.

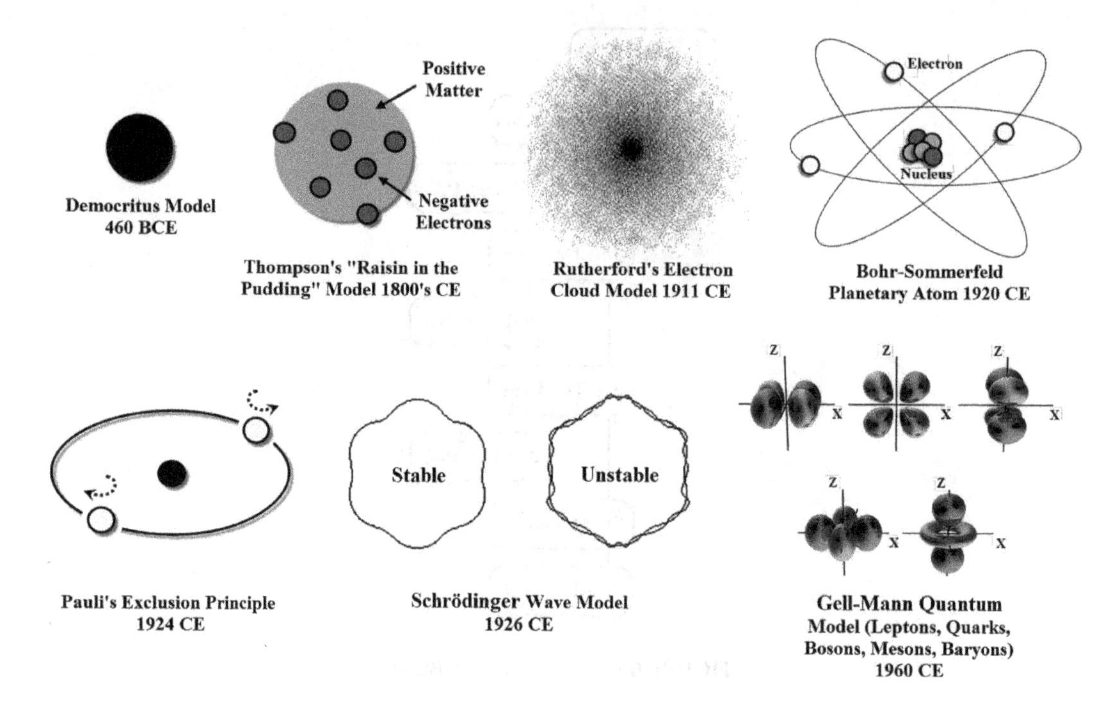

FIGURE 6.10 Atomic theory evolution.

Chapter 1 provided an introduction of the subject from an environmental or biological perspective. It is an easily understood way of looking at the subject. Yet it is only part of the story. In a similar way, chemistry students are often introduced to the atom starting with Thales of Miletus and Democritus before moving deeper into the subject by exploring the contributions of John Dalton, J. J. Thompson, Max Planck, Albert Einstein, Ernest Rutherford, Niels Bohr, Arnold Sommerfeld, Wolfgang Pauli, Louis de Broglie, Werner Heisenberg, Erwin Schrödinger, Max Born, James Chadwick, Paul Dirac, Hideki Yukawa, Murray Gell-Mann, and Yuval Ne'man. The evolving theoretical view of the atom is shown in Figure 6.10.

Moving from the simplistic to the complex, let us delve deeper into CM as it applies to the marketplace and performance optimization. Perhaps it is better understood if we look at the automotive market segment. A summary of work done by Takahiro Fujimoto is explained subsequently.[25] Quotations from Fujimoto's work are indented.

A company's capability is an organization's overall capability to evolve competitive routines even in highly episodic and uncertain situations. These include learning from mistakes, making good decisions, and grasping the competitive benefits as well as capitalizing on unintended consequences and resultant synergies.

> An evolutionary framework is particularly useful when it is obvious the system was created as more than someone's deliberate plan for survival.

Systems emergence can occur through many different paths or a combination of them.

Random trials: Outcomes are not predictable, so you might as well try everything.
Rational calculations: Objectives are deliberately chosen to satisfy or maximize organizational objectives.

[25] Fujimoto, T. (1999). *The Evolution of a Manufacturing System at Toyota.* Oxford University Press, New York, NY.

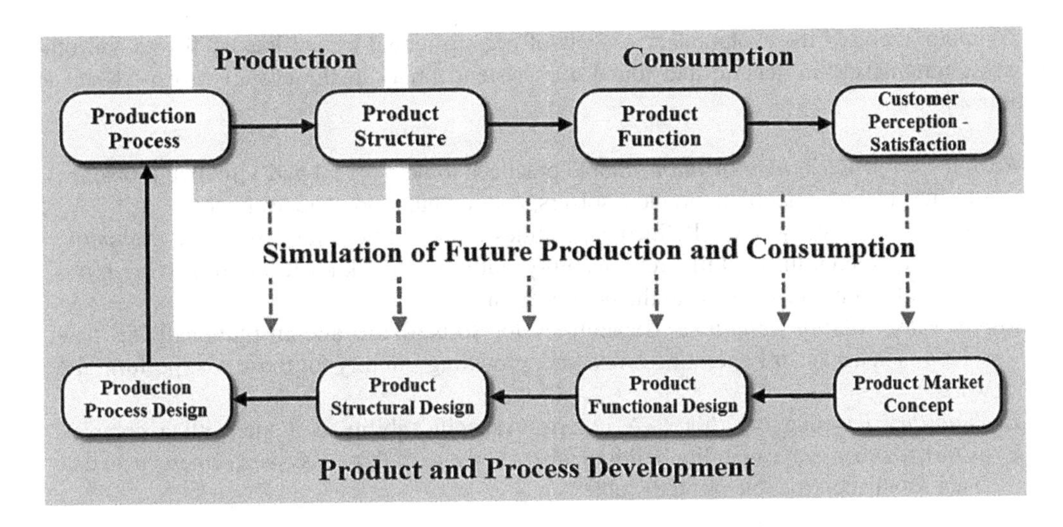

FIGURE 6.11 Product development model.

Environmental constraints: Internal (managerial perceptions) and external (laws and regulations, market pressure, etc.) constraints may in fact enhance competiveness.

Entrepreneurial vision: The vision of a gifted leader may trigger actions that break down or remove constraints.

Knowledge transfer: May be either push (driven by the source organization) or pull (adopter–imitator takes the initiative).

This product development evolutionary model can mean different things to different people. It negates the fact that systems evolve not through natural selection but through conscious effort working in tandem with firm, industry, and market factors with the resultant synergies of improving technologies. One key to the process is a total systems informational perspective of the entire manufacturing system and product life cycle (Figure 6.11).

> Information is central for two reasons: first, the notion of organizational routines as informational patterns—something like the gene unit of DNA "information"—is a logical consequence of evolutionary thinking; second, information is the common element that flows through the product development, production and supplier systems.
>
> Information is broadly defined as intangible patterns of materials or energy that represent some other events or objects, rather than tangible objects themselves. By this definition, knowledge, as well as skills, is also regarded as a kind of human embodied information asset. In the present framework, the information that runs through a manufacturing system is considered value-carrying information, meaning that it is expected to attract to attract and satisfy customers when embodied in the product and delivered to them. In this context a manufacturing routine is nothing but a stable pattern of information flow and assets embedded in a manufacturing system.
>
> Take, for example, a successful product like the Toyota Camry. When a customer buys this car and uses it every day, what he or she consumes is essentially a bundle of information delivered through the car rather than the car as a physical entity. Each time the consumer drives it, he or she is aware of the Camry's style, visibility from the cabin, feel of the seat, sound of the engine, feel of acceleration, road noise, squeaks and rattles of the dashboard, and so on. The customer also remembers the price paid for the car, and how economically it operates, its breakdowns and repairs, praise of the car from friends, even reactions of people on the street. Therefore, he or she continuously interprets and reinterprets all of these informational inputs, which translate into overall customer satisfaction or dissatisfaction.

Fujimoto analyzed the evolutionary aspects of organizational capabilities at Toyota and other Japanese automakers in general and found a consistent thread in the evolution of performance optimization.

Routines for problem identification: Stable practices to help reveal and visualize problems, transfer that information to problem solvers, and maintain problem awareness.
Routines for problem solving: Refined capabilities for searching, simulating, and evaluating alternatives; coordinate skills, responsibility, authority, and knowledge for solving problems and communicate those to the organization.
Routines for solution retention: Capabilities for formalizing and institutionalizing new solutions in a standard operating procedure, providing stability for those who perform the work.
Evolutionary learning capability: A company specific ability to cope with a complex historical process of capability building (also known as multipath systems emergence) that is not totally controllable or predictable.

The combination of these elements results in a hybridization of capabilities inherent from the factory floor, to marketing, to product design/development, to strong partnerships with vendors through leveraged innovation that incorporates innovations in work design and technologies.

Starting with the NATO CM Symposium Group discussions regarding a CI and comparing with the Fujimoto observations, we find that it does not go far enough regarding supporting information. By necessity, CM in a market performance optimization environment requires a still broader interpretation of what constitutes a CI, such as that proposed by Geza Papp.[26]

A configuration (configured) item is defined as any product that requires formal release and control of supporting documentation prior to acceptance by the user; where the supporting documentation is detailed with sufficient precision for all internal and external stakeholders to understand. A configuration item may be assigned relationships with other configuration items.

This requires a simplified definition of data or information and its management.

Control and availability of all information necessary to the improvement of product and company market viability and company compliance and awareness of environmental constraints.

It also results in proactively managing change with a single point of document dispersal in a customer-driven rather than customer derived requirements environment. Different market segments implement this in unique ways. One implementation example is that of Nokia Networks[27] who state:

Bringing together our NetAct Optimizer and Configurator with customized tools and services, we'll help you bring your management and optimization into a single solution.

You'll get a centralized view of your operations so you can easily manage your optimization and configuration program.

You'll also be able to reduce your integration and maintenance costs with all your management operations in a single solution.

---

[26] Wessel, D. (2013). *What Is a Configuration Item and Which Consequences Does the Selection of CIs Have to Your Organization?* Consultant on NATO standardization to OSD, ATL. LinkedIn NATO CM Symposium Group moderated by Dirk Wessel started on July 1, 2013. Reprinted with the permission of the author.

[27] Nokia Solutions. (2014). *Solutions*. Retrieved June 10, 2014, from http://nsn.com/, http://nsn.com/portfolio/solutions. Reprinted with the permission of Nokia Networks, formerly NSN, is the world's specialist in mobile broadband, providing network infrastructure software, hardware, and services to the world's largest operators. Nokia Networks is one of Nokia's three businesses.

> Every day a quarter of the world's population connect using NSN infrastructure and solutions. Our Multi-Vendor Configuration Management and Optimization solution comes with best in class functionality, built on our deep understanding of access and core technologies.

To others, it might be how well data files are cross-connected in a database or how many clicks it takes to find information in a PD/LM implementation. It may be the elimination of specified elements of the entire scope of the CM process if they are not applicable to the product or product line. In the case of many smart devices, the first time the unit is tested is when it is turned on by the buyer.

Personal experience and industry surveys provide a persuasive case that CM implementation directly impacts the bottom line, and sound decisions relative to management of the configuration of an item can make or break the product, its market share, and the company's longevity.

6.3.3.9.2 CM and Cost Optimization

Cost optimization in an environment where production runs in the million units is much different than cost optimization in a program driven by a single customer with limited quantities. The fundamental difference lies in the end objective and how it must be approached. The dollar savings through minor improvements over a million units can be substantial. The same improvement in a production run of a single unit may not pay for the cost of processing the change.

Consider the differences in savings involved in manufacture of the bell housing discussed in Section 1.5.2.3.1. Machining is the preferable method when making up to 10 units, at which point forging becomes more economical with savings of 50% per unit. The breakeven point considers lead time as well as material and production savings. This differs slightly from the 67% savings realized at unit number seven cited in the Forging Industry Association literature.[28] Once the breakeven point is reached, additional cost savings result due to the continued amortization of the nonrecurring cost spread over the entire production run until such time as forged part quality is reduced due to die wear which accounts for 70% of die failure.[29] Replacement dies can then be obtained at a reduced cost as there is no recurring design. The predicted die replacement can be determined using finite element modeling. This reduces in machine set-up. When combined with workflow modification (optimized press speed, die temperature, die materials, work piece material, friction, lubrication, and resultant heat transfer due to thermal softening) in a just-in-time environment, significant increases in die life result. If the production run is large enough, dramatic savings can be realized in energy costs and materials alone by moving to powder metallurgy technologies. Powder metallurgy allows production of components with near net shape, intricate features, and close dimensional precision that are finished without the need of machining.

CM influences on cost optimization take other forms as well. Staying in sync with market and technological trends can result in company-wide savings while increasing customer satisfaction. Often this takes the form of automation of status and accounting functions and data management capabilities. In some industries, it is possible to track a semi-customized order of a commercial product through the entire production flow and know the status of every component starting by keying in the order number. In other market segments, the actual cost to produce each component is so well known that informed decisions can be made regarding the effect of changes of outwardly appearing minor modifications. One instance discovered in our research showed that the simple well-intentioned act on the part of a lead design engineer to increase the diameter of a through hole to ease assembly resulted in a €742,340 implementation impact to production tooling on the factory floor. The change saved an estimated €43,670, leaving the remaining €698,680 to be absorbed out of company profits.

---

[28] Forging Industry Association. (2007). *Long Lifecycle Plus High Performance Makes Forged Components Lowest Cost.* Retrieved on July 25, 2014, from www.forging.org.

[29] Groseclose, A. R. (2010). *Estimating of Forging Die Wear and Cost.* The Ohio State University, Columbus, OH.

On a major satellite program, the configuration manager reacted to a request from a customer engineer to modify the format of documentation required for a contractual status report. There was no change proposal and no change to the contractual data item description. Taking this direction from a government official with no authority to request, direct, or authorize such changes is a violation of the contract. The violation resulted in a profit reduction of more than €436,670 at contract closeout. When asked about the change, the government employee stated that to make the change through the contracting officer would have taken too long. The contractor's senior configuration manager, who acted on the direction, stated that they always did their very best to please the customer. They had no idea that what they were doing was a violation of the contract.

Malcolm Gladwell talks of this in *Blink*[30] when describing the €8,730,000 acquisition of a sixth-century BCE kouros statue of a boy by the J. Paul Getty Museum in California. After an extensive investigation of the work, the museum agreed to buy the statue and put it on display in the fall of 1986. When it was shown to experts outside of the museum—such as Federico Zeri, Evelyn Harrison, Thomas Hoving, and George Dontas—they could instantly tell it was a forgery. Such intuitive understanding is not a by-product of a person's genetic code but a by-product of a person's intimate familiarity with all elements of their chosen field. The art specialists had seen and studied thousands of statues and found elements of the Getty acquisition as distasteful to them as the scent of wine that has gone bad is to a well-nosed sommelier.

Some would call the kind of intuitive understanding described in *Blink* as common sense. As Voltaire observed, "Le sens commun n'est pas si commun."[31] This is true of CM implementation. The question many starting a career in CM ask is, "How do I gain the knowledge necessary to become a good configuration manager?" The question is a valid one and the answer is complicated by the fact that most entering the profession have no training in CM or a CM deployment across a company. They rely on what is written down in whatever standards they are told apply. A good CM implementation needs to consider CM in all its details and manifestations. The CM implementation lead needs not only to be a visionary familiar with the requirements pertaining to the product throughout its life cycle and the development of the product itself, but also to be able to recognize and orchestrate the complexities of the data associated with the product.

As engineers in the robotics field are discovering, what is written down, what is relayed verbally, what is seen visually, and what is measured empirically has left them woefully unprepared for designing a robot that can walk, run, and climb with the ease of a person. The same applies to development of an intuitive feel for the application of CM. In the final analysis, all the standards developed are no more than guideposts along the way. They offer little help if you do not understand where you have been, where you are, where you are going, and how you recognize that you have arrived. In a way, it is like a Zen puzzle; when CM implementation is in balance, you will learn to recognize it. What it takes to bring it into balance requires a deep understanding of the professional environment and a developing situational awareness of the influence of each input, constraint, and mechanism influencing the outputs.

Perhaps the single most important understanding for CM implementation is a clear grasp of how the business operates, so that informed choices based on available options can be made. Among other things, this will facilitate properly vetting changes with all stakeholders, so that a valid analysis of exactly what the impact of the change is can be made. As configuration managers become seasoned in their profession, they will develop an intuitive feel for the interrelationships and cause and effect of change.

[30] Gladwell, M. (2005). *Blink: The Power of Thinking Without Thinking*. Little, Brown and Company, Park Avenue, NY.

[31] Voltaire, F.M. (1764) quotation "Common sense is not so common." *Dictionnaire Philosophique*. p. 866. https://books.google.com/books?id=8hkVAAAAQAAJ&pg=PA866&lpg=PA866&dq=Le+sens+commun+n%27est+pas+si+commun&source=bl&ots=3zxq7YMIae&sig=8X-I6m0ol6zziN6tSTsdfCF2P3M&hl=en&sa=X&ei=dsGdVLqNH82WyASNh4KgAw&ved=0CFgQ6AEwBw#v=onepage&q=Le%20sens%20commun%20n'est%20pas%20si%20commun&f=false p886. Accessed November 22, 2013.

The second most important understanding that the CM professional can develop is a clear perception of how the tools used in the company are interconnected. The PD/LM system database structure must be understood at least conceptually to allow the tool to be used as a tool rather than simply another system you feed data into. If the tool is not understood, it is very hard to use it to best advantage. Put another way, if you tell someone that a knife is a screwdriver, how long will it be before they understand that its primary purpose is to cut things? Understanding how the tools are designed allows the knife to be used as a knife, and leads to an understanding of how to keep it sharp and optimize its design for the purpose it was intended to perform. All PD/LM systems are generic when they come out of the box. As the CM implementation is defined, its nature becomes more specific. It can then be honed to razor sharpness and wielded by the users with surgical precision.

Applicable regulations and standards need to be read with a sense of what they are about and not specifically what the words say. Look at company policy, procedures, practices, and guidelines to get a feel for how the regulations and standards are manifested in company documentation. Go back and read the regulations and standards again and trace the specific requirements back through the company documentation. Perform a second trace through the company documentation to see how company policy, procedures, practices, and guidelines are interrelated. Ask questions if you do not understand something and become an expert in the area if no other expert exists.

Let us put the above in context using the automobile as an example. The automobile—like any other machine, program, or biological entity—must be in balance for it to operate properly. Paying attention to the feel of the vehicle and its components (sounds, performance, appearance) and acting on those observations will enable the vehicle to provide years of service. It will corner properly, stop properly, and do all the things it was designed to do. The driver need not understand how to design, manufacture, or repair a suspension system to know that a shock absorber needs to be replaced or how properly inflated tires affect the performance of the vehicle. The driver should read the owner's manual to understand how each subsystem is used, where the vehicle jack is located, what is required for maintenance, and what the maintenance schedule is.

The same understanding of how regulations and standards are implemented in company policies, procedures, practices, and guidelines—when combined with a basic grasp of the external-to-the-company regulations and standards—allows the configuration manager to know how issues in one piece of the management system influence the overall performance of the entire infrastructure. This is where cost savings and profitability can be greatly influenced by:

- Management of the configuration of the systems used to create, produce, market, and improve it.
- Implementation of changes to external influences.
- Consolidation of engineering, financial, mission assurance, production, supply chain, contracts, and traditional CM information tracking systems into a single database.
- Evaluation and implementation of proposed changes to internal requirements or to the product itself.
- Retooling workflows and automation of labor-intensive repetitive tasks, such as re-entering metadata associated with an engineering order if it is contained on the affected item.

Simply understanding the workflow for granting internal access rights to a program or product file server or intranet site can have a substantial payback. In one firm, a net savings of €742,340 over three years resulted from putting in place an automated permissions request workflow and aligning server and intranet functionality. Because most CM activities grew out of and have strong ties to older paper-based systems, there is much to be gained from a careful examination of how those systems have translated into or are being influenced by PD/LM systems. *Digital fabric* CM allows the selection of the right set of CM requirements based on the context they are used in.

6.3.4 CM Process Elements

If we return to Figure 1.1 from MIL-HDBK-61, we find the representation of typical inputs, constraints, mechanisms/facilitators, and outputs associated with CM and the development of any product.

6.3.4.1 Inputs

Inputs come from many sources. CM implementation is never a case of a one-size implementation that fits all products. CM also scales by type of product and market segment. What is considered critical in the food industry in terms of traceability is much different than what is considered important in the toy industry or aerospace industry. Items of importance and the effort expended to satisfy each product's specific management are variable over the product life cycle. As a result, inputs also change over the life of a product. Events leading up to qualification and acceptance are much different than those once the product is sold when maintenance starts. The CM implementation needs to recognize not only the place in time the product exists in but the floating nature of the focus of CM activities in synergy with it. The same attention to requirements elicitation must be given throughout the product life cycle as is paid by the concept development team when formulating a product's entry into the marketplace or when building a product specific to customer-stated and derived requirements. Typical inputs include the following:

- Market evaluation
- Program initiation
- Systems-level requirements
- Analysis
- Customer needs, wants, and desires
- Stakeholder needs, wants, and desires
- Performance measurements
- Internal and external communication
- Other analysis
- Maintenance strategies

6.3.4.2 Constraints

Constraints may be unique to the product, market segment, company, city, state, or nation, and in some cases are international. They should be considered not obstacles but challenges to be addressed, resulting in opportunities for a better product or an increase in market share. Samsung's introduction of a 128G V-NAND SSD 24-layer 3D flash memory chip in 2013, sidestepping the 30 nm spacing issue required to eliminate transistor cross-talk, is an example. Xerox's drastically reducing the number of parts needed to make a copier by over 50% through re-evaluating the copier configuration in the 1990s is another. Typical constraints are as follows:

- Governmental restrictions (local, state, national, and international)
  - Includes tariffs, trade restrictions, and program priority ratings (DX, DO, etc.)
- Environmental concerns
- Resource sustainability
- Perception
- Test market
- Timing of competitor's products
- Resources (facilities, monetary, personnel, etc.)
- Time
- Inadequate planning
- Inadequate preparation

- Stockholder expectations
- Materials

6.3.4.3 Mechanisms/Facilitators

Facilitators are those management and support functions, process and quality infrastructure, facilities, and teaming agreements that not only increase the capabilities available to perform CM functions but also improve them and their effectiveness.

- Customer mandate
- Management support
- Supplier teaming and synergy
- Sales
- Facilities
- Training
- Policies, procedures, practices, and guidelines
- Tools and infrastructure
- Global technology solutions (GTS)

6.3.4.4 Outputs

As soon as all the inputs, constraints, and facilitators are balanced against each other, outputs assure the continued viability of not only the product in the market segment but also the company.

- Consistent and appropriate implementation
- Verifiable safety
- Adequate data to mitigate constraints and associated reporting
- Product image
- Stronger supplier base
- Increased quality, reduced cost, and higher earnings before interest and taxes (EBIT)
- Improvements in processes
- CM consistent with the scope of the product

6.3.5 Requirements and CM

Requirements directly impacting decisions regarding the right sizing of the CM activities come from multiple sources.

6.3.5.1 Scope Trace Through Requirements

Two common phrases used in engineering circles are "functional requirements consolidation" and "functional decomposition of requirements." These activities are tied to an approach broadly applicable to management of requirements and requirements document generation.

The process of functional requirements consolidation (Figure 6.12) starts with requirements identification, requirements evaluation, requirements allocation to documentation (SOW, specifications, MAR, contract terms and conditions, etc.), documentation generation, and verification of included data requirements tie-out.

Each requirement is assigned to a single document to establish one source of truth for that requirement. The artifact proving that each requirement is met is clearly defined in the document and the data submission is captured in the data requirements list (DRL).

The process of functional requirements allocation starts with requirements documentation (SOW, specifications, MAR, contract terms and conditions, etc.) and the allocation of each requirement to the components, mechanisms, and subsystems of the finished product. A single requirement such as a specified mean-time-between-failure (MTBF) may be allocated to every subsystem and

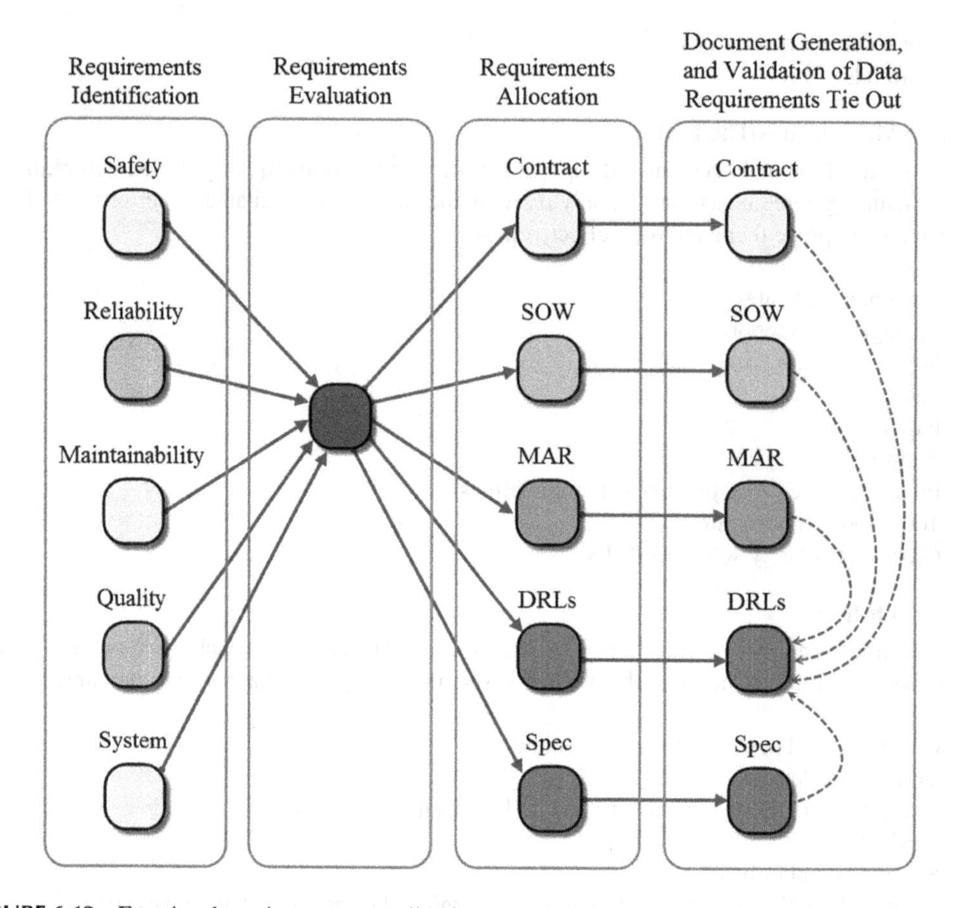

FIGURE 6.12 Functional requirements consolidation.

all electrical components and feed into the final reliability calculation. In a PD/LM environment, the component and subsystem analysis documents would form the document BOM for the final systems-level report. Systems-level MTBF antecedents are shown in Figure 6.13.

This distributive allocation of a single requirement is needed to produce elements, generate data antecedents, and consolidate that information into the final artifact to document compliance. It is

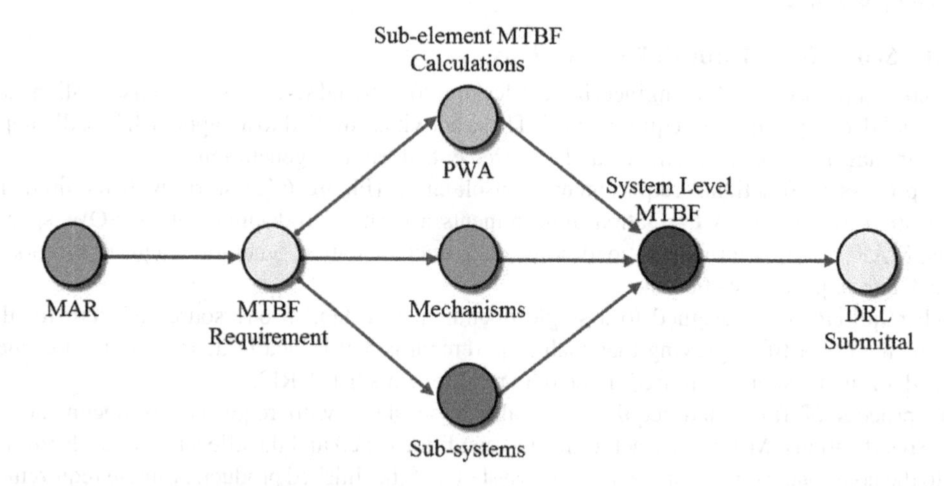

FIGURE 6.13 Systems-level MTBF antecedents.

a single digital thread. This process is repeated for each requirement and creates the warp in the digital fabric. All external and internal requirements are allocated in a comparable way.

6.3.5.2 Requirements and CM Overview

Often it becomes advantageous to reallocate requirements from one subsystem to another due to cost, schedule, technical risk, consumer mandate, or changes in capabilities or need. This would be impossible without the generation of subsystem-to-subsystem interface requirements lists and giver–receiver lists and the establishment of requirement tracking mechanisms to facilitate change management and evaluation. Such allocations change the warp on the loom we weave the digital fabric on. The lack of any one of these requirement allocation and accounting tools can lead to additional costs, as evidenced on a recent space instrument. The optics subsystem assumed that the electrical subsystem was responsible for the wiring harness used between the two subsystems. The electrical group assumed that the harnesses were being provided by the optics group. A few weeks before system integration and test, it was discovered that no cables had been designed or built. The root cause issue was found to be a failure to generate the necessary interface control drawings (ICDs) and giver–receiver lists.

On another program, the diameter of a launch vehicle was increased to accommodate a larger payload. Guidance control and other components were attached to a shelf of fixed width on the inside of the stage. The change was not properly coordinated during the configuration control board review and, as a result, was not known or understood by the entire organization. This resulted in no plans to perform another weight and center of gravity calculation or to properly flow down the impact to the structure to all organizations. The missile stage supporting the shelf was redesigned to make the structure stronger and an extra hole was added to accommodate one more fastener. The interfacing stage design team was unaware of the addition of the mounting hole. The two stages could not be assembled due to one stage having more mounting holes than the stage it mounted to. It cost hundreds of thousands to develop an adapter ring (Figure 6.14) that fit between the two stages.

A distinct trend has been observed since the 1990s of companies merging change control, data management, and status accounting functions to gain efficiencies in the CM process. This

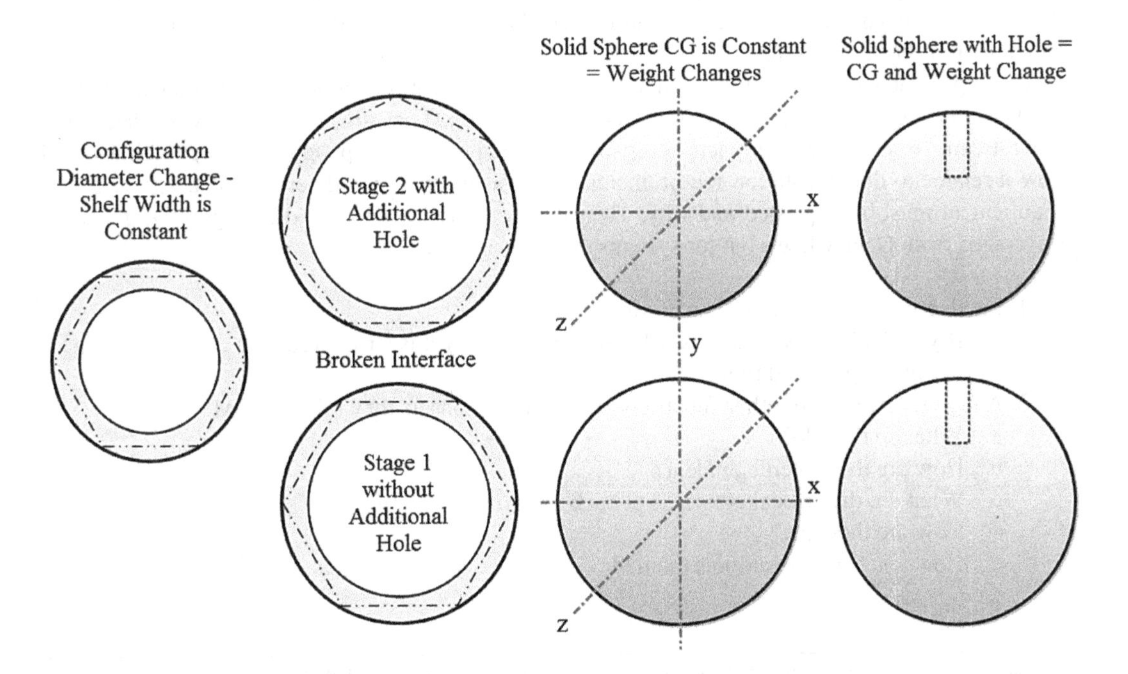

FIGURE 6.14 Launch vehicle diameter change.

is a departure from prior to 1990 when records release, status and accounting, and data management were training steps in the CM implementation. There is now also confusion between the role of systems engineering and that of CM. There need not be, as the roles are quite distinct. Systems engineering in conjunction with other program management members assesses all technical requirements and their allocation. As requirements are refined and derived requirements identified and allocated, the role of the systems engineer evolves to one of managing total system performance and operability. The CM manager manages the specific configuration and its baseline as well as controls all data associated with it, so the systems engineer has ready access to associated information.

Gentry Lee,[32] in his 2005 presentation, "So You Want to Be a Systems Engineer," at the Jet Propulsion Laboratory, states that the role of systems engineering is to understand the partial derivative of everything with respect to everything else (Equation 6.1).

$$\delta X_i \; x \; \delta X_n$$

EQUATION 6.1 The partial of everything with respect to everything else.

Their primary role is to know how every electrical component, mechanism, and subsystem affects the operation of the entire product. This ability allows the systems engineer to make informed trade-off decisions regarding design trade studies, subsystem functionality, and requirements allocation as well as to apply that understanding during test and any resultant redesign due to customer issues and feedback.

In an equivalent way, the role of CM is to understand and link the partial derivative of any piece of product information to all other pieces of product information in a way that it is easily understood and usable. There can be no self-defined boundary that constitutes CM. There is no cookbook, standard, or procedure that will define the entirety of the work scope. In many ways, CM must perform the same least-squares analysis for interrelated data that the systems engineer prepares for system-to-system connections. If you give good systems engineers one piece of data about their program, it is like injecting a dye tracer into the bloodstream. They untangle the digital thread from the digital fabric and discern how that information reacts with every other subsystem in the entire system.

If you provide good configuration managers a finite element model of a system, they know how it relates to every piece of data that was used in its creation. They also understand what data flows downstream from it; how it impacts company policies, procedures, practices, and guidelines; and how it relates to documentation requirements. As a result, they can determine if a variance to a requirement must be requested and why. The top characteristics of a good configuration manager (leveraged from Gentry Lee's list for systems engineering) are:

1. Intellectual curiosity about the product.
    a. If you don't know what it is, what it does, and how it is structured, you can't manage it or anything related to it.
2. A big picture view of other functions and organizations as they relate to CM.
    a. What do they do?
    b. How are they interdependent?
    c. What are the critical concerns of each organization?
    d. How are they met?
    e. How you can best support them?

[32] Chief engineer for the Planetary Flight Systems Directorate at the Jet Propulsion Laboratory and noted science fiction writer and co-developer of the *COSMOS* television series, http://spacese.spacegrant.org/index.php?page=videos.

3. Connections with the product—it can't be done in a vacuum.
   a. What is it?
   b. What does it do?
   c. What do other organizations need to know?
   d. Why is it important?
      i. What specific aspects are important?
      ii. How should the aspects be interrelated?
4. The ability to recognize and trace the interrelationship between all information elements and then establish meaningful data links between them.
5. The ability to recognize the impact of a change and the internal and external requirements, and perform trade-offs in how a specific dataset is formulated so all requirements are satisfied.
6. Comfort with change.
   a. All requirements cannot be defined for any product.
   b. Requirements evolve over time.
   c. Sometimes you go through the critical design phase and must go back to conceptual evaluation.
7. Comfort with uncertainty.
   a. You will only be able to define what finished looks or feels like at a point in time.
8. Self-confidence and humility.
   a. Know what you know.
   b. Know what you don't know.
   c. Don't be afraid to admit that you don't know.
   d. Ask the questions, so you can learn what you don't know.
   e. Listen so you understand the intent of what is being said and ask more questions.
   f. Don't get fixated on acronyms. Focus on what they mean in the larger sense.
9. Proper application of process.
   a. Understand why it was created.
   b. Understand how it can best be applied.
      i. Knowing only process and standards does not make a great configuration manager.
      ii. Knowing no process or standards is not much better.

6.3.5.3 Success Criteria

Successful CM implementation needs to be evaluated from a variety of viewpoints. A single project can have a very successful implementation at the expense of the product, program, or company. Projects or program management teams going rogue are not common because the results are short lived and self-serving. Going rogue is generally evidenced on small projects, where the program team makes decisions that run counter to the norm. It is often done for political advantage. One common example of this is a program manager mandating that all data deliveries and data tracking be done by general and administrative function. This spreads the costs across all programs and not as a direct charge to the project, product, or program. It causes two things.

1. The actual cost of CM is not known.
2. The project, product, or program falsely shows increased cost control and profitability at the expense of every other program in the company.

6.3.5.3.1 CM Success

Often the total impact of CM implementation cannot be measured in the day-to-day flow of a product through its life cycles. You must step back to do a meaningful assessment and ask rather simple questions.

1. Am I better off using the CM methodologies than I am without them?
2. Can I find information about anything pertaining to the product in less than 15 minutes?

3. Can I trace the design solution path from initial requirement through trade studies to preliminary design, final design integration, and test, sales, and maintenance?
4. Can I find all trade studies performed that evaluate other alternatives that may be a more viable solution due to changes in tools and technologies?
5. Can I retrospectively review the genesis of the product's current form, so that false starts and other development issues can be evaluated for process and design improvement?
6. Can I find cross-linked data to see if these same issues existed on other past and present programs to develop internal standards and training so critical internal intellectual property and design know-how is not lost? (The "Magic Sauce" is lost every 12–15 years in the space launch booster area and this is not atypical in other industries.[33])
7. Can I do all these things for programs that have been closed for 10 or more years?
8. Are all design teams on the same page or was there a breakdown in communication relative to design restraints, goals, and design evolution?
9. Are internal subsystem-to-subsystem ICDs and giver–receiver lists managed properly?
10. Do we have the proper level of change control?
11. Are we tracking the information necessary to assure success, what did we forget, why did we forget it, and what is the impact?
12. Are test and test temporary configurations managed with the same degree of control as the production unit(s)?
13. Can I easily return to the design and contractual technical requirements at any point or date in the life cycle of the program (e.g., revision of every document associated with any product event: preliminary design review, critical design review, integration and test, FCA, PCA, delivery)?
14. Do I have adequate traceability so that, at any point in the product life cycle, I can truly know all differences between what was designed and what was produced including all non-conformances and part substitutions?
15. If there is an industry-wide alert on defective parts, can I find out within 15 minutes everywhere the suspect part is used on all my products (in-production, completed, and sold)?
16. Do I know the differences between the internal policies, practices, procedures, and guidelines at the start of the product life cycle and those that exist now, so that I can do cost impact evaluations?

A company should never evaluate CM success upon strict compliance with ISO, IEEE, or other guidance documents. It is critical that it should be evaluated on its benefit to the company, customer satisfaction, and EBIT. One often stated objection to implementing CM is that it adds cost. The statement is true. What needs to be considered is not what CM implementation costs but what it saves. David Harland and Ralph Lorenz have documented the impact of inadequate CM and state the following truism, "Engineering is the art of doing with one dollar what any damn fool can do with two."[34]

The dropping of the U.S. National Oceanic and Atmospheric Administration (NOAA) N-Prime weather satellite (Figure 6.15) on September 6, 2003, was due to the undocumented removal of adapter plate bolts. It resulted in €117,900,000 of damage to the nearly completed €203,490,000 satellite. It is but one example of the impact of CM implementation. The cost to correct the resultant damage to the satellite was paid out of company profits. The damage to NASA, NOAA, the contractor's reputation, and the perception of the aerospace sector cannot be measured.

[33] Ross, L., and Nieberding, J. (2010). *Space System Development: Lessons Learned Workshop*. Aerospace Engineering Associates LLC, Bay Village, OH.

[34] Harland, D. M., and Lorenz, R. D. (2005). *Space Systems Failures: Disasters and Rescues of Satellites, Rockets and Space Probes*. Praxis Publishing Ltd, London.

FIGURE 6.15 NOAA N-Prime mishap (Image from NOAA N-Prime Mishap Investigation Final Report, NASA September 13, 2004.)

6.3.5.3.2 Project, Product, and Business Success

Angeline, a blogger at projectmanagementcommunications.com in Sydney, Australia, in response to a question on this subject (http://pm.stackexchange.com/questions/3122/definition-of-project-success) states (paraphrased):

> A project can only be successful if the success criteria were defined upfront (and I have seen many cases of projects that skip that part). When starting on a project, it is essential to work actively with the organization that owns the project to define success across three tiers, which are as follows:
>
> - *Tier 1—Project completion success:* This is about defining the criteria by which the process of delivering the project is successful. Essentially, the classic questions such as "Are we on time, budget, scope, quality?" are addressed here (adapted to whichever project management method you might be using). It is limited to the duration of the project and success can be measured as soon as the project is officially completed (with intermediary measures being taken of course as part of project control processes).
> - *Tier 2—Product/service success:* This is about defining the criteria by which the product or service delivered is deemed successful (e.g., system is utilized by all users in scope, uptime is 99.99%, customer satisfaction has increased by 25%, etc.). These criteria need to be measured once the product/service is implemented and over a defined period.
> - *Tier 3—Business success:* This is about defining the criteria by which the product/service delivered brings value to the overall organization, and how it contributes financially and/or strategically to the business. Examples include financial value contribution (increased turnover, profit, etc.), competitive advantage (x points of market share won), and so on.
>
> You can be successful on a single tier but not others. Ultimately, tier 1 matters little if tiers 2 and 3 are not met. What constitutes overall success needs to be defined and agreed as part of this exercise.

From this definition, successful application of CM activities is much more than how well a program or product perceives it. It must be viewed from the perspective of its effect in all aspects of the business. If the product and business are not a success, the CM application has failed.

6.3.5.4 Good versus Poor Requirements

A standard concept in contract law is that any document errors are construed against the preparer. For those with little training in requirements generation, it is often hard to distinguish a good requirement from a poor one. To distinguish this, you need to keep in mind the following tests:

- Is there a valid need for the requirement?
  - What happens if the requirement is not stated?
  - If the answer is "nothing," then it is not a good requirement.
- How is the requirement verified?
  - What method is used?
  - What constitutes success?
  - What artifact is required?
  - Because a requirement is a single entity … it passes or fails as one piece.
- Who or what decided the need for the requirement?
  - It needs an identifiable source.
  - Is it derived?
- Is the requirement attainable?
  - Is it technically feasible?
  - Does it fit within budget, schedule, and other constraints?
  - If it is technically feasible and you cannot afford it due to other constraints, it is not a requirement but a goal.
- Is the requirement clear?
  - Does it express a single thought?
  - Is it concise?
  - Is it unambiguous?
- Does it pass the reasonable person test?
  - Are the terms used misleading?
  - Can 10 people from different organizations read it and all come to the same conclusion as to what it says?
- Does the requirement state what is to be accomplished?
  - Requirements do not provide design solutions!
- Are you deferring?
  - Do you specify operations to be performed rather than a requirement?
  - "Clean all assemblies in accordance with the Mission Assurance Requirements Specification" is deferring and not a requirement.
- Have you made bad assumptions?
  - Have you assumed that those performing the requirement use the same tools and methodologies as you do?
- Are the requirements traceable?
  - If they are not, you need to define where the requirement came from.
- Can requirements be prioritized?
- Do requirements stand alone or are attributes or modifiers required to make them intelligible?
- Have you used a word that will drive costs?
  - The word "support"?
    - "Support" is a proper term if you want a structure to support 50 kg of weight.
    - It is incorrect if you are stating that the system will support certain activities.
  - Other words and phrases to avoid are
    - But not limited to,
    - Etc.,
    - And,

    - Or,
    - Minimize,
    - Maximize,
    - Rapid,
    - User-friendly,
    - Easy,
    - Sufficient,
    - Enough,
    - Adequate, and
    - Quick.
- Have you mixed subjects?
    - Does the requirement apply to only one system or subsystem?
        - You have defined a subsystem solution not a requirement.
        - Requirements need to be written at the highest level possible and not define or limit the solution set.
- Are the requirements single or cumulative?
    - The transporter shall withstand:
        - Crosswinds of 100 kph *and*
        - Ice loading of 28 cm *and*
        - Hail up to 9 cm
    - The transporter shall withstand
        - Crosswinds of 100 kph *or*
        - Ice loading of 28 cm *or*
        - Hail up to 9 cm
    - The solution sets for the examples are different: 9 cm hail traveling at 100 kph with an ice loading of 28 cm is a much different environment than each evaluated one at a time.
- Do you have the same or similar requirement in more than one place in the document?
    - It need only be cited once.

The process of requirements generation necessitates the understanding of certain terms and how they are used. What follows has an accepted contract law legal basis. The words "shall," "will," and "should" when used in defining requirements need to be considered terms of the art applicable to requirements writing and not as terms used in everyday speech.

At present, requirements use *shall* and each requirement applies to a single thing (e.g., each requirement must be in a single sentence). The use of *shall* is changing; see Section 5.5.8.

- *Bad*
    - The system shall accept credit cards and accept PayPal.
- *Good*
    - The system shall accept credit cards.
    - The system shall accept PayPal.
- *Bad*
    - The system shall work with any browser.
- *Good*
    - The system shall work with Firefox Quantum version 64.0.2.
    - The system shall work with Internet Explorer version 11.0.105.
    - The system shall work with Google Chrome version 71.0.3578.98.
- *Bad*
    - The system shall respond quickly to user clicks.
- *Good*
    - The system shall respond within 10 ms to any user click.

Statements of fact use *will* and are not subject to verification.

- Generally, *will* is used for things the customer provides to the supplier.
  - The procurement office will provide supplier access to the data repository for delivery of supplier documentation.

Goals use *should* and are not specific requirements. *Should* statements can be considered as nice-to-have items.

- The system should be designed to minimize return to factory for repairs.

Must has not been used in requirements as it has never been defined in a court of law and there is no documentation specifying how it is different from *should* or if it is the same as *should*. It is essential that each requirement be very simple to understand. Each requirement can usually be written in any one of the following formats:

- The System shall provide …
- The System shall be capable of …
- The System shall weigh …
- The Subsystem #1 shall provide …
- The Subsystem #2 shall interface with …

Consider the following functionality drivers from MIL-STD-490[35] and IEEE P1233.[36] Many of the following items are often overlooked.

- Communication
- Deployment
- Design constraints
- Environment
- Facility
- Functionality
- Interface
- Interfaces
- Maintainability
- Operability
- Performance
- Personnel
- Privacy
- Regulatory
- Reliability
- Safety
- Security
- Test
- Training
- Transportation

[35] DOD, U.S. (1985). *MIL-STD-490a, Military Standard Specification Practices, U.S. Department of Defense*. U.S. Government, Pentagon, Washington, DC.

[36] IEEE. (2009). *IEEE-P1233, IEEE Guide for Developing System Requirements Specifications*. IEEE, Piscataway, NJ.

6.3.5.5 Elicitation

Requirements elicitation (also known as requirements gathering) is part of the configuration implementation tool box. All requirements cannot be collected from the customer or from a single organization in the concept development phase of a new product. Requirements development hinges on inputs from the customer and all stakeholders. These inputs are combined with observations, intended use, prototyping, design teams, lessons learned, market survey, interviews, role playing, identification of known–unknowns, speculation on unknown–unknowns, and an evaluation of the total scope of the system throughout its life cycle. Christel and Kang[37] stated that the challenges come from three distinct areas.

- *Problems of scope:* The boundary of the system is ill-defined or the customers/users specify unnecessary technical detail that may confuse, rather than clarify, the overall system objectives.
- *Problems of understanding:* The customers/users are not completely sure of what is needed, have a poor understanding of the capabilities and limitations of their computing environment, do not have a full understanding of the problem domain, have trouble communicating needs to the system engineer, omit information that is believed to be *obvious*, specify requirements that conflict with the needs of other customers/users, or specify requirements that are ambiguous or untestable.
- *Problems of volatility:* The requirements change over time. The rate of change is sometimes referred to as the "level of requirement volatility."

Scope, understanding, and volatility are much like the real-time evolution of a natural disaster. You need to switch gears quickly while still retaining the capacity to adjust to react to the informatics involved to survive. As fast as technologies are evolving, the minimalistic relational mindset common in older PD/LM systems paradigms is a distinct disadvantage. The CM related PD/LM infrastructure must evolve to an ontological framework (philosophy concerning the overall nature of what things are) just to keep up with the informatics. Truth is not opinion, and holding on to the way things were done or perceived 20 years ago or even two years ago may no longer be valid. This is never truer than with autonomous and semi-autonomous systems.[38]

Requirements elicitation is one reason why having a digital twin of the product is gaining traction. We do not believe that at this point a true digital twin can be crafted. You may have one for fluid dynamics, another for thermal properties, and still others for assembly flow to mechanical interference. We have yet to see a demonstration where all of these are combined. There is simply no way to interrelate all aspects of a product without making an experience-based "informed" decision that certain things are insignificant. Early weather and ecological models suffered from such informed decisions. Weather models are a math heavy digital twin of the Earth. Digital twins of any stripe are subject to the "butterfly effect" (i.e., trivial things can have non-linear impacts on a complex system).[39] Remember that in a digital twin environment your GTS organization and PD/LM system will need to accommodate terabytes of data for each model just to capture the change iterations. Companies will also need to establish the provenance of each model derivative to certify that the neutral step or xml file is equal to the model and the machine control conversion is also equal.

---

[37] Christel, M., and Kang, K. C. (1992). *Issues in Requirements Elicitation, Technical Report CMU/SEI-92-TR-012, ESC-92-TR-012.* Software Engineering Institute Carnegie Mellon® University, Pittsburgh, PA.

[38] Durak, D., et al. (2018). *Advances in Aeronautical Informatics Technologies Towards Flight 4.0.* Springer International Publishing, Part of Springer Nature, Gewarbestrasse 11, 6330 Cham, Switzerland.

[39] Lorenz, E. U. (2015). Predictability: Does the Flap of a Butterfly's Wings in Brazil Set Off a Tornado in Texas? *Indiana Academy of Science Classics*, Volume 20, Issue 3, March 2015, pp. 260–263.

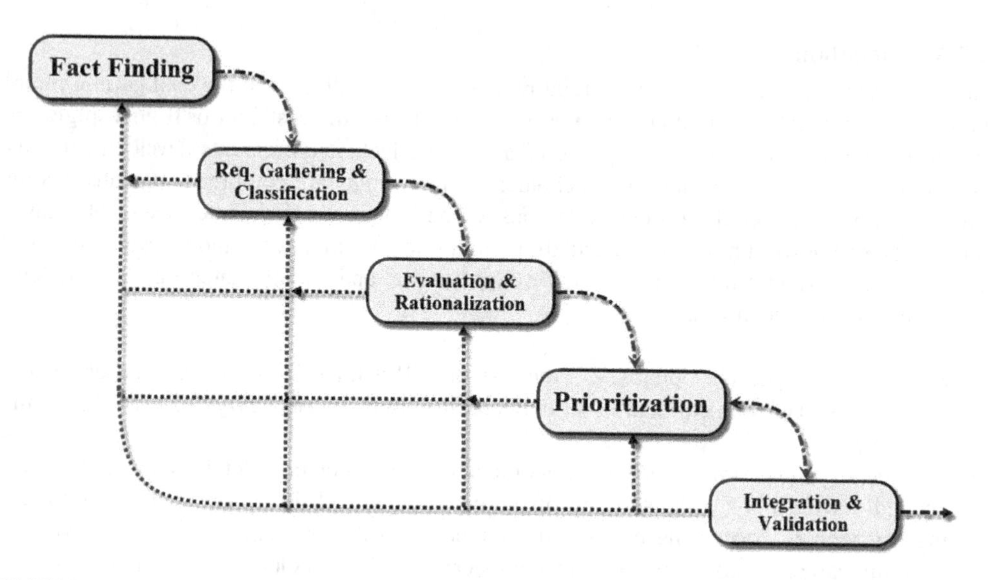

FIGURE 6.16 Christel and Kang requirements elicitation model.

Rzepka decomposes the requirements engineering process into three activities.[40]

1. Elicit requirements from individual sources.
2. Ensure that the needs of all users are consistent and feasible.
3. Validate that the requirements so derived are an accurate reflection of user needs.

It was found that in many market segments and in many companies no attempt was made to differentiate between requirements, design solutions, and non-requirements (desirements). This was simply due to a lack of understanding of the requirements elicitation process combined with a lack of understanding of the end goal of the process and the preferences of the individuals involved. Once the requirements elicitation process is clearly stated, such issues would be resolved by themselves.

Several useful views of the requirements elicitation process are defined in the following paraphrased information from Christel and Kang, whose developmental model is shown in Figure 6.16.

Issue-based information system (IBIS): A structuring method that allows the rationale underlying requirements to be organized and tracked. IBIS has several advantages. It is easy to learn and use. An indented text system was used in an early implementation of a tool for IBIS, and it was well accepted because the users were already familiar with the components (personal computers and text editors) used for the tool.

Domain analysis: The definition of features and capabilities common to systems in advance of development, providing a means of communication and a collective understanding of the domain.

- Facilitates communication
- Defines system boundaries
- Defines partitions, abstractions, projections
- Allows for opposing alternatives marking conflicts to facilitate resolution
- Makes it easier for the analyst to modify the knowledge structure

[40] Rzepka, W. E. (1989). A requirements engineering testbed: Concept, status, and first results. Proceedings of the 22nd Annual Hawaii International Conference on System Sciences, Kailua-Kona, January 3–6, pp. 339–347.

Joint application design (JAD), an IBM Corporation trademarked item: JAD's main theme is to bring together representatives with management authority and accountability into a structured workshop to foster timely decision making. JAD consists of five phases:

- Project definition
- Research
- Preparation for the JAD session
- The JAD session
- The final document

Controlled requirements expression (CORE): A requirements analysis and specification method that was developed in 1979 and refined during the early 1980s. It has not been used very much in practice in the United States. The work of Rome Laboratory (previously Rome Air Development Center) on the requirements engineering environment (REE) includes the CORE method. CORE can be considered as a paradigm in which the requirements specification is created by the customer and the developer, not one or the other. It is based on the principle of first defining the problem to be analyzed, breaking it down into viewpoints to be considered, and gathering and documenting information about each viewpoint. This data is then further analyzed and structured, an enhanced graphical representation of the data for each viewpoint is created, and then the viewpoints are examined in combination rather than in isolation. The last phase of the method deals with constraints analysis, in which non-functional requirements such as cost and time windows are identified, and the earlier phases are reviewed with respect to these non-functional requirements.

Quality function deployment (QFD): "An overall concept that provides a means of translating customer requirements into the appropriate technical requirements for each stage of product development and production." The initial steps of QFD can be described as "simply a system of identifying and prioritizing customer needs obtained from every available source." QFD is a concept that applies well to requirements elicitation, especially to an elicitation model where the customer's voice is the driving force behind requirements creation.

Soft systems methodology (SSM): It is the essence of a methodology as opposed to a method or technique in that it offers a set of guidelines or principles that, in any specific instance, can be tailored both to the characteristics of the situation in which it is to be applied and to the people using the approach. Users of SSM have to discover for themselves ways of utilizing it, which they personally find both comfortable and stimulating. Such is the variety of human problem situations that no would-be problem-solving approach could be reduced to a standard formula and still manage to engage with the richness of situations. Flexibility in use is characteristic of competent applications of SSM, and the reader should not look for a handbook formula to be followed every time. SSM is applicable to "messy, changing, ill-defined problem situations."

User skills task match (USTM): A user-centered approach to requirements expression. USTM is structured into three stages of description (describing a product opportunity), analysis (identifying a high-value solution), and decision making (delivering a business solution). It is a collection of techniques and methods designed for use by the key stakeholders in the development of initial requirements for *generic* systems. Generic systems are defined as those that are designed to satisfy the needs of many different customers or markets.

Paisley: Views the requirements specification as an executable model of the proposed system interacting with its environment. Paisley models a system as a series of asynchronous interacting processes but is criticized as being "a textual notation that is difficult to understand." Paisley specifications are *operational* in that they are implementation-independent models of how to solve a problem, rather than just being statements of what properties a solution should have. More recent papers on Paisley focus more on specifications than on requirements elicitation and analysis.

Scenario-based requirements elicitation (SBRE): Features the parallel development of requirements and a high-level design, the use of scenarios to communicate the behavior of a design, and an evaluation function to assess the suitability of the design. SBRE also recognizes the importance of an issue base with which the issues that arise during the elicitation process are maintained. Scenarios are used to structure the early interaction between users and designers in order to quickly develop a set of initial requirements. The author suggests that because scenarios have low cost and limited expressiveness, they "seem most appropriate for communicating specific system features in situations of high uncertainty."

Jackson system development (JSD): Involves specifications consisting mainly of a distributed network of sequential processes. It has been used primarily on data-processing applications. The major criticisms of JSD are that it is a very different *middle-out* approach, that the design is performed in a very fragmented manner, and that it may not apply well to large systems as there is no hierarchical representation or any overall view. JSD can be used to structure elicitation but is often used at a later development stage.

Structured common sense: A formalized outgrowth of CORE and contains the following steps:

- Agent identification (based on CORE's viewpoint hierarchies)
- Physical data flow analysis (derived from structured systems analysis)
- Action tabulation and description (derived from CORE's tabular collection)
- Causal tabulation and special case analysis

Regardless of the methodology chosen, the goal of requirements elicitation is to capture all requirements. Often requirements may not be understood by everyone making input to the elicitation process. Equal consideration must be made in the elicitation process for determining data fields in a PDM or PLM system. Do you have a requirement to report on distinct classes of procurements? If so, that data needs to be stated as a requirement for PD/LM system content associated with vendor items (specifications, statements of work, control documents, etc.). Metadata fields need to be created to capture data pertaining to sources of supply so that the information can be mined later.

Metadata fields also offer a convenient way for a company to give access into the PD/LM system to its suppliers and customers.

- A customer may be given rights to see everything associated with a program name. This includes program supplier data.
- A supplier may be given access to everything associated with that supplier but nothing else on the program(s) they are working on.
- Customers and suppliers are not given access to data not associated with them.

In each case, the PD/LM access is tied to a specific metadata field. Using this method, suppliers and customers can only see their data and are barred from accessing any other data in the PD/LM system.

6.3.5.6 Prioritizing and Tracking

Determining what is important to prioritize and track is not unlike management of your schedule. While the outward aspects appear to be very similar, once you take a closer look, you will find that the schedule is in a constant state of flux; it is an intricate dance between your goals for the day and the environment with which you interact. The same is true of CM implementation. Once the requirements are defined, there is a progression of the product through the stages of the product life cycle. The elements are prioritized and tracked. They change as the product moves through each phase. The focus of CM also changes in each phase.

Prioritizing and tracking of the right informational elements at any point in time is not unlike the job of a sommelier when pairing wine with a master chef's creations. Just as the sommelier takes the time to understand the sense of taste and what flavors and scents go together to provide a pleasure for the palate, the CM implementation must understand the sense of what is required throughout the life of a program and present a solution set that seamlessly melds with those needs.

This may sound like a philosophical discussion, but it is not. Anyone with the proper attitude can follow a recipe and make a meal. The difference between a cook and a chef is that the chef can walk through the market and intuitively understand how to blend the available products to create something new and fitting for that moment in time. They do not need or use a recipe book except as a reference. The chef is like the program's manager. The configuration manager is like the sommelier. The chef and the sommelier are a team, as are the engineering manager and the configuration manager.

If you want to see what this sense of cooking rather than the traditional textbook approach looks and feels like, a good start is to view a few episodes of the BBC series *Two Fat Ladies*.[41] Once you have done, refer to any recipe booklet for the same dish. You will immediately discover that truly understanding something is vastly different than knowing it. One can cite chapter and verse and still not understand what it means and what it implies.

Just as the sommelier is guided to the right wine pairing, the configuration manager is guided by the complexity and understanding of the product, the stated requirements, the derived requirements, and internal and external regulations, policies, and procedures. So how does the aspiring CM professional learn these skills? What follows are a selection of observations on just that gathered over a period of decades:

- Teach yourself to see the interrelationships between things. One way to open your mind to doing this is to play the game Tangoes[42] once a day, whenever you are on an airplane or whenever you are looking for an answer that eludes you.
- Bring large work issues down to size and relate them to something you know.[43]
- Develop critical thinking skills and recognize that it is easier to do a minor course correction once started than it is to start in the first place.
- Perform or participate in an N-squared analysis (Table 6.1) on every new program until you can do it intuitively.
- Determine what your educational acumen is and shore up the soft spots.
  - Understand what you aren't good at and develop a network of people who can help you in those areas.
- Seek to understand everything you can about the product, how it works, and what subsystems drive others.
- CM is not limited to the release of changes and delivering data to a customer. It is not following process 1 then process 2 and then process 3. Only by knowing the product and what you want to do with it throughout its life and applying the correct process elements in the correct way do you achieve the desired results. There is a great deal of difference between the paint by number picture and the original.
  - Procuring a complete technical data package (TDP) if you never intend to have someone else make another one is not reason enough to expend the money and time to do so. You must understand why you want the TDP and what end result it supports.
- Add value. Do not simply keep the peace by trying to accommodate everyone. Push back with sound reasoning if you know a request is not within the scope of the program or management objectives.

---

[41] Llewellyn, P., Geilinger, S., and Field, A. (Directors). (1996). *Two Fat Ladies*. [Motion Picture], BBC, London.

[42] Based on the ancient Chinese Tangram puzzle. Also see Read, R. C. (1965). *Tangrams: 330 Puzzles*. Dover Publications, Mineola, NY.

[43] Angier, N. (2009, March 30). *The biggest of puzzles brought down to size*. The New York Times.

TABLE 6.1
N-Squared Activity

CM Analysis	*External Regs*	*Internal Regs*	*Contract*	*SOW*	*Specification*	*MAR*	*DRLs*	*Mechanical*	*Electrical*	*Etc.*
External Regs	*External Regs*									
Internal Regs	Update G2.5.9	*Internal Regs*								
Contract		Dev req to G3.2.1	*Contract*				H.52 & I.27 not covered			
SOW			Attach A	*SOW*						
Specification			Attach B	Reference 2.1	*Specification*		23 TBDs inpact SE-4			
MAR		Alternate procedure	Attach C	Reference 2.2		*MAR*	5.2.1 to 5.2.5 not included			
DRLs		Dev req to G3.2.19	Exhibit 1	Reference 2.3			*DRLs*	ME-1 to 23	EE 1-18	
Mechanical								*Mechanical*		
Electrical							Dev req to G7.2.14	I-ICD	*Electrical*	
Etc.										*Etc.*

- Do the very best you can today and learn to sleep well knowing that tomorrow you will have to reprioritize and occasionally go back to the beginning and start over.
- Understand that you and others will make mistakes. Understand the risks and have a backup plan in place to mitigate those risks in case the approach you take does not work.

6.3.5.7 Relationship Between Requirements and CM

There is no set of rules that can be established for a specific product or a single phase of the product. There is no set of CM requirements that defines everything that the configuration manager must be aware of. There is no set of requirements that can be written that defines everything that a system, project, and product must do to be successful. At times, the stated requirements are poorly defined, overlooked, or missing.

Every requirement is associated with information that requires some degree of CM. The more critical the information is to the product, the higher the level of management associated with it. Well-written requirements will have a single information artifact. Poorly written requirements may have multiple or nested information artifacts. All written requirements when taken together are a subset of the information associated with CM implementation.

There is a difference between cited requirements and derived requirements. It is in derived requirements that most CM activities fall short. Most managers and engineers do not recognize the total scope of the activity and think that CM ends with the delivery of the product or in some cases the last delivery of informational data associated with a DRL. That is simply not the case. CM goes well beyond these events and in some cases a program's CM implementation influences events for hundreds if not thousands of years.

Have you ever been to Europe? Have you ever wondered what the purpose is of the ball on the steeple of many of the churches in Switzerland and Austria? On the first look, it appears to be part of a lightning strike mitigation system like the lightning rods commonly used on most large structures. While that may be the secondary purpose, the real purpose in many communities is to hold the architectural drawings for the church itself, so that if it is ever damaged or destroyed, it can be replicated with an identical copy. This represents the implementation of an implied requirement for disaster recovery as well as sound CM and European religious community records retention.

A requirement to provide a vacuum bake-out plan leads to the derived requirement to have a vacuum chamber large enough for the purpose that can be heated, hold a vacuum, and be cleaned, as well as have the requisite uninterruptable power supply, some form of instrumentation, and perhaps feed-through plates for added instrumentation. The requirement to deliver a reduced thermal model in a specific format has a derived requirement that your suppliers provide component and subsystem models that can be integrated into that format in compatible formats.

As an exercise, assume that you are responsible for evaluating a vendor's CM system and the only criterion in your internal policies, procedures, and practices is the following statement.

> Suppliers shall have an ISO configuration and data management implementation compliant with AS9100-2009, Quality Management Systems—Requirements for Aviation, Space and Defense Organizations.

- What is the real requirement and where do you start?
- How extensive is the task at the vendor?
- What specifically do you look for?
- Is it as simple as asking to see the results from an independent certification organization like the British Standards Institute for ISO or the Smithers Quality Assessments for AS9100-2009?
  - Is something more required?
  - Is something less required?

The same thought process is inherent in assessing requirements and ensuring they are met.

6.3.5.7.1 Requirements Allocation Within CM

In systems engineering, after elicitation is completed, the requirements are allocated to the design, manufacturing, mission assurance, and other functions. Understanding the functions within a configuration allows the same kind of allocation to take place regarding those functions in the CM implementation. Some requirements will align with change control, identification, status accounting, and data management. Further graduations will occur based on what the requirement pertains to (e.g., project completion success, product/service success, or business success). This is where an N-squared analysis is very helpful. The exercise will firmly establish where each requirement comes from and the planning that is associated with it.

Requirements originating internal to the company because of compliance with legal statutes naturally fall to business success. Requirements originating internal to the program because of the program plan or the contract and its attachments naturally fall to project success. Requirements related to a specific product naturally fall to product/service success. The time and attention paid to each will vary over the life cycle of the product. Strategic capabilities fall in the company success realm. Tactical capabilities relate to product/service success.

6.3.5.8 Logistics

Activities that culminate with product sales are simply a well-orchestrated precursor to the operational phase. In some industries, this is known as the "logistics phase." If the configuration has not been properly managed, it can put the company in a loss position relative to customer satisfaction, market share, and profitability. This is where a great CM implementation is easily distinguished from a not-so-great one.

One current definition of logistics is:

> Logistics is the management process that assures resources of any description are available when needed. Items managed by a logistics function can include energy, communication, networks, information, physical items (food, materials, equipment, and parts) and people. It may involve material handling, production, packaging, inventory, transportation, warehousing, and security. Logistics activities can be modeled and optimized using simulation software. The goal of logistics is to minimize the need for excess resources of any kind. The operation of a just-in-time manufacturing process relies heavily on the logistics function. In such a manufacturing scenario; materials and other resources arrive in the production flow minutes before they are needed so that parts inventory is minimized and factory through-put is maximized.

Logistic activities pertain to procurement, production, distribution, after-sales support, disposal, green technologies, and global and domestic activities. Tied up with this are product warranties and warranty and license pass-through from component and subsystem providers. The management of the information associated with each of these activities as well as operations and maintenance of the current configuration happens all around us. In the commercial sectors, electronic devices and database prompts are established to facilitate the maintenance of purchased items and to gather data regarding customer satisfaction with those products. Examples of this are as simple as a red engine light on a vehicle dash to indicate recommended service. The email you receive from a medical provider reminding you of an appointment is a logistics activity, as is the satisfaction survey you complete once any service has been completed.

Logistics combines elements of project completion success, product/service success, and business success. In a fielded product, if the design did not build in the capability to self-verify the need for a logistics activity, that information must be tracked by an individual. Self-verification is most often associated with electronic or electromechanical devices. The low battery indicator and the signal reception indicator on a personal communications device are built in. The activity associated with evaluating the version level of the software suite you have on your computer and notifying you a newer version is available is a self-verification activity. The specific configuration of a one-off or limited production run automobile or other transportation device may not have such embedded logistic elements.

Depending on the complexity of the product and its price point, the degree of logistical activity involved will vary. The more expensive the item is, the more support is generally available and planned for. During the early years, the costs of components were more expensive than the labor costs to repair a radio or television. A wide network of certified repair specialists existed. Repair shops were resplendent with oscilloscopes, tube testers, and replacement vacuum tubes (in American English, tube; or thermionic valve or valve in British English). Repair components were common and could even be purchased in larger convenience stores. Today, entertainment devices have better reception, higher reliability, and a lower purchase price when adjusted for inflation. Many are considered disposable items, not worth the cost to repair. No logistic provisions have been made to support them other than in the accessory market. Radio receivers like those produced by AOR, Atlas, Collins, Drake, Eaton, Elecraft, Grundig, Hiberling, Icom, JRC, Kaito, Persues, Tecsun, Ten-Tec, and Yaesu tend to have higher prices and higher levels of logistics support.

This reflects a continuance of the trend that started with logistic support activities during World War I. If the cost to repair is higher than the cost to replace stock, a new assembly is ordered instead of repair parts being ordered. The exception was limited to cases where the item is repairable in the field by the user and critical to the mission objective. Ground troops were issued a repair kit that included firing pins and replacement springs for their firearm but not replacement caps for their canteens. Logistics continues to evolve, as demonstrated by the Plateforme d'Échange Normalisée et Centralisée d'Information Logistique (PENCIL) initiative. PENCIL will establish a new way to share logistics data between the French army and industry based on PLCS/STANAG 4661.

How a company determines logistics for repair or replace and warranty activities is dependent on company culture and philosophy. Companies such as Timex have a very simple approach. They may repair your watch by installing new or reconditioned components or replace your watch with a similar model. It may also be mandated by regulation or be specific to a contract. Regardless of the source of the requirement or its application and what it entails, CM implementation is key and must consider what is required at the start of a new project as part of CM planning.

Perhaps the greatest challenge facing the maintenance of fielded systems is efficiency of repairs and maintenance status. The International Society of Automation (ISA) estimates that manufacturers lose €565,050,000,000 globally every year to downtime. To combat this cost drain, companies are moving away from reactive and preventative maintenance and embracing predictive maintenance that is made possible for 5G technologies.

Recent innovations in the field of virtual reality and artificial reality will change how operations and maintenance activities are done. The use of virtual reality (VR) and augmented reality (AR) to assist maintenance operations (Figure 6.17) is on the rise. Instead of carefully written time

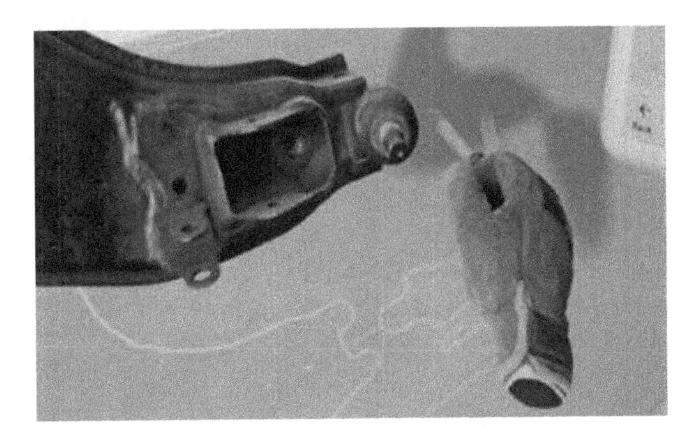

FIGURE 6.17 F-35 maintenance team explores VR, AR for training needs (U.S. government).[44]

[44] Sexton, D. Maj. (2018). https://www.arnold.af.mil/News/Article-Display/Article/1719912/f-35-maintenance-team-explores-vr-ar-for-training-needs/. Accessed December 21, 2018.

compliance technical orders (TCTOs) detailing every step required to remove, replace, and check out the installation of components; maintenance staff are now able to put on a VR headset and see the entire maintenance sequence floating with voice-over instructions beside the unit they are working on. The VR instruction sequences can be backed up and replayed if necessary. An added advantage of VR and AR technologies is the complicated assembly sequences for manufacturing or test of a product can be performed in VR before the production hardware is touched.[45]

[45] Christiansen, B. (2018): *The Use of AI and VR In Maintenance Management*, Engineering.Com, 2018-12-03, https://www.engineering.com/AdvancedManufacturing/ArticleID/18100/The-Use-of-AI-and-VR-In-Maintenance-Management.aspx. Accessed December 4, 2018.

7 Configuration Management Support of Functional Resources

7.1 QUESTIONS ANSWERED IN THIS CHAPTER

- Identify the areas of manufacturing most critical to be under configuration management (CM) control.
- What is the relationship between inventory control and CM?
- Demonstrate how CM interfaces with
  - Engineering (new products)
  - Engineering (adaptations of existing products)
  - Manufacturing
  - Command media
  - Contracts

7.2 INTRODUCTION

CM implementation as it applies to other organizations is overlooked to the detriment of project or program. This aligns CM and DM activities with support functions. The activities assigned to each of the functional resources are typical and not all inclusive. Functional resources include contracts and are intentionally centric to contracted work in many areas. A contract or program can be with the public at large, with an internal organization, or in a contractor or subcontractor/vendor relationship. What follows should be read with a broad interpretation.

7.3 CM AND PROJECT MANAGEMENT

CM activities tie to program management and vary from program to program. They are based on variations in regulatory compliance, market sector, product mix, customer base, and product market sector. CM is centric to ensuring that the product is documented to the degree necessary to ensure reproducibility, acceptance, and post-acceptance activities and communications.

7.3.1 Acceptance

CM activities include preparation of acceptance documentation including certificates of conformance (COC) or other documents such as the U.S. government's DD Form-250, Material Inspection and Receiving Report (MIRR). One often overlooked CM acceptance activity found on U.S. government programs is the need to document formal acceptance of all data deliverables that required approval during the life of the program. Instructions for DD Form-250 contained in the U.S. Federal Acquisition Regulations (FAR) DOD FAR Supplement (DFARS) Appendix F specify in F-301(b)(2)(iii)(B) that the shipment number for the last DD Form-250 is designated by adding a Z to the end of the shipment number. This is generally referred to as a Z-Code DD Form-250. Data acceptance is often the last activity to take place after other contract requirements have been met. The DD Form-250 was supplanted by the Wide Area Workflow Receiving Report (WAWF RR). Use of DD Form-250 is at present the norm for some other U.S. government agencies such as NASA. Both DD Form-250 and WAWF RR are multi-use vehicles as described subsequently.

7.3.2 DD Form-250 Use[1]

The WAWF RR and DD Form-250 are multipurpose reports used:

1. To provide evidence of government contract quality assurance at origin or destination.
2. To provide evidence of acceptance at origin, destination, or other.
3. For packing lists.
4. For receiving.
5. For shipping.
6. As a contractor invoice.
   a. The WAWF RR or DD Form-250 alone cannot be used as an invoice, however, the option exists to create an invoice from the RR or a combination (invoice and RR), both of which minimize data entry.
7. As a support for a commercial invoice.

7.3.3 CM and Project Integration Management

As companies move toward domestic and international collaboration, a very high degree of transparency is required between all stakeholders. CM is the primary means to ensure that all leveraged innovation elements are united. This is different from project communication.

CM ensures that the correct data are disseminated and that the individual collaborators working to bring a product to market understand what is required of each other. Collaboration can take the form of giver–receiver lists, interface control drawings, specifications, statements of work, and other forms of information.

In one of the firms studied, the task was to apply intellectual property (IP), Technical Assistance Agreement (TAA), and U.S. International Traffic in Arms Regulation (ITAR) legends using the PD/LM report writer. If they could apply legends with the help of a pull-down menu, it would allow a single document to be used on multiple programs and transmitted to multiple customers and vendors. A team of the top talent in the company was pulled together to ensure that the implementation was clean.

The solution was rolled out and everyone liked it. Unfortunately, there were issues. Markings could not be applied to book-form documents. A book-form document is comprised of a set of pages in vertical format that are all identified as sheet 1 and numbered 1.001 and up. These are followed by sheets 2 and up in vertical format numbered sheets 2 and up. Interestingly, this document format was known but not considered. It has still not been resolved two years later. Standard sheet sizes are shown in Figure 7.1.

CM is associated with not only the management of configuration and data associated with those products but also the management of the infrastructure required to ensure that the product can be managed. This includes the management of information distribution, information description, information scheduling, progress reporting, the escalation (resolution) process, and the associated administrative closure.

7.3.4 CM and Scope Management

Figure 9.2 shows how requirements are allocated from the product level (often referred to as the "functional baseline") to associated systems, subsystems, and components to create the allocated baseline. Scope management ensures that technical requirements are understood and that uncertainties are eliminated. The faster that all items classified as "to be determined" (TBD),

[1] U.S. DOD FAR Supplement (DFARS) Appendix F, F-103 Use.

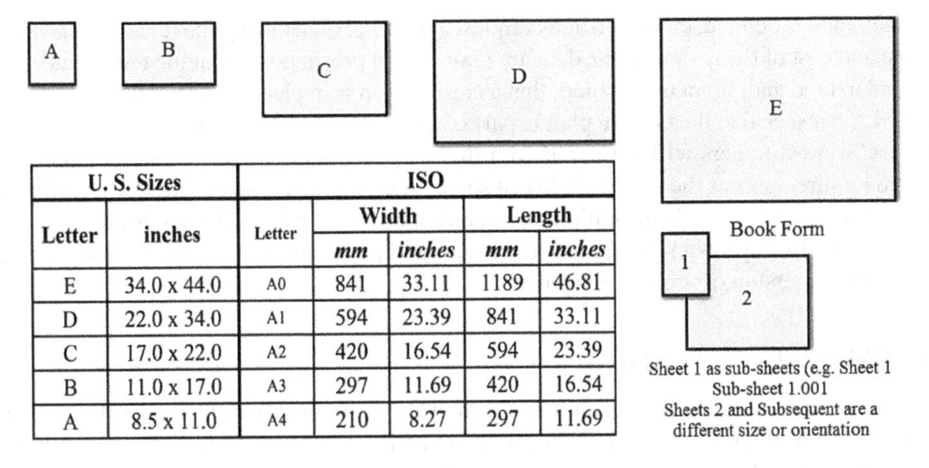

U. S. Sizes		ISO				
Letter	inches	Letter	Width		Length	
			mm	*inches*	*mm*	*inches*
E	34.0 x 44.0	A0	841	33.11	1189	46.81
D	22.0 x 34.0	A1	594	23.39	841	33.11
C	17.0 x 22.0	A2	420	16.54	594	23.39
B	11.0 x 17.0	A3	297	11.69	420	16.54
A	8.5 x 11.0	A4	210	8.27	297	11.69

FIGURE 7.1 Standard sheet sizes in the United States and metric.

"to be specified" (TBS), "to be reviewed" (TBR), and "to be negotiated" (TBN) are closed, the firmer the technical requirements become.

- TBD means a variable to be defined by the proposer.
- TBS means to be specified by the customer.
- TBR means to be reviewed and subject to revision.
- TBN means to be negotiated between the customer and the contractor.

There are other requirements in the product life cycle that must also be allocated in the same way. Reviews such as those shown in Figure 4.9 fall into this category. These reviews are most often delineated in the work statement. Typically, they are also tied to development program and cost events. On programs with milestone billing, the contractor is paid when the milestone is completed. Below is one such milestone payment description:

> Payment shall be made to the contractor upon successful completion of the Preliminary Design Review defined in Statement of Work (Section 3.3.2.1).

A review of statement of work (SOW) Section 3.3.2.1 revealed that there was a tie-out to the associated data item delivering the preliminary design review (PDR) presentation but no description of what *successful completion* means. The lack of a definition is effectively another TBN in the documentation often overlooked until after the review is held. This places the contractor at considerable risk as the payment could be held open for months after the event took place. Successful PDR—in this case, completion—could mean any of the following:

- The PDR was held.
- All customer and contractor action items are closed.
- All contractor action items are closed.
- All customer and contractor action items mutually agreed upon as critical (e.g., must be resolved prior to continued work) are closed.
- The customer wants to follow tasks to resolution that should be addressed during the critical design review and will hold payment for PDR until that is completed.

The question arises, "Who is responsible for finding these other undefined requirements and getting them resolved?" Ultimately, it is the program manager and, until what constitutes a successful

PDR completion is defined, customers may request additional data during the review. This increases the scope and cost of the review. If the data are managed, it presupposes that the requirement for the data is understood and, if not understood, that a closure plan is in place to ensure that understanding is reached. That says that the closure plan is part of the CM implementation.

Many CM practitioners will not agree with this view and see the identification and closure of undefined requirements as the responsibility of some other organization. However, they fully accept that they play a significant role in configuration control boards (CCBs), whose roles involve evaluating changes for all impacts including risk to closure and payment. This includes taking active and participatory ownership of scope management.

7.3.5 CM and Time Management

Perhaps the most recognized proponent in the United States for process efficiencies through motion study is the team of Frank Bunker Gilbreth Sr. and Lillian Evelyn Moller Gilbreth. They were acutely aware that if the process was evaluated to eliminate wasted motion, the workflow was more efficient. Lillian continued the pioneering work running the consulting firm Gilbreth, Incorporated, long after Frank's death in 1924, outliving him by 48 years. The Gilbreths' methodology differs from other kinds of efficiency management, such as Fredrick Winslow Taylor's. Taylor's focus was reduction of process times rather than improving system process. Time management efficiencies through process in the CM area directly impact the entire program life cycle.

This may be exemplified by looking at the web-based Oracle's AGILE© PD/LM system. As typically implemented, Oracle's AGILE© requires a document object to be created before it can be released. The document object contains metadata fields. Oracle's AGILE© also requires an engineering order (EO) for all releases of documents and any associated parts. Oracle's AGILE© even allows the creation of an EO from the document object itself. What does not happen is the auto-population of all the metadata fields from the document object to EO. These metadata fields are generally company defined during implementation and PD/LM rollout. They may include:

- Document number
- Document name
- Program name
- Work authorization (WA) of root charge number
- Configuration item identifier (CII) number

This oversight means that the EO originator must fill these metadata fields again prior to the EO being moved to the review or approval states in the workflow. Companies using the Oracle's AGILE© PD/LM system can choose to accept this or, as the release process owner CM, can choose to circumvent the time required by each user to accomplish this activity by creating a generic EO, document object, and associated parts for each program and correctly filling out the metadata fields. The generic EO will have the document object and the associated part objects on its "affected items" tab. By then providing a link to the EO in the CM plan, those doing releases have a prototypical EO. As the EO, document object, and associated parts can be saved with a new identifier, as much as 10 minutes per item can be saved. This is a Gilbrethian CM process improvement.

Looking at CM implementation across all functional resources from this perspective forms much of the grist in the process improvement mill. It helps companies to maintain a competitive edge. Ultimately time management is directly associated with cost management.

7.3.6 CM and Project Communications Management

Information distribution, information description, information schedule, and progress reporting are all CM-centric and DM-centric activities. CM acts as the program manager's tool for maintaining information access and storage as well as permission structures for program servers, intranet sites such as SharePoint, and site interlinks to ensure a single point of truth on the site and that data links are maintained.

7.3.7 CM and Project Cost Management (Earned Value Management)

We evaluated the Oracle's AGILE© PD/LM system against implementing the Gilbrethian CM process improvement described earlier for anticipated cost savings on a program valued at €27,000,000. It results in a positive impact of over €131,001.75 to the program bottom line.

7.3.7.1 Assumptions

- Average billable engineering rate of €174.67/hour
- 1,000 documents (90% of the metadata carries forward from revision to revision)
- 3,000 parts (90% of the metadata carries forward from revision to revision)
- Five EOs per drawing (90% of the metadata carries forward from revision to revision)

7.3.7.2 Calculation of Cost Avoidance

Cost avoidance is calculated using Equations 7.1–7.8.

$$1{,}000 + 3{,}000 = 4000 \text{ items revised}$$

EQUATION 7.1 Number of items revised.

$$4{,}000 \text{ items revised} + 1{,}000 \text{ EOs for initial release} = 5{,}000 \text{ objects each revision}$$

EQUATION 7.2 Number of objects each revision.

$$5{,}000 \times 10 \text{ minutes} = 50{,}000 \text{ minutes}$$

EQUATION 7.3 Minutes spent inputting metadata for first release.

$$50{,}000 \text{ minutes} / 60 \text{ minutes} = 833.3 \text{ hours}$$

EQUATION 7.4 Minutes spent inputting metadata for first release.

$$833.3 \text{ hours} \times €174.67 \text{ per hour} = €145{,}552.51$$

EQUATION 7.5 Cost savings on first release.

$$€145{,}552.51 \times 90\% = €130{,}997.25$$

EQUATION 7.6 Cost savings per EO after initial release.

$$€130{,}997.25 \times 4 = €523{,}989$$

EQUATION 7.7 Cost savings for four EOs after initial release.

$$€523{,}989 + €145{,}552.51 = €669{,}541.51 \text{ cost avoidanc}$$

EQUATION 7.8 Calculation of cost avoidance.

7.3.7.3 Total Program Cost

Total program cost (TPC) must be defined in terms that everyone can understand. It is not a concept that those outside of program management are used to thinking about or implementing. CM as practiced in the U.S. military industrial base has sequestered itself from those it supports across the industry and at the companies they are part of (the company). As a result, many tools utilized by these CM practitioners are of limited use to both internal and external customers. Understanding everything that CM entails and implementing sound CM practices have universally been shown to reduce TPC while increasing reliability, quality, and safety.

The TPC approach means that CM must look at the tools with a broader perspective and understand how others use them, how CM uses them, and the cost benefits of their continued use. If a single click or mouse movement for those using the tool is deleted, it generates a large cost savings over the life of any program and results in a more satisfied customer base.

7.3.7.4 Value to the External Customer

Value to the external customer in its simplest form can be summed up as satisfying internal and external customer requirements:

1. Making external customers succeed.
2. Providing good hardware, software, and services below cost and on schedule.
3. Meeting the minimum requirements of the contract.
4. Establishing and maintaining a congenial and candid working relationship with the customer's contracts and program management and their DM staff.
   a. Listening to their needs.
   b. Evolving the structure of how data are tracked and reported to maximize use.

Anything done beyond this is extra, unless required to meet regulatory requirements or needed to keep International Organization for Standards and other certifications (e.g., AS9100) active.

7.3.7.5 Value to the Company

Value to the company (VTC) must be defined as maximizing EBIT and bottom-line profitability. Program decisions must meet the company's core initiatives such as diversity, sustainability, and safety while being the first-choice provider of its goods and services.

VTC also requires making hard decisions and broadening CM implementation across traditionally drawn lines. Being proactive instead of reactive and protecting the company's interests through ethical decisions is a must. This pays off in combination with a sound management schema of product structure and the data structure and regulatory, contractual, and technical requirements.

7.3.8 CM and Project Quality Management

7.3.8.1 Protection from Audit

Protection from audit is the component of the CM implementation that ties directly into a sound management schema of not only any associated product structure but also the data structure and contractual and technical requirements. It includes things such as

1. Having a single-source of truth
2. Maximized use of current tools
3. Maintenance of a library of all contract documents in their conformed state (current contractual state) along with copies of all lower-level versions or revisions throughout the life of the contract

7.3.9 CM and Project Risk Management

Risk management is a multifaceted approach to either eliminate or reduce risks over the life of a program. Some elements related to CM implementation have been discussed as they specifically relate to time management, communications management, and project cost management.

Perhaps one of the most important aspects of risk management is associated with CM tools. Many organizations do not have a truly integrated GTS infrastructure either closely tied to CM implementation or reporting through a CM change control system. Without these, GTS decisions can be made regarding the company's *tool suite* that do not support ongoing programs or projects.

In response to the *end of service* announcement for selected International Business Machines (IBM) rational products shown in Table 7.1, one company surveyed reported that GTS in conjunction with the software engineering department determined there was no impact, as the company was moving away from this product. This determination led to a decision of not migrating to the newer versions of the software, and plans were made to deactivate existing licenses internal to the company, forcing an earlier migration to the new tool suite.

Unfortunately, several satellites having a remaining mission life up to 15 years relied on the IBM rational tools to maintain the on-orbit software. The impact assessment performed by GTS and software engineering had not considered anything beyond new development in their decision tree, nor did they socialize the withdrawal date of the software with stakeholders outside the two departments. Information instead filtered back to the company via a software CM forum on LinkedIn.

A similar error occurred at a different company studied. The firm's business was centric to a specific customer who mandated the type and version of the software used for documents and drawings in all its contracts. Contracts with the customer totaled more than €7,160,000,000. Company GTS decided that it was in the company's best interest to move all the computers to the latest version of available applications and rolled out the implementation during a Saturday upgrade. Unfortunately, the new software suite was not fully downward compatible. Although users could open files from previous versions, they had to save in the newer version. Data deliveries to the customer on all programs were impacted. Data submittals were rejected as not compliant with contractual requirements and payments to the contractor were withheld. The effort to reinstate the previous version of the software suite to be compliant with contractual requirements ended up costing the company more than €131,000,000 once customer fines were added in.

7.3.10 CM and Action Item Lists

Starting with the later 1990s and carrying through to the present, emphasis was placed on managing the typical work day using pre-printed daily, weekly, and monthly planners. The concept was touted as the next step in personal time management and marketed by a variety of companies under names such as Day-Timer©, Action Day Planner©, Day Designer©, and others, many of which are still

TABLE 7.1
End of Service IBM Rational Selected Products 2015. VRM, Virtual Resource Manager.

Program Number	VRM	Withdrawal from Support Date	Program Release Name
5724-X73	7.1.0	April 30, 2015	Rational Synergy
5724-V66	7.1.0	April 30, 2015	Rational Synergy
5724-X80	5.2.x	April 30, 2015	Rational Change
5724-V87	5.2.x	April 30, 2015	Rational Change

available today. The idea was that if your day, week, and month were organized, your life would be organized as well and that would leave more time to concentrate on the things in life that mattered.

On a large fixed price program, the manager issued daily, weekly, and monthly planners to the entire staff and mandated their use. Employees were instructed to enter events in the monthly planner, expand them in the weekly planner, and further expand them in the daily planner. The mandate included all activities being performed. Employees were further informed that all actions should be prioritized and that the daily, weekly, and monthly planners would be audited Monday of each week.

Six weeks into the program it was behind schedule and over cost. A team of experts was hired to find out what was going on. They discovered that each employee was spending more than 32 hours a week keeping the planners updated to pass the program manager's time efficiency audits. That time was charged directly to the program. The experts' recommendation was to stop the insanity and to move into an integrated action item list. The vice president of operations agreed. The program manager was let go. His obsession with the planners had cost them more than €20,079,014. The program never fully recovered from the cost hit.

7.4 CM AND SYSTEMS ENGINEERING

A long-time and highly respected CM practitioner, when asked what CM did for engineering, replied:

> CM manages the engineering infrastructure, infrastructure improvements and documentation providing configuration status accounting data as required so that engineering can do its job.

Although this is an oversimplification, it is not far from the truth in many of the highly integrated organizations we studied. In retrospect, our question should have been, "How does CM implementation support systems engineering?" We asked this new question while researching this second edition and the reply was:

> CM implementation weaves the various digital threads together creating a digital fabric. That fabric interconnects data nodes providing systems engineering the information needed to make informed decisions and trade-offs between subsystems to optimize total system performance while maintaining requirements compliance.

Translated for systems-speak, it means that CM implementation lets the systems engineer understand the partial derivative of everything with respect to everything else (which we expressed earlier in Equation 6.1).

7.5 CM AND PRODUCTION

In a CM audit in 2011 against AS9100 requirements implementation, the auditor asked three critical questions:

1. How do you control the configuration of models and files in the CAD databases?
2. How do you control the configuration of the files provided to the production floor for use in your numerical control machines and additive manufacturing?
3. How do you verify that only the current models and files are used during the production process?

These questions point out an aspect of CM implementation that is occasionally overlooked. If you do not know what must be managed, you may not be managing it. The discovery of what must be managed presupposes a basic understanding of the tools used by every functional resource and how each output from the CM implementation interfaces with those resources.

It is not enough to follow a checklist. CM implementation must consider why the items are on the checklist to begin with and evaluate the currency of those items against the current company environment. Once designs are released to the factory floor, the information in them must be available in the form needed by the software managing the machine itself. If it isn't, you have entered a man-in-the-middle situation. So how does this play out in a real-world environment? In the early days of computer aided design/computer aided manufacturing, drawings were done on vellum or paper. Copies were provided to the factory and the production staff would code drawing parameters into a machine that produced a Mylar tape. The tape was fed into the machine and it controlled the milling or lathe operation. This process evolved into one where the production staff now loads files into a gcode interpreter that converts files into computer numerical control (CNC) machine instructions. If the files are not captured in the PD/LM system at time of release, they are not being managed by the PD/LM system and change control is lost. CNC, electrical discharge machines (EDM), and additive manufacturing operations have not yet moved to a standardize file format. Typical file types that need to be captured in the PD/LM system are:

- CNC mills and lathes
    - .gcode via a gcode interpreter using
        - .DXF in polyline format
        - .eps
    - EDM wire and ram (die sinking)
        - .STEP
        - .pdx (parasolid)
- Additive manufacturing (3D printing)
    - .STL
    - .OBJ
    - .gcode
    - .3MF
    - .X3G
    - .AMF
    - Collada
    - VRML97/2

7.6 CM AND FINANCE

The term "finance" as used here relates to company finance rather than to program fiscal management. Each year, the finance department prepares justification for tax write-offs for technological improvements associated with research and development efforts. Technological improvements are also inherent in all firm-fixed price work. The customer (internal or external) is paying a firm-fixed price for the design, development, fabrication, test, and post-acceptance activities associated with a product. The contractor provides whatever talent, facilities, and innovation is required.

The data required to support tax write-offs are gathered over the operational life of each program and segregated by each innovation. CM implementation tracks and allows reporting on technological improvements using metadata fields in the PD/LM system.

7.7 CM AND LEGAL

The legal department ensures compliance with all regulatory requirements as well as the protection of the company's IP rights. CM implementation is critical not only to disseminating IP plans and markings but also to ensuring that only properly marked data are provided to customers, vendors, and strategic partners. CM implementation also ensures compliance with national data rights protection such as the U.S. ITAR or specific marking requirements associated with TAAs with international partners or customers.

7.8 CM AND THE CONTRACTS DEPARTMENT

As pressures increase to keep non-direct charges down, significant pressure is placed on contracts departments to reduce indirect costs. Only direct costs generate indirect dollars. Contracts department staff reductions do not equate to contract requirements reductions. This places a greater reliance on a CM implementation that generates PD/LM data to meet contract required DM activities. These reports were manually generated by less experienced contract professionals in times past.

Contract department CM and DM tasks include:

- Maintaining a conformed (cut and paste) version of the entire contract and contract technical requirements.
- Ensuring that change proposals include the proper forms, formats, and proposed language that represent the intent of the change. This includes proper wording of new requirements, so that undocumented TBS requirements are eliminated.
- Preparing acceptance paperwork, such as the MIRR (DD Form-250), certificates of completion, and so on.
- Interfacing between contracts and property management to ensure that all customer and contractor property is identified and managed.
- Auditing risk mitigation activities.

A contract is an agreement for a lawful purpose that consists of an offer and acceptance with specific terms made between two or more competent parties in which there is a promise to do or refrain from doing a certain thing and an acceptance of that promise in exchange for some form of valuable consideration.

The recent move in the United States to allow electronic signatures on real estate transactions is not the norm. Most of the market segments still rely on a manually signed contractual instrument except in those cases where the transaction is over the counter. A signature is also required for the purchase of prescription medications. Although most dispensers of prescription drugs do not have a CM department by name, they still have very sophisticated DM activities often networked across the region's druggists and medical providers to ensure patient safety. In such cases, the patient can be a configuration item or configured item.

A typical U.S. government contract has four parts and 13 sections. The order of precedence is as follows:

- Part I takes precedence over Parts I, II, and IV.
  - Contract Sections A through H
- Part II takes precedence over Parts III and IV.
  - Section I
- Part III takes precedence over Part IV.
  - Section J
- Part IV takes precedence over nothing.
  - Sections K through M

The intent of the parts and sections in U.S. government contracts is included in most contracts for goods and services in commercial and international contracting. Sections are as follows:

A. Solicitation/Contract Form
B. Supplies or Services and Prices/Costs
C. Description/Specs/Work Statement
D. Inspection and Acceptance
E. Packaging and Marking
F. Deliveries or Performance

G. Contract Administration Data
H. Special Contract Provisions
I. Contract Clauses
J. List of Documents, Exhibits, and Other Attachments
K. Representations, Certifications, and Other Statements of Offerors
L. Instructions, Conditions, and Notices to Offerors
M. Evaluation Factors for Award

7.9 CM AND SUPPLY CHAIN MANAGEMENT

The relationship between a company and its suppliers is much the same as the relationship between the customer and the contractor.

CM implementation should include maintenance and release of the conformed subcontract and attachments to ensure that the company has documented the current agreement with each subcontractor. Doing this is critical to program discussions with the vendor and to answering program audit questions.

Maintenance of the conformed subcontract may appear to be at odds with the advent of automated CM tools. Generally, it was found that CM tools were not used. Like the contract between the company and the customer, the subcontract along with the exhibits, attachments, and appendices were found to require manual rather than electronic sign-off. The *legally recognized* contract was the paper copy, not its electronic equivalent.

Supply chain management is responsible for ensuring that subcontract requirements are met. This includes the following:

- Proposal management to and from the vendor
- Management and evaluation of supplier data submittals
- Source-selection activities
- Vendor site surveys to evaluate the acceptability of their internal functions including CM
- Management of stockrooms and inventory
  - Indoor tracking of assets in 2018 was still problematic due to inaccurate technology and changing environment. At best, indoor trackers are good only to a few meters.[2]

Everything that a program transmits to the customer, the supplier provides to the contractor. CM functions are reversed in the supply chain relationship. Under a subcontract, the company evaluates the acceptability of a product after its submittal by the supplier to the company. Under a contract, the acceptability of a product is reviewed prior to company submittal to a customer. CM is integral to evaluating the acceptability of products provided by the vendor and ensuring they are properly evaluated and accepted using change control.

CM implementation also ensures that suppliers are completely vetted before they become inseparable players in the company's market strategy. An example of this is the recent acceptance of FACC (Ried im Innkreis, Austria) joining SPACE Deutschland e.V. (Berlin, Germany) as a full member.[3] FACC specializes in design, development, and production of advanced aircraft components and systems.

7.10 CM AND FACILITIES

The configuration of the facility infrastructure is as important to the function of the company as the understanding of the product produced is to the customer. This includes the certification and maintenance status of all company-owned facility infrastructures as well as the management of

---

[2] Leverage, IOT for All (2018): *Overview of Indoor Tracking*, https://www.iotforall.com/overview-indoor-tracking-technology/?utm_source=newsletter&utm_campaign=IFA_Newsletter_Dec20_2018. Accessed December 17, 2018.

[3] Francis, S. (2018). FACC joins aerospace industry association SPACE. *Composites World* e-zine. https://www.compositesworld.com/news/facc-joins-aerospace-industry-association-space; accessed December 31, 2018.

the associated provenance of each item. Modern CM implementation often includes the creation of digital twins for building parking and entrance, meeting rooms, manufacturing layout, receiving inspection, and stock rooms to optimize facility traffic flow.

7.10.1 CM and Property Management

Many of the functions CM implementation provides to contracts manifest themselves in an interface with property management. This includes coordination of receipt, use, marking, and tracking of customer-provided and subcontractor-provided property, stockroom inventory, stockroom property transferred to a program/project, and property assigned to employees. CM implementation also manages the *transfer of accountability* and shipment of finished products to the customer.

7.11 TIME-PHASED CM ACTIVITIES

What follows is a set of recommended CM implementation activities generally associated with company functional resources. In many cases, questions such as what must be done as well as why it is important have been addressed. It is not meant to be all inclusive but rather typical of the generic CM implementation. It represents the best practices we observed. In many cases the activities fall outside of what was considered the norm in companies evaluated.

It is written assuming a contract relationship exists between the contractor and a U.S. government agency. It can be applied conceptually to the *quasi-contractual* relationship existing within a company and between different subsidiaries of a company. As with all CM activities, a CM department may not be the functional organization in any one company performing the tasks despite the tasks being CM in nature.

7.11.1 At Award

7.11.1.1 The Contract

Working in concert with the program contract manager, the program configuration manager needs to obtain a complete copy of the contract and all associated contract documents. Each document needs to be carefully examined for contractual CM requirements, data submittal requirements not contained in the data requirements list (DRL), disconnects between contract parts or documents, and a watch list that if not socialized to those supporting the program will result in unallowable costs. This evaluation also includes a CM analysis of open TBD, TBS, TBR, and TBN items that expose the program and company to risk. The contract is kept under configuration control with all changes being managed using a formal change evaluation and implementation process.

7.11.1.2 Contract Provisions and Reference Documents

7.11.1.2.1 What Has to Be Done

The program configuration manager needs to find full-text copies of every referenced provision that is cited in the contract and put them in one or more documents.

- Full-text FAR clauses need to be collected into a single editable/searchable document such as MS Word, so the clause can be found quickly.
  - Clauses should in order by number and not segregated by contract section.
  - The contract section where the clause is located can be appended to the number and description for reference.

- FAR agency supplement clauses such as the DOD, NASA, and Air Force supplements each need to be separately identified by number and not by contract section.
- The same thing must be done for each non-FAR provision such as those involved with U.S. Uniform Commercial Code (UCC) or international contracting.
- If a contract clause says to do something in accordance with another cited clause or FAR, UCC, or other provision that exists in the contract, copies of those documents also need to be found and put in the program document repository.

It does not stop there. As all the contract attachments (including the SOW, specifications, mission assurance requirements, and DRL requirements) are implemented by the contract, full-text copies of these documents as well as any compliance documents and reference documents listed in them also need to be gathered. Should a compliance document in turn have requirements, documents and reference documents need to be gathered. It is recommended that this process continue until the lowest-level source documents are reached.

7.11.1.2.2 Why It Has to Be Done

This is the first line of defense against customer and internal audit findings. Often, auditors will find a clause in the contract, pull the latest version from their sources, and audit against the latest requirements. Requirements evolve over time and the clause they pull may not be the clause that is in your contract. If you do not have a full-text copy of the contract clause to prove that the program is compliant, you have lost the battle and may have to spend scarce resources to comply with the more stringent requirements.

Often, requirements implemented in specifications and other attachments are buried as low as level 12. These will also cause contractors compliance issues if you don't have a full-text copy. It is a sound CM practice to work with contracts to add a special provision to the contract that limits contractor compliance to the third level of indenture. Referenced documents below the third level of indenture should by specified in the contract as guidance.

7.11.1.3 Hidden Data Submittal Requirements

7.11.1.3.1 What Has to Be Done

Every contract document must be carefully reviewed to find and plan for any data submittal requirements not part of the DRL. These are identified from the top down and tied out from the bottom up. This means that the review of the contract, its provisions, and each contract section needs to be carefully read, and submittals must be documented. The list of hidden data submittal requirements then needs to be compared to the DRL list for possible duplicates. It is recommended that a no-cost proposal be submitted to the customer to add the additional items, so that all deliverables are captured in a single document.

Culprits include but are not limited to incident reporting, reporting damage to government property, Federal Information Processing Standards, GTS security plans, submittals of *new-technology* rights requests, submittal of a final scientific or technical report, open ended requirements associated with reviews, and so on.

7.11.1.3.2 Why It Has to Be Done

Traditionally, the draft contract is not reviewed with CM implementation in mind. Without proper identification and planning for submittal of all contractual data, the company risks being not compliant with the contract. This results in the following:

1. Loss of goodwill on the part of the customer
2. The extra expense of responding to audit findings
3. Withholding of payments and lower award fee resulting in a hit to EBIT

7.11.1.4 Disconnects Between Contract Parts or Documents

7.11.1.4.1 What Has to Be Done

During the review of documentation, the CM implementation needs to consider disconnects between contract sections. This is one of the worst trip wires that a contractor can experience. Disconnects generally always exist between DRL delivery requirements, deliverable items list (DIL) requirements, and packaging and marking requirements.

Note that some contracts do not have a DIL but instead use Sections B and F of the contract for deliverable items and their delivery date. To list everywhere we found disconnects would expand this document by separate chapters dedicated to each buying agency and sub-chapters dedicated to buying offices within the agency. Let it suffice to say that no contract we looked at was clean in this area.

Special attention needs to be given to any reference to the company documents in customer documents. One program was written up for having references in the customer-provided SOW to documents that required the company to use a paper release system. The paper system had been retired when the company moved to a PD/LM system in 2003. The paper system requirement had been cloned forward from a 1996 program. Since there was no CM implementation review, the requirement was not noticed until after award. It was a costly mistake requiring a paper release that had to be duplicated in the PD/LM release. The customer stringently enforced the paper requirement.

7.11.1.4.2 Why It Has to Be Done

After source selection activities are finished, contractors often find that the actual contract they sign doesn't accurately reflect what they assumed had been negotiated. Contractors who fail to do a complete contract review are blindsided by missing or conflicting contract requirements. This has resulted in:

- Products being delivered to the wrong place
- "Ship to" and "Mark for" addresses differing between various parts of the contract
- DRL and data item descriptions not matching
- Shipping instruction with freight on board (FOB) source specified in one place and FOB destination specified for the same item somewhere else.

7.11.1.5 The Watch List

7.11.1.5.1 What Has to Be Done

During the review of documentation, the CM implementation needs to consider contract provisions that, if not socialized, may cause trouble during program performance. Many of the examples that follow were gleaned from lessons we learned on programs we looked at. They include:

- Failure to pay attention to clauses prohibiting flying on other than U.S. flagship carriers in NPR 6000.1, *Requirements for Packaging, Handling, and Transportation for Aeronautical and Space Systems, Equipment, and Associated Components.*
- Limits on overtime, banked time, and pay scales.
- Time limits on delivery of inspection of systems records.

Each item found needs to be documented in the program plan, released in the PD/LM system, and socialized. In many cases, the program manager may request CM assistance in generating a risk mitigation plan for each item. As additional lessons learned items come up, they should be added to the watch list.

7.11.1.5.2 Why It Has to Be Done

A few hours spent at the start of the program can save hundreds of thousands of dollars and valuable schedule time later. Trouble for the contractor can take many forms. Two common ones are:

1. Ambiguous language on what constitutes acceptance for any given deliverable or review.
2. Costs such as flying on a non–U.S. owned airline being unallowable.

These affect the cost and schedule and have a direct impact on EBIT. They can result in weeks if not months of internal and external action items to clear up if found too late or to generate root cause documentation.

7.11.1.6 Contract Status Worksheet

Sound management of the configuration of the deliverables is tied directly to the sound management of the contractual technical requirements. There are many ways to do this. Perhaps the easiest is to create and maintain over the life of the contract a spreadsheet that tracks the contract sections and attachments in column headers against a row for each authorized change to the contract.

A second tab on this spreadsheet can be used to track each task order (TO) on the contract.

At the top level, the contracts manager maintains traceability of contract documents against proposals and contract modifications through the work authorization system.

The contracts manager generally does not maintain traceability of contractual documents against proposals and contract modifications. This is a task that CM implementation needs to address. CM is uniquely fitted to do this as part of the proposal synergy through the CCB process.

7.11.1.6.1 What Has to Be Done

At program award, you need to implement a method for tracking the revisions and changes to all contract requirements against the change proposals and customer contract change requests, and implementing contract modifications. You also need to create a tracking method for making sure that all TOs under a contract are identified, the price, schedule, and deliverable items are tracked and delivered, and the TO is closed.

Experience has shown that a single spreadsheet is suited for this activity and, if properly configured, provides graphical status to the program. In some cases, minimal reporting is needed other than to ensure that the program understands where they are and where they have been. If you have a modern PD/LM system, that is where this needs to be done as part of the digital fabric.

7.11.1.6.2 Why It Has to Be Done

Loss of control over the contract technical requirements is a disaster. Once the technical requirements are lost or muddled, there is no way that a contractor can stand before the customer and unequivocally state that the item being delivered meets contractual requirements. If you don't know what the requirement are, you cannot certify that you have met them. Although others may be responsible for generating a requirements verification matrix, it is the configuration manager who has the prime responsibility for ensuring that the program staff understands the genesis of each change to the current and evolving contractual technical documents.

7.11.1.7 Setting Up a Conformed Contract and Contractual Data Repository

One of the greatest services a comprehensive CM implementation can provide to the program management team is the maintenance of a conformed contract. A conformed contract is defined as a copy with every change documented in it. It is essential in several areas, including the following:

1. Knowing what the contract says at any point in time.
2. Identifying which modification added each requirement and what that requirement is.

3. Tying requirements to bid assumptions and customer acceptance of those assumptions to avoid scope creep from the customer and volunteerism by program personnel.
4. Being able to evaluate change proposals against the contract.
5. Tying out TBD, TBN, TBS, and TBR items in the contract baseline.
6. Being able to properly plan and work on the program, ensuring contractual requirements are met.
7. Being able to push back against customer personnel who are looking at an earlier version of the requirements.

It is wise to always provide the customer with a copy of the conformed contract kept by the contractor. Customers appreciate this and rarely have resources to make one themselves.

7.11.1.7.1 What Has to Be Done

At contract award (CA), a location must be established as the historical document repository for the program. This should be identified as "Contract at Award." Every document that was gathered as part of Section 7.11.1.2 must be included. Once this repository has been created, the entire directory will have to be cloned as the conformed contract. It is this set of documents that will be used to reflect the current contractual requirements. The current contract should also contain all contract modifications and TOs as well as approved major and critical variances (deviations and waivers). Down revisions of superseded documents are kept for the mission life of the program in a data archive. Doing this in the PDLM system as part of the digital fabric is optimal.

The CM implementation should consider maintaining a hard copy of the conformed contract and attachments for the program manager to assist with customer discussions and negotiations as well as for audit support. This may appear to be at odds with the idea behind digital threads and digital fabrics. Until the customer base moves to all electronic signatures via workflows, contracts along with the exhibits, attachments, and appendices require manual signatures. This makes the hard copy the *legally recognized* contract, not its electronic equivalent. The PD/LM version is identical and much more useful for data analytics and compliance analysis.

7.11.1.7.2 Why It Has to Be Done

Few contractors can have the ability to stand before a body of auditors and state with any certainty that the company understands and has properly managed the contract technical baseline, and that the hardware meets all contractual requirements. This capability has been lost over time as companies move toward matrixed organizations to save costs.

Systems engineering may track elements of the technical requirements and the program manager may believe they are tracking it all. The contracts manager may have copies of all the contract modifications in one file, a copy of the awarded contract in another, and copies of some parts of the contract somewhere else. Without a conformed contract, no one has the complete picture.

7.11.2 Critical Activities Not Related to Product Performance

Every contract requires several things to happen as part of the infrastructure management at program start. Are there special contract requirements that must be planned for and tied out? How are resources going to be managed? How is your CM implementation going to approach each? What are the findings, tie-outs, and next actions?

The configuration manager should approach every contract from day one with the view to understanding exactly what it is needed to close out the contract. Part of what CM brings to the program management team is an understanding of how to track closed items and how to tie out open items so they become closed. This process of identifying something that is an open requirement and its tie-out is essential.

Examples include the following:

- The company did not recognize that there was a contract requirement for it to buy and provide launch insurance for an Earth observation satellite. No mitigation plan was put in place until a week before launch. At that point it was too late to broker a cost competitive agreement with the insurance providers. The mistake cost the company over €1,046,232.
- The statement of work (SOW) stated that the contractor needed to come to an agreement with the customer on what constituted "acceptance" six months prior to the acceptance review. No mitigation plan was enacted and the requirement was forgotten about until the day the review took place. The mistake cost the company over €17,437.
- The company missed a requirement to provide an emergency reaction plan for the life of the program. The mistake cost the company over €298,176.

The CM implementation needs to make sure that everything is identified and tied out.

7.11.2.1 What Has to Be Done

Corporations can no longer afford to stand up on a program an army of CM professionals with distinct specializations like specification writing, vendor SOW preparation, supply chain audit support, vendor survey for compliance with CM standards and methodologies; change managers; and a document subject matter expert (SME) for each document invoked by the contract.

The task of looking over the documentation with an eye toward closing out the contract still exists; it just now must be done with limited staffing. Every CM implementation should assign a point of contact to read and retain a working knowledge of the content of the following documents:

1. SOW
2. Customer specifications
3. Mission assurance requirements
4. DIL
5. Government/customer-furnished equipment/materials/property/data list
6. Contract provisions
7. Interface requirements documents
8. Organization conflicts of interest plans
9. GTS security plans
10. Program management plans
11. Mission assurance plan
12. Environmental verification requirements
13. Special calibration requirements
14. Award fee and milestone payment plans
15. Integrations and test plans
16. DD Form-254 security plan or equivalent

If the CM implementation states that CM personnel should do DM in the broadest sense, then part of the task is to identify, assign, monitor, and complete the status of items that need to be tied out.

7.11.2.2 Why It Has to Be Done

Unfortunately, we all know about someone who forgot to do something like lock the front door when they left on vacation. They returned to find all their possessions (or perhaps only a single item of immense value) missing.

- Every open item that is not tracked to closure is an unlocked door and the things we lose are very valuable.
- The company loses credibility in the eyes of the customer; recovery is very expensive; and the company also loses self-respect and industry respect as well as future business.

7.11.2.3 Resource Assessment and Management

Expert craftspeople are masters at understanding and applying customer needs in the most beneficial and economical way possible. If this is not done, it limits the chance of success, drives the cost and/or schedule, and often leads to an inferior result.

Resources take the form of

1. *Budget:* Time and money.
2. *People:* A talent assessment weighed against item 1 and the cost to train them.
3. *Practices:* Sifting through mandatory and non-mandatory past practices to find those that may apply and tailoring them for the job at hand.
4. *Physical tools:* Computers, desks, file cabinets, server space, and local area networks.
5. *Electronic tools:* Spreadsheets, process automation, PD/LM systems, manufacturing/material resource planning systems, anomaly, and corrective-action tracking systems.
6. *Mentors:* People who can guide someone through the requirements.

7.11.2.3.1 Budget

Budget is much more than the program schedule, negotiated price for the job, and how much of it is allocated to various departments. It includes the schedule and cost allocation for CM implementation. This allocation varies due to current and proposed labor rates and overheads, withholding of management reserve, and the grade level of people assigned to perform the task. We have seen CM implementation budgets vary greatly depending on which business unit a program is assigned to. It is not uncommon to see G&A and OH rates in one business unit much higher than in another business unit. When a program is moved into the lower G&A and OH rate business unit, the contracted price for the work does not change. The lower rate does, however, free up more of the available dollars to be spent on performing work (Equation 7.9).

$$\text{Higher G \& A and OH rates} = \text{Fewer direct dollars}$$

$$\text{Lower G \& A and OH rates} = \text{More direct dollars}$$

EQUATION 7.9 G&A and OH relationships.

Budgets are generally assigned and allocated to CM implementation based on the experience of the program manager. If past CM implementations have not shown value for the money spent, CM may be viewed as a necessary evil and be underfunded. If CM implementation improves product safety and quality, worth is measured according to a different scale and it is properly funded.

The business unit and program expect people to do all the necessary tasks without waiting to be assigned to them. This places a great responsibility on CM, as it requires a larger system view of what is required by the contract as well as refined situational awareness and a very proactive stance when it comes to implementing the program.

Many CM professionals interviewed struggled with this concept. They were used to waiting until the PD/LM system sends an email or until someone calls prior to taking any action. Companies can no longer afford CM implementation to be inward looking and reactive. Competitiveness mandates that CM implementation be outward looking and proactive. Making the organizational transition is often difficult for some seasoned employees. Dozens of individuals interviewed stated that they could not embrace the paradigm shift, saying, "I've been getting pay increases and promotions for over 20 years doing what I'm doing. Do not expect me to do more by being proactive for the same pay."

7.11.2.3.2 What Has to Be Done

We tend to think of non-recurring cost budgets in the form of employee head count. This may be convenient; but the fact remains that two employees with the same training and labor grade are not equal in all respects. An employee who is an excellent match to a program with a specific contract structure and customer may not be the best match for the same task with a different customer or contract structure. Employees with a vertically integrated program management background may not be the best fit for programs with a horizontally integrated management structure.

Employees assigned to a single program or set of programs for a long time may do more harm to the company in the long run. It tends to freeze the employee into a *one size fits all* mode of CM implementation. Twelve years after moving to a PD/LM system, one company surveyed still has a single program using paper release methodologies because that system is contractually mandated. CM personnel working exclusively on such a program will only understand that release system although their department peers will have evolved to different CM methodologies. It is hard for them to see the digital thread when they have spent years following a paper trail.

Change is a good thing; change is constant. Globally, CM tools and implementation are evolving rapidly to leverage new tools and processes while remaining true to basic CM principles. Budgets need to include employee training. The CM implementation must remain current with changes in regulatory and guidance documents. Staying up to date fosters continual improvement in knowing what we are doing, why we are doing it, and how we are doing it. This knowledge must be tempered with an assessment of how best to apply it to CM implementation on current programs and of how to leverage it to CM implementation on projects yet to be started.

The key for every CM professional is to understand what they don't know and ask for help. A person assigned to a program should (1) honestly evaluate their own capabilities against CM implementation requirements, and (2) identify their areas of weakness so that a proper mix of talents can be applied. The positive result is that CM implementation is not using limited budgets unnecessarily and failing CM requirements.

Training is critical to the success of both the program and the individuals. It also means that if a point-in-time issue is encountered, which only happens once or twice in every program, it is more cost effective if a subject matter expert (SME) comes in and does that task and provides a few hours of mentoring. This is opposed to the program/project CM spending weeks trying to do something they do not fully understand and then, when there is a crisis, calling in the SME to clean up the work that was done.

It is critical for the program configuration manager to work closely with the program business analyst (BA) to establish and document the approach to be taken to add CM hours to changes. Generally, CM effort equals 5–7% of the engineering base hours on all engineering change proposals, contract change proposals, and TOs negotiated on the program. If the bid is for 100 added hours, CM should be added in at an additional 5–7 hours. Often, this does not occur, and CM budgets are eroded over the life of longer programs whereas the budgets of organizations supported increase during the same time frame, as does the scope of the CM effort.

7.11.2.3.3 Why It Has to Be Done

Part of the job of the leads and managers is to move people away from their comfort zone after a proper assessment of each individual's capabilities. This is done not to force someone to become great at something that they do not have the aptitude for, but rather to develop the gifts they do possess while making them aware of their own limitations. Such challenges hopefully engender a deep appreciation for their weaknesses and sensitize them to the need to ask for help when confronted with these limitations. Leads and managers must remember that you can't put into a person what is missing: you can only develop what is already there. A favorite quotation of ours from Albert Einstein[4] is, "Everybody is a genius. But if you judge a fish by its ability to climb a tree, it will live its whole life believing that it is stupid."

---

[4] Pettigrew, T. (2013). https://www.macleans.ca/education/uniandcollege/why-we-should-forget-einsteins-tree-climbing-fish/, accessed January 1. 2019.

If a CM professional is doing the same job for more than three years, then it is time to think about a different assignment. Three years is the average time before a company's internal infrastructure evolves. When a configuration manager is assigned to a program that lasts from four to thirty years, (let's refer to it as program Alpha), they implement their configuration management activities based on the requirements and tools current at that point in time. The tools and implementation remain in place until the contract is completed.

Over time, improvements in configuration management tools and methodologies are made, but program Alpha is not required to adopt them as it would impact the company's relationship with the customer. Within a period of four years, the differences are so great that much of what program Alpha is doing is no longer applicable to CM implementation on later bids and awards. The program Alpha configuration manager's knowledge of how configuration management is implemented is out of date and they must be retrained before being assigned to a new program. We have seen excellent employees who were not able to make the transition from 15- to 30-year-old technology to the digital fabric mindset.

Budget increases for CM implementation due to cost-bearing changes are often overlooked during the bid process. An advocate needs to be in place if CM implementation budget increases are going to be properly bid, negotiated, and allocated. One company surveyed decided that bid and proposal funds could be stretched if they used past actuals in CM to cost new business. After an analysis of recent programs for CM charges (and software CM charges), one program stood out as having CM charges equaling 3% of the engineering base hours. An executive challenge was made to bring in a new fixed price program at the same percentage.

The bid was successful. After three months, it became apparent that CM had been underscoped. A later evaluation of the 3% program showed that CM spent 20 hours a week of unpaid and undocumented overtime for its entire life. It had been running (had CM been funded) at almost 9% of the engineering base hours. By the time the error had been identified, six other programs had been awarded based on the same cost model with the same built-in budget issue and profit loss.

7.11.2.4 IP Plan

When assessing IP requirements, each contract must be evaluated against the contract requirements that drive it. Part of a contract may be under the U.S. government FAR Part 12 rules for commercial contracting, where the IP belongs to the contractor. Another part of the contract may be under FAR Part 15 rules for contracting by negotiation with restricted and limited rights.

7.11.2.4.1 What Has to Be Done

CM implementation needs to include the contracts manager and the technology and IP department to propose, document, disseminate, and implement a clear IP making schema that protects the company.

The task of dissemination and implementation can be daunting if the CM professional is uncomfortable with making reasonable determinations or unclear which documents are to receive what form of IP marking.

7.11.2.4.2 Why It Has to Be Done

FAR Part 15 requirements for data marking are straightforward. If the data are not properly marked, the customer may remove all markings and use the data any way they like. This includes providing it to competitors. Data that are not marked properly will erode a company's competitive edge, and the technological differentiators that make the company the obvious choice disappear.

Working in the FAR Parts 15 and 12 commercial and international product sectors is challenging. Someone outside of the CM will not find it easy to grasp all implications of these sectors. The CM implementation must include providing lots of guidance in this area. Releasing an IP plan is a significant help.

7.11.2.5 DD Form-254 and "For Official Use Only" Marking

The U.S. government defense programs contain what is known as the DD Form-254, DOD Contract Security Classification Specification. This form may contain a section pertaining to "for official use only" (FOUO) markings. This section is very often misunderstood by companies, their customers (if a subcontractor to the prime contractor on a U.S. government contract), and in some cases the U.S. government contracting officer and their security representatives located at contractor facilities.

FOUO markings are simply a reminder to U.S. government personnel that they need to take a closer look at the document prior to releasing if it is requested under the Freedom of Information Act (FOIA).

FOIA requests have nine exemptions due to issues of sensitivity and personal rights. They are listed in Title 5 of the U.S. Code, Section 552, which states

(b) This section does not apply to matters that are:

1. (A) Specifically authorized under criteria established by an Executive Order to be kept secret in the interest of national defense or foreign policy and (B) are in fact properly classified pursuant to such Executive Order;
2. Information related solely to the internal personnel rules and practices of an agency;
3. Information specifically exempted from disclosure by statute (other than section 552b of this title), provided that such statute (A) requires that the matters be withheld from the public in such a manner as to leave no discretion on the issue, or (B) establishes particular criteria for withholding or refers to particular types of matters to be withheld; FOIA Exemption 3 statutes;
4. Trade secrets and commercial or financial information obtained from a person and privileged or confidential;
5. Inter-agency or intra-agency memoranda or letters, which would not be available by law to a party other than an agency in litigation with the agency;
6. Personnel and medical files and similar files the disclosure of which would constitute a clearly unwarranted invasion of personal privacy;
7. Records or information compiled for law enforcement purposes, but only to the extent that the production of such law enforcement records or information (A) could reasonably be expected to interfere with enforcement proceedings, (B) would deprive a person of a right to a fair trial or an impartial adjudication, (C) could reasonably be expected to constitute an unwarranted invasion of personal privacy, (D) could reasonably be expected to disclose the identity of a confidential source, including a state, local, or foreign agency or an authority or any private institution that furnished information on a confidential basis, and in the case of a record or information compiled by a criminal law enforcement authority in the course of a criminal investigation or by an agency conducting a lawful national security intelligence investigation, information furnished by a confidential source, (E) would disclose techniques and procedures for law enforcement investigations or prosecutions, or would disclose guidelines for law enforcement investigations or prosecutions if such disclosure could reasonably be expected to risk circumvention of the law, or (F) could reasonably be expected to endanger the life or physical safety of any individual;
8. Information contained in or related to examination, operating, or condition reports prepared by, on behalf of, or for the use of an agency responsible for the regulation or supervision of financial institutions; or
9. Geological and geophysical information and data, including maps, concerning wells.

Because DOD contracts are issued subject to FAR requirements and the FAR is a federal statute, much of what is associated with them is excluded from FOIA requests under Exemption 3 above.

In the final analysis, FOUO markings at the contractor level only apply to Exemption 4, trade secrets and commercial or financial information obtained from a person and privileged or confidential:

- Contractor's progress, status, and management report
- Cost performance report
- Integrated master schedule
- Contract funds status report

They are deemed to be confidential business data. Should the program manager and contracts department declare these as not being confidential business data, then nothing on the program must be marked with an FOUO legend. FOUO marking instructions need to be added to the program CM plan as part of item identification.

7.11.2.6 CM Plan

CM plan content is covered in many guidelines and requirements documents and may be specific to a commodity type. Generic plan content, as outlined in the GEIA HB-649, *CM Handbook*, needs to be tailored to that commodity type with different specific planning if the product is hardware, software, firmware, models, support services, analysis services, and so on.

Occasionally, it may be advisable to develop a single CM plan for all programs of a certain type. This single plan is then supplemented by a unique program addendum. This approach presents several advantages over having a unique CM plan for each program. It allows for the incorporation of modifications to the CM plan template for the company to be made once for all programs under the umbrella of the business unit plan rather than updating one plan per program. It maintains consistency of approach to CM implementation for a unique set of customers and assists in managing audits and limiting findings. It also facilitates efficiencies as all programs falling under the single-plan approach are managed the same way, using the same workflows and the same signature authorizations.

7.11.2.6.1 What Has to Be Done

Occasionally, it may be advisable to develop a single CM plan for all programs of a certain type (program types may be by commodity or by contracting method such as firm fixed price, fixed price incentive firm target, etc.).

1. This single plan is then supplemented by a program unique addendum for each program the plan applies to:
   a. Contract number
   b. CII numbers
   c. Top assemblies for each contract hardware, software, or other deliverable items along with all interface control drawings
   d. Program CM contacts, flight and non-flight collection numbers
   e. Configuration baselines

2. Product data management (PDM) system workflow for a standard EO
3. Document approval requirements and distribution
4. Document review criteria and instructions
5. IP markings
6. Data right assertions
7. ITAR-designated official and requirements
8. Links to the SharePoint program site
9. Server file path and who to contact for access

10. IP plan number, contacts, and guidance
11. Contract data requirements list management and distribution
12. Cover letter requirements
13. Deliverable item preparation requirements
14. Product classification
15. Contract documents in the PDM system
16. All deviations to the program CM plan
17. Hardware quality assurance document number
18. Links to document templates for SOWs, specifications, drawings, system engineering reports, and special instruction regarding program unique abbreviations
19. CCB
    a. Acceptance documentation and its preparation[5]
20. Unique material requirements

The CM plan addendum is the first-level communication document for CM implementation with all program personnel. It should include all the CM information any person assigned to the program requires.

7.11.2.6.2 Why It Has to Be Done

The number one complaint we received on programs working in a horizontally aligned engineering structure was a lack of consistency in the CM implementation approach across program implementation and start-up. Multiple requests were made for a single document or presentation that provided most of the information someone joining the program would need to start work, which could be used as a reference point for the life of the program. The more distributed the support functions became, the greater the communication issues were. This did not appear to be an issue in vertically aligned integrated product team environments.

7.11.2.7 DRL Tracking System

Status accounting and DM is critical to program success. It includes tracking and delivery of data to the customer as well as tracking of DRLs received from program vendors. Many companies are moving toward relational PD/LM-centric data tracking where multiple deliveries of one DRL can be scheduled, reminders can be sent, and automated workflows can be used to facilitate delivery to the customer.

This was not observed to be the norm across many business sectors. Often, it was found that spreadsheets were used to track deliverables due to lack of more efficient tools. One program studied used a PowerPoint format to provide DRL performance status to the GSFC and NASA management chain. GSFC found this approach helpful as the format allowed it to be incorporated into presentations to NASA headquarters.

7.11.2.7.1 What Has to Be Done

CM coordinates the customer DRL submittals with the program and they collectively determine the best way the company can meet the data-reporting requirements with the minimum of expended effort. Many new tools are available for assisting with DRL tracking, due date notification, and status.

One particularly attractive approach is to release the customer's DRL requirements document in the PD/LM system after creating each DRL as a document object and including these in the DRL attachment BOM in an unreleased state for the length of the program. These DRL objects act as

[5] A major disconnect was found in most of the programs evaluated as to who is responsible for preparing the COC, DD Form-250, or other acceptance document. Except for perhaps small-study efforts, it is recommended that this be the responsibility of the configuration manager.

collectors for released documents and other artifacts that are generated to meet the DRL requirements. As DRLs are released and submitted to the customer under their unique item identifier, they are added to the DRL collector under the DRL attachment. You can think of this as having a bowl for each DRL requirement and thinking of the individual submittals as jelly beans. If a single report (e.g., a purple jelly bean) meets requirements in three DRLs (bowls), a purple jelly bean then ends up in each of the three bowls.

Vendor DRL collectors are likewise released in the PD/LM system, forming the BOM of the SOW sent to the vendor. As individual documents meeting DRL requirements are received from the vendor, they are put into the PD/LM system for change control review and added to the DRL collector(s) under the SOW to the vendor. One method of accommodating vendor DRLs in a PD/LM system is shown in Figure 7.2.

7.11.2.7.2 Why It Has to Be Done

The lack of attention to DRL submittal requirements will delay payment of the final invoice on a program. This may result in rejected invoices against milestone billing events or award fees and erosion of a company's credibility with its customer base.

As all DRLs are captured in the PD/LM system, indentured BOM reports can be generated from the system either in portable document format (pdf) or as a spreadsheet form. This can be submitted for final data acceptance with a Z-code DD Form-250, or equivalent. It is the last acceptance document approved prior to the program entering the growth and maturity phases of the life cycle. In the U.S. government programs, this may be referred to as the operations and maintenance (e.g., logistics) phase.

7.11.2.8 DRL Cover Letters

Many contracts require that a cover letter accompany each data submittal. With the advent of electronic media delivery, this can be an expensive luxury. It is another holdover from paper systems not yet purged from procurement guidelines. CM implementation should strive to delete the cover letter requirement during the bid and proposal activities. When this is not possible, the next option is to negotiate a monthly or quarterly email summary of data transmittals.

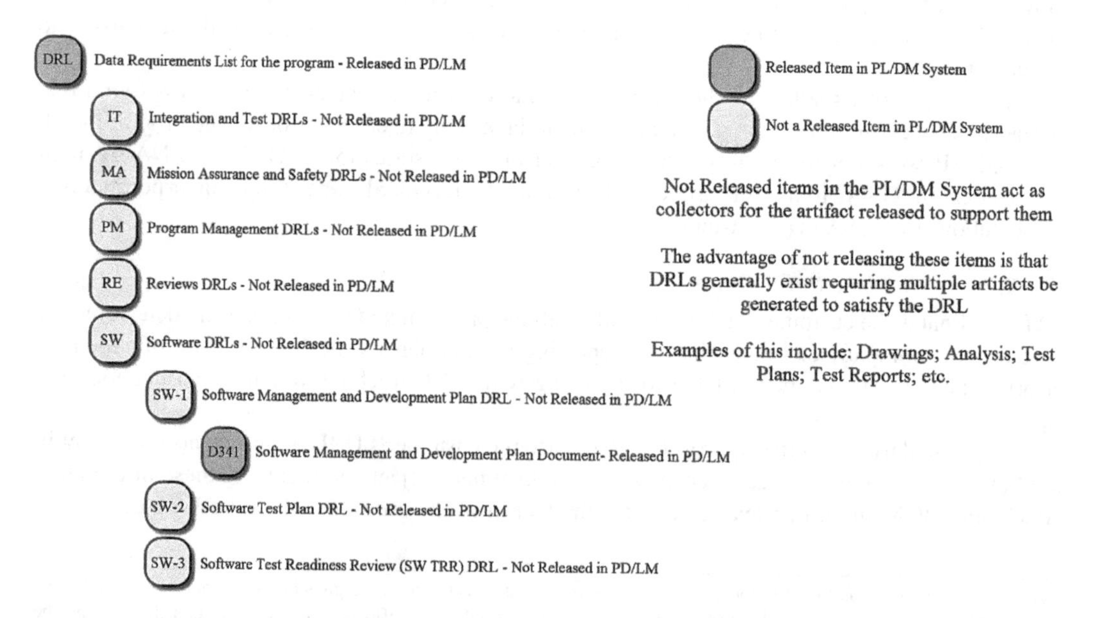

FIGURE 7.2 DRLs in a PD/LM system.

Evidence of e-delivery is manifested by the date time stamp on the site or in the system to which data are delivered. The date time stamp is the official record receipt in the customer's DM system. The transmittal is often accompanied by an email to the customer.

Customer mandated transmittal cover letters must be individually prepared, signed, scanned, and distributed via email with a follow-up hard copy delivered using overnight courier. We found cases where scans of a signature were inserted into cover letter electronic files and the file was then distributed. These signatures were copied and fraudulently applied to other documents not originating from the company.

7.11.2.8.1 What Has to Be Done

Once a bid decision is made, all draft contract requirements must be reviewed against CM implementation to ensure that they are adequately scoped and costed. A CM implementation requiring hard copy transmittal letters when electronic delivery of data is also specified does not make sense. Inconsistencies of this kind are costly if not found and mitigated. At the very least, the customer should be asked why the disconnects exist and requirements that make sense should be discussed. If the requirement cannot be changed prior to award, the bid must reflect the additional costs.

We found across industries that data requirements were not consistently reviewed during the new business bid and proposal process. This resulted in two programs with similar technical requirements having CM implementation being bid at the same price, ignoring the fact that one program had 2,500 data submittals and the other had 150.

7.11.2.8.2 Why It Has to Be Done

The cost of creating, signing, scanning, tracking, and maintaining files for transmittal letters over the life of a program is not trivial. The cost of preparing 500 letters alone (not unusual for a small program) is estimated at 125 hours or about €21,833.75, if you assume a billable rate of €174.67 per hour. Larger programs may have as many as 10,000 separate data transmittals and as a result 10,000 separate cover letters that must be written, coordinated, signed, transmitted, and tracked. This is 20 times the amount used to calculate the impact noted earlier and equates to an expenditure of approximately €436,675. By eliminating the requirement, the company bid becomes much more cost competitive and implementation after award less prone to audit findings.

7.11.2.9 Standardized Data Distributions Internal and External

In an environment based on electronic approvals, we found that little thought had been given to standardizing the flow of critical information to internal and external customers via electronic notification. Often, data deliveries were done by a professional with a strong understanding of DRL distribution statements where very limited distribution was the norm instead of the exception. In the evolving nature of distributive program infrastructure, wide distribution is the norm and a broader range of internal and external people receive notification that data have been delivered. Distribution groups (Figure 7.3) are easy to create in most email systems, with the advantage that

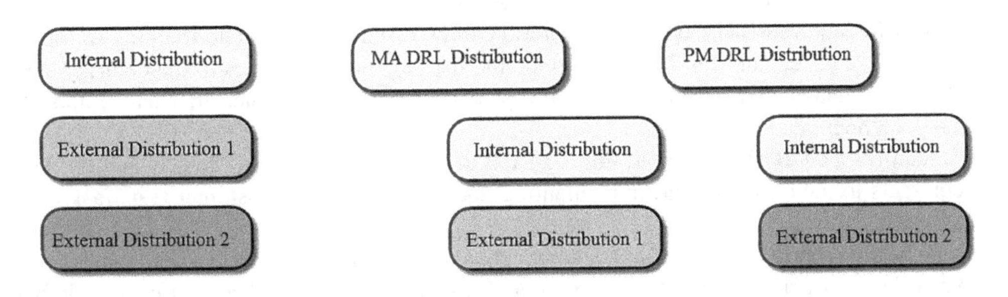

FIGURE 7.3 Distribution groups.

larger distributions can be created as a combination of a standard distribution group plus a unique-to-a-DRL subject (such as "program management").

As a result, new groups can be created that require less maintenance than one group for each DRL. Changes to customer personnel or internal personnel need to be made only in one group and, as that group is a member of a larger distribution, the change propagates to those other groups. User groups can also be generated in PD/LM systems to notify key departmental representatives of every PD/LM release. The same methodology can be used.

7.11.2.9.1 What Has to Be Done

The CM implementation office (after working with the program team and customer) generates a distribution list of internal and external personnel who will be notified of every DRL transmittal. Program DRL distribution groups are created in the email system and used for each DRL delivery to the customer.

The CM implementation office (after working with the program team) generates a list of key company personnel who should be notified of every PD/LM release on the program. A program EO distribution object is then created in the PD/LM system as a global group. User groups should be named simply so anyone can find them. If the program name is "Rohit," the user group name could be "Rohit_EO_Dist."

7.11.2.9.2 Why It Has to Be Done

Part of the failing of any organization is its lack of communication, which is manifested in many ways. One of those is its inability to adequately distribute information so everyone is working to the same documentation set and document revisions.

Information is power, but only when it is freely shared among all parties who have a need to know. If this information is sitting on a server or in an engineering notebook, it is useful only to the person who knows it exists. Once distributed to those who need it, information can change the effectiveness of the entire organization.

7.11.2.10 The CCB

CM implementation requires CM to act as the secretary to the CCB (e.g., like the secretary for a corporation rather than the administrative assistant). Without a good set of CCB minutes, the program technical requirements are soon lost. Only three organizations exist for the life of the entire program: program management, finance, and CM. CM by necessity often takes on the role of program knowledge manager. It is in this role that the CCB needs to be viewed. Notes taken in the moment regarding decisions must be recorded in the context of program impact and not the moment.

7.11.2.10.1 What Has to Be Done

A meeting should take place between CM, program manager, financial analyst, lead systems engineer, and contracts manager to determine CCB structure, attendance, and meeting requirements.

CM participation in the CCB process is necessary so proposals can be properly evaluated for changes to CM requirements, completion criteria can be discussed, adequate tracking to the contract technical requirements can be maintained, and the negotiated changes to the contract can be properly disseminated and socialized with the product managers, product area leads, and other program personnel.

This requires a CCB digital thread linking CCB actions to the proposals and transmittal letters from contracts to the customer and implementing documents from the customer (e.g., task orders, contract mods, etc.).

A copy of the proposed contractual document changes should also be kept so that, when a modification to the contract is made, the company and the customer know if a modification is being made "out of the proposed sequence". This happens more often than many would believe. Often, contractors have as many as 25 proposals in the customer's review process at one time. All of these could impact the same document paragraph. When later proposals are approved before earlier proposals, the language in the contract modification can become dangerously muddled.

Assume that you have a requirement to deliver 5 sandwiches on contract. You have submitted two proposals both adding an additional sandwich. After both proposals are approved you should have a total of 7 sandwiches on contract.

Proposal 1: Modify paragraph 3.4.1 from "5 sandwiches" to "6 sandwiches."
Proposal 2: Modify paragraph 3.4.1 from "6 sandwiches" to "7 sandwiches."

If proposal 2 is approved before proposal 1, since you do not know what the language is predicated on, you don't fix the originally proposed language to reflect the out of sequence approvals before it becomes contractual. That results in the following:

Modification 1 for Proposal 2: Modify paragraph 3.4.1 from "6 sandwiches" to "7 sandwiches."

Proposal 1 is then approved after the proposal 2 language change. You don't fix the originally proposal language to reflect the approval of the proposal 2 change. That results in the following:

Modification 2 for Proposal 1: Modify paragraph 3.4.1 from "7 sandwiches" to "6 sandwiches."

In this scenario, the out of sequence approval deleted one sandwich because you should have had seven on contract after the approval of both proposal 1 and proposal 2.

Knowing that the modification language is not consistent with the proposal order allows you to correct the modifications before they are signed by both parties. Below is what the language modifications should have been changed to due to the out of sequence approvals:

Modification 1 for Proposal 2: Modify paragraph 3.4.1 from "5 sandwiches" to "6 sandwiches."
Modification 2 for Proposal 1: Modify paragraph 3.4.1 from "6 sandwiches" to "7 sandwiches."

7.11.2.10.2 Why It Has to Be Done

If you don't know where you are going, and you don't know when you are going to arrive, you never know where you are! To effectively manage a program from the CM perspective, the contractor can never lose the contractual technical baseline. If they do, there is no way that the company can stand up in front of any customer or auditor and state that the thing that was built meets the contractual requirements.

It is incumbent on the configuration manager to make sure that the baseline is managed and that the company knows what each modification does, how it ties back to what was bid, and what that bid was predicated on. In many cases, this management requires that change language in an already submitted proposal must be updated to account for out-of-sequence approvals of changes submitted later.

7.11.3 Hardware Quality Instruction and Software Quality Instruction

Programs often issue what are known as a "hardware quality instruction" (HQI) and "software quality instruction" (SQI) documents providing critical information regarding process inspections, documentation reviews, customer flow-downs, and so on. These are released in the PD/LM system. CM implementation must ensure that the HQI and SQI contain no errors relative to item identification, quality levels, referenced document numbers and descriptions, or PD/LM system collection identification that may tie into a non-integrated information DM system. Such errors can cause major performance issues during program performance.

7.11.3.1 What Has to Be Done

CM implementation ensures that the items produced and the services being provided are safe and meet mission assurance requirements. Part of the implementation is to review any document that flows down requirements to the program to ensure that the requirements are correct.

7.11.3.2 Why It Has to Be Done

Frequently, hardware and software quality personnel multitask across all programs and the HQI and SQI are created from a boilerplate template. We have observed disconnects between CM implementation requirements and released HQI and SQI documents. The disconnects often do not surface until physical audit prior to delivery. By that time, it is too late to take corrective action.

7.11.4 Program Data Repository Basics

CM implementation requires that data repositories be set up and managed. Each type of repository is purpose specific and may be a subset of the entire data repository structure for a program. The three main data repositories the CM uses are program specific: intranet, servers, and program-related PD/LM structure based on program metadata. Each repository has its own strengths and weaknesses. They will be addressed in turn.

7.11.4.1 Program Intranet Site

Many companies are moving toward intranet technologies to manage documents not requiring strict change control. If implemented, versioning control is a distinct improvement over server structures and will become more common across all industries over time. Not everything can be migrated from company file servers to intranet sites at present. Both intranet and servers have distinct advantages and disadvantages. CM implementation needs to strike a balance between the two.

7.11.4.1.1 Strengths and Weaknesses

Intranet sites offer the following advantages over servers:

- The ability to digitally create a link to data contained elsewhere on the intranet, allowing for a single point of truth instead of multiple points of truth.
  - A server may contain several copies of the same document with the same file name that are not identical. Intranet use may not eliminate this, but it is significantly reduced as items can link to other intranet pages or provide a direct link to the PD/LM system.
- Check-out and check-in capabilities exist with built-in versioning control.
  - Assuming someone remembers to turn it on for the site being used.
- Easily maintained permission structures.
- Quick access to data with searching capabilities.
- The ability to have multiple people editing the same document without conflicts.

Many intranet implementations have the following drawbacks:

- File sizes are much smaller than those accommodated by servers.
- Certain file types (extensions) are not allowed.
- Many offending characters exist that cannot be used in site, file, and folder names.

7.11.4.1.1.1 Site Name Restrictions

- Site names cannot contain the following characters: \ / : * ? " < > | # {} % & " ~ +
- You cannot start a site name, subsite name, or site group name with an underscore character (e.g., _) or with the period character. (The authors recommend avoiding the underscore in site names.)

- You cannot use the period character consecutively in the middle of a site name.
- You cannot use the period character at the end of a site name.

7.11.4.1.1.2 File Name Restrictions

- File names cannot contain the following characters: “ # % & * : < > ? \ / { | } ~
- You cannot use the period character consecutively in the middle of a file name.
- You cannot use the period character at the end of a file name.
- You cannot start a file name with the period character.

7.11.4.1.1.3 Folder Name Restrictions

- You cannot use the following characters anywhere in a folder name or a server name: ~ # % & * { } \ : < > ? / | “
- You cannot use the period character consecutively in the middle of a folder name.
- You cannot use the period character at the end of a folder name.
- You cannot start a folder name with the period character.
- A limit of single directory/file name length—128 characters (including file extensions).
  - This is the same limitation on the company servers.
- Use of underscores in intranet file names is not recommended.
  - This is the opposite of recommended file names on a server.

7.11.4.2 Program Servers

At one time, program servers were the data repository of choice and supplemented the hard copy files used for document release and control. Since the 1980s, this has been changing. Today, with the advent of very sophisticated PD/LM systems, expanded database tools, and infrastructure, only a mix of all the tools available will meet the growing immediacy of having information readily available.

7.11.4.2.1 Strengths and Weaknesses

Servers offer a way to store data files and organize them using an indentured directory structure. They are available at a relatively inexpensive price point. Seldom accessed files can be stored on servers with longer access times and lower cost at the expense of such niceties as hot swappable drives and redundantly mapped and synced server structures.

They are, however, limited as listed below:

- A limit of single directory/file name length including file extensions.
- No versioning controls.
- Directories and files are easily deleted or moved if the user is not careful.
- File lack of availability on items not accessed regularly due to system issues, upgrades, and GTS backup pressures to not keep backups after certain time periods.
  - Some GTS organizations were found to maintain no backups older than three months.
  - Seldom accessed data may become corrupt with no capability to restore it as a backup no longer exists.
- Corruption of PowerPoint and other files happens due to issues with archiving software at the GTS backbone level or due to other issues.
- Propagation of multiple uncontrolled copies of the same document is possible.
  - These are not identical but rather are “works in progress” with the risk of people working to an unreleased copy.

- Inadvertent creation of new subdirectories with permissions that are different than the directory it resides under causes issues with data access.
  - Generally, data permission is a string progression downward. This means the user has access to higher directories.
  - This is different from non-nodal access where data can be accessed directly without relying on a string of permissions stemming from the top directory.
- Complicated permissions structures.

7.11.4.2.1.1 File Path Length Limit The entire file path length limitation is easily exceeded because of the trend toward making file names as descriptive as possible. Unfortunately, the GTS backbone is not able to provide the server user a warning that file names are exceeding the maximum path length as they are loaded. This results in what looks like a reasonable set of nested directories and subdirectories and files with no ability to access them later as the path is longer than the file length limitations of the server. This is more common on some server software platforms than others. Many times, file lengths are imposed by the server software provider and not due to limitations of the server itself (e.g., the server software artificially imposes the restriction).

7.11.4.2.1.2 No Versioning Control The lack of versioning control means that without research you cannot tell what the current version is. There is no single point of truth. Workarounds must be set up to clearly establish the version within the file name itself. This is often accomplished by adding the date in the file name. Unfortunately, there is no standard for doing this across industry. As a result, users are unable to quickly evaluate changes between versions. Often, two copies of the same file exist in different subdirectories that are identical except for a different date in the file properties. By putting the entire date in YYYY-MM-DD format, the ASCII formats used on the server will sort the file by date. Unfortunately, this cuts into the available total file path length.

7.11.4.2.1.3 Files Being Moved or Deleted It is extremely easy to click on a directory, subdirectory, or file (thinking you have let all the pressure off the mouse button) and inadvertently drag it into another directory or delete it completely from a server. There is no safeguard against this. CM implementation can play a critical DM role as the focal point for finding lost directories and files or working with GTS to recover deleted ones. The first step is always to search on the entire directory for the missing file name or directory name. Only if it is not found and moved back to its proper place in the directory structure should GTS be contacted for a server restoration. Server restoration of even a single directory is never easy. File updates or new files added to the server since the last back-up will need to be uploaded a second time.

It is recommended that GTS provide a weekly text listing of the complete directory structure on program servers. This will allow rapid searches for the directory name and content to assist in recovery of deleted or moved files or directories. Experienced CM professionals have stated that most often directories and files have been moved into other directories rather than deleted. Without a listing of the server directories, it is often not easy to know the real name of the directory being searched for.

7.11.4.2.1.4 File Corruption Unfortunately, files do become corrupted. Files that in the past have lent themselves to becoming corrupted (due to either external vulnerabilities or server software upgrades) are .jpeg and.ppt. This is something very hard to recover from since photos are leveraged forward to document heritage design and presentations for major reviews are leveraged forward from older programs to newer ones. Photos and packages from major reviews are also used to evaluate product and service issues. It is recommended that a zip file of all the individual presentation files be created and stored in the PD/LM system with the release. Similar steps should be taken for critical photos.

7.11.4.2.1.5 File Propagation Many in the engineering community like to save their documents on their computer. They may also keep a copy in their department subdirectory on a server. This presents issues on several fronts.

- There is no single source of truth for the document's released or approved version state, which is an audit issue.
- Changes to the document that were made at the time of PD/LM release, distribution, and implementation may not be captured on the desktop or functional subdirectory copy.
  - Typographic errors, formatting issues, IP rights statements, etc., are often modified before release.
  - If a server or desktop version is used to make a later revision, the release changes are lost.
- Changes to the desktop or functional subdirectory copy for modification and use on another program may not be using the latest document.

GTS can run a search on the program server that identifies all cases where two or more copies of a document with the same name exist. CM implementation should require that this search be run once a quarter. When duplicate files are identified, the duplicate copy should be deleted from the server and a shortcut to the source document put in its place.

7.11.4.2.1.6 Differing Permissions Access to directories and files on a server is controlled in two ways. Access can be granted directly or by creating a permissions group. It is easy to tell which of these two methods is used on the server by going to the root directory, opening the permissions for the subdirectory in question, and opening the security tab.

Permissions that are granted to one directory are generally propagated down through the entire subdirectory string as each new directory is created. Any directory that has a limited permission access is referred to as *locked down.*

Occasionally, other directories are intentionally moved into a locked down directory and directory access issues arise. Because these moved directories inherited permissions from the directory under which they were created, those permissions move with them. Permissions need to be identical between the gaining directory structure and the losing directory structure or they can no longer access the moved data.

7.11.4.3 PD/LM Systems

The technologies used to perform change management and some configuration status and accounting are evolving. PDM systems will become product life cycle management (PLM) systems integrated to fulfill the ever-growing need for cost control, supply chain interface, data mining, and reduction. There are concerns that a single system will not be robust enough to meet the needs of all users. The last three attempts we made to map PLM informatic nodes proved fruitless. Interdependencies were too complex to trace the digital threads. This resulted in company customized reports yielding incomplete data.

Between 1996 and 2012, a certain company moved from an IBM mainframe-based records and release system to C-Gate to Oracle's AGILE© user-based interface and eventually to Oracle's AGILE© web-based interface. In each instance, the ability to perform change releases was improved while status and accounting and DM suffered. By the end of 2018, CM was equated with change analysis instead of planning, identification, change control, status accounting, review, and audit. CM implementation misconceptions due to tool changes are taking place in most companies. Selecting a PL/DM solution without looking at the scope of CM requirements degrades the company's ability to meet contractual mandates.

7.11.4.3.1 Strengths and Weaknesses of PD/LM Systems

PD/LM system strengths include the following:

- Track and manage all changes to product-related data
- Spend less time organizing and tracking design data
- Improve productivity through the reuse of product design data
- Enhance change collaboration
- Generally keyed into or integrated with supply chain management

These strengths can be leveraged in many ways and CM can help make or influence intelligent decisions in their use.

PD/LM system weaknesses include the following:

- Data mining is dependent on the quality of the data being input—"garbage in, garbage out" holds true.
    - Create a company-wide PD/LM data metadata methodology and stick to it.
- Data entry is case sensitive.
- Searches do not always yield what the user expects due to the user not adding the wildcard "*" prior to the number they are searching for.
    - If you follow ASME drawing standards and use prefixes, searches on the root number without the wildcard will result in a null set (e.g., no information will be found).
- Searching for historical data may not yield the results the user is looking for due to a lack of consistency in how the files were migrated from previous systems.
    - "SYS200" is not the same as "SYS-200".
- Data access can easily be locked down by the user, leaving them with no access.
    - Generally, this is due to user access being tied to the object's program name.
    - If the wrong program name is entered and saved, the file creator may not be able to access the PD/LM object they created because they are not a member of that program.
- Multiple ways of doing the same thing can lead to confusion.
    - Configuring the PD/LM implementation should ensure that there is only one way to create objects.
- Datasets in pull-down menus need to be constantly monitored against the controlling standards to keep them in sync.
- A user can add a proxy to sign for them when they are out, but proxy assignments will not agree with the sign-off matrix of the program for which the work is being approved.
    - Unless proxy assignments are authorized in the change control section of the CM plan, it will result in an audit finding of non-compliance with released program documentation.
- The PD/LM system may allow the use of I, O, Q, S, X, and Z in production release revisions.
    - This is a violation of the company's drafting manuals if it is predicated on lettering prohibitions utilized in older paper-based systems.

7.11.4.4 PD/LM Product Structure

The PD/LM product or BOM structure forms the roadmap for supply chain, stockrooms, manufacturing, and test. It must be complete from the lowest piece part through the top assembly for the item being produced or the service being performed.

7.11.4.4.1 What Has to Be Done

Relationships must be established in the BOM of each part, assembly, or document to create a digital thread that can be exploited to weave the digital fabric.

7.11.4.4.2 Why It Has to Be Done

Incomplete BOM structure causes breaks in supply chain activities. Without the structure in place, parts and materials cannot be ordered. Incomplete SW BOMs impact the ability of the SW to be merged. Incomplete document BOMs impact on the ability to rapidly access critical data.

7.12 THE FIRST 90 DAYS AND BEYOND

This is the most critical period of CM implementation and future program performance. Within the first 90 days, major agreements and decisions are made defining working relationships between the customer and the company. This is also when much of the background infrastructure is set up to ensure that the program has the tools to support unique needs. CM implementation should view each new program as an opportunity to move the state of the art forward. This requires an implementation team that is knowledgeable, trained, and proactive. A firm's competitive edge is only maintained in this way. Simply doing what you did last time is comfortable; but it comes at the cost of building in the same mistakes and workarounds.

7.12.1 The 90/90 Rule

Discussions with program support personnel and program managers have clearly shown that 90% of the decisions that drive program costs are made in the first 90 days of a program. These decisions include a determination of whether management tools are or are not developed, and clear channels of internal and external communication are or are not established; and what role each organization will or will not play. These are critical to CM planning and identification.

The program discussions relative to these decisions need to maximize CM implementation across the program infrastructure. Clear definitions and interfaces must be defined and documented.

7.12.1.1 Setting Expectations

Each member of the program management team and their staff influence the company's ability to perform on a program. Their influence on the shape of the CM implementation fabric is based on:

- What worked for them in the past.
- What did not work for them in the past.
- How it could have been better.
- How they view documenting interfaces and giver–receiver lists.
- Past interactions with other support functions.
- Their concerns relative to the program, cost, and schedule.

After information is obtained, the needs, wants, and goals of the program need to be balanced with CM requirements, lessons learned, and the available CM tool suite to arrive at a best-fit scenario for the program CM implementation.

Once this has been done, key players can be called together for a discussion outlining how the CM implementation addresses the concerns and how those needs are met. The following question needs to be addressed: "By doing what you are proposing, are you negatively impacting cost, schedule, or risk?"

7.12.1.1.1 Reducing Cost

Cost can be defined in two ways and CM implementation influences both.

1. Cost reduction: "Are you doing things so well that it costs less to do them?"
2. Cost avoidance: "Are you doing things so efficiently that it gives you time to do something else that needs to be done?"

By addressing the entire product life cycle, an implementation solution can be executed that mitigates multiple risks for a greater positive contribution across the program. DRL tracking methods and delivery status or other report formats may be so complicated that it takes the non-CM user 10 minutes to find useful information. This can easily add 10,000 hours in wasted time on a program.

Eliminating a single click on a server or intranet site also has tremendous payback. Using industry standards instead of someone's favorite home-grown implementation can pay high dividends. The cost of one highly paid senior engineer to perform a data reduction task equals the salary of three less experienced specialists.

7.12.1.1.2 Reducing Schedule

Schedules are always changing, and each program functional lead has their own. These roll into the program's consolidated schedule. CM implementation can either reduce or increase schedule time. One of the most frequent complaints we heard involved lack of a consistent progression in CM implementation. In a single company, one CM implementation would result in the highest level of excellence by exploiting to full advantage the capabilities of the PD/LM system. On another program that was awarded later, the CM implementation minimally used the PD/LM system, relying instead on manual spreadsheets discontinued in the 1970s in the rest of the company. We each have our "hot buttons"—things that we insist on, not because they are necessarily right or the best, but simply because they are familiar. Those implementing a CM solution on the second program were not willing to embrace process change. PD/LM was viewed as a fad that would soon fade. Generally, archaic practices were accompanied with comments like, "You can't expect everyone to be computer literate and the company is going to be glad we are doing it this way in the long run."

At another firm, we looked at requests for government security ratings (classified, secret, top secret, etc.) reviewed and logged prior to submittal to the government for processing. The company was concerned that clearances were taking an excessive length of time to be approved. We found 200 days of unsubmitted requests in a box on the responsible employee's desk. When asked about the box, the employee smiled and said, "I was hired in 30 years ago and I use the first-in, first-out system. When I get a new request, I put it at the end of the queue and then process the oldest one. If it looks all right I log it in and send it to the government. Sometimes I don't get a new request for weeks. That backlog of 200 requests in the box, my friend, is my job security!" Extrapolated company records showed that during that 30 years, 9,834 requests had been processed. This averaged just over 8 hours per request over 30 years. The average at other companies was 1.3 hours per request. Their backlog of unprocessed requests averaged less than 7. The 200-request backlog impacted the company in two ways.

1. Work on classified programs was delayed by months due to lack of cleared personnel.
2. The daily loss to the company (assuming the 1.5-hour average) in clearance processing time was 81.24% per day or 1,690 hours per year.

CM is an activity done by everyone at the company. Signing a security request and passing it on to the next step in the workflow is no different than any other form of change. Signature approval of 9,834 engineering changes over 30 years would be intolerable in a program environment that sees that number of approvals per person per year.

7.12.1.1.3 Reducing Risk

Risk reduction takes many forms. There are five main types:

1. Cost
2. Schedule
3. Safety
4. Quality
5. Audit

Listing them may be repetitious, but everything involved with CM implementation affects each of these in many ways. Every signature on an EO in the PD/LM system at one company surveyed cost €43.67. Adding one signature to all EOs may appear trivial until it is put into perspective. A single added signature resulted in an impact of €209,616 at one company (3,400 drawings plus 1,400 system engineering reports times €43.67). The signature was not required. This impact did not include changes to documents other than drawings; when they are included, the true impact was three times higher.

There are also schedule hits and audit risks due to an extended sign-off loop. Rarely is the number of signatures mandated by a company or specified by contract. Analysis of over 15,000 PD/LM EO approvals indicated that after the fourth signature is obtained, additional signatures add little to ensure that the change has captured all impacts. Tabletop reviews prior to starting the EO in the PD/LM system were found to be less costly, resulting in an efficiency gain with an associated cost reduction of over 30%.

7.12.2 Proposals and CM Budget

CM implementation hours are traditionally reduced during the proposal phase for the initial award to make the cost number fit in the anticipated *winning bid* cost window. When we asked about this, we were told that CM implementation was a necessary but expensive evil and the less it was funded the better. CM implementation increases were also not generally bid on change proposals once a contract was awarded. The CM function is a critical part of the program infrastructure and hours need to be increased as the program price grows due to proposal activities. CM implementation is part of the *marching army* costs on any program. Increases in scope drive CM implementation scope and, as a result, cost and schedule. Historically, the CM effort on added scope from change proposals equals 5% of the engineering base. This can be adjusted upward to 7% if the proposal is in the high-risk category. If CM budget is not increased with each change, the program will not be properly funded.

7.12.3 CM Interfaces

CM implementation creates a positive working relationship with the business unit, program management staff, and program employees if it is proactive and includes finding innovative ways to accomplish the CM tasks set forth in the contract. The relationships are supplemented by knowing the internal and external CM customers, anticipating their needs, and knowing how best to fill those needs within the limitations of the tools the implementation has at its disposal.

7.12.3.1 Internal CM Interfaces

Each organization and each employee has special needs. A few things to watch out for during CM implementation are as follows:

- Does the employee use the PD/LM tool daily (mechanical, electrical, I&T)?
- Does the employee only release documents sporadically throughout the program life?
- Do they want a lesson, or do they want to know only one thing?
- Never point the employee to a document and tell them to read it.
  - Seeing someone do it may be all it takes.
  - Everyone learns differently.
- Never quote rules and regulations at employees. They called CM because they read the rules and regulations and do not understand them.
  - *Talking to* and *conversing with* are two different things.
- Sit down with employees instead of sending emails.
  - If an email string is longer than four replies, it is time for a meeting.
- Be consistent—most people in a program organization work several different programs, and CM implementation should not change every time a new program starts.
  - If the implementation rules are changing, provide employees with a gap analysis explaining what has changed and why.

  - If CM is not sure how something was done on the last program someone worked on, reach out to the configuration manager on that program.
  - You cannot do a gap analysis if you do not know how wide the gap is.
- Be flexible and always learn how the latest CM implementation was crafted.
  - Always move the state of the art forward by using the latest CM implementation as the starting point.
  - The older the CM implementation is, the more it cost.
    - Programs in the 1970s employed a CM staff of 20 to 30 people.
    - Current programs may use 15% of that number.

In one company studied, CM had made system-wide decisions that affected CM implementation each year for 18 years. Every improvement reduced overall cost or resulted in cost avoidance. These improvements were not retroactively implemented (flowed across all active programs), but simply phased in when a new program started. One program still existed using an 18-year-old CM implementation. When the program ended, CM staff members were reassigned. They had to be extensively trained in the current suite of CM tools and processes. The sudden transition from using a paper system to a PD/LM system was so great that many quit the company in frustration.

Continual improvement in CM, once started, must flow across the entire department and all programs. The approach taken by the company being discussed resulted in disenfranchising those on existing programs whenever an improvement was implemented. Do not be afraid to fail when making suggestions for improvements. Every promising idea can be used—perhaps not at the time and place it was originally thought of, but downstream when business conditions allow it. Proposals may impact other organizations. It takes time to socialize things across a horizontally organized company.

Improvements that are quickly accepted across the enterprise are based on improvements making it easier, faster, and less expensive for everyone, not just for one department or activity. We have seen many risk and opportunity submittals to process improvement boards claiming that the change will reduce costs. When analyzed, the change often reduces the time to do a repetitive task by two hours in one area while increasing the cost to all other organizations by 40 hours.

7.12.3.2 Know Your Internal Customer

Some organizations only release a few documents once each program. They will never become PD/LM experts. By the time they use the PD/LM system a second time, several years may have passed and they need to be trained again. Such organizations will willingly allocate a portion of their budget to have CM to generate their changes, document objects, and BOMs in the PD/LM system, as it saves them and the company money in the long run.

If assigned a new program, call a CM peer review to discuss how the department has traditionally worked with an individual or an organization. It will enable you to work effectively with the individual(s) and go a long way toward maintaining or improving how CM is perceived by the business unit and program.

Do this before reacting to any request for support believed not to be a CM responsibility. Oftentimes, CM can accommodate non-traditional support requests by first thinking about them and creating a CM implementation plan modification. If this is not done, it is easy to lose track of the scope of CM implementation on a program. CM implementation needs to balance use of available tools intelligently by collectively applying critical thinking skills developed within the CM organization.

No company pays employees to be automatons. Distribution of data by the CM conveys that the configuration manager agrees with the content. Bad information can be distributed with apparent CM concurrence if the data is not reviewed for correctness prior to distribution. Sitting silently in a meeting and not speaking up when you see others going down the wrong path constitutes CM approval of the wrong decision. CM silence may result in the company investing in doing something or going down a path the CM representative knows is wrong. CM implementation requires all employees to shout out if they see something that isn't right.

7.12.4 Program Dos and Don'ts

Each program can be successful or not, depending on the actions of those working on it. There is no cookie-cutter approach that will fit every program's needs and no minimum set of things that CM implementation needs to encompass to make it a success.

With each new program start, the company has the chance to make better-informed decisions and to create an environment where CM implementation within the company becomes more closely aligned with how CM is performed outside of the company. This synergy between knowledge management and CM is critical.

The following CM implementation notes are based on decades of experience.

DO

- Spend time understanding critical issues and helping people with those issues.
- Listen with the view of understanding the intent of the conversation rather than just the words.
  - A person saying that they understand what is being said does not imply they agree with it or that they will act because of that understanding.
- Seek to understand the goals and needs of the program management team.
- Pay attention to other people's hot spots.
  - When they reach out to you, their problem is the highest thing on their priority list.
  - Put CM priorities on the back burner for the good of the program and give extra help when needed.
- Recognize that the CM implementation developed on the last program may not be good enough on the next project.
- Look at the program from the program manager's perspective.
  - The more you understand the programmatic needs, the better you will understand how CM can meet those needs.
- Recognize that every day will be a learning experience and CM department capabilities will grow from those collective experiences.
- Reach out and ask questions in areas you are unfamiliar with.
  - Learning and sharing that learning grows the entire company and makes for a better CM implementation.
  - It also avoids serious career limiting errors.

DO NOT assume that:

- Everyone knows what program and enterprise CM implementation entails.
  - Much of what CM implementation is and CM professionals do they will never see.
- Everyone understands why CM must do certain things.
- CM implementation will be greeted with open arms.
  - You cannot expect to teach everyone how to do things in the PD/LM system and have it stick.
  - If you do, you will become frustrated when they call for help.
  - Remember that if sausage lovers understood how it is made, many of them would soon lose their liking for it.
- CM can live on past virtuous deeds and accomplishments.
  - That is like saying, "I don't have to eat today because I ate last year, last month, last week, or yesterday."
  - You need to "move with the cheese."[6]
- CM knows better than anyone else what the program needs.
- CM knows better than anyone else how the company works.

[6] Johnson, S. (1998). *Who Moved My Cheese*, G. P. Putnam's Sons, 375 Hudson St. New York, NY 10014.

- Program CM does not have to understand the interface between enterprise CM implementation and program CM implementation.
- If a process is not documented, it is prohibited.
  - Process improvement is ongoing.
  - New processes are always tested on programs prior to enterprise implementation.
- If something is not specifically authorized by process, it is prohibited.
  - Procedures and processes don't work that way.
  - You can always do more than is documented.
  - You can always do something less or differently than what is documented if you have an approved process variance.

DO NOT let yourself believe that CM implementation can be accomplished:
- Without understanding as well as the contract manager the parts of the contract that apply to CM.
- Without interfacing with your peers and program personnel.
- Without being part of the program management team.
- Without reading change requests it to see if they make sense.
- In the same way it was on the last program and be successful.
- Without doing whatever it takes to provide a product that makes the customer successful while maximizing profitability.
  - This may include having someone send a document and request that CM format it, apply IP statements to it, and get it into the PD/LM system to release.
- Without weekly interface with the program functional leads and other program critical functions.
- Without contract and contractual requirements documents being managed under CM.

7.12.4.1 CM as a Service Organization

CM implementation should never change processes to make it easier for CM while making it more difficult for everyone else. The only exception to this is in those cases where adding a little extra work for everyone results in greater paybacks than the total effort expended.

A general definition of CM as a service organization is

> CM implementation provides support to others in a way that is seamless, comprehensive and eases the way, so they can complete specific tasks needed for program performance while meeting program or project commitments levied through internal or external contracts. It anticipates the needs of those it supports before they arise and makes significant contributions to the overall goals of the organization by being proactive.

7.12.5 The Many Aspects of CM

CM professionals wear different hats depending on program needs.. In the eyes of many departments, the perception that something is a CM task leads to a non-standard CM implementation.

- CM can either step into the role that no one else is doing or grow the responsibilities of the CM department or
- CM can submit the issue to the risks and opportunities board for adjudication.

7.12.5.1 Speaking with One Voice to the Customer

A company should always speak with one voice to the customer. In some cases, the company may have several programs with the same buying office. This makes the one-voice concept harder to apply. CM implementation plays a very important role in maintaining consistency.

It is recommended that CM employees on programs with the same buying office stay in constant coordination. Issues surfacing on one program will soon filter across other programs. If possible, use the same formats for reporting DRL status and other status to the buying office. It makes it easier for the customer and allows people to back each other up with less transition time.

7.12.5.2 Domestic ITAR Marking Inconsistencies

The contract and the designated ITAR officials should never disagree relative to domestic ITAR legends. Should the designated ITAR official and the contract manager disagree, CM, legal, and the designated ITAR official need to sort it out before including it in the CM implementation. The purpose of this meeting is to reach consensus between the parties and ensure that program ITAR training is provided.

7.12.6 The Property Management Interface

The more a CM professional knows about property management, the better it is for the company. Traditionally, CM prepares DD Form-250 or COC and property management prepares the associated DD Form-1149.

CM implementation must be tied into the property management process and document the interface with the contracts department in the identification section of the CM plan. CM must coordinate the unit price to put on DD Form-1149 and DD Form-250 or COC. The contracts department, property management, and CM interface also ensures that residual property is properly transitioned off the program when it ends.

7.12.6.1 DD Form-250, Material Inspection and Receiving Report

On many programs, it is CM's responsibility to complete DD Form-250 (Figure 7.4) and provide it to contracts management. Contracts management coordinates it with the customer's buying office to ensure that the "Ship to," "Mark for," billing office, and other information is acceptable. Once this has happened, CM needs to provide a copy of the document to property management, so that they can prepare DD Form-1149. To prepare DD Form-250, CM needs to work in concert with contracts and the financial analyst to ensure that the unit and extended prices are correct.

There are nuances to DD Form-250. These can get the company into trouble if not understood.

- Shipment numbers in block 2 must match the contract DIL or contract line item numbers applicable to the program.
- Forgetting that block 2 on the last DD Form-250 on the program has a "Z" appended to the shipment number (e.g., if there are 21 shipments, the very last one reads 0021Z not 0021).
- Improper date formats in block 3.
- Improper acceptance points in block 8.
- Improper prime contractor information in block 9.
- Improper "administered by" information in block 10.
- Improper payment information in block 12.
- Improper "shipped to" information in block 13.
- Improper "marked for" information in block 14.
- Improper item number in block 15.
- Improper stock/part number information in block 16.
- Improperly filled out blocks 21a and 21b.
- Improper use of block 22.
- Incomplete information in block 23.

MATERIAL INSPECTION AND RECEIVING REPORT

Form Approved
OMB No. 0704-0248

The public reporting burden for this collection of information is estimated to average 30 minutes per response, including the time for reviewing instructions, searching existing data sources, gathering and maintaining the data needed, and completing and reviewing the collection of information. Send comments regarding this burden estimate or any other aspect of this collection of information, including suggestions for reducing the burden, to Department of Defense, Washington Headquarters Services, Directorate for Information Operations and Reports (0704-0248), 1215 Jefferson Davis Highway, Suite 1204, Arlington, VA 22202-4302. Respondents should be aware that notwithstanding any other provision of law, no person shall be subject to any penalty for failing to comply with a collection of information if it does not display a currently valid OMB control number.

PLEASE DO NOT RETURN YOUR COMPLETED FORM TO THE ABOVE ADDRESS.
SEND THIS FORM IN ACCORDANCE WITH THE INSTRUCTIONS CONTAINED IN THE DFARS, APPENDIX F-401.

1. PROCUREMENT INSTRUMENT IDENTIFICATION (CONTRACT) NO. | ORDER NO. | 6. INVOICE NO./DATE | 7. PAGE | OF | 8. ACCEPTANCE POINT

2. SHIPMENT NO. | 3. DATE SHIPPED | 4. B/L TCN | 5. DISCOUNT TERMS

9. PRIME CONTRACTOR CODE | 10. ADMINISTERED BY CODE

11. SHIPPED FROM *(If other than 9)* CODE | FOB: | 12. PAYMENT WILL BE MADE BY CODE

13. SHIPPED TO CODE | 14. MARKED FOR CODE

15. ITEM NO.	16. STOCK/PART NO. DESCRIPTION *(Indicate number of shipping containers - type of container - container number.)*	17. QUANTITY SHIP/REC'D*	18. UNIT	19. UNIT PRICE	20. AMOUNT

21. CONTRACT QUALITY ASSURANCE

a. ORIGIN

☐ CQA ☐ ACCEPTANCE of listed items has been made by me or under my supervision and they conform to contract, except as noted herein or on supporting documents.

DATE — SIGNATURE OF AUTHORIZED GOVERNMENT REPRESENTATIVE

TYPED NAME:

TITLE:

MAILING ADDRESS:

COMMERCIAL TELEPHONE NUMBER:

b. DESTINATION

☐ CQA ☐ ACCEPTANCE of listed items has been made by me or under my supervision and they conform to contract, except as noted herein or on supporting documents.

DATE — SIGNATURE OF AUTHORIZED GOVERNMENT REPRESENTATIVE

TYPED NAME:

TITLE:

MAILING ADDRESS:

COMMERCIAL TELEPHONE NUMBER:

22. RECEIVER'S USE

Quantities shown in column 17 were received in apparent good condition except as noted.

DATE RECEIVED — SIGNATURE OF AUTHORIZED GOVERNMENT REPRESENTATIVE

TYPED NAME:

TITLE:

MAILING ADDRESS:

COMMERCIAL TELEPHONE NUMBER:

** If quantity received by the Government is the same as quantity shipped, indicate by (X) mark; if different, enter actual quantity received below quantity shipped and encircle.*

23. CONTRACTOR USE ONLY

DD FORM 250, AUG 2000 PREVIOUS EDITION IS OBSOLETE.

Reset

FIGURE 7.4 DD Form-250, Material Inspection and Receiving Report.

- Not getting DD Form-250 signed off in at least block 22 and DD Form-1149 signed by both the losing and the gaining contracting officer prior to shipment.
- Because major variances (major and critical deviations and waivers) are a departure from the technical requirements baseline, all approved major variances need to be attached as an addendum to the DD Form-250 for that item along with full-text copies of the approved documents.

In some cases, acceptance cannot take place until some event has taken place. On a spacecraft program, this may be on orbit acceptance. Typically, it is after instrument dry-out in vacuum, subsystem-by-subsystem checkouts after thermal balance is achieved, and then instrument checkout, as well as a complete set of functional on orbit tests, are complete. Such acceptance is documented in block 22b of DD Form-250. This should not preclude the contracting officer from signing block 23.

7.12.7 DD-1149, Requisition and Invoice/Shipping Document

Once DD Form-250 is completed and socialized with the contracting officer, CM provides a copy to the property administrator with a request that DD Form-1149 be prepared. When the DD Form-1149 is received back from property management, it needs to be checked against DD Form-250 for accuracy. This is part of the checks and balances that CM provides to property management and the program.

Once DD Form-1149 is reviewed and corrected, both DD Form-250 and DD Form-1149 must be provided to the contracting officer through the company contracts department. After the documents are signed off by the customer, the shipment is prepared.

When delivering an item from one U.S. government program to another, the customer's contracting officer on the losing program and the contracting officer on the gaining program both need to sign the DD Form-1149 before the shipment can take place.

7.12.8 Shippers and DIL Deliveries

There are often mandatory inspections and container marking requirements that must be met before shipment. CM implementation is a key to making sure that these take place. Each delivery should be planned, coordinated, and moved through the system by a team that includes production and CM. Points at which you stop and take assessment should be included throughout the process. The following questions always need to be answered:

- Is DD Form-250 signed?
- Is DD Form-1149 signed?
- Does the shipper have notes in it that copies of the signed DD Form-250 and DD Form-1149 must be placed in the shipping container (with the originals in the contracts department's hands and a copy of both documents delivered to CM)?
- Does the shipper have notes that contracts, production, and CM be notified as the shipment moves from the contractor's facility to the destination?
- Does the shipper require documentation that shipment receipt and acceptance at the destination has taken place?
- Are all documents captured in the PD/LM system and a digital thread established that links them to the item being delivered?

7.12.9 Certificates of Conformance

The COC content is much the same as general information in a DD Form-250. The COC identifies the customer, item contracted for, item identification (e.g., CI, model number, service description, etc.), quantity, unit price, unit of measure, extended price, and all approved variances. It is signed by the company prior to being submitted for signature by the customer.

7.13 SUMMARY

Good CM implementation practices are critical to the survival of an organization. CM implementation activities can be performed by one department or allocated across many departments. It is as important in the maintenance of the GTS infrastructure as it is in maintenance of the configuration of a facility, a piece of capital equipment, or a system or service sold to a customer.

How CM is implemented may embrace all aspects of CM envisioned by the regulatory and guidance documents, may embrace a subset of them, or may be transcendent—evolving CM faster than regulations can keep up. CM implementation is entirely dependent on the allocation of responsibilities for performance of CM activities to one or more departments or the assumption of other department responsibilities by the CM department.

To truly implement in the broadest sense the meaning behind the words "CM is chartered to perform configuration planning and management, configuration identification, configuration change management, configuration status and accounting, configuration verification and audit" is a far-reaching goal. It presupposes that the CM practitioner has a firm grasp of the financial implications of the tasks they perform and the decisions they make. It also presupposes they understand the risk posture of the program and the direct impact that they have on meeting program risk and profit goals.

8 Functional Interfaces

8.1 QUESTIONS ANSWERED IN THIS CHAPTER

- What program elements should fall under change control?
- Should enterprise CM implementation include change control of all outward facing documents?
- What are the implications of poor CM implementation during the manufacturing process?
- Should CM implementation include command and control media?
- What is a change control board?
- What is the connection between verification and specification requirements?
- What is the connection between project scope and project closure?

8.2 INTRODUCTION

> Design engineering companies face increasing market and competition pressure forcing them to offer innovation in ever-shorter cycles for ever-lower prices. In their struggle to keep profit margins up, companies more and more must broaden their horizon for resource acquisition—both human and material—to a global level. Their pursuit for the best conditions has started with globally distributing and outsourcing production of less complex parts and products a few decades ago, but has reached product development and design engineering workload in the last years of the twentieth century.[1]

Globalization stretches functional resources beyond the norm. It also results in a CM implementation at the enterprise level with a much broader scope than CM implementation on a program. Implementation starts at the enterprise level and spreads downward through the organization and outward across the supply chain. While a hand holding a piece of paper is not the paper itself, the hand is involved, and so if any element of CM is involved then CM is present. This argument is gaining wide acceptance internationally. This is certainly true of SAE/EIA 649-C. The trick is to tailor each of the elements that make up CM to the task at hand. In the case of financial records, although a receipt for petty cash may not require subsequent change or version control, the date on the receipt acts as the first (and only) release date. Assessing the CM implementation requires a rudimentary understanding of what functions in a typical organization do. We have attempted to provide that in this chapter.

8.3 PROJECT MANAGEMENT BASICS

Project management has existed since the times of the pharaohs and pyramids (2700 BCE) and is seen in every other wonder of the ancient world. Although activities such as the building of the pyramids may seem more production oriented, there are multitudes of supporting activities tied to project management: concept selection, scope and change management, human resources, and support infrastructure.

[1] From the International Conference on Engineering Design, ICED'07, August 28–31, 2007, Cite des Sciences et de L'Industrie, Paris, France: Cross-Cultural Collaboration in Design Engineering—Influence Factors and Their Impact on Work Performance.

The modern incarnation of project management is often associated with the post–World War II reconstruction in the 1950s. Project management is not a homogeneous discipline but a collection of knowledge areas. These areas are clearly defined by the Project Management Institute, Project Management Body of Knowledge,[2] as:

1. Integration management.
2. Scope management.
3. Time management.
4. Cost management.
5. Quality management.
6. Communication management.
7. Risk management.
8. Procurement management.
9. Stakeholder management.

8.3.1 Project Integration Management

According to the Project Management Institute, project integration management must include the processes and activities needed to identify, define, combine, unify, and coordinate the various processes and project management activities.

The specific areas of interest to CM implementation are:

- Development of the project charter.
- Development of the project management plan.
- Directing and management of project execution.
- Monitoring and control of project work.
- Performing integrated change control.
- Closing of a project or phase.

These processes and activities allow the project to be managed effectively through coordination of planning and execution to meet the defined scope while maximizing return on investment. The nature of these items requires that changes to them be formally evaluated and dispositioned. This ensures that proposed alterations are tied to a change in scope or in project direction. Any change will require review and a mechanism for articulating the changes through the project team.

8.3.2 Scope Management

In project management parlance, *scope* is the work that must be done to deliver a product according to the product's required functions and features defined in the contract. Reaching this objective requires an expenditure of money and an investment in time, talent, and material. Tasks to achieve this objective are created and monitored. Changing the product's required functions and features as defined in the contract impacts the activities the team should undertake. Accepting direction that is not contractual (e.g., from customer representatives with no authority to do so) happens so frequently that it is commonly called "scope creep." If the change is driven from within the program team, it is called "gold plating." Failure to control scope creep and gold plating will significantly impact the probability of success. In some circles, this is referred to as program "death by a thousand lashes." We have seen programs where the functions and features requirements being

[2] Haughey, D. (2013, July 9). *Project Management Body of Knowledge.* http://www.projectsmart.co.uk/, http://www.projectsmart.co.uk/pmbok.php. Accessed August 4, 2013.

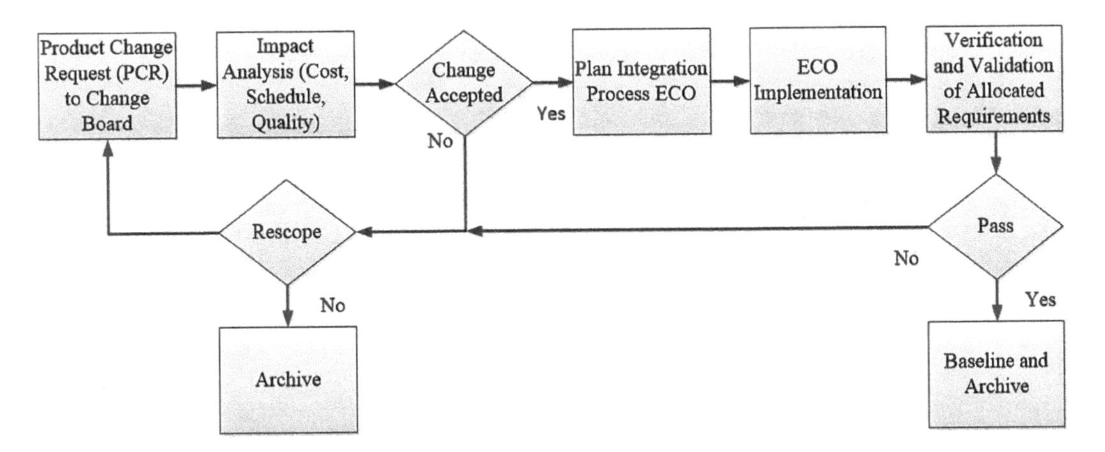

FIGURE 8.1 A typical change assessment engineering change order (ECO).

performed to are so different from those on contract that the product or service failed functional and physical configuration audits.

> It is here the configuration management is heavily involved as this is the embryonic first phase of the product lifecycle. It is here that configurations are identified from a user, contractual and enterprise business framework.[3]

Several tools are available to define the project scope.

- Project charter.
- Specifications.
- Interfaces.
- Statements of work.
- Schedules.
- Work breakdown structure.

Project management controls the scope definition documentation, which is updated only with approved changes and subsequently communicated through the project team. This traceability of scope to approved changes ensures no *gold plating* or unaccounted-for *scope creep*. The scope documentation is managed throughout the program/project life cycle. Strict change control is essential. Experience suggests no project will go through its life cycle without incurring change. These changes move through a change management system, much like the one in Figure 8.1. Prior to change approval, the risks associated with the change are evaluated by the stakeholders. This includes the risk of the change not being approved, often overlooked during impact analysis of the change.

8.3.3 Time Management

Time management employs tools to quantify the staffing and delivery aspects of the project. To do so, activities and all dependencies are scheduled. Dependencies are the interconnections between tasks and objectives. An example of a dependency is the relationship between software delivery and software testing. A finish-to-start dependency results as software must be

[3] Dirk Wessel, Chief CM Section NCI Agency input during text review.

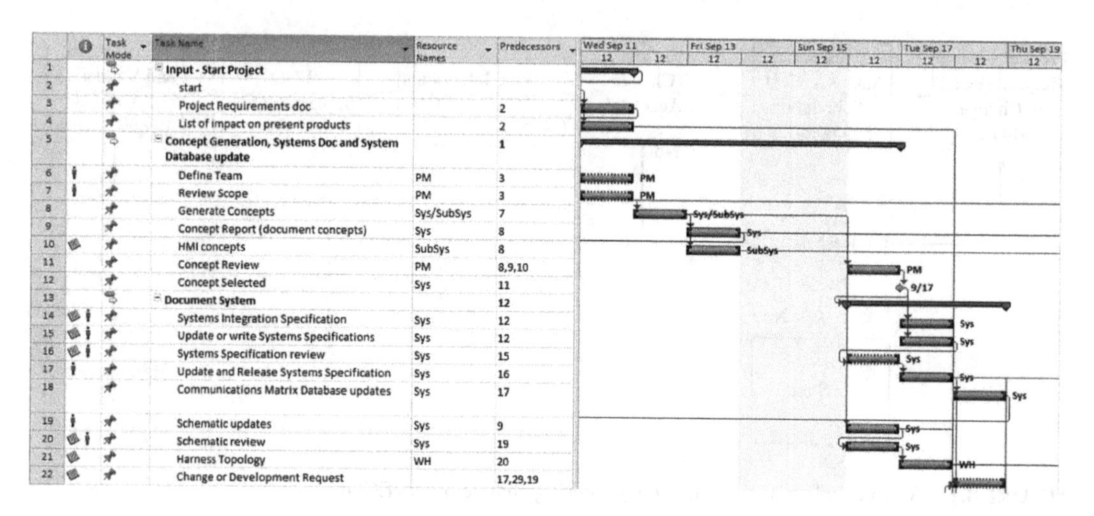

FIGURE 8.2 Gantt chart.

complete before software testing can begin. A Gantt[4] view (Figure 8.2) may be useful. Planned performance is tracked against actual performance in the schedule. This allows the project manager, team, and executive management to understand actual vs. planned progress visually. If the dependencies are correct, the impact of change to the project implementation schedule can be assessed.

8.3.4 Project Communications Management

A communication plan will help in managing and maintaining expectations. The communication plan is under change control throughout the project life cycle. Engineering change orders may be required by the customer, enterprise re-organization, or a change in key stakeholders or project sponsors. The typical contents of the communication plan are from *Project Management of Complex and Embedded System.*[5]

8.3.4.1 Information Distribution

The communications plan should include a chart detailing communication responsibility. International Organization for Standardization (ISO)/TS 16949, "Quality Management Systems Particular Requirements for the Application of ISO 9001:2000 for Automotive Production and Relevant Service Part Organizations," requires the inclusion of customer-specific documents.

8.3.4.2 Information Description

The communications plan should include a description regarding the type of information to be distributed, its format, its content, and its level of detail.

8.3.4.3 Information Schedule

The communications plan should include a communications schedule. The communications schedule defines the method for assessing the rates of information delivery (e.g., as required, weekly, every two weeks, monthly, etc.).

[4] A kind of bar chart developed by Gantt, H., in 1910.

[5] Pries, K. H., and Quigley, J. M. (2009). *Project Management of Complex and Embedded Systems*. Auerbach Publications/Tailor & Francis Group. p. 26. Reprinted with permission from the Publisher.

8.3.4.4 Progress Reporting

The communications plan should define the method used for collecting and reporting project status or progress reports.

8.3.4.5 Communications Plan Revision

The communications plan should define the method for revising, updating, and otherwise refining the communication management plans as the project progresses.

8.3.4.6 Escalation Process

The communications plan should define the escalation process for the project. Team members may be divided in implementation paths or how requirements are interpreted. Define in advance the escalation process regarding how to break such *log jams* when they occur, reach consensus, and meet schedule.

8.3.4.7 Administrative Closure

The communications plan should define generating, gathering, and distributing information for phase formalization or project completion. It consists of documenting project results to formalize acceptance of the proposed or completed products by the customer. It includes collecting the digital threads for project records, ensuring that they reflect the contractual requirements, analyzing project success, recording effectiveness and lessons learned, and archiving for future use.

8.3.5 Project Cost Management (Earned Value Management)

Customer driven measuring of cost and schedule performance is known as earned value (EV). It originated in the U.S. DOD and directly influences the tools the contractor needs to put in place. EV management (EVM) centers on budgeted cost of work scheduled (BCWS), actual cost (AC) of work performed (ACWP), and budgeted cost of work performed (BCWP).

Successful implementation of an EVM system requires:

- Connection between specific work packages and the cost accounting system.
- Appropriate and quick time-reporting mechanisms.
- Measurements of the present state of tasks with respect to completion.

Meeting these criteria enables mapping both schedule and cost for the project. The specific measurements of interest are listed subsequently.

8.3.5.1 Schedule Performance Index

Schedule performance index refers to the ratio of work performed to the value of the work planned.

8.3.5.1.1 Schedule Variance

Schedule variance measures how much ahead or behind the project is compared to its schedule (Equation 8.1).

$$\text{Schedule variance (SV)} = \text{BCWP} - \text{BCWS}$$

EQUATION 8.1 Schedule variance calculation.

A positive SV indicates the project is ahead of schedule, whereas a negative SV indicates the project is behind schedule.

8.3.5.2 Cost Performance

Cost performance is the ratio of earned value to actual cost.

8.3.5.2.1 Cost Variance

Cost variance measures how much ahead or behind the project is compared to its budget. It can be measured on a cost basis using the planned budgeted work against the actual planned work (Equation 8.2).

$$\text{Cost variance (CV)} = \text{BCWP} - \text{ACWP}$$

EQUATION 8.2 Cost variance calculation.

A positive CV indicates the project is under budget, whereas a negative CV indicates the project is over budget.

Unaccounted-for changes to the scope of a project will affect cost monitoring and control actions. Making the most of these progress measurements requires that the baseline be updated to reflect the expected performance due to scope creep and gold plating activities. Negotiated as well as contractor proposed revisions submitted in accordance with the contractual "Changes" can add or delete tasks, change cost and schedule estimates, change material costs and task duration, and create additional interdependencies that must be accounted for.

8.3.5.3 Other Elements Associated with EVM

Our discussion regarding cost and schedule variances will not make you an expert. It serves only to introduce the terms, so they are familiar. Other terms and equations that you will need are shown in Equations 8.3–8.6.

$$\text{Budget at completion (BAC)} = \text{Baselined effort} - (\text{Hours} \times \text{Hourly rate})$$

EQUATION 8.3 Budget at complete calculation.

$$\text{Estimate at completion (EAC)} = \text{AC} + \text{ETC}$$

EQUATION 8.4 Estimate at completion calculation.

$$\text{Variance at completion (VAC)} = \text{BAC} - \text{EAC}$$

EQUATION 8.5 Variance at completion calculation.

$$\text{Estimate to complete (ETC)} = \text{EAC} - \text{AC}$$

EQUATION 8.6 Estimate to complete calculation.

8.3.6 Project Quality Management

Dave Garwood states, "Quality can best be defined as conformance to expectations[6]." Project quality management involves the actions, processes, and activities performed to deliver on these expectations. Expectations can come in the form of international, national, enterprise, and contractual

[6] Garwood, D. (2004), *Bills of Material for a Lean Enterprise*, Dogwood Publishing Company, Inc. 111 Village Parkway Building 2, Marietta, Georgia 30067. Fourth printing April 2010.

requirements. Organizations may have quality handbooks and quality policies as part of an enterprise mission assurance implementation. Evaluating quality is a comparison of a product or service to an idealized goal and the ability to meet those idealized target measurements.

To provide this assessment and to adjust the project direction to achieve the ultimate quality targets requires measurements, analysis, and course corrections. The course corrections can take the form of change to:

1. Product
2. Product development processes
3. Product manufacturing processes
4. Project management processes

Item identification ties directly to quality management activities in all hardware, software, and firmware disposition during change review and approval. Figure 1.18 presented one aspect of disposition decisions based on differing levels of cleanliness. Other disposition decisions may pertain to material choice, material testing and vendor certification activities, and monitoring.

8.3.7 Project Risk Management

Project risk management activities are needed to avoid or mitigate the impacts of the anomalies or anticipated eventualities on the project. This risk mitigation is recurring throughout the project. Risks are logged in a risk register for analysis and mitigation. The probability (likelihood) and impact (severity) of the item on the program success is categorized. Actions required to alleviate the risk are identified and assigned to individuals. This living document is modified as the project moves through the design, development, and production phases. Risk assessment will be discussed in Chapter 13.

8.3.8 Action Item List

An action item list is used to articulate what tasks need to be done, when each task needs to be completed, and who is responsible for the task across interfacing departments or with supply chain companies. The list changes over the life of the program as older actions are closed and new ones emerge. Experience suggests the action item list needs to be managed like any other document using change control. If change control is not used, the management method must be included in the CM Management and Planning section of the CM Plan.

8.4 ENGINEERING BASICS

"Engineering" is a general term that comes from the Latin *ingenium*, meaning "cleverness" and *ingeniare*, meaning "to contrive or devise." Engineering can be broken down into subdisciplines like aerospace, automotive, chemical, civil, configuration management, Earth systems, efficiency, electrical, environmental, mechanical, nano, parts, production, quality, software, and systems. Each of these has multiple subspecialties, including thermal, stress, and reliability. Marketing identifies opportunities. Engineering brings products to life that can take advantage of those opportunities. Product development engineering requires a tightrope walk between tightly and loosely controlled structures to allow creative room for engineers to explore design possibilities. This does not mean zero control but relies more on the engineer to act in accordance with good development techniques.

Defined processes for product development work address this need in a *closed-loop* system. Typically, the system is the set or subset of processes to which the organization works to design and

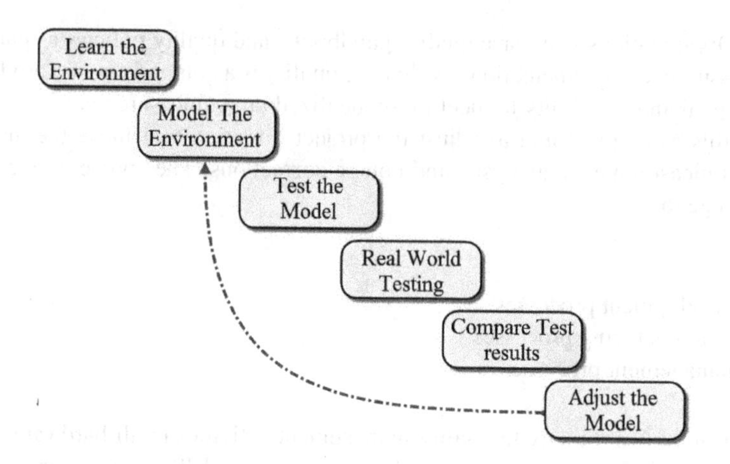

FIGURE 8.3 Scope, systems requirement, and component requirement.

develop the product. The processes are consistent with Software Engineering Institute,[7] Capability Maturity Model® Integration®, or AS9100 standard(s).

8.4.1 Design and Requirements

CM implementation includes requirements allocation, which is the catalyst for the development process. A business case without requirements will not result in a product or service. Requirements without a business case will not be funded. Engineering allocates technical requirements to systems and subsystems to implement. Figure 8.3 depicts how requirements and requirement changes are tied together.

Requirements always flow from the highest level to the lowest level in a "parent-child" relationship. Proof that the requirement has been met through analysis, test, inspection, or demonstration always flows from the lowest-level requirements. The most detailed requirements are traceable to the next higher level, which is in turn traceable to the next higher level in an indentured or "child-parent" relationship. Figure 8.3 shows the parent-child relationship. A good requirement allocation is:

- *Clear, concise, and valid:* If it is not, it should not be a requirement.
- *Complete:* All information necessary for achieving the requirement.
- *Consistent:* No contradictions between requirements.
- *Uniquely Identified:* One requirement ID per requirement.
- *Traceable:* Traceable all the way back to the scope of the project.
- *Verifiable:* Able to perform objective analysis/test to confirm objectives are met.
- *Prioritized:* Level of urgency associated with project objectives.

Figure 8.4 shows the digital thread as requirements are allocated downward and provides a simple requirement numbering schema. Starting with product requirement 1.1, the sub-tier requirements are subsequently broken down into smaller pieces that are easier to verify. This process of allocating requirements to the lowest level is also referred to as a requirements functional decomposition. As artifacts are created that prove each requirement has been met, another digital thread connects the lowest-level artifact upward to the product requirement 1.1. When all digital threads are in place, requirement 1.1 is verified.

[7] Christel, M., and Kang, K. C. (1992). *Issues in Requirements Elicitation*, Technical Report CMU/SEI-92-TR-012, ESC-92-TR-012. Software Engineering Institute Carnegie Mellon® University, Pittsburgh, PA.

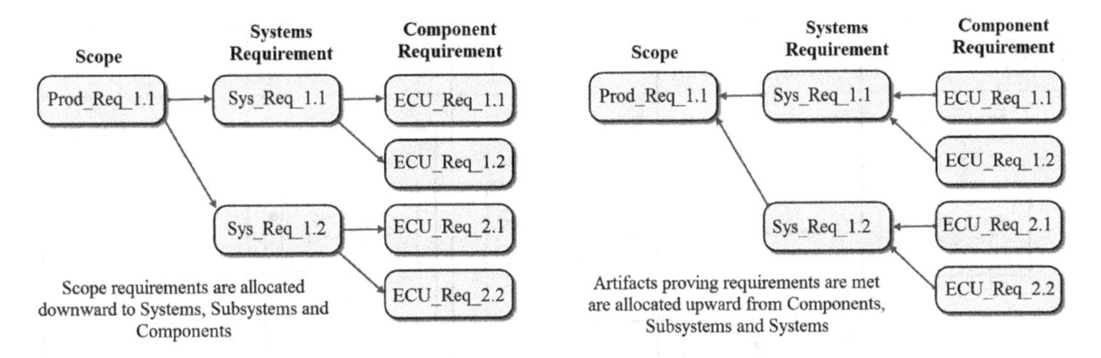

FIGURE 8.4 A typical requirements allocation.

8.4.2 Simulation and Verification

Simulation may be used to explore unique design solutions without making the actual parts. Simulation requires a lower investment in time, money, and opportunity cost. Development risks are reduced by using a digital twin. Different digital twins can exist for evaluating interfaces, structures, thermal characteristics, controllers, and such. Test models generated from the digital twin need to accurately represent operational constraints. Test simulations are run after the model is created to determine the model's alignment with the real-world environment and measured test data. Once the test model accurately represents the real world, it can be utilized to predict system behavior. Model development, as shown in Figure 8.5, is an iterative and multiple loop process of managed building, testing, learning, and model adjustment.

8.4.3 Bills of Materials and Bills of Documentation

Bills of materials (BOMs) are typically from either a design-centric view or a manufacturing-centric view, as depicted in Figure 8.6. The BOM is the first collection of data that supply chain management utilizes to assess the probable cost of the product, and manufacturing engineers use to set up the workflow through the production centers. If the product includes embedded items, the BOMs may consist of a microcontroller/microprocessor and associated code, making a sub-assembly.

BOMs are not the only type of bill associated with products. Other kinds of bills include:

- Bills of documents (BOD) identifying all documents in hierarchical order like a document tree.
- Bills of resources (BOR) listing people, machines, and material required for capacity planning.
- Bills of lading (BOL) for shipment of goods.

These BOD, BOR, and BOL items are captured in the PD/LM system and form digital threads necessary to weave the digital fabric.

8.5 PRODUCTION BASICS

Eli Whitney's "uniformity system" is cited as the source of many engineering disciplines in the United States and abroad. CM implementation is one of them. Many students in the United States during the 1950s, '60s, and '70s were taught that Whitney was the first to demonstrate interchangeable parts. We now know this was not true. Whitney's part interchangeability system (also known as the "armory system" and "American system of manufacturing") was perfected in the U.S. by Captain John Hall and Simeon North. Whitney staged a demonstration of part interchangeability

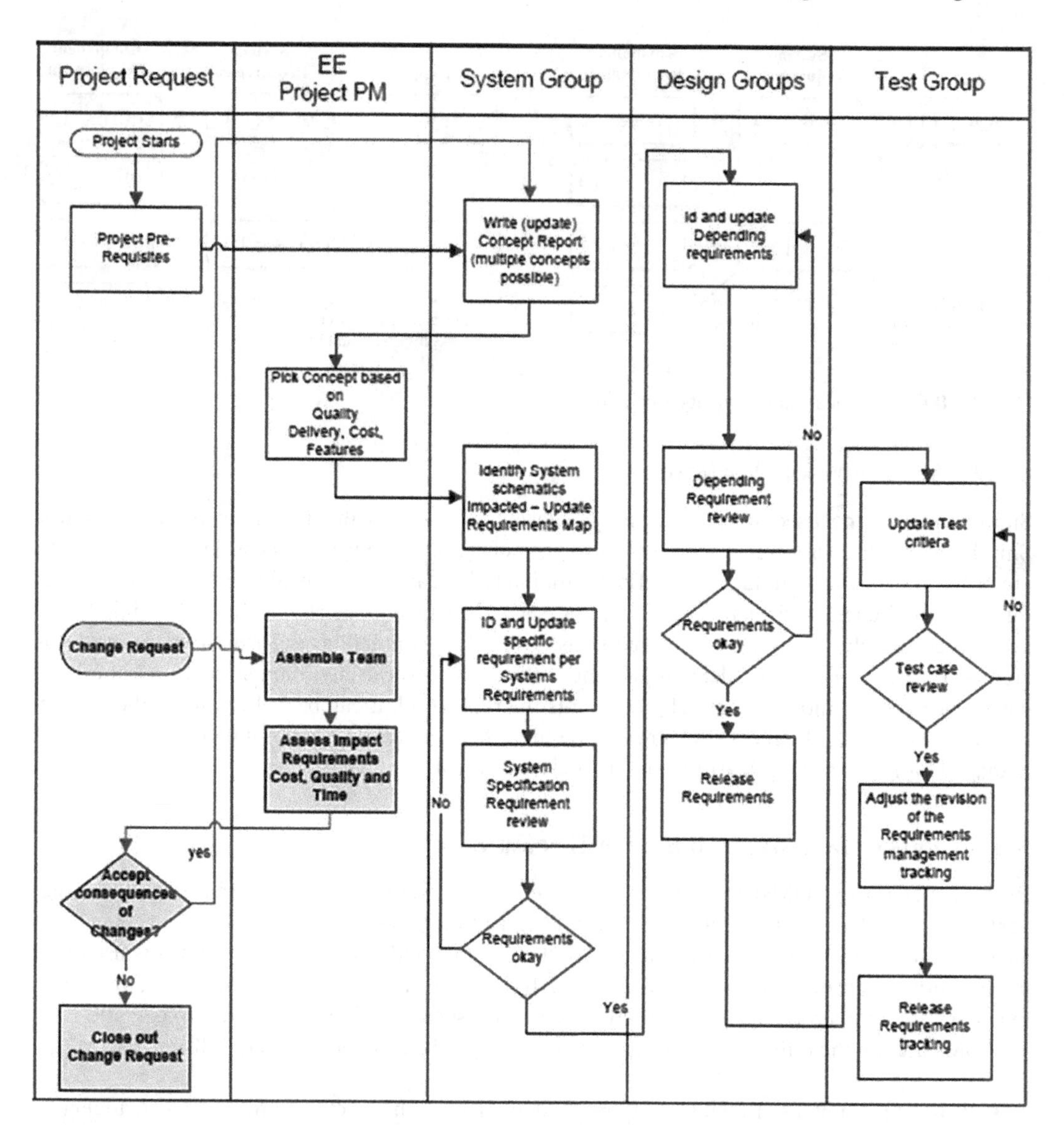

FIGURE 8.5 Test requirements management system.

before members of the U.S. Congress and newly elected President Thomas Jefferson in 1801. It is now widely regarded as a hoax using carefully selected and meticulously crafted parts and not a demonstration of the production capability of Whitney's armory. Whitney's armory did not achieve interchangeability until after his death in 1825.

Jean-Baptiste Vaquette de Gribeauval, Honoré Blanc, and Louis de Tousard are credited as being the true pioneers of parts interchangeability,[8] as we see it today. There is evidence that while Jefferson was ambassador to France he was aware of the efforts of de Gribeauval, Blanc, and de Tousard. He was impressed and obtained a pamphlet describing the benefits of part interchangeability and a description of how it was being accomplished. Jefferson passed the document on to Henry Knox the American Secretary of War. Knox gave the pamphlet to Whitney. When Jefferson returned to the USA he worked to fund interchangeability development.

[8] Althin, T. K. (1948). *C. E. Johansson, 1864–1943: The Master of Measurement.* Stockholm, Sweden: Nordish-Rotogravyr.

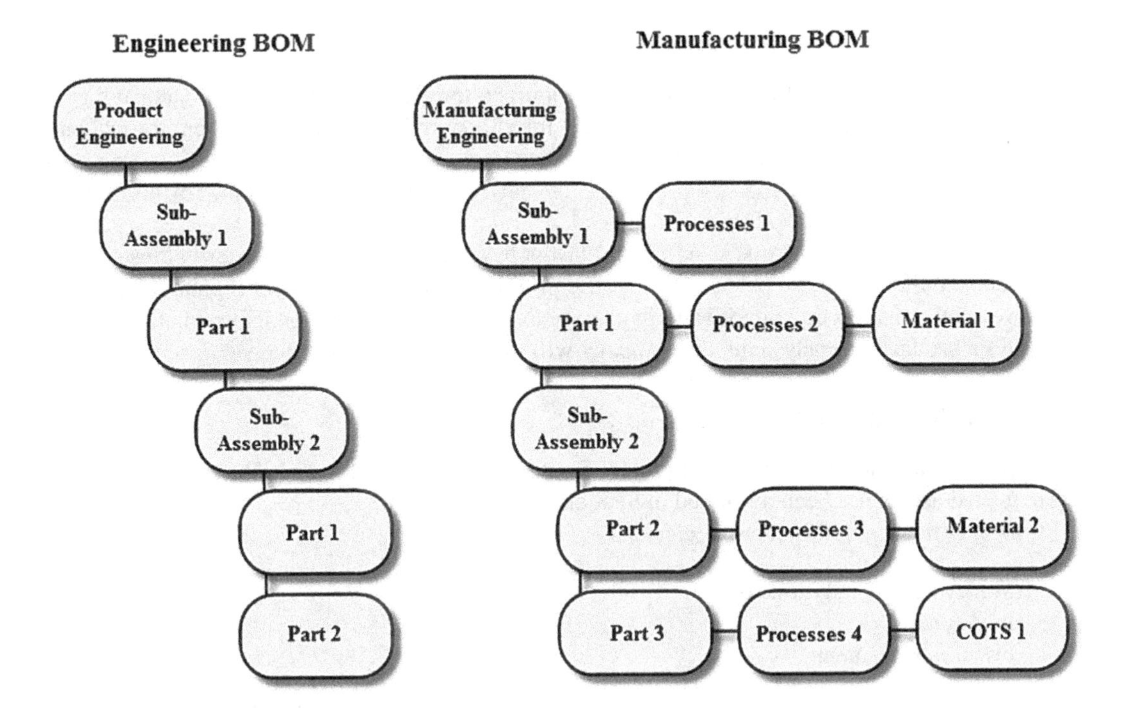

FIGURE 8.6 Engineering vs. manufacturing BOM.

The pamphlet itself may have been authored by Blanc.[9] At the time the pamphlet was written, Honoré Blanc was making 10,000 muskets a year with interchangeable parts for Napoleon Bonaparte well before Whitney's demonstration took place (Figure 8.7). There is also evidence suggesting that Jefferson tried unsuccessfully to entice Honoré Blanc to move to America. Production has advanced considerably from the days of Honoré Blanc. CM implementation has led to standardized parts and sub-assemblies usable on more than one product.

8.5.1 Production Setup

As a product develops, the tools, processes, workflows, and production infrastructure also evolve. Manufacturing personnel need to be included in the design process to ensure that the design is something that can be economically made. Design needs to be consistent with production methods, tools, and techniques. Design capabilities also evolve, and the use of additive manufacturing synergies will reduce the time and money associated with getting the product to market.

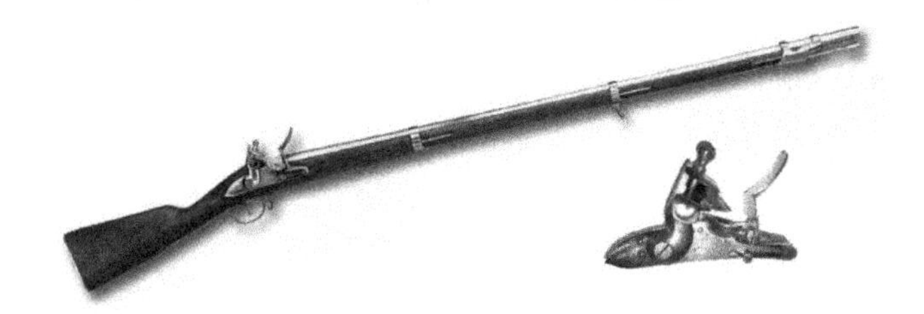

FIGURE 8.7 Honoré Blanc model 1776 musket.

[9] Roe, J. W. (1916). *English and American Tool Builders*. New Haven, CT: Yale University Press.

8.5.2 Production Test and Verification

Various tools have been utilized over time in the journey toward attaining the manufacturing goal of 100% product acceptance, lowered component inventory, and increased component throughput. These have gone by zero defects, total quality management, just-in-time manufacturing, lean production, Kaizen, Six Sigma, and so on. *The Goal*, by physicist Eliyahu M. Goldratt,[10] explores these themes in an easily understood story narrative.

When the production processes and factory layout have been optimized, a battery of tests is run on products built from these processes to assure product accuracy against Six Sigma and process improvement goals. Process modifications may be required before these goals are met. If testing indicates a production-ready state, the company will ramp up production.

8.6 FINANCE BASICS

Finance falls into two areas: program/project finance and company finance. The project management related aspect has been discussed in Section 8.3.5.

Company finance in this context refers to:

- Money and capital markets.
- Investments.
- Fiscal management.

Company finance produces financial statements for the governing board and stockholders:

- Balance sheet, a statement of financial position.
- Income statement, a statement of comprehensive income, and a statement of profit and loss or a profit and loss report.
- Statement detailing cash flow relative to operating, investing, and financing activities.

Larger corporations may require the inclusion of statement notes, executive analysis, and associated management decisions. The International Accounting Board[11] states:

> The objective of general purpose external financial reporting is to provide financial information about the reporting entity that is useful to present and potential investors and creditors in making decisions in their capacity as capital providers.

Should anything in the ambient environment change, then the reports will need to be adjusted. Some of the areas of concern are as follows:

- Current ratio
- Acid (quick) ratio (i.e., the ability to meet short-term liabilities with short-term assets)
- Inventory turnover ratio
- Days sales outstanding
- Fixed assets turnover ratio
- Operating capital ratio
- Debt ratio
- Profit margin on sales
- Return on assets
- Return on equity

---

[10] Goldratt, E. M., and Cox, J. (1992). *The Goal*. North River Press, Great Barrington, MA.

[11] IASB. (2010, September). *Conceptual Framework for Financial Reporting 2010*. http://www.iasplus.com/, http://www.iasplus.com/en/standards/other/framework. Accessed April 12, 2014.

Corporate finance is also responsible for payment of debt, allocation of costs to the proper cost pools and objectives, and preparation and payment of taxes.

8.7 LEGAL BASICS

New product development work requires analysis from a legal department perspective to ensure regulatory compliance. This starts at the local level (city) and progresses upward. An organization whose product is built or consumed internationally will have to comply with those laws as well. Legal department staff is trained to handle an ever-changing regulatory and contractual climate. If the final product is unable to meet the claims made for that product, then customer dissatisfaction may involve customer remuneration or product replacement through legal action.

The legal department also manages intellectual property marking, new-technology reporting, and marking compliance. In the international setting, this includes the implementation of trade restrictions pertaining to identified technologies such as the U.S. International Traffic in Arms Regulations. The legal department investigates competing or existing intellectual property and patents to ascertain the uniqueness of any company-generated intellectual property. The legal department bridges the gap from technical language to legal language when filing patents. This may include managing any remuneration to employees.

Unless the enterprise is specifically engaged in providing goods and services to a government, it may not have a specific department managing contracts and contractual issues. Should a separate contracts function exist, it is involved with contract interpretation and compliance with the customer community and not with the company's contractual relationship with its employees.

8.8 CONTRACTS DEPARTMENT

Many companies have a standalone contracts department. Each contract has associated risks. Depending on the contract type, the risk is weighted in favor of either the contractor or the customer. This is shown in Figure 8.8. The contracts department performs all functions relating to the bid and proposal contractual review as well as management of the contractual relationship between the company and its customers. The similar relationship between the contractor and its suppliers is

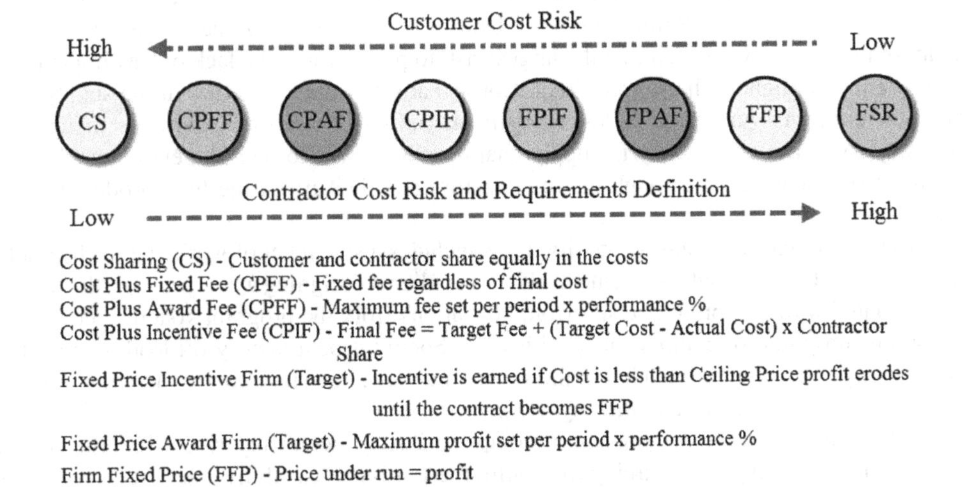

FIGURE 8.8 Customer versus contractor risk profile.

administered by the supply chain management in close coordination with the contracts department to assure mandatory requirements are flowed down to suppliers. In some cases, this area may be referred to as "purchasing." In other situations, the purchasing department is a sub-function under supply chain management that specializes in buying commodities (COTS items) whereas subcontract administrators specialize in major procurements.

The discipline of contract management includes the following:

1. Providing contractual documents that define the scope of the program.
2. Negotiating terms and conditions.
3. Agreeing and managing changes and amendments.
4. Monitoring performance.
5. Confirmation of contract closure.

A contract is a written or oral legally binding agreement between competent parties for a legal purpose. It may be a promise on both sides to do something for something else of perceived value or it may be a promise not to do something in exchange for something of perceived value. Contracts are binding agreements between identified parties with an agreement. Contract life cycle management (CLM) seeks to minimize the risks and maximize profits. CLM is one aspect of program life cycle management.

> Nearly 65% of enterprises report that contract lifecycle management (CLM) has reduced exposure to financial and legal risk.[12]

8.9 SUPPLY CHAIN MANAGEMENT BASICS

Supply chain management involves all the activities associated with the acquisition of materials, goods, and services from suppliers. The supply chain itself consists of service providers for analysis, test, and evaluation activities as well as raw material suppliers and sub-assembly suppliers, and extends through the delivery of finished products, supplies, and services to the customer. The supply chain network is shown in Figure 8.9. At each level of the supply chain, starting with the raw materials, value is added to the item being produced. With value comes documentation relative to the product and thus some form of CM implementation. At a meeting in Eindhoven that included the authors, DAF Trucks, ASML, Philips, and Prof. Dr. A. (Ton) G. de Kok of the Technical University of Eindhoven (TU/e), it was stated that the largest risk to product was the lack of CM implementation consistency throughout the supply chain. Consistent CM implementation in the supply chain is not a new concern. During the 1800s, many industrialists like Andrew Carnegie thought that the best way to ensure consistency in supply chain products was to own the entire supply chain from mining of raw material through the manufacture and delivery of the final product (vertical integration).

Supply chain activity for non-commodity items includes a statement of work, data deliverables, and a flow-down of contractual requirements. If the supplier is designing a new product, the requirements associated with that product result from allocations from the program system specification, or a program subsystem or component specification. Should it be a factory off-load, one or more control drawings may also be included in the subcontract (e.g., interface control drawing, procurement control drawing, vendor item drawing, etc.).

As products and services become more complex, a single company rarely has the internal resources to do everything. Outsourcing to acquire these missing capabilities has become the new business norm. Some industries are starting to develop dedicated supply chain relationships with

[12] Raj, A. (2014). *Best Practices in Contract Management: Strategies for Optimizing Business Relationships*. Retrieved April 14, 2014, from http://www.academia.edu/, http://www.academia.edu/5102810/HR_Assgt.

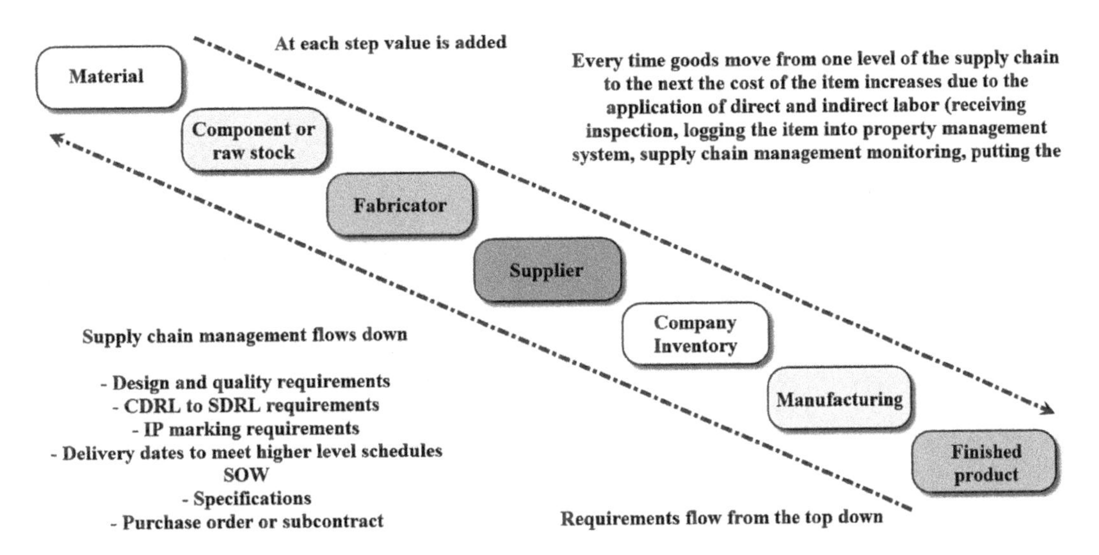

FIGURE 8.9 Supply chain network.

other highly regarded firms to assure overall company goals. This symbiotic relationship strategy works well if suppliers devote the necessary funds in research and development. A single supplier has been known to place the entire product line in jeopardy. Recent product recalls for this very reason include cell phone batteries supplied to Samsung and car airbags supplied by Takata Corporation, Japan.

8.10 FACILITIES BASICS

If the organization is of a *brick and mortar* type, it will need to manage those facilities. Facility management is the coordination and maintenance of the physical things that keep the organization running and working at peak performance. Enterprise-level CM implementation is centric to addressing facilities concerns. This includes the utilization of space and layout, manufacturing, test equipment, and machinery; organizational infrastructure and equipment to keep our personnel safe and productive; management of property and metrology to assure item certification for use is current; and compliance with regulations and assorted codes (e.g., health, safety, and fire codes). Facilities also manages the maintenance of equipment, including repair to or replacement of capital assets such as buildings, vehicles, and equipment. Maintenance work includes preventive maintenance work as well as *repairs as fail* efforts. Activities to head off those untimely failure events (such as inspecting the equipment) also fall within this area. One critical facilities department subfunction is that of company and customer property management. An approved property management system includes activities associated with receiving inspection, stock room control, proper allocation of materials, segregation of customer property from company property, and bonded storage of company stock transferred to programs.

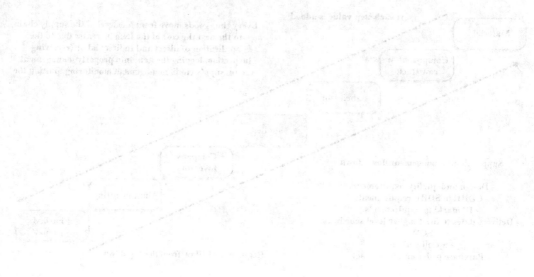

9 Configuration Management Baselines

9.1 QUESTIONS ANSWERED IN THIS CHAPTER

- How is product growth planning accomplished via development baselines in an incremental and iterative development methodology?
- What mechanism drives a delta to the functional baseline?
- What mechanism drives a delta to the allocated baseline?
- What mechanism drives a delta to the product baseline?

9.2 INTRODUCTION

There is a lot of confusion about what a configuration baseline is. A baseline is simply a snapshot of where something is at any point in time. Once a baseline is established, you can measure changes to it. A program may have hundreds of things referred to as "baselines." You will often hear about cost, schedule, contract, performance, and other baselines at your place of employment. When it comes to CM implementation, a configuration baseline has a specific meaning.

This broad definition differs from how the term is applied to CM implementation. Only three baselines are often defined in a contract. These are identified as:[1]

1. Functional baseline
   a. A definition of the required system functionality describing functional and interface characteristics of the overall system, and the verification required to demonstrate the achievement of those specified functional characteristics.
   b. The definition exists in the contract requirements at time of award.
   c. Subsequent contract changes do not create a new functional baseline. They create a delta to the functional baseline.
2. Allocated baseline
   a. A definition of the configuration items making up a system, and then how system function and performance requirements are allocated across lower-level configuration items (hence the term "allocated baseline").
   b. The allocation of functional baseline requirements as they existed in the contract at time of award.
   c. Subsequent contract changes creating a delta to the functional baseline do not create a new allocated baseline. They create a delta to the allocated baseline.
3. Product baseline
   a. Documentation describing all necessary functional and physical characteristics of a product; the selected functional and physical characteristics designated for production acceptance testing; and tests necessary for deployment/installation, operation, support, training, and disposal of the product.

[1] AcqNotes, http://acqnotes.com/acqnote/careerfields/configuration-baselines. Accessed January 21, 2019.

The following are the critical concepts to remember about a configuration baseline:

- It is a snapshot in time.
- It is a commitment point.
- It is a control point, also known as the point of departure.
- It is a way to measure where a product really is.
- It is controlled externally by a customer or it is established internally, depending on where it occurs and on the individual content of the review or the context of that review.

We decided it might be interesting to express what a delta looks like from a datum shift perspective. Just as a datum is a set of values to define a specific geodetic system, with differences being known as a "datum shift," east and west map coordinates of the Earth are established using lines known as "longitude." These lines pass through the axis of rotation at points known as the "north" and "south" poles. The Earth's magnetic poles are not aligned with its axis of rotation. To read a map of the Earth's surface, a datum shift known as a "declination correction" is made. Declination corrections are indicated on maps. The declination corrections for several locations on March 8, 2014, are shown below.[2] As can be seen, declination corrections are not static due to shifts in the magnetic fields of the Earth.

- *Afghanistan:* Kabul 2° 52′ 25″ E changing by 2.1′ E per year
- *Argentina:* Buenos Aires 8° 18′ 21″ W changing by 9.8′ W per year
- *Australia:* Canberra 12° 18′ 0″ E changing by 0.5′ W per year
- *Austria:* Vienna 3° 38′ 56″ E changing by 6.1′ E per year
- *Belgium:* Brussels 0° 27′ 22″ E changing by 7.7′ E per year
- *Botswana:* Gaborone 15° 29′ 10″ W changing by 2.0′ W per year
- *Brazil:* Brasilia 3° 38′ 56″ E changing by 6.1′ E per year
- *Canada:* Ottawa 13° 31′ 26″ W changing by 2.6′ E per year
- *China:* Beijing 6° 36′ 49″ W changing by 3.2′ W per year
- *Costa Rica:* San Jose 1° 11′ 53″ W changing by 7.5′ W per year
- *Denmark:* Copenhagen 2° 59′ 54″ E changing by 7.6′ E per year
- *Egypt:* Cairo 4° 3′ 33″ E changing by 5.2′ E per year
- *England:* London 1° 7′ 26″ W changing by 8.5′ E per year
- *Fiji:* Suva 12° 17′ 45″ E changing by 0.6′ W per year
- *Finland:* Helsinki 8° 10′ 19″ E changing by 7.7′ E per year
- *France:* Paris 0° 0′ 15″ E changing by 7.7′ E per year
- *Germany:* Berlin 3° 7′ 24″ E changing by 7.0′ E per year
- *Iceland:* Reykjavik 14° 51′ 18″ W changing by 16.8′ E per year
- *India:* New Delhi 0° 53′ 40″ E changing by 1.4′ E per year
- *Italy:* Rome 2° 33′ 50″ E changing by 5.7′ E per year
- *Japan:* Tokyo 7° 6′ 20″ W changing by 1.0′ W per year
- *Kenya:* Nairobi 0° 32′ 47″ E changing by 3.3′ E per year
- *Korea, South:* Seoul 8° 2′ 13″ W changing by 2.0′ W per year
- *Mauritius:* Port Louis 18° 16′ 18″ W changing by 3.2′ E per year
- *Mexico:* Mexico City 5° 5′ 40″ E changing by 6.7′ W per year
- *New Zealand:* Wellington 22° 16′ 7″ E changing by 1.5′ E per year
- *Norway:* Oslo 2° 24′ 26″ E changing by 8.9′ E per year
- *Russia:* Moscow 10° 32′ 49″ E changing by 6.3′ E per year
- *Senegal:* Dakar 7° 51′ 36″ W changing by 7.0′ E per year

[2] Calculated using NOAA. NOAA, U.S. (2014, July). *Magnetic Field Calculators.* http://www.ngdc.noaa.gov/geomag-web/#declination. Accessed July 2, 2014.

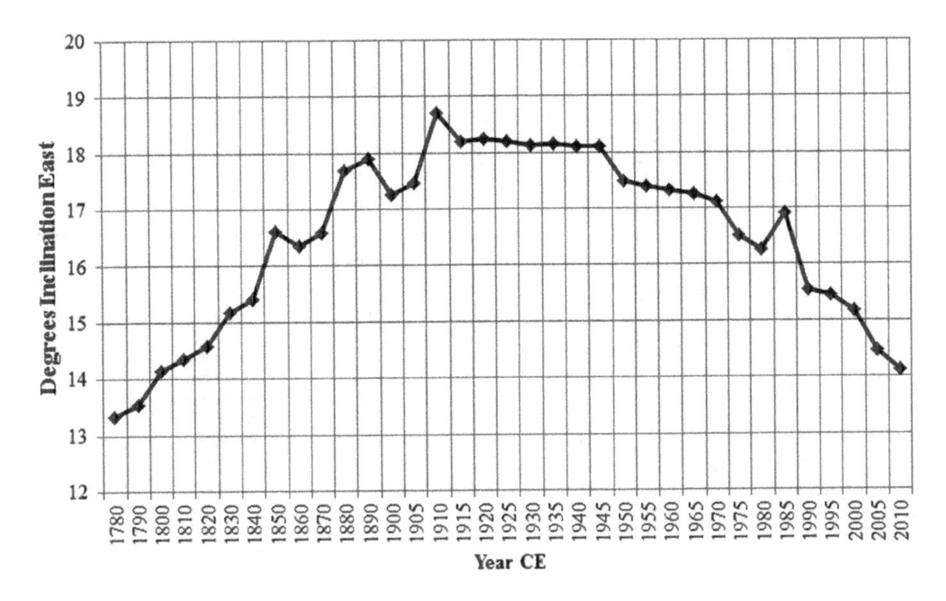

FIGURE 9.1 Inclination shift in Carson City, Nevada.

- *South Africa:* Pretoria 17° 44′ 26″ W changing by 4.2′ W per year
- *Spain:* Madrid 1° 14′ 29″ W changing by 7.4′ E per year
- *Sweden:* Stockholm 5° 18′ 5″ E changing by 8.0′ E per year
- *Switzerland:* Bern 1° 30′ 40″ E changing by 6.7′ E per year
- *Turkey:* Ankara 5° 16′ 36″ E changing by 5.3′ E per year
- *United States of America:* Washington DC 10° 53′ 23″ W changing by 0.9′ W per year
- *Venezuela:* Caracas 11° 58′ 13″ W changing by 5.7′ W per year
- *Zimbabwe:* Harare 8° 12′ 45″ W changing by 0.4′ E per year

The degree of inclination shift eastward calculated over time for Carson City, Nevada, is shown in Figure 9.1. Configuration baselines defined by CM guidance documents and regulations also experience shift over time (Figure 9.2). IEEE Standard 828-2012[3] states:

> A baseline provides a logical basis for comparison. A specific version of a single work product by itself, or a set of work products together, can be established as a baseline.

It goes on to state that:

> During the course of product development, a series of baselines is established, enabling assessment of the evolving product's maturity at different points in time.

The program technical documentation at contract award forms what is known as the "functional baseline." Contract technical requirements can be very dynamic over the life of a program; but changes to technical requirements after award **do not** create a new "functional" baseline. Changes to the program technical documents occur for a variety of reasons. Often specifications contain to-be-determined, to-be-reviewed, or to-be-specified requirements. They may also contain to-be-negotiated paragraphs or items. As the program progresses, these items are replaced with hard

[3] IEEE. (2012). *IEEE-828-2012—IEEE Standard for Configuration Management in Systems and Software Engineering.* IEEE, Piscataway, NJ.

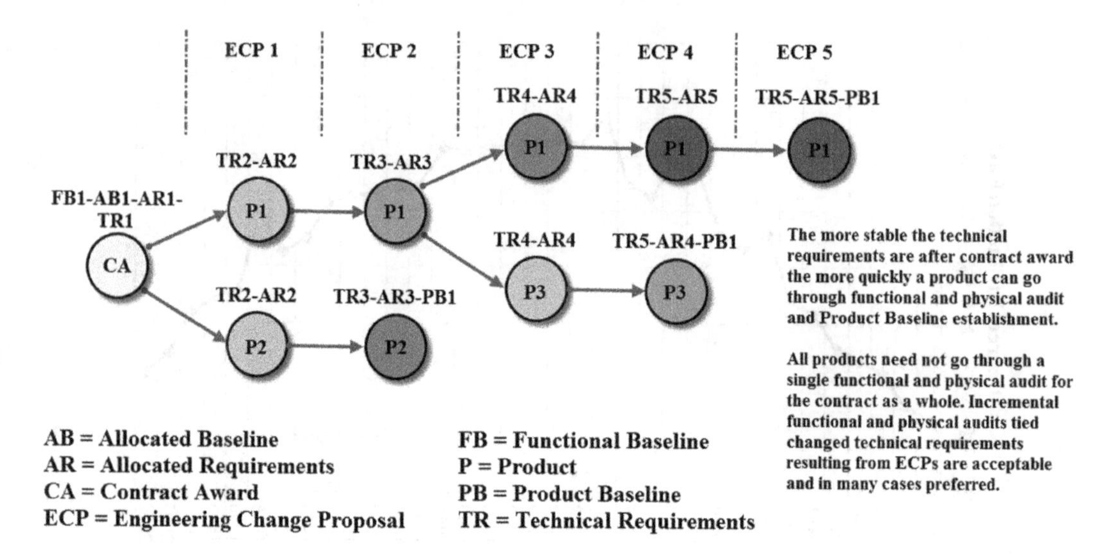

FIGURE 9.2 Requirements drift versus baseline.

values or ranges that pertain to performance. Referenced documents in specifications or mission assurance requirements (MARs) may be modified to accommodate changes to statutory requirements. Interface requirements may change due to the capabilities of other contractors if the customer is the systems integrator.

Once the technical requirements in the awarded contract are evaluated, those requirements are allocated to various systems and subsystems. This creates what is known as the "allocated" baseline. Subsequent changes to contract technical requirements ripple through the contractual documentation and may require allocations of new or delta requirements to accommodate them. New or changing requirements **do not** create a new "allocated" baseline.

A weather satellite was recently required to comply with 58 different technical documents at contract award. These formed its functional baseline. Over its decade-long developmental life, more than 238 changes occurred to the technical requirements. Each of these changes rippled through the developmental effort. Some of the requirements documents changed as many as 26 times. This program volatility resulted in an updated requirements allocation. Some requirements remained static over the life of the program, as others drifted rapidly before becoming static. Still others drifted continuously until months prior to test for a system or subsystem.

If we look at commercial firms, we find integrated CM implementation approaches that weave digital threads into enduring digital fabrics that government sector firms can only dream of emulating. Commercial practices for the maintenance of products such as automobiles are incredible. OEM parts are available for products produced decades earlier.

CM implementation requires that CM planning accommodate technical requirement evolution and that the accommodation be documented in the CM plan. Programs with a logistical component for operations and maintenance (O&M) need to define field- and depot-level maintenance interfaces as well as parts sourcing and storage. Leveraging the engineering and manufacturing development portion of Figure 4.9 forward, we arrive at Figure 9.3. If the program is developing many different deliverable items, each item will have its own functional, allocated, and product baseline (e.g., a jet engine and the jet engine transport and positioning cart are not subject to the same technical requirements).

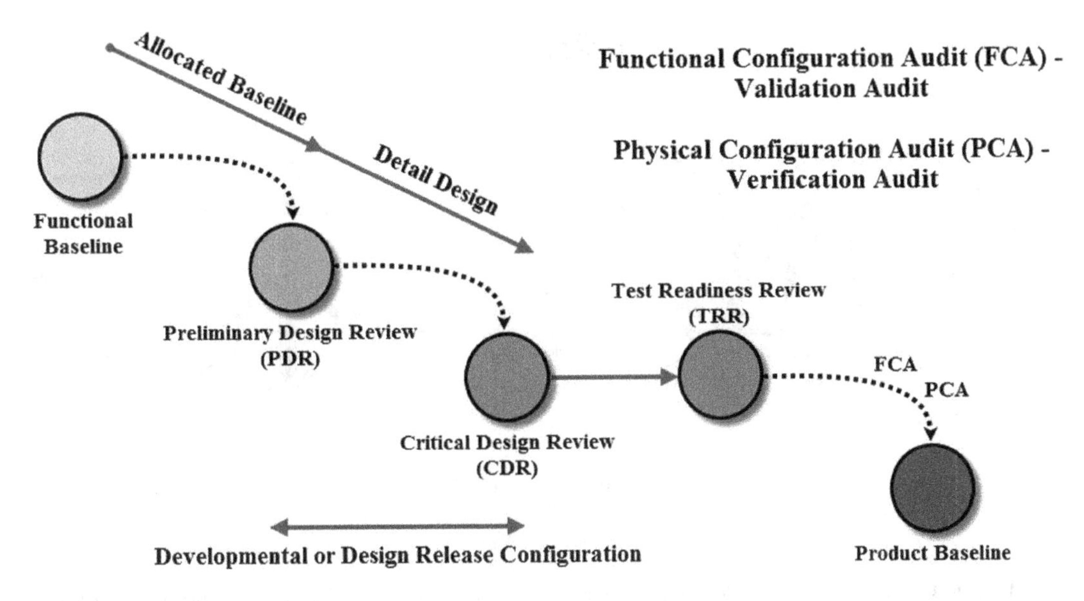

FIGURE 9.3 Engineering and manufacturing development.

From a commercial perspective, O&M determinations start with a predetermination of what portions of the final product are deemed cost effective to repair or replace. These items fall into two broad categories:

1. Parts that can be ordered and replaced by the owner (also known as "field-level maintenance").
2. Parts that can be ordered and replaced only at a repair facility (also known as "depot-level maintenance").

We have chosen the carbide miner's lamp (which produced acetylene gas from calcium carbide and water) as an example of O&M parts selection. Thomas Wilson developed the first commercially economical method of producing calcium carbide in 1892 using lime and coke. Calcium carbide when mixed with water produces acetylene gas, as shown in Equation 9.1.

$$CaC_2(s) + 2H_2O(l) \rightarrow Ca(OH)_2(aq) + C_2H_2(g)$$

EQUATION 9.1 Calcium carbide when mixed with water produces acetylene gas.

Wilson's U.S. patent was subsequently acquired by the Union Carbide Corporation in 1895, and it found immediate application in the commercial lighting, mining, automotive, and bicycle markets. Our example will focus on the Auto-Lite Carbide mining lamp manufactured by the Universal Lamp Company of Springfield, Illinois. The Universal Lamp Company was organized in 1913 by Jacob S. Sherman, who filed his first patent in 1914. The company produced Auto-Lite lamps until 1960. It had reduced competition in 1922 when it purchased Shanklin Manufacturing Company and its Guy's Dropper miner's lamp line. Auto-Lite was chosen because, during its 47-year span, it maintained its distinctive look despite configuration changes that included:

- Six changes to the top design
- Nine changes to the top markings

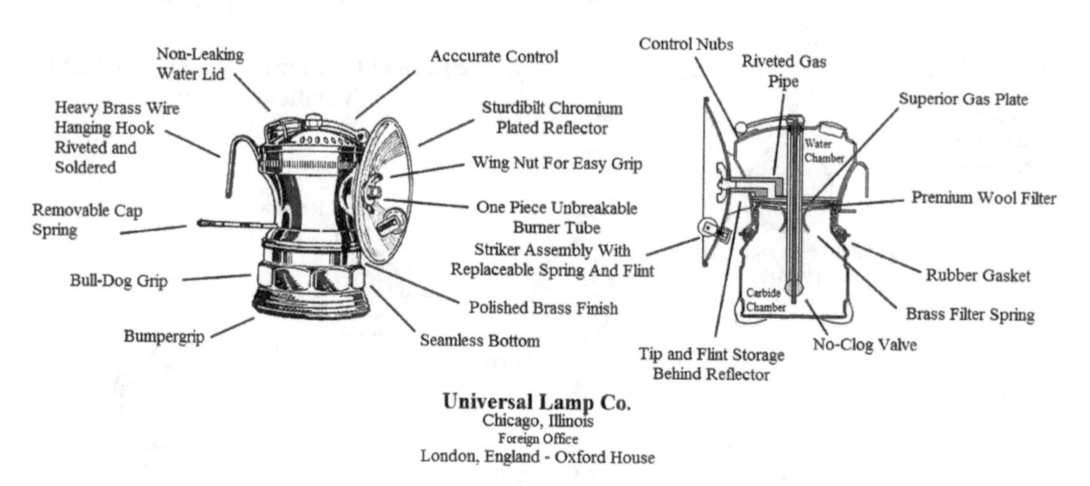

FIGURE 9.4 Auto-Lite™ carbide mining lamp. (Courtesy of Universal Lamp Co., Springfield, Illinois; Foreign Office: London, England—Oxford House.)

- Four changes to the bottom design
- Seven changes to the bottom markings
- Six variations of cap mounting hooks
- Three changes to the water valve
- Three changes to the water door
- Two changes to the gas tube

Figure 9.4 shows the lamp design as it existed in 1926.

Replacement parts for the Auto-Lite miner's lamp included those O&M items needed to keep the lamp performing properly. These were available at any location the lamps were sold. Stores stocked the "Universal Supply Box" (Figure 9.5), which contained an assortment of repair parts:

- Bumper grip
- Ceramic and brass gas tips
- Superior gas plate
- Removable cap sprint
- Lamp brass bottom replacement
- Lamp brass filter spring
- Chromium reflectors (multiple sizes)
- Steel gas tip reamer
- Striker assemblies
- Striker flints
- Striker back nut
- Striker cap
- Tip protectors
- Rubber gasket
- Wing nuts
- Wool felt filters

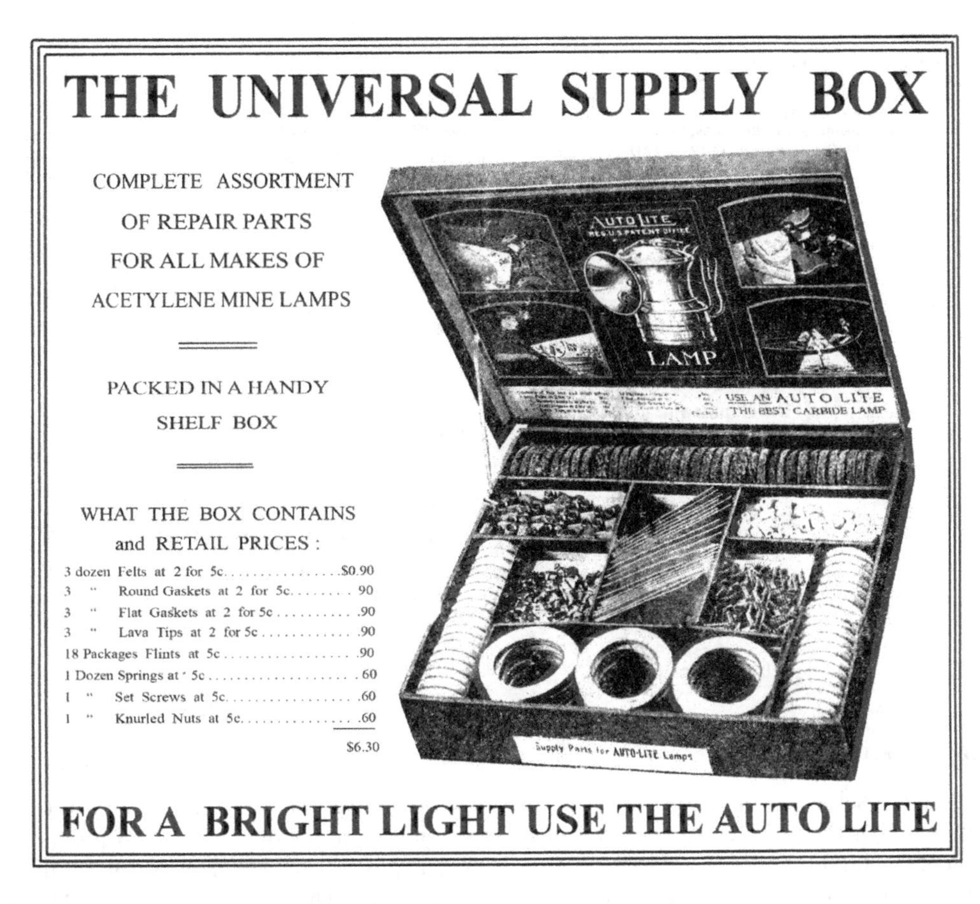

FIGURE 9.5 Logistical aspects of the Auto-Lite.

9.3 THERE ARE MANY BASELINES

Section 5.2.5 of SAE/EIA 649-B-2011[4] provided accepted guidance regarding the functional, allocated, and product baselines as well as developmental configuration or design release baseline. Baseline relationships are shown in Figure 9.6. SAE/EIA 649-B-2011 definitions are going to be used for the baseline descriptions because SAE/EIA 649-C-2019-02 was crafted to encompass an international, commercial, and services view of CM requirements instead of focusing on government, aerospace, and defense.

Events such as a verification audit or a validation audit are known by many different names in different industries and in different countries. In an international collaboration, becoming adept in ascertaining the sense of what something is called and applying that root-level understanding in crafting the CM implementation is important. Do not fixate on specific words or acronyms used to describe a CM function. If the terms are not familiar to everyone, they need to be defined in the Identification section of the CM plan (refer to SAE/EIA 649-C-2019-02 Table 2 for neutral terms and equivalencies). Seeking to understand and reach consensus may include comparison of the product life cycle phases described in Chapter 1 against how they are defined by stakeholders in the collaboration. As CM guidance documents such as ISO and EIA gain acceptance, terms will become standardized.

[4] SAE/EIA 649-B-2011, SAE Standard: Configuration Management Standard, SAE International, Warrendale, Pennsylvania.

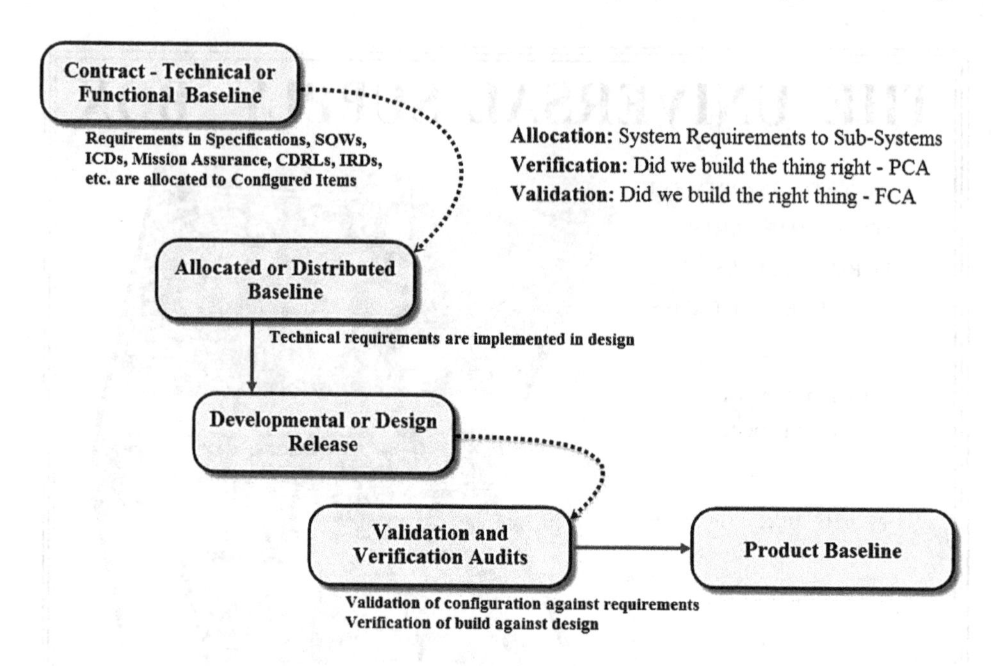

FIGURE 9.6 Baseline relationships.

9.3.1 Functional

Functional baseline is an integral part of the development phase of the product life cycle. It defines what the product is supposed to do, the regulations involved, the cost points and schedule, the target market, and other requirements such as testing and qualification requirements as well as acceptance and delivery/distribution criteria. The SAE/EIA 649-B-2011 definition read:

> A functional baseline is established for a product when the contract or order is placed and represents the originating set of performance requirements for the product.

9.3.2 Allocated

Continuing with the SAE/EIA 649-B-2011 definitions, we find:

> Allocated baseline(s) are established for major components of the product (e.g., configuration items) where the performance requirements of the product are "allocated" down via separate specifications and often establish the functional baseline for suppliers.

The functional and allocated baselines are often blurred in commercial development, with the allocation of requirements to systems and subsystems occurring as part of concept development. This implementation causes a steep learning curve.

9.3.3 Product

Returning to SAE/EIA 649-B-2011 definitions, we find:

> Product baseline is established after completion of design, development, and testing such that the original performance requirements have been proven to be successfully achieved and designated physical parameters confirmed; thus, leading to commencement of limited or full production.

The standard points out that the product baseline can be owned by either an internal or external customer. This is a very important distinction, as it attempts to include products developed to internal requirements for an external market segment as well as products developed entirely for an external customer (such as for government use).

9.3.4 Developmental

SAE/EIA 649-B-2011 says:

> As design information is released, it becomes part of the design release configuration controlled by the developing activity.

9.3.4.1 Design Implications

Each change processed after the initial release of the design modifies the design release configuration. Configuration status accounting reports on the current design configuration of any part, subassembly, assembly, or product (these words are used loosely and include models and software modules as well as 3D printing image files) will yield different results depending when the report is generated until the product baseline is established. Such reports are simply a snapshot of the design at a point in time (e.g., at the year, month, day, hour, minute, and second the report is run).

Many of the companies acknowledge the fluidity of the design release activities and choose not to identify functional, allocated, or product baselines within the PD/LM system. They reason that—while the functional, allocated, and product baselines have specific contextual meaning—the design configuration is intrinsic to the design process itself and not subject to "artificially set" baselines or reviews of the design itself. Capturing design release configuration can be performed *ad hoc* (e.g., on demand) at any point in time after the release of the first piece of engineering. Only the current configuration status could be retrieved.

9.3.4.2 Change Implications

Where the moving design release configuration data comes into its own is in the analysis of change impact. Configuration status and accounting reports, if the digital thread is maintained, do facilitate an evaluation of design interdependencies. This requires very careful attention not only to the product structure but also to the document structure. Such design interdependencies can be improved if the digital threads establish not only familial relationships of the affected item but also non-familial relationships of the related item external to the direct parent–child hierarchy, as shown in Figure 9.7. If all relationships are established between items in the PD/LM system, chances improve dramatically of assessing the true impact of a proposed change and capturing all impacts of the change. Despite best efforts, impacts are often overlooked if the digital threads are not present.

The impact error rate in a horizontal functional organization has been observed over a period of 37 years. We found a 20% greater change impact error rate (e.g., impacts not identified) in horizontally vs. vertically oriented program structures. This is not to imply that vertically oriented program structures are error free or that horizontal program structures are error prone. Instead we draw the conclusion that both program structures benefit if the digital fabric is woven as the program progresses, so interdependencies can be quantified.

9.4 TRIGGERS FOR MOVING FROM ONE BASELINE TO THE NEXT

The events that trigger the establishment of a baseline and the movement of a design from one baseline to the next are defined by the customer and nature of the product. In the case of the Auto-Lite miner's lamp, updates mandated by the marketplace based on customer feedback were required to

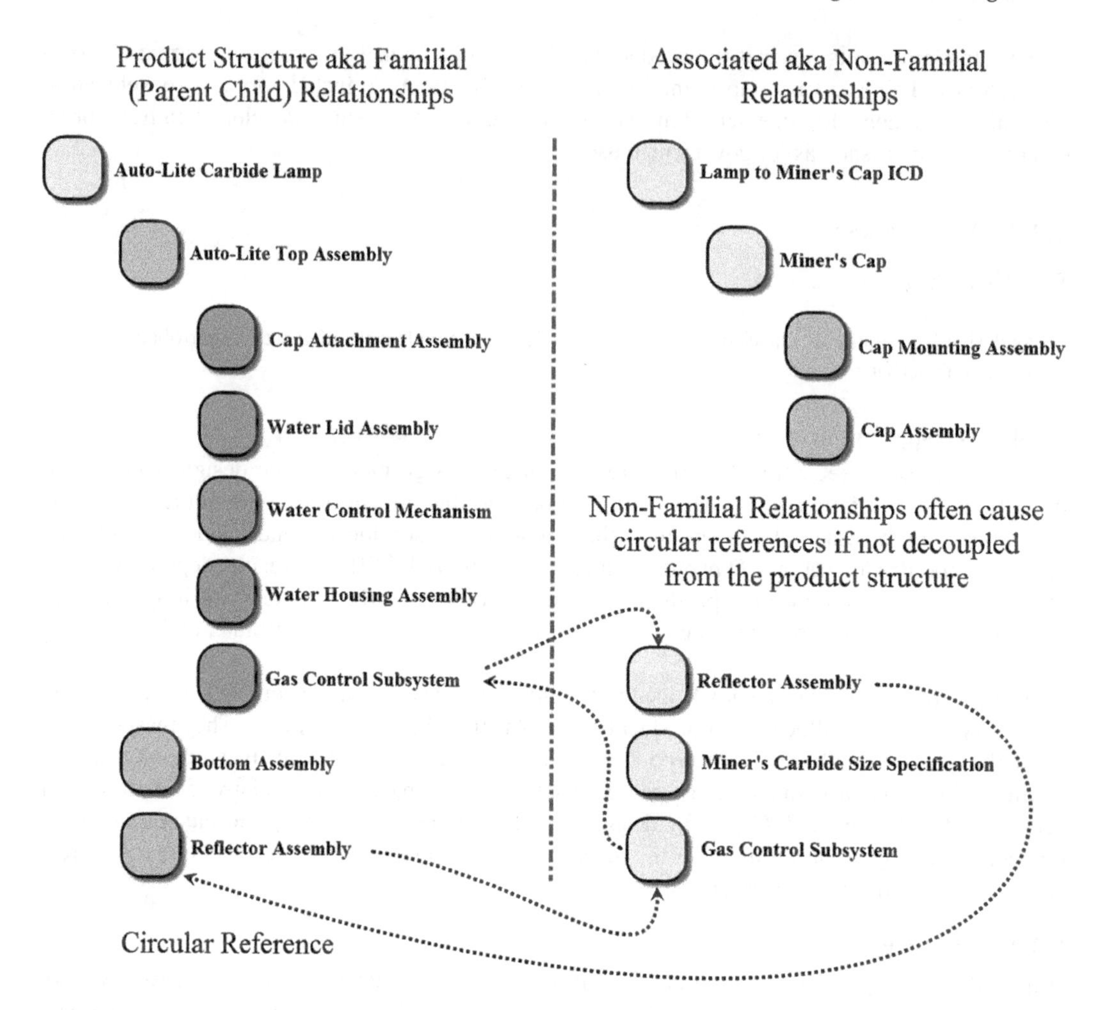

FIGURE 9.7 Familial and non-familial PD/LM structure.

maintain market share. The rise in popularity of spelunking (cave exploration) artificially kept the market vibrant long after the lamp's use for mining started to dwindle.

A contractual relationship, such as with a government or as a subcontractor, defines events that determine the movement from one baseline to another. These are often codependent on a review or an audit that drives design modification to meet technical requirements.

9.4.1 CM Implications of a Product Baseline

Once an item has been validated against requirements and the product comparison to engineering has been verified, the product baseline is established. The design is said to be frozen and the design authority is often transferred from the contractor to the customer. Minor or class II changes become moot at this point. *Customer* approval is required for any design modification. This includes changes due to part substitutions and design updates due to part obsolescence.

Validation and verification are required on all programs; but a formal FCA or PCA implies multiple quantities of the same item will be built. Such a PCA may include complete disassembly of the first or second unit produced (the first unit may have been used for item qualification activities). Each physical part is then compared against drawing requirements for tolerance of manufacturing and surface coatings as well as reassembly using released assembly drawings. One PCA we are aware of took six weeks to accomplish these activities.

There is no engineering reason for a government customer to mandate a formal functional configuration audit (FCA) or physical configuration audit (PCA) to establish a product baseline on a contract for a one-off unit. Discussions with many customers indicate that FCA and PCA are mandated, sometimes partway through performance, if the customer is not satisfied that a contractor is not managing requirements or design. Customers may also mandate an FCA and PCA if it is contained in their program management or systems engineering plan boilerplates. If this is the case, the impact drives costs that could have been put to better use.

We are also aware of cases where government contractors that used a redline engineering system did not incorporate these redlines, and as a result not only failed to disclose the fact that redlines existed but also delivered a technical data package (TDP) to the customer that was impossible to build to. This negated the contractual requirement for a TDP to maintain the industrial base. This double bookkeeping was cited by several government personnel as a reason for more stringent FCA and PCA activities. Cheating of this kind is detrimental to all contractors as well as to those in harm's way who depend on the product and its documentation to keep them safe. Being in harm's way is not limited to first responders or the warfighter.

STS-41-C was NASA's 11th space shuttle mission. One objective of the mission was to repair the malfunctioning Solar Max satellite. A special capture tool—the trunnion pin acquisition device (TPAD)—was developed to help capture the satellite. Unknown to the TPAD developers, redlined engineering changes adding multilayer insulation attachment points were not incorporated into the released product baseline documentation. These added grommets made it impossible for the TPAD to function. After three failed attempts to use the TPAD, mission specialist George D. "Pinkie" Nelson attempted to wrestle the satellite to a standstill using the manned maneuvering unit and failed. Eventually Solar Max was commanded from the ground to enter a slow spin, allowing the "Canadarm" remote manipulator system arm to capture it. After repairs, the satellite was redeployed.

The lack of clean configuration documentation resulted in the near loss of a mission specialist, a one-day extension to the STS-41-C mission, and a near critical expenditure of the orbiter's reaction control system fuel. One beneficial outcome of the Solar Max experience and similar events is a marked trend by industry to limit redlined engineering to test procedures with a requirement that redlines be captured in the PD/LM system.

9.5 BASELINES AND CONFIGURATION IDENTIFICATION

So far, we have discussed the essence of a configuration baseline. What remains is the relationship between baselines and configuration identification. On the surface, configuration identification is rather basic. It simply implies that a product is identified in a way that distinguishes it from every other product produced by not only your company but also every other company producing comparable products. We can call this configuration granularity. Other forms of granularity exist in CM, such as the data granularity associated with medical coding shown in Figure 5.2. For CM implementation purposes, granularity is defined as the degree of identification required to provide the level of understanding necessary to distinguish one thing from another. In the case of medical coding evolution from ICD-9 to ICD-10, the level of granularity increased to more clearly identify procedure and injury location.

Proper configuration identification facilitates post-delivery support and maintenance of a product in the production baseline as well as ensuring that the correct part configurations are used during final assembly, test, and qualification activities. Traditional practices in the government sector—such as assigning a single CI to a family of printed wiring boards, despite board-unique white wire changes modifying board functionality—are giving way to utilization of item identification for each unique configuration. Tracking item lot date codes for an

TABLE 9.1
Phase-Based Part Re-Identification

	Prototype	*Final Design*	*Fabrication*	*Test Complete*	*Acceptance*	*Phaseout*
Mark 1	1234567-101P	1234567-101	1234567-101	1234567-101T	1234567-101TA	1234567-101TAP
Mark 2*	1234567-201P	1234567-201	1234567-201	1234567-201T	1234567-201TA	1234567-201TAP
Mark 1	1234567-700	1234567-001	1234567-001	1234567-101	1234567-201	1234567-301
Mark 2**	2234567-700	2234567-001	2234567-001	2234567-101	2234567-201	2234567-301

* The first update after first sale is distinguishable as being part of the part family in the dash number.
** The first update after first sale is distinguishable as being part of the part family in the root number.

In all cases, part designs that are scrapped or discontinued are replaced with a new out of family item identification

electrical component to the installed location on a wiring board is increasingly mandated in government MARs.

9.5.1 Phase-Based Item Re-Identification

Item re-identification, as it moves between one baseline and the next, is more sophisticated in some business sectors than in others. Re-identification of parts as they move from prototype, to product development, to fabrication, to test, and to distribution is viewed as the normal flow. Item re-identification is also more sophisticated in software CM than in hardware CM. One approach to a phase-based part re-identification system is shown in Table 9.1. Suffix designators are used to denote items as they are tested, accepted, and phased out to assure that applicable items are properly tracked and not commingled in inventory.

Without planned phase-based re-identification, multiple subsystem updates that increase functionality by 5% or more could not be implemented in concert. Such block updates are a necessity when the implementation of multiple changes at the same time is additive to achieve an overall improvement in the performance nodes.

9.6 VALIDATION AND VERIFICATION IN NON-GOVERNMENTAL SALES

The most sophisticated implementation of configuration principles is always found in those industries where human health is involved. On April 10, 2014, Food Seminars International offered a seminar with the title "Validation and Verification 101: What you need to know." To those who are governmental sales centric, this may appear to be an odd topic for the food industry. On closer examination, the need to understand the differences between the U.S. Department of Agriculture and the Codex Alimentarius Commission, established by FAO and the World Health Organization in 1963, as a product moves along its life cycle is part of product CM. Perhaps another factor in the advances where health is involved is that safety validation throughout the process with a distinct focus on product risk (microbiological, physical, chemical, or allergenic) as well as process capability is not optional.

Real-time need to ensure that products meet rigorous standards, although using the same basic CM principles, takes on unfamiliar forms. It results in processes that are much more flexible and capable than those used in industries reliant on governmental sales. Even firms with a small-to-medium governmental sales reliance as part of their entire product offering have CM implementation that is farther reaching because their survival depends on it.

9.7 PRODUCT ACCEPTANCE

The purpose of CM as a profession culminates in the acceptance of the product or service by a customer. In the commercial sector, for one-off or limited quantity sales, and in the governmental sales sector, two primary forms of acceptance are often utilized: the DD Form 250 and the certificate of conformance. Goods come in two types: durable and non-durable. Non-durable goods are consumed upon or shortly after sale; food products are non-durable. Products such as mechanical devices are durable goods.

10 Configuration Control

10.1 QUESTIONS ANSWERED IN THIS CHAPTER

- What happens when there is insufficient control over the configuration?
- How do changes during development impact the contract closure phase of the project?
- Can we relax configuration control requirements during development?
- How do excessive controls impact the prototype process?
- What activities are associated with configuration status accounting?
- Does configuration status accounting control the configuration?
- What activities are associated with configuration verification and audit?
- Does configuration verification and audit control the configuration?
- What is the objective of the functional configuration audit?
- What is the objective of the physical configuration audit?

10.2 MANAGEMENT OF A CONFIGURATION

Our research has shown a sharp division between market segments regarding not only the scope of CM implementation but also the depth in which items are managed. Established companies with a long history of government sales trended toward using CM released processes to "make sure we never make that mistake again" rather than eliminating the root cause of the mistake in the first place.

During a FCA audit, an employee was tasked with transporting paper build documents to the audit team from another location. They loaded up a document cart and wheeled it to their car. The amount of documentation was large and initial attempts to get everything into the vehicle's back seat didn't work; everything had to be taken out of the car and rearranged. During this process the employee put several of the files on top of the vehicle. The CM lead, noticing that loading was taking longer than anticipated and being fearful of holding up the audit process, ran up and took over. Since everything had been removed from the document cart, the employee was directed to collapse the cart and put it in the car's trunk while the lead loaded the car. The lead closed the car door and the employee drove off. The documents on the roof were overlooked and later found in the company parking lot. Remembering to put everything in the vehicle before driving away isn't part of CM implementation, but cautions about it were added to the entire suite of CM processes to "make sure we never make that mistake again."

We found a similar instance in a company in the commercial sector, not so reliant on government business. Their response to items being left in the parking lot was more pragmatic. They put up signs near building exits that read, "Protecting intellectual property is everyone's business. When transporting items between buildings in your own vehicle, slow down. Take time to look around, under, and on top of the vehicle to make sure you have everything." They also ran a short story on their intranet. It was viewed as an issue of having too many people trying to speed up loading of documents into a car rather than a CM issue.

This leads to the question, "What is the purpose of CM?" At its simplest, CM is a way of thinking and managing information and change critical to product/program, enterprise, and regulatory aspects of anything. Judith Blaustein's definition,[1] "CM acts as the DNA of systems," is eloquent. It ties CM implementation to the core of what it means to be human.

[1] Blaustein, J.. (2013). *What Is the Purpose of CM?* Judith Blaustein of Blaustein and Associates. https://www.linkedin.com/groups/What-is-purpose-Configuration-Management-4393863.S.263480123?trk=groups_most_popular-0-b-ttl&goback=%2Egmp_4393863. Accessed November 13, 2013.

10.3 CM IS CRITICAL TO QUALITY AND SAFETY

CM implementation is critical to both quality and safety, and comes in many disguises. Sometimes, even CM professionals fail to recognize it. Sometimes, it is distributed to other organizations. Sometimes, it is consolidated in one organization. CM implementation always exists in some form regardless of who is doing it. It is all around us.

CM implementation includes establishing rules crucial to communication and requirements traceability. The cost savings associated with having developed a standardized statement of work, specification, mission assurance requirements document, and interface control drawing format for use on every program is staggering. The key to success is not in creating the generic template but in never deleting a paragraph header. If a paragraph is not required, the paragraph number and heading must be maintained, and the next line is identified simply as "Reserved." This means that specifications will always have the same requirement in the exact same paragraph (e.g., within all specifications 3.2.4 is assigned to "Reliability" and addressed in the same place in every specification created).

At the start of the program, functional requirements are laboriously allocated to systems and subsystems. These are input into a requirements management tool. Every "shall" statement is assigned a separate identifier in the tracking system. This kind of requirements management is shown in Table 10.1. Anytime a paragraph numbering is deleted, the data in the tracking system must be modified to account for the renumbering. If paragraph numbering is not altered there is still work in the tool to delete sub-paragraphs under the one being deleted but subsequent numbers are not impacted (e.g., if 3.5.6 is changed from "Wind environment" to "Reserved," 3.5.1.1 and lower paragraphs are deleted but paragraphs 3.5.7 and higher are not impacted). Tremendous cost savings result when requirements traceability is preserved, as shown in Figure 10.1 and Table 10.2.

10.4 THE CM TOWER OF BABEL

Etymology is the study of words, where they originated and how they changed over time (not to be confused with entomology, the study of insects). In Australia, nearly 400 different languages were spoken prior to European influences. These may have originated from a common root language or set of root languages with divergence occurring as the continent was peopled. Pidgin (simplified) languages may also have developed because of trade and took on a life of their own, adding to the linguistic base. Today, the indigenous linguistic base has collapsed to approximately 70 distinct forms, of which 40 are endangered and 15 are still widely used.

This offers insight into what we observed in CM terminology. Distinct divergence in CM-related terminology can be traced to each introduction of technologies and industry subtypes resulting

TABLE 10.1
Typical Requirements Tracking in DOORS

Customer ID	DOORS ID	Unit Specification, 1-543A-78B-6 Revision A	Verification Method
	PSS-35	**3.2 Lifetime and Reliability**	
	PSS-36	**3.2.1 Mission Life**	
Unit_Spec_010	PSS-37	The mission life of the unit **shall** be no less than ten years.	Analysis
	PSS-38	**3.2.2 Storage Life**	
Unit_Spec_010	PSS-39	The unit shall not be degraded during 2 years of service.	Analysis
	PSS-40	**3.2.3 Storage Compatibility**	
Unit_Spec_010	PSS-41	The unit **shall** be designed to be compatible with ground storage requirements.	Analysis
	PSS-42	**3.2.4 Reliability**	
Unit_Spec_010	PSS-43	Unit reliability **shall** be greater than 0.985 after 7 years of operation plus 2 years of storage prior to use.	Analysis
	PSS-44	**3.2.5 End of Life**	
Unit_Spec_010	PSS-45	Unit Reliability at End of Life (EOL) at 10 years plus 2 years of storage prior to operational use **shall** be greater than 0.975.	Analysis

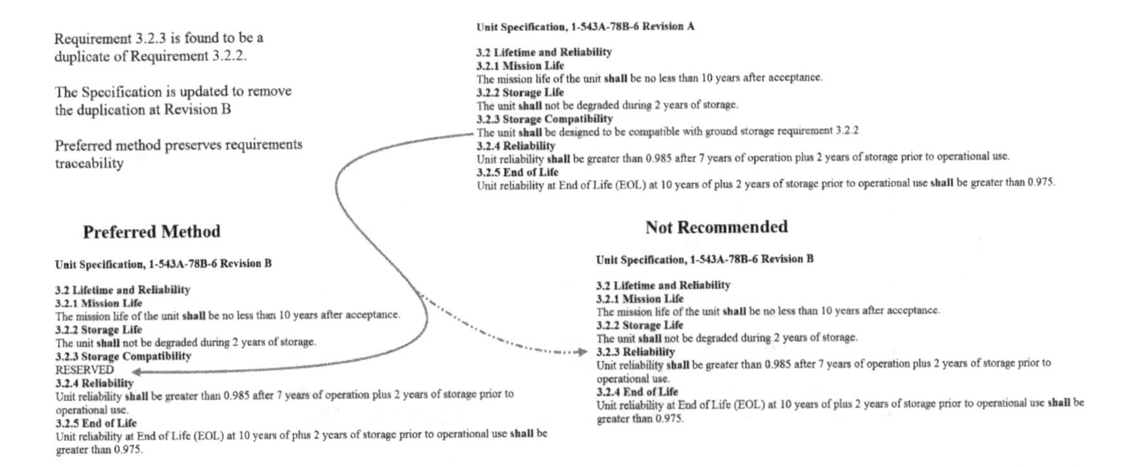
Requirement 3.2.3 is found to be a duplicate of Requirement 3.2.2.
The Specification is updated to remove the duplication at Revision B
Preferred method preserves requirements traceability
Unit Specification, 1-543A-78B-6 Revision A
3.2 Lifetime and Reliability
3.2.1 Mission Life
The mission life of the unit shall be no less than 10 years after acceptance.
3.2.2 Storage Life
The unit shall not be degraded during 2 years of storage.
3.2.3 Storage Compatibility
The unit shall be designed to be compatible with ground storage requirement 3.2.2
3.2.4 Reliability
Unit reliability shall be greater than 0.985 after 7 years of operation plus 2 years of storage prior to operational use.
3.2.5 End of Life
Unit reliability at End of Life (EOL) at 10 years of plus 2 years of storage prior to operational use shall be greater than 0.975.
Preferred Method
Unit Specification, 1-543A-78B-6 Revision B
3.2 Lifetime and Reliability
3.2.1 Mission Life
The mission life of the unit shall be no less than 10 years after acceptance.
3.2.2 Storage Life
The unit shall not be degraded during 2 years of storage.
3.2.3 Storage Compatibility
RESERVED
3.2.4 Reliability
Unit reliability shall be greater than 0.985 after 7 years of operation plus 2 years of storage prior to operational use.
3.2.5 End of Life
Unit reliability at End of Life (EOL) at 10 years of plus 2 years of storage prior to operational use shall be greater than 0.975.
Not Recommended
Unit Specification, 1-543A-78B-6 Revision B
3.2 Lifetime and Reliability
3.2.1 Mission Life
The mission life of the unit shall be no less than 10 years after acceptance.
3.2.2 Storage Life
The unit shall not be degraded during 2 years of storage.
3.2.3 Reliability
Unit reliability shall be greater than 0.985 after 7 years of operation plus 2 years of storage prior to operational use.
3.2.4 End of Life
Unit reliability at End of Life (EOL) at 10 years of plus 2 years of storage prior to operational use shall be greater than 0.975.

FIGURE 10.1 Preserving requirements traceability.

from the technologies. There is a non-linear expansion of not only CM terminology, but what appears to be a single-minded divergence of CM methodologies tied directly to technological innovation (Figure 10.2).

This has left many CM professionals rather adamant that there is "old CM" (i.e., old school) CM and something called "new CM." LinkedIn discussion groups have pronounced that four main CM types exist:

1. *Old CM:* Hardware-related CM
2. *New CM:* Software and firmware CM
3. *GTS CM:* GTS CM of servers and computing assets
4. Knowledge management CM

Terms like "old CM," "new CM," "GTS CM," and "knowledge management CM" are meaningless. The specific linguistics used to describe CM implementation will change with the market sector and are in vogue for a fleeting time in many areas due to the current state of the tools used in each market sector. They are not driven by the CM implementation but by tool users. They should not be used as differentiators between potential employees or between new CM implementation tools. The principles behind CM implementation do not change over time.

TABLE 10.2
Requirements Change Preferred DOORS

Customer ID	DOORS ID	Unit Specification, 1-543A-78B-6 Revision A	Verification Method
	PSS-35	**3.2 Lifetime and Reliability**	
	PSS-36	**3.2.1 Mission Life**	
Unit_Spec_010	PSS-37	The mission life of the unit **shall** be no less than ten years.	Analysis
	PSS-38	**3.2.2 Storage Life**	
Unit_Spec_010	PSS-39	The unit shall not be degraded during 2 years of service.	Analysis
	PSS-40	**3.2.3 Storage Compatibility**	
Unit_Spec_010	PSS-41	Reserved	
	PSS-42	**3.2.4 Reliability**	
Unit_Spec_010	PSS-43	Unit reliability **shall** be greater than 0.985 after 7 years of operation plus 2 years of storage prior to use.	Analysis
	PSS-44	**3.2.5 End of Life**	
Unit_Spec_010	PSS-45	Unit Reliability at End of Life (EOL) at 10 years plus 2 years of storage prior to operational use **shall** be greater than 0.975.	Analysis

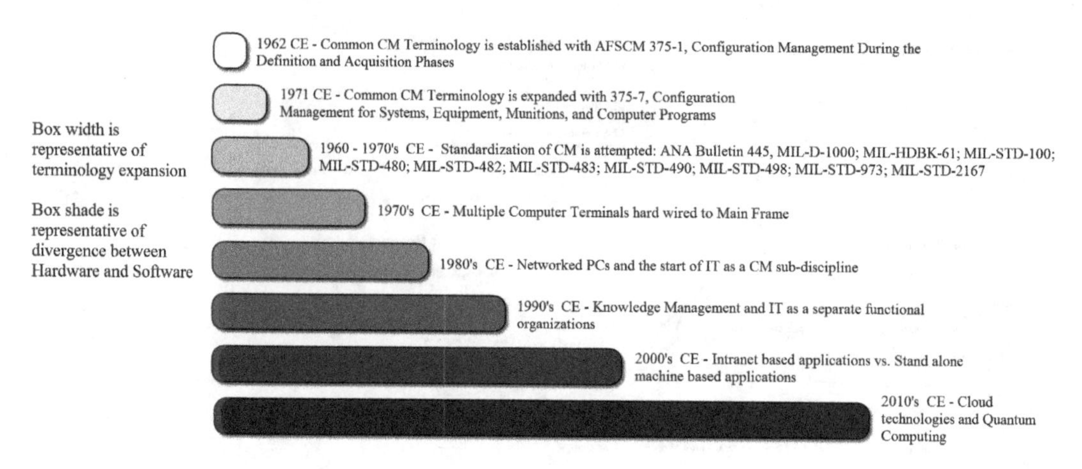
1962 CE - Common CM Terminology is established with AFSCM 375-1, Configuration Management During the Definition and Acquisition Phases
1971 CE - Common CM Terminology is expanded with 375-7, Configuration Management for Systems, Equipment, Munitions, and Computer Programs
Box width is representative of terminology expansion
1960 - 1970's CE - Standardization of CM is attempted: ANA Bulletin 445, MIL-D-1000; MIL-HDBK-61; MIL-STD-100; MIL-STD-480; MIL-STD-482; MIL-STD-483; MIL-STD-490; MIL-STD-498; MIL-STD-973; MIL-STD-2167
Box shade is representative of divergence between Hardware and Software
1970's CE - Multiple Computer Terminals hard wired to Main Frame
1980's CE - Networked PCs and the start of IT as a CM sub-discipline
1990's CE - Knowledge Management and IT as a separate functional organizations
2000's CE - Intranet based applications vs. Stand alone machine based applications
2010's CE - Cloud technologies and Quantum Computing

FIGURE 10.2 Events tied to CM language divergence.

It is the premise of this book that management of things is ingrained in the human experience. It has existed since hunter–gatherer times and is critical to the present. As societies moved from a nomadic to a pastoral existence, the steps required to manage the surrounding world became more complicated. Livestock were improved through selective cross-breeding with the hope of animals with greater physical bulk or better-tasting flesh or greater yield of milk or eggs. Grains were selectively harvested for larger seed yield, resulting in the plants developing a narrower viable ecological niche and becoming dependent on human intervention for their survival.

During the period of the great guilds, apprentices learned all aspects of product management associated with their trade. A master smith could source raw materials and make everything required to set up a profitable business in any village or town large enough to support the trade. Configurations of the tools, materials, processes, testing, and bringing product to market were understood and could be replicated or modified with a clear understanding of what was being done and why. A master tradesman was a living treasure and the physical embodiment of their craft. A smithy could stand unmolested by warring factions at any crossroads for generations. The body of knowledge and capabilities of the smith were respected and protected in an environmental niche that transcended shifting political boundaries.

As societies grew, markets consolidated. Small firms supplying superior product succumbed to larger firms providing lesser product at a lower cost and higher uniformity. Although one could still purchase specialty items, the cost of doing so outweighed the advantages. There was little need to pay for porcelain dinnerware displaying the family crest when, for lesser cost, you could dine on dinnerware identical in every detail to that used by royalty.

Public and political backlash against inferior products or products deemed dangerous to government resulted in regulatory control. Court favoritism manifested itself in trade monopolies, and government-to-government antagonism was demonstrated with trade sanctions. The economic engine—often spurred on by government—continued. Product became more complex and the management of product required new theories, and, in some cases, old theories repackaged in current terminology. This is where we find the scope of CM implementation today.

10.4.1 Use Drives the CM Planning and Management Systems Design

Section 1.5.2.1 explored adaptive radiation in engineering. This can be used as a metaphor for what is currently observed in the world marketplace as it is related to CM practice. Most market segments are driven to produce product acceptable to a large purchasing base (this product type will be referred to as "multiple build end items" to simplify the discussion). A subsector of the sales in all market segments are low-volume highly specialized items (this product type will be referred to as "one-off end items" to simplify the discussion).

10.4.1.1 Multiple Build End Items

Interviews and online discussions have culminated in a very interesting finding. The closer the industries were to the commodity market, the better was their overall grasp of the full spectrum of managing configurations and the applicability of CM to their product. CM implementation at each product level relies on sound CM implementation of the items sourced from the level below. The more automated the production process, the more reliant the producer is on the consistency of those sourced items and the more catastrophic the results of poor CM implementation.

During the early days of chemical formulations for scientific research and subsequently for mass production of medicines, the purity of the chemicals used was paramount. Only with the use of chemicals that were untainted by other elements or compounds could the effectiveness of medicinal compounds be quantified and replicated. This lack of CM implementation was one premise behind the inability of Dr. Robert Campbell to replicate a proven cancer treatment in the movie *Medicine Man.*[2] Production of chemicals, chemical compounds, and other raw materials formed at the lowest level of the product supply chain mandates CM implementation at that level.

Multiple build end items are sold without customization for a customer's unique needs. Automobiles fall into this category, as do computers and almost anything the public can purchase. Any configuration an individual can order from all available options can also be ordered by someone else. Some multiple build items may have limited product runs. Purpose build fly rods are an example. Due to the rapidly developing composites technologies, a company may decide to build 1000 units before changing technologies. The fact that the purchaser can have their name imprinted on the fly rod does not remove it from the multiple unit fabrication classification. Any other purchaser can have that same name imprinted on another copy of the 1000 unit run.

10.4.2 Six Sigma

What makes CM implementation particularly challenging on multiple build end items is CM of processes and acceptance criteria to ensure the suitability of all products sold. This leads us to a movement toward what is known as "Six Sigma." Six Sigma methodology was originally developed by Motorola in 1968 and is deeply rooted in probability analysis. A calculation called "root mean squared" is used to determine how far something is (standard deviation) from the mean value of all things being evaluated. The math is relatively simple and can be dispensed with.

Let's assume that once the final grades for a class are calculated, we find that we have two A's, eight B's, thirty-two C's, eight D's, and two E's. Plotting the grades as data points on a graph creates a distribution that is bell shaped and for that reason the plot is called a bell curve. The grades form what is known as a normal distribution because the number of A's and E's are the same and the number of B's and D's are the same. C is the mean grade. Grades B and D are one standard deviation (or sigma) away from the mean grade. Grades A and E are two standard deviations (or two sigma) away from the mean grade.

Six Sigma is six standard deviations away from the mean. If a data distribution is approximately normal (Figure 10.3), then about 68% of the values are within one standard deviation of the mean, about 95% of the values are within two standard deviations, and about 99.7% lie within three standard deviations. This is known as the *68-95-99.7 Rule*, or the *Empirical Rule.*

Processes that operate with "Six Sigma quality" over the short term are assumed to produce long-term defect levels below 3.4 defects per million opportunities. A survey of multiple industries found that many firms are claiming Six Sigma capabilities that have never built anywhere near a million items.

[2] McTiernan, J. (Director). (1992). *Medicine Man.* [Motion Picture].

zσ	percentage
1σ	68.27%
1.645σ	90%
1.960σ	95%
2σ	95.450%
2.576σ	99%
3σ	99.7300%
3.2906σ	99.9%
4σ	99.993666%
5σ	99.99994267%
6σ	99.9999998027%
7σ	99.9999999997440%

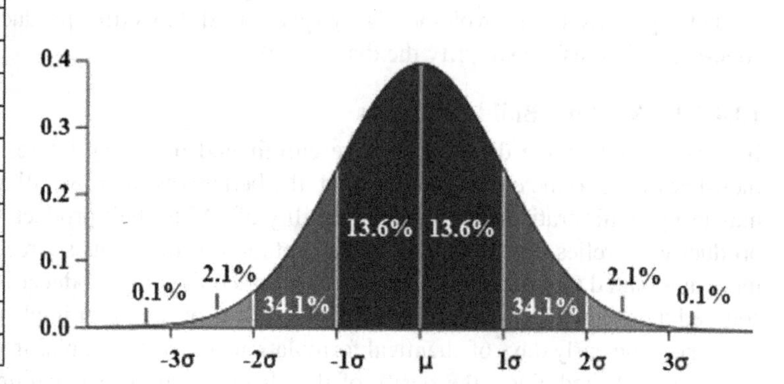

FIGURE 10.3 Rules for normally distributed data.

Their claims are process based rather than product based. What they are saying is that the process is written so well that, if you follow it a million times, people will only make a mistake less than 3.4 times.

10.4.3 Multiple Build End Item and a Product Baseline

Historically, product baselines were established for those items that were entering the U.S. government inventory in sufficient quantities to ensure uniformity of product. Design authority was removed from the product developer and design changes were elevated from minor to major (e.g., class II to class I). The design was frozen, and no changes could be made unless mandated by the government. The product baseline is often established concurrent with functional configuration audit (FCA) and physical configuration audit (PCA) activities.

There is a long history of non-conforming product in the annals of every nation. Establishing a product baseline was one way to limit and perhaps eliminate non-conformance after the international conflict of 1918 referred to as "World War I." What followed eventually resulted in what can best be described as a form of administrative insanity. Product specifications were written for almost everything purchased by the U.S. military, including items such as MIL-F-3897 – *Military Specification Fruitcake Bar* and MIL-C-1161 – *Collar, Dog, Leather.*

In the United States during the post–Korean conflict years (1950–1953), it was discovered that the disadvantages of managing all of the product specifications outweighed the advantages. Many of the specifications were made inactive as better or equal products were commercially available for a much lower cost. The advent of the International Organization for Standards (ISO) and its standards has resulted in further savings on most multiple build items used by the U.S. government.

10.5 ONE-OFF END ITEMS

One-off end items were the norm prior to the industrial revolution and are now found only in high-end products.

Items developed specifically for government use fall into this category. Except for firearms, ammunition, and other defensive and offensive items carried by military personnel, most items rarely see production runs above 10,000 units. This has led to an interesting phenomenon relative to Six Sigma at some companies that runs like this:

- I am only producing one item like this.
- I do not get paid if it does not meet acceptance criteria.
- If it passes, then 100% of production units are accepted.
- Therefore, the product when accepted is produced at higher than Six Sigma levels.

Although the first three statements may be true, the fact remains that only a one-off capability has been demonstrated and the product is produced in a less-than-one-sigma environment.

A second phenomenon observed is that even though only one such item will ever be produced (as is the case with most satellites), many contracts contain requirements for formalized FCA and PCA events and product baseline establishment and submittal of a technical data package (TDP). It is understandable that an operational asset requires physical and functional evaluation. The cost associated with formalized audits and a TDP on a six-to ten-year development effort that allows a different company to produce another unit isn't warranted. Electronic-based technologies refresh on about a three-year cycle. Some refresh in as little as nine months. Given an average three- to five-year design and build cycle for a single-instrument spacecraft, the technology being flown is outdated before ignition of the booster carrying it into space.

Parallels exist across all market segments and all customer bases. In most cases, the producer does a much better job of maintaining documentation required to reproduce one of its products than the consumer. Consumer electronics is often in the forefront with updated owner's, operations, and repair manuals; parts; and upgrades available via web-based support sites.

10.6 MANAGEMENT OF CHANGE

Unmanaged change abounds in the natural world as flora and fauna adapt to environmental stresses around them. Certain species of rattlesnakes have been hunted in the United States to such an extent that the percent of snakes that fail to develop rattles has risen over time.

Unmanaged change in a company can have similar unplanned consequences. Some form of engineering order (EO) was found for engineering changes. Most included both a *was* and an *is now* comparison to assist in evaluating the change and as a means of communication with internal and external stakeholders. The temptation to make changes to company policies, procedures, practices, and processes updates was so great that a lack of *was* and *is now* information prevailed in some companies. This pointed to a lack of CM implementation at the enterprise level.

A move to use an environmentally friendly process for applying conformal coating to printed wiring assemblies (PWAs) was made at one firm studied. Everything looked good on paper and the change was released. No testing was done to verify the cross-compatibility of materials. No coordination of the change had taken place with the programs needing conformal coated PWAs. Some weeks after the new process was implemented, PWAs using the new process experienced the conformal coating pulling away from the PWA surface. All PWAs coated during the period had to have the conformal coat stripped away. Components mounted on the PWA were damaged during the cleaning operation, further driving costs. Root cause analysis showed an incompatibility between the chemical composition of the coating and the chemical composition of the resin used to bind the fiberglass board substrate. Everything they needed in order to do the analysis was found on the producer's material data sheet (MDS); it simply wasn't looked at.

10.6.1 Engineering Change Request

An engineering change request (ECR) or any form of change request (e.g., software change request, process change request, action request, risk and opportunity request, etc.) is a documented lien against a released item. These liens can be against a single item or against multiple items.

Change requests are written to mitigate eight situations:

1. Design or other issues that surface through use or test.
2. Component obsolescence, unavailability, or industry alerts such as government and industry data exchange program (GIDEP).
3. Product enhancements.

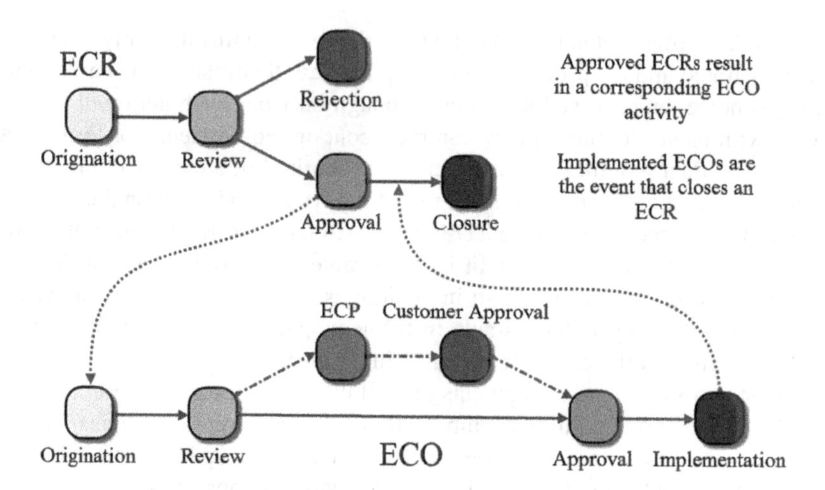

FIGURE 10.4 ECR–ECO interdependencies.

4. Cross-functional incompatibilities.
5. Administrative clean-up or typographical errors.
6. Process changes.
7. Regulatory changes.
8. Requests from management.

There is an interdependency (digital thread) between change requests and the changes processed to implement them. This is shown graphically in Figure 10.4.

10.6.2 Engineering Change Orders

Changes to documentation (e.g., drawings, analysis, specifications, policies, procedures, practices, and processes) as well as manufacturing and documenting bills of materials (BOMs) are made in most product data/life cycle management (PD/LM) systems using engineering change orders (ECOs) or similar mechanism. Changes authorizing a contractor to start work use a contract change order associated with an engineering change proposal (ECP) submitted by the contractor to the customer in response to a request for proposal.

The current trend in CM guidance documents is to classify these as major and minor.

Major changes impact

- Specified contractual requirements or product technical attributes.
- An established functional, allocated, or product baseline.
- Product safety.

Minor changes are those that impact anything else.

The definitions in MIL-STD-973, CM, are more concise.[3] What follows has been modified only to add in definitions for the acronyms used.

> 5.4.2.2.1 Classification of engineering changes. An engineering change shall be classified as Class I or Class II by the preparing contractor in accordance with this standard. Class I ECPS shall be referred to the Government for approval or disapproval. Classification disagreements shall be referred to the Government for final decision. A proposed engineering change to a CI, or to any combination or

[3] DOD, U.S. (1992). *MIL-STD-973, Configuration Management*. U.S. Government, Section 5.4.2.2.1 and 5.4.2.4.

discrete portion thereof, shall be determined to be Class I by examining the factors below, as contractually applicable, to determine if they would be impacted as a result of implementing the change. The change shall be Class I if:

a. The Functional Configuration Documentation (FCD) or Allocated Configuration Documentation (ACD), once established, is affected to the extent that any of the following requirements would be outside specified limits or specified tolerances:
   1. Performance.
   2. Reliability, maintainability or survivability.
   3. Weight, balance, moment of inertia.
   4. Interface characteristics.
   5. Electromagnetic characteristics.
   6. Other technical requirements in the specifications.

Note: Minor clarifications and corrections to FCD or ACD shall be made only as an incidental part of the next Class I ECP and accompanying Specification Change Notice (SCN) or Notice of Revision (NOR), unless otherwise directed by the Government.

b. A change to the Product Configuration Documentation (PCD), once established, will affect the FCD or ACD as described in 5.4.2.2 paragraph 1a or will impact one or more of the following
   1. Government Furnished Equipment (GFE)
   2. Safety
   3. Compatibility or specified interoperability with interfacing CIs, support equipment or support software, spares, trainers or training devices/equipment/ software.
   4. Configuration to the extent that retrofit action is required.
   5. Delivered operation and maintenance manuals for which adequate change/revision funding is not provided in existing contracts.
   6. Preset adjustments or schedules affecting operating limits or performance to such extent as to require assignment of a new identification number.
   7. Interchangeability, substitutability, or replaceability as applied to CIs, and to all subassemblies and parts except the pieces and parts of non-reparable subassemblies.
   8. Sources of CIs or repairable items at any level defined by source-control drawings.
   9. Skills, manning, training, biomedical factors or human-engineering design.

c. Any of the following contractual factors are affected:
   1. Cost to the Government including incentives and fees.
   2. Contract guarantees or warranties.
   3. Contractual deliveries.
   4. Scheduled contract milestones.

5.4.2.4 Class II engineering changes. An engineering change which impacts none of the Class I factors specified in 5.4.2.2.1 shall be classified as a Class II engineering change.

Confusion existing in the bidding and implementation of major (class I) changes when it comes to PD/LM systems is primarily due to a lack of understanding on the part of engineering support organizations of contractual technical/functional requirements management.

Consider the following customer-requested requirements change when the contractor is working to an allocated baseline:

Vehicle gross combination mass (GCM) shall be reduced from 2,500 kg to 2,200.

The resulting proposal is reviewed internally. A proposal is submitted to accommodate the change request. As there was no change to any of the vehicle-performance criteria, a lower weight solution was arrived at, which included changes to vehicle structural design as well as changes to materials. None of the changes could be implemented prior to customer approval. After approval, the new set of technical/functional requirements is established, and contractor engineering changes to implement the changes are within the general scope of the modified contract. As a result, all the changes that are processed to implement the weight reduction are minor changes approved by the contractor because the contractor is the design authority for minor changes.

This is very different when a change occurs after the product baseline is established. Changes to the product baseline are major changes as the customer is now the design authority.

Curiously, despite all the advantages of PD/LM environments, it was found that many standards, customers, and companies are reluctant to commit to full digital functionality. What follows is a list of the most common paper system rules found to have been leveraged forward in PD/LM implementation.

- Restricting the number of unincorporated EOs.
  - Incorporating advance EOs is an area of document control that requires latitude, which is hard to put into a black and white policy.
  - Rules such as a hard policy of advance EO incorporation; having five outstanding advance EOs is not advisable because in some cases the very first advance EO is so complex that it should be incorporated immediately.
  - In many cases, the advance EOs are needed in a hurry. If they pile up on a design that will never be produced again, it is not worth the cost of incorporating them after the fact.
- Prohibiting the use of revision letters I, O, Q, S, X, and Z.
  - This is a paper system holdover from the days when hard copies of drawings and other documents were stored in filing cabinets. It was believed that I, O, Q, S, X, and Z could be mistaken for the numbers 1, 0, 0, 5, 8, and 2 and this would lead to documents being misfiled.
  - It is not required if all the records are held in the PD/LM system.
- Including the PD/LM system metadata information on released engineering.
  - Program, prepared by, responsible engineer, revision history, contract number, U.S. contractor, government entity code (CAGE), Dun and Bradstreet Data Universal Numbering System (DUNS) number, next assembly, and so on are all common metadata fields.
  - Repeating the information in the document released in the PD/LM system eats into the non-recurring cost of the program.
  - It is also a potential source of audit findings.
- Hard copy signatures.
  - In a PD/LM environment, electronic signatures are used to release all documents that are not part of the contract between the customer and the contractor.
  - In many instances, even the contract is signed electronically.
  - The requirement to physically sign the face page of a document that is controlled in a PD/PL system using electronic approvals as part of the work flow is redundant.

10.6.2.1 Review Changes

Change review is a weak area in many organizations and the review differs depending if you are the customer or the contractor. Turning to MIL-HDBK-61A, *CM Guidance*, the following figures are found to be representational of similar functions in all customer–contractor relationships[4] (Figures 10.5 through 10.7). Inherent in CM implementation is change control as well as management of temporary departures from the technical requirements for a limited number of deliverable units. These departures are called variances (deviations and waivers).

Figure 10.6 is representative of the customer's role in driving change. When multiple contractors are developing products that form part of a larger assembly or system, changes made by one of the contractors can dramatically impact what the other contractors are designing. For example, say you are building the chassis for a vehicle that will be delivered to the customer for final assembly. The chassis must

[4] DOD, U.S. (2001). *MIL-HDBK-61a, Military Handbook: Configuration Management Guidance*. U.S. Government, Figures 6.1 through 6.3.

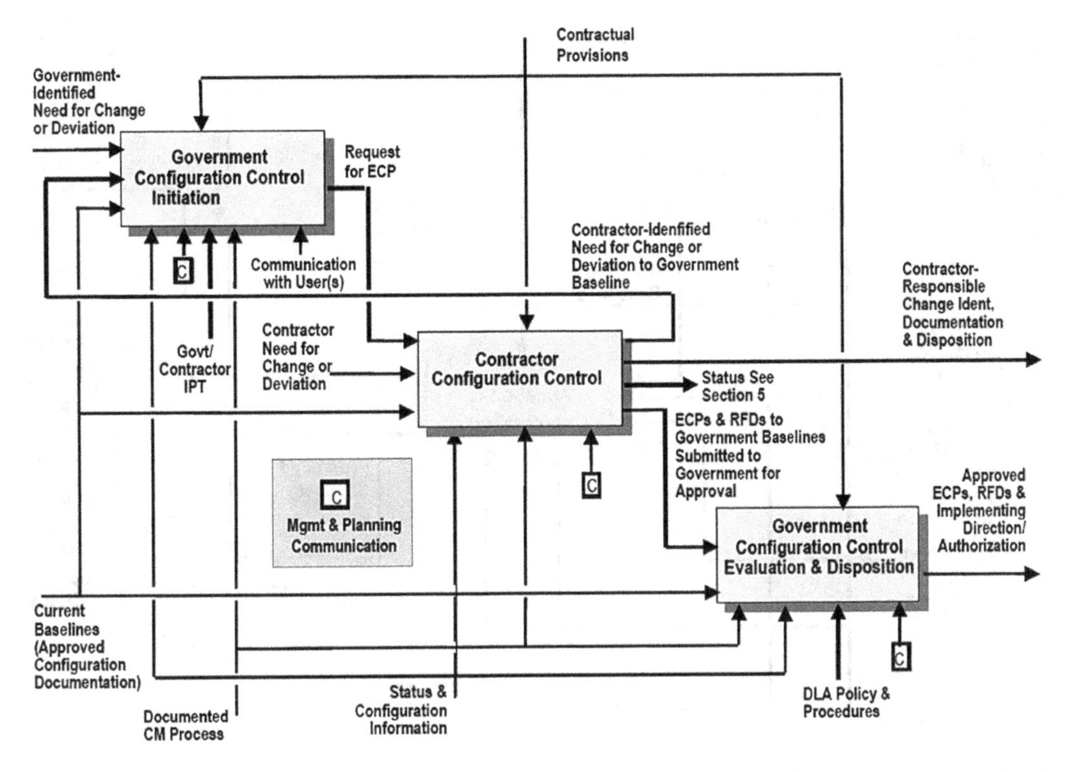

FIGURE 10.5 Configuration control process. (Data from DOD, U.S., *MIL-HDBK-61a, Military Handbook: Configuration Management Guidance*, U.S. Government, Figures 6.1 through 6.3, 2001.)

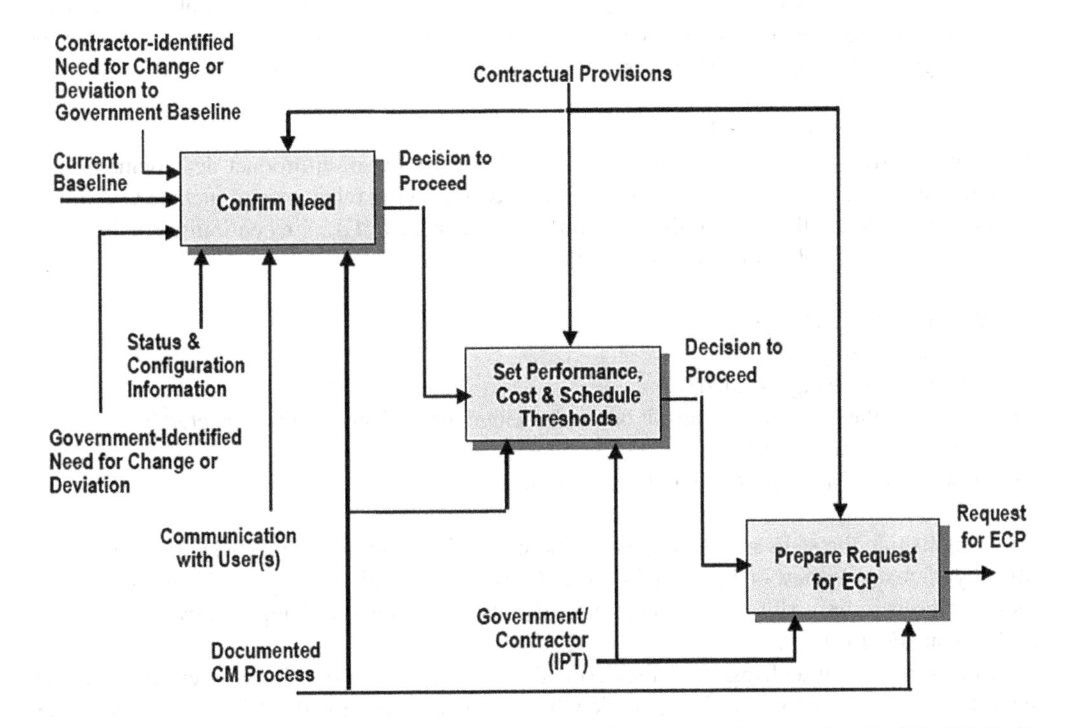

FIGURE 10.6 Government configuration control change initiation. (Data from DOD, U.S., *MIL-HDBK-61a, Military Handbook: Configuration Management Guidance,* U.S. Government, Figures 6.1 through 6.4, 2001.)

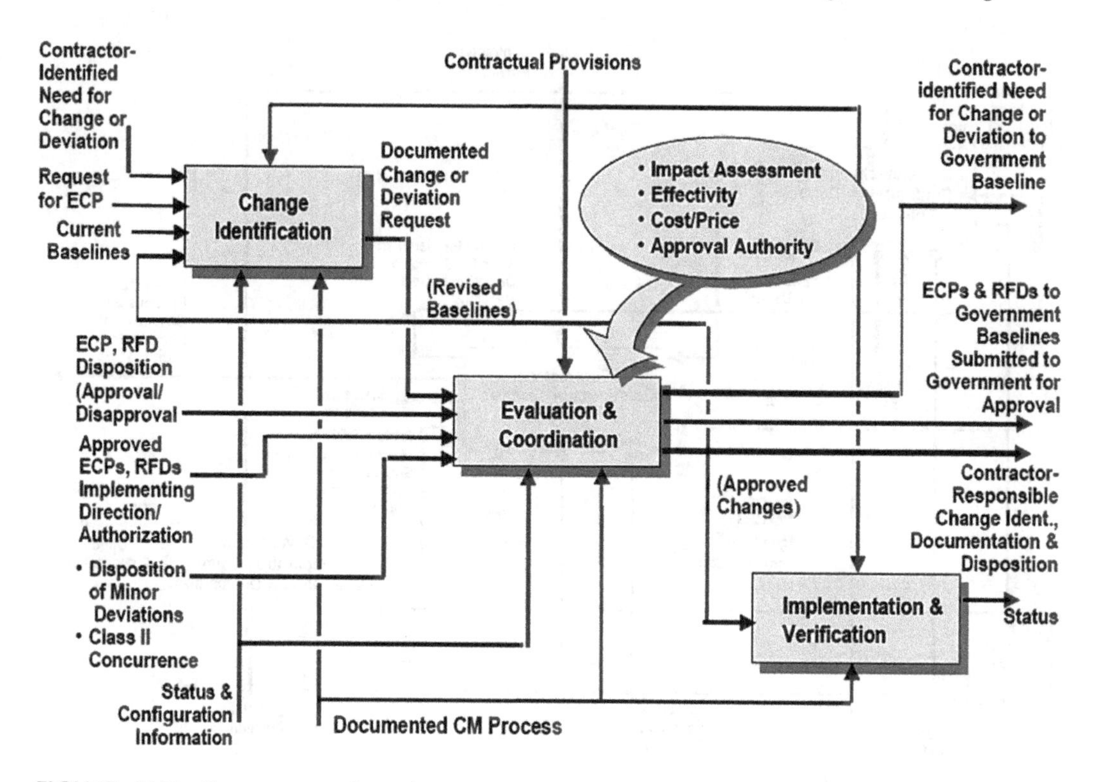

FIGURE 10.7 Contractor configuration control. (Data from DOD, U.S., *MIL-HDBK-61a, Military Handbook: Configuration Management Guidance*, U.S. Government, Figures 6.1 through 6.5, 2001.)

accommodate the engine, transmission, suspension, exterior and interior trims, and other subsystems from other contractors. Should the engine mounting location change due to an associate contractor, the chassis design will be impacted. Therefore, an interface control working group is critical.

10.6.2.1.1 Roles for the Review

Organizational decoupling of CM implementation from the program/product development team results in muddled change control. When coupled with a growing reluctance to include CM implementation throughout all phases of the program life cycle, a lack of design continuity results.

The basic elements of change review are relatively simple.

- Who (needs the change?)
- What (is the impact?)
- When (is the change needed?)
- Where (in the product line must it be implemented are delivered units impacted?)
- Why (is it needed then?)
- How much (of an impact is it and what will it cost?)

Accomplishing the steps needed to answer these questions can be a little tricky. Overlaying the criticality of change review on Figure 1.3 (cost and influence of critical design decisions plotted over time) highlights a sharp rise in the degree of criticality of a complete change review very early in development (Figure 10.8).

Adequate review of a change becomes critical very early in the program life cycle. How then should the roles for conducting change review be assigned? Certainly, there is a change manager, stakeholders, the originator of the change document, an individual responsible for the design of the item(s) being changed, and a change release manager.

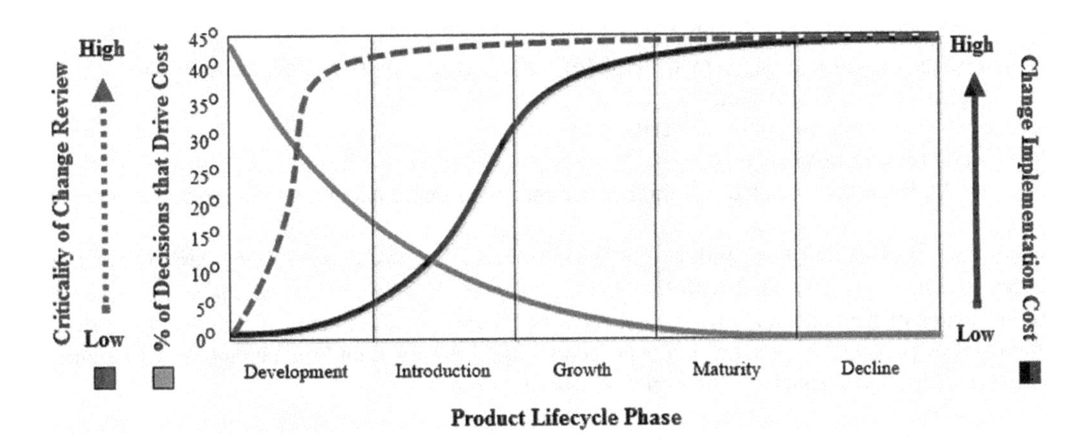

FIGURE 10.8 Increasing criticality of change review over time.

The change manager should be the individual on the team who has the overall responsibility for identifying the first-level and second-level impacts of the proposed change. The change manager works closely with the originator of the change document (ECO or equivalent) to ensure that all stakeholders are identified prior to the change entering the review process. This change impact assessment relationship is shown in Figure 10.9.

- First-level impacts
  - The item being changed
  - The next higher level of assembly where the item is used
  - The previous level of assembly (items in the BOM)
  - Any contractual impacts
  - Any vendor impacts
  - Internal subsystem to subsystem interface control drawings (ICDs)

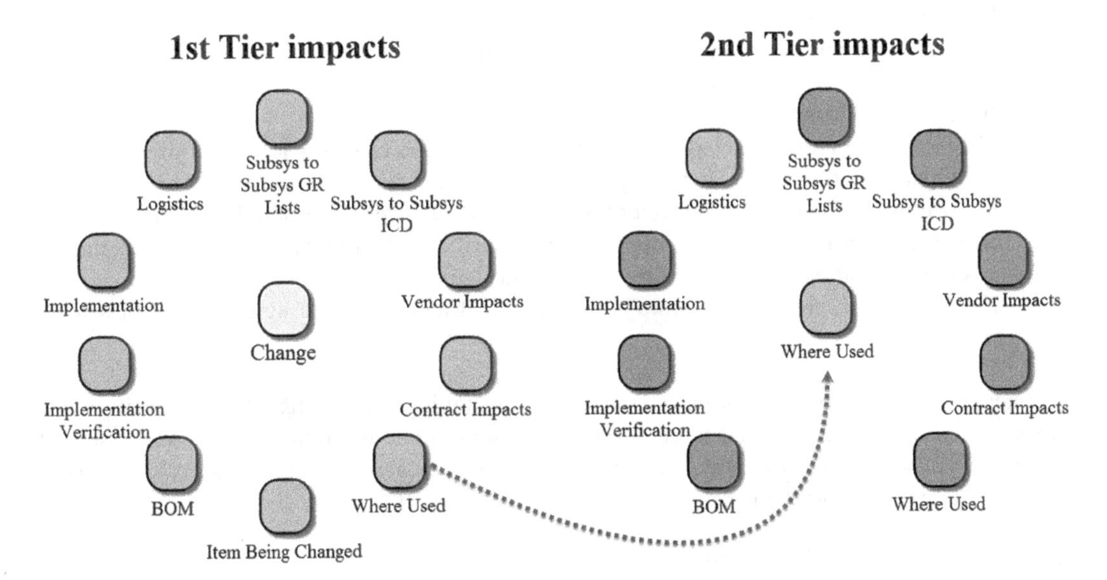

FIGURE 10.9 Change impact assessment.

    - Internal subsystem to subsystem giver–receiver lists
    - Logistics impacts on previously delivered items and items waiting to be shipped
    - Implementation strategy
    - Implementation complete verification
- Second-level impacts
    - All impacts associated with higher-tier and lower-tier items in the total product structure

The owners of all first-level and second-level impacts become change stakeholders. They in turn analyze the impacts associated with the change and provide feedback to the change manager and the originator of the change to ensure that all impacts are identified prior to change approval. The individual responsible for the design of the item(s) being changed and the change release manager should be consulted throughout the initial assessment process.

10.6.2.1.2 How to Conduct the Review

The tendency to exchange information in a transactional environment is the text equivalent of a "sound bite." Text messaging, chat, email, and other forms are increasing, and not all of it finds its way into the change control process. This means that some stakeholders have more information than others and the change impact assessment is fractured. Although some kinds of information transfer can be expedited in this way, it is recommended that the initial evaluation of first-level and second-level impacts as well as the coordination of those impacts with the stakeholders be held in a meeting environment. This is the only way the third-level and fourth-level impacts surface. Meetings should be chaired by the change manager and co-chaired by the originator of the change document.

The synergy achieved in a meeting environment allows comments made by one stakeholder to influence how others in the meeting view the impacts and leads to better informed decisions as well as complete analysis and implementation strategy development. Interviews with professionals involved with change review have generated the following set of guidelines:

- The meeting should be chaired by the change manager.
- The co-chair acts as the secretary and takes notes regarding impacts and actions assigned.
- All one-tier and two-tier items are identified and provided to the stakeholders at least 48 hours prior to the meeting.
- Silence on the part of any individual indicates agreement.
- The chair will call on one stakeholder at a time and open discussions will take place after the stakeholder has finished.
- All stakeholders are expected to view the entirety of the change and not self-limit their participation to the specific item they are responsible for.
- For example, a change to a connector type may drive changes to on-board computers, wiring boards, thermal, structure, test equipment, logistics, integration, test, verification, and existing analysis as well as the wiring harness design.
- The change will not be placed into formal review if there is any disagreement on how to proceed.

International collaborations often see these meetings taking place using teleconferencing. This frees stakeholders from the need to be physically present in one location and allows inclusion of associates in the meeting who may later be tasked to work on action items or research impacts to related systems and subsystems.

10.6.2.1.3 Capturing All Affected Items

Use of the process outlined in Section 10.6.2.1.2 will drive out impacts to most affected items. It is dependent on the level of experience of the combined change review team and their knowledge of the entire product life cycle, including the infrastructure supporting product design, manufacture, assembly, test, delivery, and maintenance. It may be necessary for the results of the stakeholder review to be independently assessed by a smaller team not closely tied to the program/project should the magnitude of the change be high enough.

Every change manager has a blind spot. It may be due to lack of knowledge or unfamiliarity with the area impacted by a change, or simply be an oversight due to overwork. It creates a risk that some affected items may be missed, resulting in an incomplete impact assessment. The risk is mitigated through stakeholder review early in the process, which increases the likelihood that all impacts are identified and understood.

10.6.2.2 Approval/Disapproval

After a review of the many automated workflows used in PD/LM systems, the following best practices were documented.

- Change approval or approval with comments was preferred over disapproval; however, it was incumbent on the change document originator to ensure that all comments were incorporated prior to notifying the change release manager that the change was ready either for release or for transmittal to the customer for their approval prior to release.
- Documents associated with the change are checked out from the PD/LM system and updated, and the updated documents are checked back into the PD/LM system to preserve comment incorporation traceability. Most PD/LM system designs allow for reviewers to see previous versions of the document revision.
- Rejection of a change for any reason required the change to be returned to the pending state and to start the review and approval workflow over from the beginning. For this reason, rejection was not preferred. Approval with comments rather than rejection is preferred.
- If an individual believes they should not be on the list of approvals, a simple email or phone call to the change originator is acceptable.

A review conducted when these practices were evident prior to the change entering the approval workflow found approval rejection incident rates at near zero levels (0.00032%).

10.6.3 Variances, Part Substitution, and Defects

10.6.3.1 Variances

Variances are departures from requirements discovered prior to acceptance. All variances require customer approval prior to acceptance. Variances cannot be for all units produced. Should this need arise, the associated technical requirements need to be changed so a variance is not required.

10.6.3.1.1 Critical Variances

Critical variances are temporary departures from requirements involving safety or when the variance is against requirements defined as critical.

10.6.3.1.2 Major Variances

Major variances are temporary departures from performance, interchangeability, reliability, survivability, maintainability, or durability (item and its spare parts). They can be known (planned, also known as deviations) or unknown (unplanned, also known as waivers). In both cases, the departure is for a select number of units only.

Planned variances occur when a requirement is evaluated and it is determined that the requirement cannot be met prior to design start and fabrication. Unplanned variances surface during validation and verification activities when it is discovered that the design does not meet the technical requirements.

10.6.3.1.3 Minor Variances

Minor variances are any variances not determined to be critical or major. Minor variances are associated with materials (e.g., a milled part that has a surface scratch on it). Minor variances are dispositioned by a material review authority as scrap, rework, or use as is.

10.6.3.2 Part Substitutions

Authorized part substitutions do not constitute a variance or a defect. Authorized part substitutions may occur at several levels (company, program, associated design facility). Many PD/LM systems are not set up to account for part substitutions beyond the drawing level. Part substitutions are equivalent or better than the part defined in the design (e.g., a 6 TB solid state drive when a 3 TB drive was ordered).

10.6.3.3 Defects

Defects are non-conformances discovered after acceptance. There are two kinds of defects: latent defects and patent defects. Most contracts contain provisions regarding seller and buyer responsibilities under inspection, acceptance and warranty, or product liability clauses. The definitions below are from the U.S. Federal Acquisition Regulations.

- *Latent defect* means a defect that exists at the time of acceptance but cannot be discovered by a reasonable inspection.
  - A computer program has the option to print on any printer, but if printers other than the default are chosen, it results in an error message. This is a latent defect.
- *Patent defect* means any defect which exists at the time of acceptance and is not a latent defect. The rule of law is *caveat emptor.* The buyer alone is responsible for checking the quality and suitability of goods before a purchase is made.
  - Discovering after purchase that one of the lug screws holding a tire was missing is a latent defect.
  - It could have easily been discovered at the time of purchase.

10.7 CONFIGURATION VERIFICATION

Each approved major change may result in a revised set of functional requirements. These in turn are allocated to systems and subsystems and each allocated requirement needs to be verified. The overlying assumption is that if each allocated requirement is verified, then the entire system is verified.

10.7.1 Functional Aspect

Functional requirements are traditionally verified using four methods applied singly or in combination. They are analysis, inspection, test, and demonstration. Simplified examples of each method follow.

10.7.1.1 Analysis

Analysis as a form of evaluating an item's capabilities is a valuable tool when other verification methods cannot be used. A design element can be digitally modeled (e.g., a digital twin is created) to verify predicted performance. Operational conditions of spacecraft may exist that cannot be tested except in the space environment. Instead, they are evaluated against what is known of the environment from spacecraft in similar orbits or missions.

10.7.1.2 Inspection

Inspection is a method of direct quantitative measurement. A nine-bolt attachment interface may exist that is represented by a master gage. Assemblies are inspected against the master gage to verify they meet the requirement. Parts made using subtractive technologies are inspected in a dimensioning and tolerancing laboratory. Parts made using additive technologies use other methods like X-ray computed tomography or micro CT scanning.

10.7.1.3 Test

Test is a method of verification of a single functionality that often uses simulators rather than production parts. A life test may be run to prove that an actuator is able to meet a duty cycle requirement of 125,000 cycles. The actuator is placed in a test fixture and a minimum of 125,000 cycles are run. Often a test will continue until component failure to evaluate expected component life and establish the recommended replacement cycle of the item based on time in service.

10.7.1.3.1 Test Temporary Configuration

A product may be required to withstand a set of vibration frequencies and durations. Mass simulators are used instead of actual production components during the test. Once the test is complete, the simulators are removed and the product moves to the next assembly operation.

10.7.1.3.2 Test Fixtures

Test fixtures are used to assist with testing at the component, unit, and assembly levels. A fixture must be representative of both sides of an interface.

10.7.1.3.3 Simulators

Simulators (digital twins) are often the electronic equivalent of a test fixture, and may be representative of the final assembly that software will be operating on. They allow software to drive a digital twin of the final product in all the scenarios mandated by the requirements allocation to the software system.

10.7.1.3.4 Test Anomalies

During the testing process, anomalies can happen. An anomaly is simply a result that differs from expectations. Sometimes, the anomaly may be due to poor assumptions or faulty calculations regarding how the item will perform. In other cases, it is not so simple.

Voltage transience in electronic circuits may be observed. Each instance must be traced to its root cause and dispositioned. Voltage transience may be attributed to test equipment, power supplies, driving the circuit in a way other than intended, or several other sources. If the observed behavior is repeatable, it helps with root cause analysis. If the observed behavior happens only once, such anomalies are said to be observed but unverifiable. Root cause analysis may indicate design flaws that must be corrected prior to product acceptance.

10.7.1.4 Demonstration

Demonstration is a form of validation utilized when there is no other way to prove that a technical requirement is satisfied. A requirement may exist for an aircraft to accommodate 196 pieces of luggage with a maximum size of 22 cm × 35 cm × 56 cm, including handles and wheels, in the overhead compartments. A demonstration is conducted with 196 pieces of rigid foam blocks to see if they all fit. Often demonstrations are used when requirements are specified using minimum values, such as that the aircraft must fly faster than 500 knots (880–926 km/h) at 12,192 m.

10.7.2 Aids to Functional Aspect Verification

Aids to functional verification differ depending on the nature of the product itself. Verifiers may be used as a static analysis tool in very large-scale integration design, bridging the gap between layout syntax and some form of behavioral analysis.[5] This provides an adjunct to rule checkers used for

[5] Rubin, S. M. (1994). *Computer Aids for VLSI Design*. Retrieved May 19, 2014, from http://www.rulabinsky.com/cavd/text/chap05-4.html.

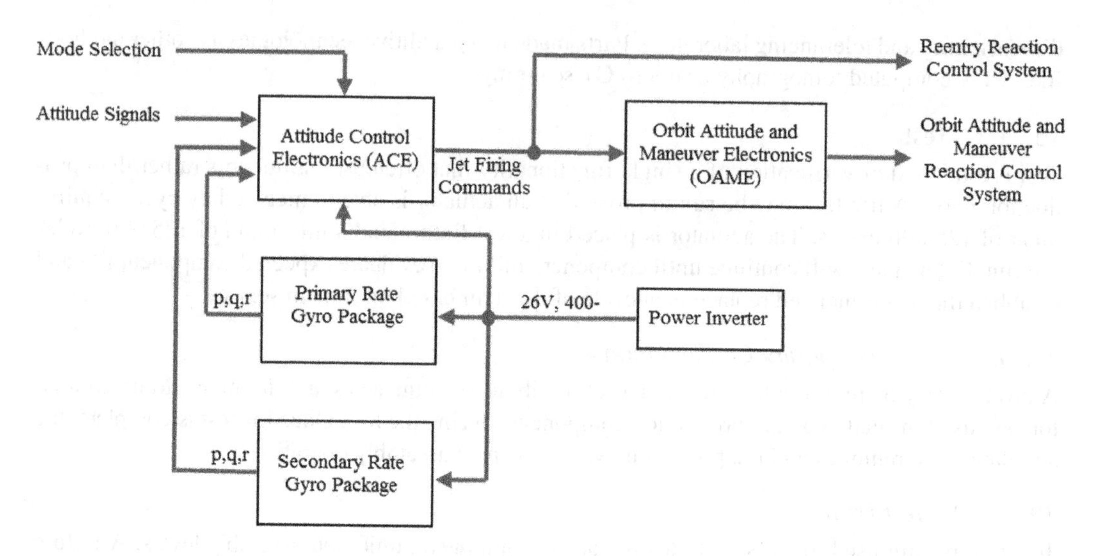

FIGURE 10.10 Gemini spacecraft attitude control block diagram.

the verification of low-level environment-dependent correctness conditions. Use of timing verifiers in lieu of complex sample protocols for independent computing environment circuit simulators is faster and provides much the same information.

Functional block diagrams of systems, subsystems, and component families such as cabling have been found useful not only in analysis but also in resolution of test anomalies. Figure 10.10 is the functional block diagram of the attitude control and maneuvering electronics system of the U.S. Gemini spacecraft circa June 1962.

10.7.2.1 CM and Test Execution Trials

Testing part of a configuration prior to completion of the entire deliverable often leads to early identification of design flaws. In an analogous way, the execution of dry runs of the test itself will pinpoint errors in the test setup and testing assumptions. In either case, management of the configuration of the test parameters as well as the test setup, and controlled release of the iterations of the test parameters as well as the test setup and test results, ensure that the test is repeatable and that all observed results and test procedure redlines are captured.

It is highly recommended that the drawing for the item being tested, the test procedure, the as-run test procedure, and the test report be linked together via a digital thread in the PD/LM system. Relationships function outside of change control, allowing the digital thread to be established without the need for a released change. In a multiple unit build environment, using an EO for establishing the digital thread between the procedure and the as-run procedure for each serial number made is a costly burden that adds nothing to the integrity of the documents and units under test.

Countless failures exist due to not incorporating test redlines and releasing them immediately after the test is complete. On a U.S. government program, the lack of documentation (1) relative to the redlines made to procedures and (2) of the complete test setup (including versions of the Mathcad and other setup parameters) resulted in the inability of the original manufacturer to replicate the testing on the second unit build. It cost hundreds of thousands of dollars in lost schedule to stand up a new test program.

Similar issues frequently occur in software and firmware development. In another instance, documentation for software loaded into programmable read-only memory (PROM) chips for multiple build test equipment indicated a released version 1.0. When several *qualified* test units were examined, the code versions on the PROMs inside were found to range from V 2.5 to V 11.0. There was no record of the changes after V 2.5, nor could evidence be found that any functional qualification

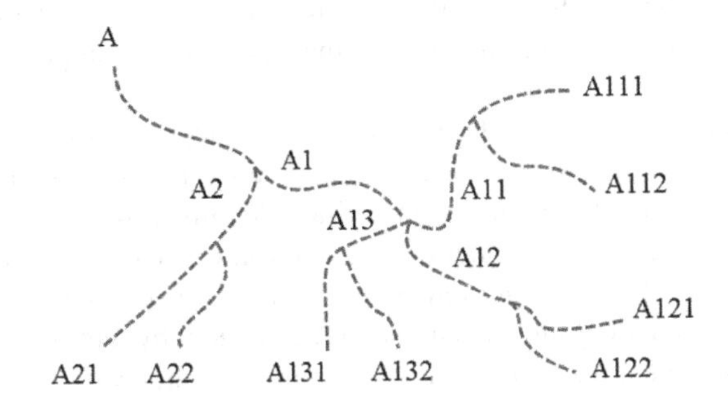
As requirement A is allocated to lower and lower subsystem elements branching of the solution set occurs
A
A1
A2
A11
A111
A112
A13
A12
A121
A122
A21
A22
A131
A132
Only when all branches are verified can requirement A be said to have been verified

FIGURE 10.11 Requirements allocation branching.

testing had taken place. No version description documents or associated software code for the various code releases could be found other than what was loaded on each PROM. To make the next PROM code updates, 110 units had to be recalled. The code itself was reverse engineered prior to a new set of code with a different software identification being loaded on the recalled units.

10.7.2.2 Subsystems (Product Sub-Element) Verification

It is easy to imagine the functional baseline as a hydra. Each requirement is a hydra head that must be fed. Each time a requirement is allocated to multiple subsystems (such as those having a hardware and software component), additional subrequirements are established. This is known as "requirements allocation branching" and is shown in Figure 10.11.

A single requirement may branch to the third or fourth level. Each time a requirement branches, it creates two subrequirements—just like cutting the head off the mythical hydra caused two heads to grow in its place. Only when all heads have been fed and the complete set of allocated requirements have been met can the system functional requirements be said to have been verified.

CM implementation needs to address verification. Many interviewees stated that systems engineering handles requirements tracking as well as monitoring the status of the artifact used to satisfy each requirement branching. While that may be true in some market segments, it does not negate the fact that it is a function necessary to the management of the configuration. In other market segments, this activity is the responsibility of a single department that may be called "configuration management" or "requirements management."

10.8 SYSTEMS (PRODUCT) VERIFICATION

What, then, is "systems verification" or, more properly, "product verification"? It is that point in the development of a product where the producer feels satisfied that all requirements necessary to bring the product to market have been satisfied.

Assume that the product being developed is a long-stemmed red rose with

1. Superior blossom shape.
2. Blossom size larger in diameter than 7.62 cm.
3. A straight stem longer than 40.64 cm.
4. A superior scent, which will last for three weeks after cutting.

Only when all four requirements have been met can it be said that the product requirements have been verified.

While the verification of the breeding requirements for a new species of cut rose may appear to be simple, the details associated with getting everything right are as complex as those associated with bringing any other product to market.

An updated version of SolidWorks software for 3D development of a vehicle such as the Wiesmann MF5 roadster may have thousands of discrete subsystem requirements. Each must be separately verified. Each verification requires an artifact documenting that the requirement has been met. Each artifact must be released under change control. Each artifact must be subjected to a gap analysis confirming that all of the requirement has been met. If the requirement has been met, all is well. If not, another branching must occur, so you can show by another means that you met the open item. If you cannot meet the leftover bit of the requirement, a variance must be submitted.

10.8.1 Live-Fire Exercise

Many complex systems are put through a form of testing known as a *live-fire exercise*. This is any set of realistic simulated scenarios using the product in its intended environment. The terminology stems from weapons testing and, like many terms associated with management of product configurations, it has become common throughout many market sectors. In aerospace, the term *live-fire exercise* is often called *test as you fly*. Realistic simulated scenarios may take the form of beta testing in the software environment, some form of total systems functionality testing in a hardware environment, or a service demonstration in a services environment. The use of a "test mule" in the automotive industry is a form of live-fire exercise.

10.8.2 Analysis, Inspection, Test, and Demonstration Master Plan

Digital threads are created as the functional requirements have been allocated and a verification method has been determined (if not stipulated by the customer). These are captured in a software tool (use of spreadsheets for this purpose is not recommended). Items specified to be verified by analysis, inspection, test, and demonstration are assigned tracking identification numbers, and document numbers are pre-assigned to each artifact to be generated. Actual vs. planned delivery of each item is statused in a program completion schedule that becomes part of the analysis, inspection, test, and demonstration master plan with a digital thread linking it back to the top functional requirement. The master plan is a living document, updated as requirements allocation branching occurs.

Artifacts supporting verification reviews and audits need to be delivered to the customer as they are released. This incremental delivery is critical as PD/LM systems often do not have the bandwidth to output the entire verification digital fabric in a 24-hour period. If they can output the data. the GTS communication pipelines aren't robust enough to allow it to be transmitted, and neither do companies have enough staff to make the deliveries. This may change when 7G and 8G technologies arrive and data deliveries are made directly from some future PD/LM system.

10.8.3 Requirements and Closure

Should a program/project have no requirements changes during the development phase, the digital thread linking allocated baseline to analysis, inspection, test, and demonstration will remain static over the program life cycle.

Each requirement change adds complexity. If the change review process fails to capture everything during the impact assessment, the program may never know if the verification goals have been met until they surface at verification reviews and audits. CM implementation activities need to encompass tracking the genesis of requirements as well as closure artifacts (Figure 10.12). This is where change control pays for itself many times over.

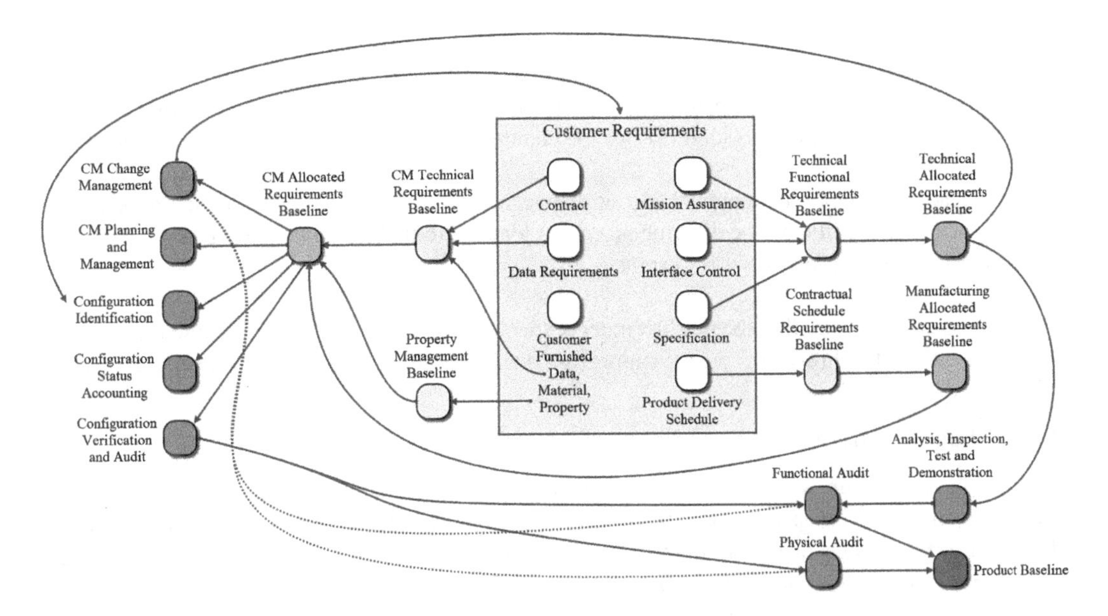

FIGURE 10.12 Requirements genesis and closure.

10.9 CONFIGURATION AUDIT

Configuration audit is ongoing throughout the product life cycle. It exists in two distinct forms: in-process audits and formal audits.

1. In-process audits
   a. Generally tied to change control of data error management of PD/LM input for a specific program.
   b. The effect of implementing in-process configuration audits is cumulative.
   c. They ensure that a result better than 50% is obtained when searching for data (refer to Section 5.3.1).
2. Formal audits
   a. Generally physical and functional.

10.9.1 In-Process Audits

Ongoing audits are referred to as "in-process audits." In-process audits performed by CM or in conjunction with CM include audits of:

- Allocated design against requirements.
- Changes to design against requirements.
- Internal approval of major changes against those authorized to approve prior to submittal.
- Internal approval of minor changes against those authorized to approve prior to implementation.
- Change incorporation against affected items.
- Analysis, inspection, test, and demonstration results against expectations.
- Data prepared for submission against data item descriptions.
- Customer acceptance of data against data approval timelines.
- Vendor data against vendor data requirements.
- Vendor product against functional and physical requirements.
- Contract changes authorized by the customer against what was proposed.
- Changes were reviewed and approved by the proper contingent of people.

- Changes were only released after a complete review was conducted.
- Changes were released by someone having the authority to do so.
- Associated metadata is complete for all fields in the PD/LM system prior to release.
- The responsible engineer is identical for the change and all affected items (e.g., documents and parts).
- The description fields are identical for the change, all affected items, and the document being released (e.g., the title on the drawing is carried forward to the EO and the PD/LM objects).
- The document approvers list is current.

In-process audits ensure that the product meets expectation, the configuration is known and maintained, and the digital thread linking all supporting documentation is controlled and maintained.

10.9.2 Formal Audits

Formal audits can occur as a set of iterative events or a single all-inclusive event. Product entering government inventory consists of two distinct audits, covering the areas of functional compliance (FCA) and conformance of the product to product documentation (PCA). MIL-STD-3046 (U.S. Army), *Department of Defense (DOD) Interim Standard Practice: CM*, provides the best direction for the U.S. government procurements as of 2018.

If an external customer is involved, the buying organization will

- Provide co-chairperson.
- Review audit agenda.
- Prior to audit, provide contractor with the name, organization, and security clearance of each government participant.
- Participating in audit activities, identifying discrepancies, and determining their criticality.
- Review daily minutes to ensure that they reflect all significant government inputs.
- Provide formal acknowledgments to the contractor of audit results (e.g., after receipt of audit minutes, the contractor will be informed if the audited area met with approval, contingent approval, or disapproval).

If an external customer is involved, the selling organization will

- Provide co-chairperson.
- Accommodate participation by subcontractors/suppliers as appropriate.
- Establish a location.
- Establish whether incremental audits will be conducted.
- Provide an audit plan.
- Provide an agenda.
- Provide conference room(s).
- Provide applicable documents (specifications, drawings, etc.).
- Provide test results.
- Provide minutes, including action item audit.
- Provide tools and inspection equipment.
- Provide access to facilities, security clearances.
- Provide support personnel.
- Isolate item(s) to be audited.

Key audit personnel are shown in Figure 10.13. It is highly recommended that a pre-audit or audit dry-run be performed prior to the formal audit. This will be addressed more fully in the discussions that follow.

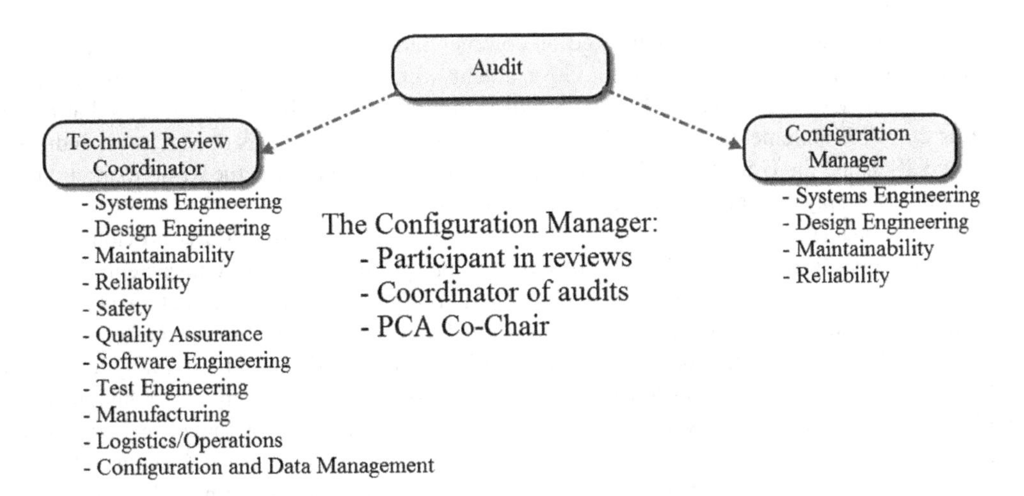

FIGURE 10.13 Audit key personnel.

10.9.2.1 Functional Audits

The product functional audit evaluates the allocated requirements against the functional requirements. Each of the functional requirements is validated to ensure that the right thing was built. For U.S. government programs, this is called the functional configuration audit (FCA). As defined by MIL-STD-3046:

> The objective of the FCA is to verify the CI's performance against its approved functional and allocated configuration documentation. Test data for the FCA shall be from test of the configuration of the prototype or preproduction article. If a prototype or preproduction article is not produced, the test data shall be that collected from test of the first production article. Subject to prior acquiring activity approval, the FCA for complex items may be conducted in increments. In such cases, a final FCA may be conducted to ensure that all requirements of the FCA have been satisfied.

Putting this in non-governmental terms, it is an audit of the proof presented by a company that the product functions in accordance with requirements. If the product contains components not previously qualified, it also includes the qualification test results.

Perhaps one of the most extreme recent cases of functional requirements allocation and verification took place in the development of the Bugatti Veyron Super Sport 16/4. The vehicle pushed the envelope of what was possible in automotive design. It is possibly the most sophisticated automotive achievement to date, with attested top speed of 431.3 km/h at Volkswagen Group's test facility at Ehra-Lessien, exceeding the design goal of 425 km/h. This is faster than Formula One cars. The W16 engine produces 984.8 kW. The seven-gear manual transmission works with torques up to 1250 N m, allowing the car to go from zero to 100 km/h in 2.5 seconds. It can stop in 2.3 seconds from that speed. To maintain operational environments, the vehicle requires 12 radiators. All this sounds impressive and, to put it into perspective, it must be remembered that the top speed of 431.3 km/h requires that the vehicle stays on the ground at that speed.

Below are the takeoff speeds for several aircraft by comparison:

- Cessna 150: ~100 km/h
- Bombardier 45: ~271 km/h
- Boeing 747 and Airbus A340: ~290 km/h
- Concorde: ~360 km/h

The embodiment and synergy of the technology in the Super Sport version of Bugatti Veyron mandated a comprehensive CM implementation with tight configuration and data management control to standards higher than mandated on commercial imaging satellite programs. This is not unusual in commercial market sectors. Government sectors adhere to preconceived definitions and limitations leveraged forward from U.S. government standards from the 1960s. What it indicates is that CM implementation—while meeting the minimum guidelines of Electronic Industries Alliance, ISO, and the U.S. DOD—has advanced to the enterprise level in the world marketplace. It has done this simply to survive in the commercial environment.

The functional audit may be combined with a "functional systems audit" (FSA), previously known as a "formal qualification review." The FSA ensures that system functional requirements not evaluated at the product level have also been demonstrated. You may say, "Wait a minute, where did that requirement come from?" The entire configuration and all the documentation associated with it are encompassed by the CM implementation. It is an enterprise-wide activity and not simply relegated to a few items that someone decides to call out for an audit of functional or physical compliance. The FSA looks at how the products you make play together with other items you make, and items associate contractors make, and items the customer makes. Put in automotive terms, does the automobile tire you designed fit on the tire rims on our entire fleet of cars, or only on a rim made by you?

Pre-functional audit activities include the following:

- Identify configuration(s) to be audited, audit participants, and facilities required.
- Prepare audit agenda.
- Prepare audit checklist.
- Provide verification cross-reference checklist.
- Provide list of all deviations/waivers.
- Collect artifact information (e.g., test plans, test procedures, as-run test procedures, reports documenting analysis, test, demonstration, and inspection results).

Post-functional audit activities include the following:

- Prepare and publish audit minutes.
- Record audit results in product status accounting reports.
- Close out corrective actions.

10.9.2.2 Physical Audits

A physical product audit evaluates the product against the engineering that defines it (*physical* includes such intangible items as models, software, and other representational engineering products used to produce the end product). Each of the design elements is verified against the finished product to ensure that the product was, in fact, built to that engineering. For U.S. government programs, this is called a *physical configuration audit* (PCA). Turning again to MIL-STD-3046:

> A Physical Configuration Audit shall be conducted for each CI and related component as required by the contract. The PCA shall verify the design documentation is a clear, complete and accurate representation of the product. Satisfactory completion of a PCA is required to establish the formal product baseline. The PCA shall include a detailed audit of engineering drawings, MBD datasets, specifications, technical data, design documentation, and associated listings which define each CI/component. The PCA shall also determine that the quality assurance testing and acceptance requirements of the product configuration documentation are adequate to ensure end item performance. The PCA shall not be started unless the FCA for the item has already been accomplished or is being accomplished concurrent with the PCA. Subject to prior acquiring activity approval, the PCA for complex items may be conducted in increments. After successful completion of the audit and the establishment of a product baseline, all subsequent changes to the product configuration documentation shall be done by formal change control only. Additional PCAs may be performed during subsequent production if required (e.g., if there has been a long gap in production or if a new manufacturer for the product is established). For software, the product specification, Interface Design Document, and Version Description Document (VDD) shall be a part of the PCA.

Physical product audits can include complete disassembly of the product, inspection/measuring each manufactured part and subassembly against the released engineering, and then reassembly as defined by the engineering. This is more likely in cases where a complete TDP has been prepared. The TDP is self-contained and does not rely on a single manufacturer's processes. A TDP can be provided to multiple technically qualified suppliers capable of manufacturing the product.

The physical product audit is often co-chaired by the contractor's configuration manager and a representative from the customer. It is highly recommended that audit ground rules be established prior to the review addressing the following areas:

- Time and place.
- Definition of a finding versus an observation.
  - Finding.
    - Item does not match the engineering.
      - No variance is in place.
    - Approved changes are not incorporated.
  - Observation.
- Engineering is not in error but could be clearer to facilitate reproducibility.
  - Finding resolution timeline.
  - Constitution of a successful review.
  - Factors such as number of findings.

Pre-physical audit activities include the following:

- Identify item(s) to be audited and depth of drawing review.
- Identify audit personnel, required facilities, and scheduling of audit activities.
- Provide list of all variances (deviations/waivers).
- Prepare audit checklist.
- Identify any difference between the physical configurations of the selected production unit and any developmental unit(s) used for the audit.
- Certify or demonstrate to the customer that these differences do not degrade the functional characteristics of the unit(s) selected for the audit.

Post-physical audit activities include the following:

- Receive customer notification of audit approval, audit contingent approval (requiring correction of specific discrepancies), or audit rejection and requirements for re-accomplishment.
- Prepare and publish audit minutes.
- Close out corrective actions.

10.9.2.2.1 As-Built versus As-Designed

A critical component of the physical audit is verification that the product was made in accordance with the engineering and production documentation. At the simplest level, this can be a comparison of the product structure against the items used. Substitute parts, minor hardware anomalies, and any other variances are identified to the lowest level of the product structure. Documentation supporting the *as-built* versus *as-designed* analysis may even include receiving inspection reports, lot sampling, and MDSs for components as well as consumable items such as solvents and other cleaning agents. It may also include the lot/date information (e.g., lot/date code) of each component used on a printed wiring assembly and where each component is located.

10.9.2.2.2 Material Review Board

The *material review board* (MRB) is responsible for determining the disposition of non-conforming materials (e.g., material or manufactured items that fail inspection) and identifying corrective actions required to prevent future discrepancies. Non-conforming customer-supplied production materials cannot be modified or disposed of without customer permission. On U.S. government contracts, this is tied to the "government property" clause in the contract. Material is identified on the government-furnished property/material and data listing (most often a contract attachment). Similar provisions and lists can be found in most contracts where the customer provides property, material, and data. It is important to remember that material includes items received from a vendor as well as items produced by the contractor. MRB activities include the evaluation and disposition of minor variances, also known as "minor deviations" or "waivers."

The basic MRB dispositions are:

- Scrap the material or item.
- If a design change is required, the variance number is tied to a change in the PD/LM system.
- Rework completed assemblies.
- Transfer MRB material to stock.
- Return purchased items to the supplier (including customer-supplied production material).
- Re-grade material.
  - Rework must be performed, and the item re-inspected before it can be placed in stock.
  - Approved for "use as is" is moved directly to stock.

MRB membership often consists of the following disciplines:

- Manufacturing engineering
- Design engineering
- Materials
- Quality
- Purchasing

MRB dispositions are an element of the *as-built* versus *as-designed* documentation.

10.10 MATERIAL RECEIPT AND INSPECTION

Traditionally, the inspection process follows the following steps:

- First article inspection
  - Inspect inquiry for non-inventory parts
    - If non-conforming, the item goes to the MRB for disposition
- Documentation verification
  - Drawings
  - Specifications
  - Vendor inspection and test records
  - Identification label accuracy
  - Material certification receipt accuracy
  - Markings and packaging
  - Supplier status
  - Receipt record accuracy
  - Expiration date
  - Approved variances (deviations and/or waivers)
  - Evaluation against any active alerts (GIDEP, or other)
- First article inspection data sheet

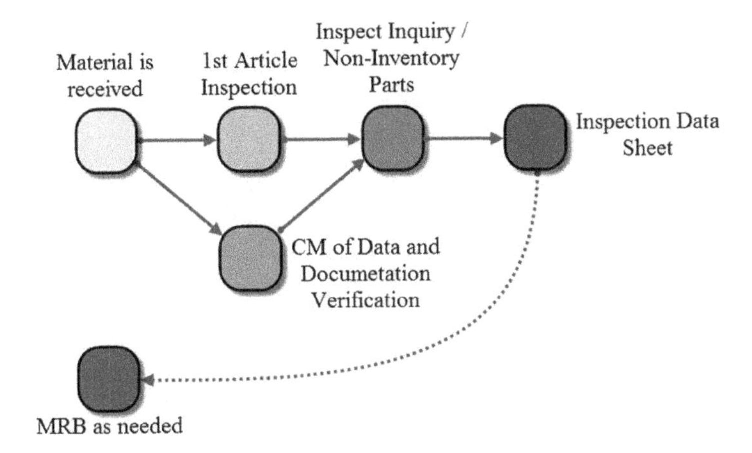

FIGURE 10.14 Material receiving and inspection flow.

Each of these steps involves CM at the vendor with the resultant generation of configuration-controlled records by the customer. This flow is summarized in Figure 10.14.

Material receipt and inspection for incoming materials applies to goods, production material, process material, and customer furnished items. What follows are the accepted definitions for each of these.

10.10.1 Goods

Material with utility and/or intrinsic value purchased by, internally produced by, and/or supplied to a company.

10.10.2 Production Material

Material that becomes part of a shippable product.

10.10.3 Process Material

Materials required in the manufacturing process that do not become part of the shippable product.

10.10.4 Customer-Furnished Material

Items provided by the customer that are consumed during the production process or items that become part of the final product.

11 When Things Go Wrong

11.1 AREAS OF CONFIGURATION MANAGEMENT RISK

The yearly quantity of mechanical, interface, electronic, and other CM implementation issues documented in the world press is staggering. The news coverage bears witness to areas where poor CM implementation or none at all has caused significant problems for the projects, products, companies, and industries. This chapter presents examples where lack of consistent enterprise CM implementation was identified as a root cause. The late Larry Bowen[1] was instrumental in identifying areas that can be tied directly to CM implementation disasters. His initial list has been expanded to include the following:

- Hardware and software integration.
  - Software and hardware changes that are not coordinated among the communities, creating problems in integration test.
- Ineffective interface control working groups (ICWGs).
  - Dealing with external interfaces especially in concurrent development programs. This keys with poor memorandum of understanding, memorandum of agreement, and associate contractors agreement that address the ICWG activity.
- Milestone slips because metrics were not used to determine exactly how a project was proceeding.
  - Needed early enough to establish workarounds to fix the problems.
- Inordinate change processing duration and number of changes.
  - Resulting in greatly fluctuating functional requirements and the inability to formulate a meaningful allocated baseline.
- Lack of change implementation after approval.
  - Resulting in mixed stock and/or items in the inventory that do not conform to engineering.
- Lack of item identification control.
  - Resulting in poorly designed items having the same item identification as their replacement parts.
- Undocumented changes or part substitution in final product.
  - Resulting in life cycle management issues.
- Lack of proper configuration control board (CCB) stakeholder review.
  - During and after development.
- High rate of scrap and repair.
  - Root cause analysis not performed.
- Maintaining software code and its associated design documentation.
  - Baselines are out of sync and incomplete.
- Lack of baseline control.
  - Resulting in misunderstood requirements, analysis, and CM implementation.
- Inadequate training to those performing CM implementation functions.
  - Result in a lack of CM control.

We have collected examples of each of these areas through personal experience, one-on-one interviews, and online discussion groups and have presented them in the sections that follow.

[1] CM Department Manager at Ball Aerospace & Technologies Corp. prior to his demise in 2009.

11.2 HARDWARE AND SOFTWARE INTEGRATION

11.2.1 Commercial Titan—Intelsat 603

Commercial Titan (also known as 6F3) was developed as a leveraged design from Martin Marietta's successful Titan 3 line of boosters. The design accommodated deploying one or two satellites into space utilizing a dual-dispenser control system. Commercial Titan first launched on January 1, 1990, carrying Skynet 4A for the U.K. Ministry of Defense and JCSAT 2 for J SAT International.

The payload manifest for the second launch on March 14, 1990, was for Intelsat 603. Due to lack of coordination between the software and hardware development teams, the hardware team connected the lower payload cable to the spacecraft and the software team sent the separation command to the upper payload cable (not connected to the spacecraft). When the satellite was ordered to free itself from the Commercial Titan vehicle, engage its apogee kick motor, and stabilize its low orbit using satellite thrusters, nothing happened. This reduced its operation life from 15 years to 12 years. Intelsat 603 was recovered by the Endeavour shuttle (Figure 11.1) on its May 7, 1992, maiden voyage (STS-49).

When Commercial Titan started, it was anticipated that a market existed to support five commercial launches a year over a five-year period. The second launch included a refly-for-free clause in case something went wrong. Due to the losses sustained with the reflight of the recovered Intelsat 603, Commercial Titan was withdrawn from service after the fourth launch on September 25, 1992.

11.2.2 *Космос*-3M—QuickBird 1

On November 20, 2000, a Космос-3M (also known as Cosmos 3M) was launched from Plesetsk Cosmodrome at 23:00 coordinated universal time (UTC). The second stage of the 3M shut down prematurely due to configuration errors and could not be restarted to place it into the proper polar orbit when the spacecraft reached apogee. The QuickBird 1 payload retrograded in the atmosphere, with reports of the bolides received from Uruguay (Montevideo and Paysandú) and Argentina (Santa Fe and Buenos Aires). Figure 11.2 shows the QuickBird 1 bolides.

The destroyed QuickBird 1 satellite was built by Ball Aerospace for DigitalGlobe, LLC. DigitalGlobe had insured the launch not only for the cost of the development, build, and launch of the satellite but also for loss of anticipated profits. The Cosmos 3M failure and the insurance payout associated with it caused launch insurance rates worldwide to increase.

FIGURE 11.1 Intelsat 603 rescue.

FIGURE 11.2 QuickBird 1 bolides over Argentina.

11.3 INEFFECTIVE ICWGs

11.3.1 Missile X Interfaces

Development of a new intercontinental ballistic missile (ICBM), the LGM-118, to replace the aging U.S. Minuteman and Titan 2 ICBMs and keep pace with Soviet technologies, was started in February 1972. It was designated "Missile X," or simply "MX." Initial operational capability was scheduled for 1985. By 1977, it was decided that the MX would comprise three solid-propellant rocket motor stages and a liquid-fueled post-boost vehicle carrying 10 multiple independently targeted reentry vehicles.

Martin Marietta was awarded the systems integration and test contract, select components, and development of associated ground support equipment. MX was designed to use a cold launch system much like the submarine-based Polaris missile. Six launch canisters were built by Hercules in Utah, and were the largest filament-wound graphite structures in the world at that time. The canister technology led to filament-wound boosters for the U.S. space transportation system (STS) shuttle boosters.

When stored in its launching canister and launched by the high-pressure gas jet, the missile was held away from the canister walls by many small *adaptor plates* mounted to the side of the missile. When the canister was cleared, these adaptor plates fell away. Due to inconsistencies in ICWG control activities, not all missile stages were the same diameter. This resulted in one solid stage having a different pad configuration than the others (Figure 11.3).

Propulsion:
First stage—Thiokol solid-fuel rocket motor; 2225 kN thrust
Second stage—Aerojet solid-fuel rocket motor; 1225 kN thrust
Third stage—Hercules solid-fuel rocket motor; 290 kN thrust
Post-boost stage: Rocketdyne restartable liquid-fuel rocket motor
Diameter: 2.34 m
Length: 21.6 m
Weight: 87,300 kg (at launch)
Speed: 24,100 km/h
Range: 10,900 + km
Service ceiling (Apogee): 800–1000 km
Armament: 10 × Mk.21 MIRV
Cost: $70,000,000
First launch: June 18, 1983
Initial operational capability: December 1986
Last launch: July 7, 2004
Retired in 2005

All figures are approximate.

FIGURE 11.3 Missile X (also known as Peacekeeper.)

FIGURE 11.4 Mars climate orbiter.

11.3.2 Mars Climate Orbiter

The Mars climate orbiter (Figure 11.4) for studying the climate of Mars was launched on December 11, 1998. On September 23, 1999, NASA lost contact.

The peer-review findings indicate that the ground-based computer software used English pound-seconds whereas the contract for spacecraft development specified metric newton-seconds. The orbiter approached Mars from an altitude of 60 km instead of 150 km. The €109,170,000 spacecraft was placed on a trajectory that brought it too close to the planet and it was torn apart in the Martian atmosphere.

11.3.3 Apollo and Lunar Module CO_2 Filters

The same Apollo 13 failure that placed astronauts Fred Haise, Jim Lovell, and Jack Swigert in jeopardy led to the discovery of an unknown ICWG failing related to the lithium hydroxide CO_2 filters. Apollo 13 had been designed to use square filters, while the Aquarius lunar module had been designed to use circular ones. This inconsistency in design never surfaced as a concern until it was too late. The oxygen tank explosion in the service module forced the crew to move into the lunar lander. This resulted in one of the greatest engineering workarounds in NASA history (Figure 11.5). The workaround to make the oblong Apollo 13 filters work in the Aquarius lunar module required liquid-cooled garment bags, hoses from the red suits, bungee cord, duct tape, and socks from the red suits.

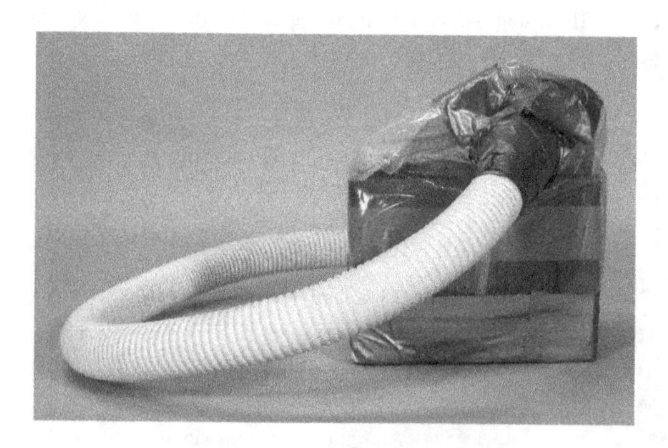

FIGURE 11.5 Apollo to lunar module lithium hydroxide filter workaround. (Smithsonian Air and Space Museum photo)

11.4 MILESTONE SLIPS BECAUSE METRICS WERE NOT USED TO DETERMINE WHERE A PROJECT WAS

11.4.1 Throwing Decisions Over the Fence

The development of a new classified instrument was being done on a very tight schedule in an integrated project team environment. The program had been aggressively bid with very low profit margins and it was critical that all changes be reviewed and approved in a less than a five-workday period. The average number of reviewers on all changes for the program was set at six by the program manager. No metrics were put in place to track ECO approvals due to the five-workday mandate. The development team was rapidly approaching a major review when it was discovered that design of several key components was stalled due to lack of change approvals.

Analysis revealed that over 30 engineering changes had been in signature an average of 58 workdays. A closer analysis of each change determined that delay was caused by an approval waterfall effect. A change in signoff for 118 days for a gasket was held for lack of agreement regarding what material should be used. The decision bounced between the designer and the materials and processes group on roughly two-month centers. Approvers would sign off "contingent" on further review by someone in another department. Over 48 people had reviewed the change and no decision had been reached by the time we were made aware of it. Some reviewers had contingently approved multiple times. The part's material characteristics drove its design and size. That size drove the next higher assembly's design and size. Without a decision, the other 29 other changes could not move forward.

Once metrics were developed and implemented, multiple other assemblies were also found to be at risk of meeting their design complete dates. These assemblies were not associated with a single design element such as mechanical or electrical but instead were found across all disciplines.

11.4.2 Company-Level Procedural Approval Holds Up Multiple Programs

To accommodate a wider variety of customers, a company decided to redefine its quality management processes to allow for multiple levels of quality, safety, and reliability in its products based on customer needs. Rather than designing and manufacturing products with only the highest quality components, the company could reduce costs on products with lower risk profiles or a shorter expected operational life.

All impacted policy and procedure changes were agreed to. A set of redlined documents was produced and engineering change requests (ECRs) were created. The generation of the ECOs was turned over to the owning departments. The bid and proposal department proceeded with bidding new work against the policy and procedure redlines in anticipation of all changes being approved and implemented within 20 workdays.

The company was awarded multiple contracts three months later for products based on a lower level of quality and the use of materials with a shorter lifespan. The contracts were approved by the contractor's vice president based on his being told that the policy and procedure changes had been made. During an external audit, it was found that one department had taken no action to start the ECO approval. Resolution of the finding that work was being done other than as documented revealed that a transition from a retiring department manager to the successor had taken place at the time of the ECR generation. The new manager was not aware that any actions were needed.

Reviews on the new awards were delayed for a period of 40 workdays while the missing ECO was approved and people were trained. The late implementation of the policy and procedure changes required documentation generated by the department on the impacted contracts to be resubmittal for customer approval prior to Heritage, PDR, and CDR reviews taking place. This caused program delays and fee reductions.

11.5 INORDINATE CHANGE PROCESSING DURATION AND NUMBER OF CHANGES

11.5.1 Low Temperature Microgravity Physics Facility

The low temperature microgravity physics facility (LTMPF) was being developed by NASA to provide a long-duration, low temperature and microgravity environment on the International Space Station (ISS) for fundamental physics investigation.

LTMPF (Figure 11.6) started in 1995 as a joint Jet Propulsion Laboratory (JPL), Ball Aerospace, and Design Net Engineering effort funded by NASA. The system as originally conceived was for a dewar that provided two experiments at liquid helium cryogenic temperatures for up to five months at a time to be mounted on the Japanese experiment module–exposed facility (JEM-EF).

During the summer of 2002, the initial flight of the LTMPF was delayed from late 2005 until early 2008 in the ISS manifest due to LTMPF budget overruns. Further delays occurred because of the Columbia accident grounding the shuttle fleet. The LTMPF continued development for a launch readiness now scheduled for early 2008.

Excessive change traffic and concept redefinition contributed to cost growth and eventual descope of the project. The obround (oblong) dewar had been completed, but much of the remaining activity was pulled back by JPL due to an experiment swap resulting in significant modifications to the LTMPF design from the previous functional requirements. On September 9, 1999, the LTMPF completed a successful requirements definition review.

Changes made included a redesign of the dewar vent line to lower the on-orbit temperature of the liquid helium bath from 1.6 K to 1.4 K, significant wiring modifications to accommodate the microwave drives needed for the superconducting microwave oscillator (SUMO), reconfigured electronics boxes, and a redesigned magnetic shield to meet SUMO's strict requirements. Additionally, work was being spent on defining how SUMO and the LTMPF would interface with the primary atomic reference clock in space experiment. LTMPF was eventually canceled due to the remanifestation of the last STS payloads. The lack of a clear set of design criteria and multiple iterations of the functional requirements drove the LTMPF costs and may have contributed to the decision to cancel and not to manifest to a different launch vehicle.

11.6 LACK OF CHANGE IMPLEMENTATION AFTER APPROVAL

11.6.1 Apollo 13 Oxygen Tanks

Apollo service module cryogenic oxygen tanks were well insulated and could hold several hundred pounds of liquid oxygen for breathing, electrical production, and water generation. The heater and protection thermostats were originally designed for the command module's 28 V DC bus. A design

FIGURE 11.6 Low temperature microgravity physics facility (courtesy NASA).

change to electronics instrument monitoring led to a voltage change for which original parts design was not compatible. The specifications for the heater and thermostat were later changed to allow use with a 65 V ground supply so tanks could be quickly pressurized. The tank contractor did not ensure that previously delivered units were modified to handle the higher voltage.

The shelf carrying the Apollo 13 oxygen tanks was originally installed in the Apollo 10 service module but was removed to fix a potential electromagnetic interference problem. During removal, the shelf was accidentally dropped about 5 cm because a retaining bolt had not been removed. The oxygen tank appeared to be undamaged, but a loosely fitting filling tube was apparently compromised. Photographs suggested that the close-out cap on the top of the tank may have hit the fuel cell shelf. After the oxygen tank was filled for ground testing, it could not be emptied using the drain line. To avoid delaying the mission by replacing the tank, the heater was connected to a 65 V ground power to boil off the oxygen. Jim Lovell signed off on this procedure. It should have taken a few days at the thermostat opening temperature of 27°C. However, because the tank had not been modified to accept 65 V, when the thermostat opened it fused the relay contacts in the closed position and the heater remained powered.

This raised the temperature of the heater to an estimated 540°C. A chart recorder on the heater current showed that the heater was not cycling on and off, as it should have been if the thermostat was functioning correctly. No one noticed it at the time. Because the temperature sensor was not designed to read higher than 27°C, the monitoring equipment did not register the true temperature inside the tank. At 540°C the gas evaporated in hours rather than days.

The sustained high-temperatures also melted the Teflon insulation on the fan power supply harness, causing the wires to be exposed. When the tank was refilled with oxygen, it became a bomb waiting to go off. During the Apollo 13 *cryo stir* procedure, fan power passed through the bare wires apparently shorted, producing sparks and igniting the Teflon. This in turn boiled liquid oxygen faster than the tank vent could remove it (Figure 11.7). The investigation board did recreate the oxygen tank failure; it did not report on any experiments that would show how effective the cryogenic malfunctions procedures were to prevent the system failure by de-energizing the electrical heater and fan circuits.

Oxygen tank 2 was not the only pressure vessel that failed during the Apollo 13 mission. Prior to the accident, the crew had moved the scheduled entry into the lunar module forward by three hours. This was done to get an earlier look at the pressure reading of the supercritical helium tank in the lunar module (LM) descent stage, which had been a suspect since prelaunch. After the abort decision, the helium pressure continued to rise and the mission control predicted the time that the burst disc

FIGURE 11.7 Apollo 13 service module after separation (NASA photo).

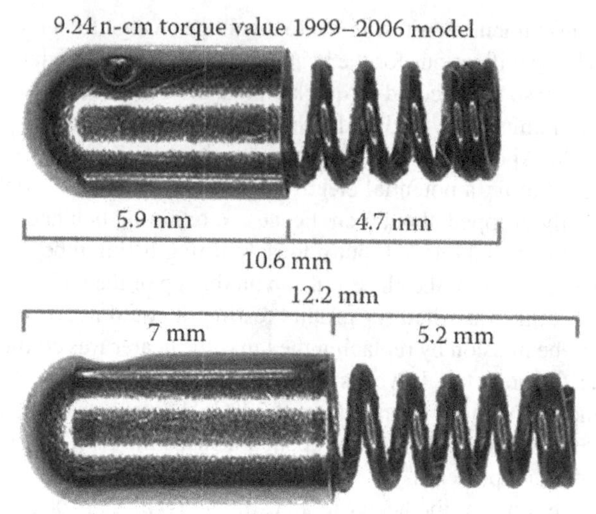

FIGURE 11.8 General Motors' detent plunger for switch 10392423.

would rupture. The helium tank burst disc ruptured at 108:54, after the lunar flyby. The expulsion reversed the direction of the passive thermal control roll (nicknamed the "barbecue roll").

11.7 LACK OF CONTROL OF ITEM IDENTIFICATION

11.7.1 General Motors Corporation Switch Recall

On June 30, 2014, General Motors announced a massive worldwide recall for ignition switches in millions of vehicles manufactured during the previous five years. The lack of adequate detent plunger force could cause the ignition key to rotate from the run to the accessory position. This problem was known as early as 2001.

The switch in question was part number 10392423. The force exerted by the detent plunger pressure at 9.24 n-cm was found to be insufficient for its intended purpose. A service advisory was issued in 2005 and a replacement detent plunger increasing the torque value to 17.57 n-cm was designed in 2006. The design change was approved and incorporated into all new switches manufactured, but the part number of the switch was not changed. Unfortunately, as with the Apollo 13 oxygen tanks, no retrofit activity took place on the previously produced switches and as the switch part number did not change, there was no straightforward way to determine which vehicles had been made with the more robust switches. Over a dozen people died as a result (Figure 11.8).

The issue ties directly to a lack of proper part identification. No two parts should ever have the same item identification if they are different, and differences in part identification roll up to the level of interchangeability.

11.8 UNDOCUMENTED CHANGES OR PART SUBSTITUTION IN FINAL PRODUCT

11.8.1 Solar Maximum Mission

In January 1981, three fuses in the solar maximum mission's (SMM) attitude-control system failed due to a voltage spike. This necessitated the SMM to rely on the magnetorquers to maintain attitude. This use of the magnetorquers prevented the satellite from being used in a stable position and caused it to wobble in its orbit.

FIGURE 11.9 Solar Max and STS-41-C (NASA photo).

The April 6, 1984, STS-41-C mission successfully captured the SMM, replaced the entire attitude control system module and the electronics module for the coronagraph/polarimeter instrument, and installed a gas cover over the X-ray polychromator (Figure 11.9).

It was not without issues. A special capture device called the "trunion pin attachment device" mounted between the hand controllers of the manned maneuvering unit (MMU) was built to mate with the trunion pin on the spacecraft. It failed when its trigger could not be activated due to interference caused by an undocumented grommet put on the spacecraft prior to launch in February 14, 1980. This jeopardized the STS-41-C mission timeline and astronaut safety as the spacecraft had to be manually captured.

11.8.2 Drop-In Replacement Sometimes Not So

We have witnessed instances where form, fit, function, and quality requirements were violated for what was advertised as a drop-in replacement. Failure analysis indicated insufficient configuration identification and change control. Form, fit, function, and the quality of any substitute part must be thoroughly vetted against the part specified in the engineering. The more critical the subassembly or component, the closer the scrutiny. The automotive industry uses a production part approval process (PPAP) to successfully launch vehicle projects and to manage the changes to the product. The PPAP contains a list of 19 documents or activity areas:

1. Design records.
2. Engineering change notes.
3. Engineering approval.
4. Design failure mode effects analysis.
5. Process flow.
6. Process failure mode effects analysis.
7. Control plan.
8. Measurement systems analysis.
9. Dimensional results.
10. Material performance tests.
11. Initial sample inspection.

12. Process study.
13. Laboratory certifications.
14. Appearance approval report.
15. Sample production parts.
16. Master (golden) sample.
17. Checking aids.
18. Customer-specific demands.
19. Part submission warrant.

Not all of these apply to every engineering change. CM implementation should ensure that a comprehensive approach or measure of due diligence is applied. There are five PPAP submission levels.

Level 1: Warrant submission only.
Level 2: Warrant submission and physical parts with some supporting data associated with the change (e.g., product change or process change).
Level 3: Warrant submission, physical parts, and complete supporting data (all prescribed documentation).
Level 4: Warrant submission, physical parts, and any other customer-identified requirements deemed necessary to mitigate the risks associated with the change.
Level 5: Warrant submission and physical parts with complete supporting data reviewed at the supplier's manufacturing location (usually for process changes and implications).

PPAP is predicated on customer awareness of engineering and process changes. Experience suggests interchangeability issues arise even in industries where the PPAP process is well established, as the front-line engineers' level of prowess may be insufficient for success. Success is dependent on comprehensive stakeholder change reviews and interdepartment communication.

11.9 LACK OF PROPER CONFIGURATION CONTROL BOARD STAKEHOLDER REVIEW

11.9.1 Harness Support Bracket Redesign Without Harness Team Input

An automotive development project with a multinational product development team was working toward a common target. Periodic convergences using prototypes took place to ensure that interfaces were maintained. Prototype parts that did not merge well indicated a lack of stakeholder inclusion during engineering change approval. In some cases, intradepartmental communication also appeared to be lacking. Analysis revealed that design engineers at each of the sites failed to coordinate changes within their own group.

Requirements at start of design had been baselined and properly allocated. Yet, as the project moved forward, the team focus started to drift, resulting in a bracket redesign change driven by the structures team with no stakeholder review by the wire harness group. The bracket existed solely to support the wiring harness. The alteration made it unsuitable for its intended purpose (Figure 11.10).

11.9.2 Small ICBM Internal Shelf Design

The U.S. small ICBM (intercontinental ballistic missile) project (Figure 11.11) was originally conceived as a mobile ICBM in response to the Soviet development of the SS-24 rail mobile ICBM and

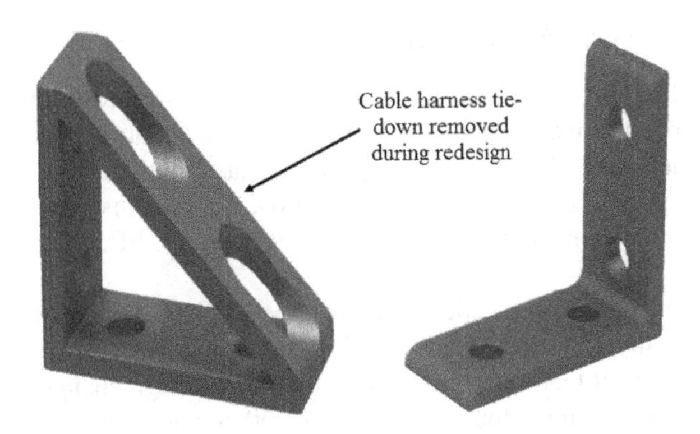

FIGURE 11.10 Harness bracket redesign.

the SS-25 road mobile ICBM. During the developmental process, it was determined that a slight increase in the missile diameter would provide the necessary geometry to accommodate a different payload configuration. The change to the missile was negotiated and the necessary redesign was started.

Inherent in the missile design was a shelf attached to the interior of the missile and used to hold guidance control and other components. An analysis performed on the larger diameter impacts by the shelf design team indicated a need to make the shelf more ridged. This was solved by increasing the number of mounting holes between the payload stage and the propellant stage below it. The design change was not coordinated with the vehicle structures and propellant stage design teams, who continued to develop a larger diameter missile with the same number of stage-to-stage mounting holes as the small diameter design.

The CM manager noticed the design disconnect before hardware had been manufactured and the program CCB was expanded to include the entire design team and not just the individual design cells. The first successful test flight took place on April 18, 1991. The program was canceled in 1992. The small ICBM graphite fiber-wound solid booster technology was leveraged forward on the side boosters of the Delta and the Orion stages of the Pegasus air launch missile.

FIGURE 11.11 Small ICBM (U.S. Air Force photo).

11.9.3 Informal Change Management

A project's technical requirements were to develop a new electronic architecture for heavy vehicles. This architecture change consisted of a suite of electronic control units (ECUs) connected via a data link sharing network. The company employed a stage gate method of project development using an iterative prototype parts and software approach. Problems surfaced when two entities delivering subsystem components were found to be working to different technical and interface requirements.

Requirements verification efforts identified that the product did not perform as expected or documented. Additionally, functions that worked in the previous iteration of the system no longer worked. One of four project team members had made small design changes, neglecting to involve team members. The parts from the four teams came together, and nothing fit or worked as planned.

11.9.4 Open Loop and Stove Pipe Changes

Issues were found during prototype fit checks and interoperability testing on a new vehicle platform. They were eventually traced back to designs not being checked out from or back into the design vault as the interfacing parts evolved. Not vaulting and merging developmental modifications into the digital model prohibited access by other teams to the latest design information. Design solutions evolved independently caused a six-week development delay and costs being spent on prototype parts that had little hope of interfacing properly when assembled.

11.10 HIGH RATE OF SCRAP AND REPAIR

11.10.1 Untested Item Substitution

In early 2009, a major electronics firm producing an environment-resistant printed wiring assembly (PWA) for a new product line suddenly started experiencing issues in its conformal coat processes at its main conformal coat facility. The issue persisted until 90% of the conformal coat was bubbling during the cure process and, after cure, the coating would detach during vibration testing. At the time, 10% of the board build and conformal coat had been offloaded to a subsidiary. All conformal coating was being done using the same processes and equipment yet only the boards conformal coated at the subsidiary passed inspection. An evaluation of process records at both conformal coat facilities showed that the processes had been adhered to, humidity levels were acceptable when the conformal coat was applied, and bake temperatures and time were adhered to. Process engineering was stumped.

It was discovered that, to save costs, the supply chain had found a second source for conformal coat material and issued it to the factory floor at both locations. The new material was administratively added to the database of authorized part substitutions without first being analyzed by parts materials and processes and validated against existing conformal coat procedures. The subsidiary still had stores of the older material in stock and had not yet started using the new material.

The estimated cost impact to rework the PWAs and tear down the main conformal coat facility to remove all traces of the new material exceeded €1,400,000. The root cause was determined early enough that the new materials were pulled from the subsidiary before it could be contaminated as well. The secondary facility was directed to return the new material to the vendor and replacement stock was ordered from the previous supplier.

11.10.2 Drug Manufacturing Issues

In 2018, several blood pressure medicines known as angiotensin II receptor blockers were recalled due to contamination errors. The recall targeted a discrete set of manufacturers. Among the drugs involved were valsartan and losartan. The recall resulted in patients' having either to change their blood pressure treatment regimen or to pay increased prescription prices due to scarcity of product from other manufacturers.

11.11 MAINTAINING SOFTWARE AND ITS ASSOCIATED DESIGN DOCUMENTATION

11.11.1 Source Code and Compile

Another piece of the CM pie is the ability to move backward and forward from revision to revision. A software engineer incrementally developed the features for the product, which were reviewed at established milestones. Software was compiled and burnt into microcontrollers. A series of tests were conducted to ascertain the level of compiled code functionality. The engineer made note of the things that worked and those that did not. The compiled code was placed in a secure area, guaranteeing the ability to return to the level of functionality that had been tested. Work continued, correcting the known bugs and adding the next phase of content. The updated source code was compiled and burnt into another microcontroller. When it was tested, nothing worked. The system did absolutely nothing.

The engineer had failed to vault the source code and had only vaulted the conversion of the binary file into assembly language. It took considerable time to perform a disassembly analysis of both compiled versions to determine root cause. It was probably a suitable self-punishment for not checking the binary file into the software CM system.

11.11.2 Merge Failure

Another problem area is merging product sub-tier elements. It appeared to be more of an issue for software products, but it was not software exclusive. Due to the multiplicity of cases discovered, only a summary will be provided. Development proceeds based on a defined set of requirements; subsystems or subroutines are developed by different teams. When requirement changes are not properly coordinated across all teams, some teams will create subsystems or subroutines that no longer interface with the rest of the development effort. This results in what is called a merge failure. CM implementation ensures that the content of subsystems or subroutines is documented and vetted across all development teams. When this is not taking place, the lack of functionality is discovered during testing or production, or when a customer states that the product does not work as advertised or as they expected it to perform.

11.12 LACK OF REQUIREMENTS CONTROL

11.12.1 Subassembly of Switch

Rocker switches are common to many automotive dashboard layouts. In one case, a switch supplier was working with the dashboard supplier to make switches with a look and feel unique to a new line of trucks. Two levels of testing were used.

1. The supplier would prove the switch could be moved from off to on and from on to off (one switch cycle) over the entire operational temperature range.
2. The customer would prove the same thing could be done when the switch was mounted to a dashboard prototype.

Supplier testing proved the switch could withstand a service life of 2,000 cycles across the specified thermal environments and shipped switches to the customer for prototype dashboard testing. A

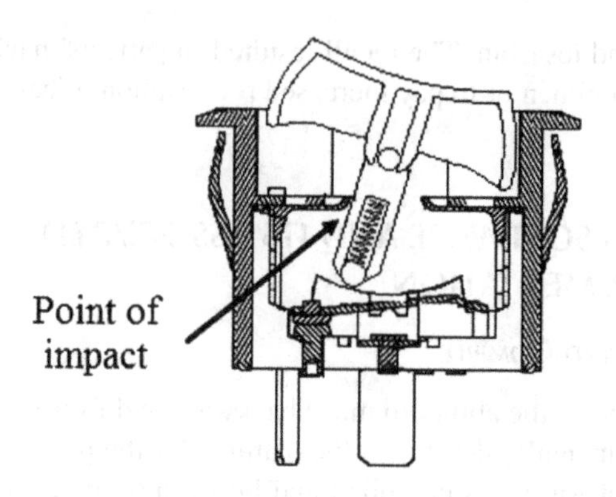

FIGURE 11.12 Rocker switch internal diagram.

standard rocker switch design was used. The switch button design included a tube on the bottom that housed the steel alloy spring, nickel-plated carbon steel roller ball (Figure 11.12).

The customer test team created a thermal test fixture that could cycle the switches while they were connected to a powered electrical circuit (e.g., under electrical load). The test setup used a dashboard prototype to ensure that the switch-activating mechanism did not push the rocker past the point where the switch clicked from the on to off position or the off to on position (e.g., past the switch point).

Thermal chamber testing started, and the switch failed in less than 24 hours. When the switches were disassembled, it was found that the spring tube broke away from the switch top, rendering the switch useless. During root cause analysis it was found that the thermal test fixture had used calculations from the previous iteration of the dashboard curvature. The concave curve of the new dash design was deeper than the last prototype tested, which resulted in the switch-activating mechanism putting too much pressure on the switch at either end of rocker travel. The excess pressure deformed in the housing each time the switch was cycled, bending the tube. The repeated bending caused the tube to break. Had the proper calculations been used, the switch-activating mechanism motion would have paralleled the new dash curvature and the switch would not have been overstressed.

11.12.2 Lack of Allocated Baseline Management

An automotive project to create a new vehicle was underway. The vehicle would build on an existing platform and the changes were largely on the interior, updated electrical and electronic systems, and an exterior facelift reflecting industry design trends. The company decided to develop all the modifications simultaneously. Unfortunately, neither the project nor the line organizations implemented robust change control for managing the requirements allocations. Instead, the allocations were documented in a manual log maintained in the lead engineer's notebook.

The lack of allocated baseline management caused great harm to project schedule, cost, and morale. This was compounded by a decision not to spend resources to create a digital twin of the components. Subtle changes in one area caused interference issues with other areas that were only uncovered during a vehicle prototype assembly. A meeting was held, and the manual log was converted to electronic format and placed under change control.

11.12.3 Design Drawing Updated From Wrong Revision

A modification to existing hardware was needed. A change to a small portion of the functional schematic defining the system needed to be made before other design groups could start their redesign activities. The systems engineering group was short on resources and was not able to make the functional schematic updates in time. Another group agreed to make the drafting changes. The change was approved. The update should have been made to the released Revision C in the PD/LM system. Unfortunately, a copy of Revision B was updated and released as Revision D. This undoing of all of the Revision C changes was not noticed for two weeks and caused further delays in the program's already tight schedule.

11.13 OPERATIONAL AND MAINTENANCE ACTIVITIES

11.13.1 Google Traffic Routed Through China

Nigeria's Main One cable company took responsibility for a malfunction that temporarily caused some of Alphabet Incorporated's Google global traffic to be misrouted through China, saying it accidentally caused the problem during a network upgrade. Main One said in an email that it had caused a 74-minute anomaly by misconfiguring a border gateway protocol filter used to route traffic across the Internet. That resulted in some Google traffic being sent through Main One partner China Telecom.[2] Similar misconfiguring of border gateway protocol filters caused multiple outages over the last decade, including cases where traffic from U.S. Internet and financial services firms was routed through Russia, China, and Belarus.

11.13.2 Airbus A380 Engine Failure

Air France flight AF66 from Paris was passing over Greenland, bound for California, when it was rocked by an uncontained engine failure. Pictures taken from inside the Air France plane showed the front cowling and fan disc of the No. 4 engine had completely sheared off (Figure 11.13). Fortunately, none of the 497 passengers was injured. The emergency had echoes of the November 2010 failure of a Rolls-Royce engine on a Qantas A380 after take-off from Singapore. That incident led to the grounding of all six A380s in the Qantas fleet for three weeks. On October 3, 2017, the Danish aviation authorities delegated the investigation to the Bureau d'Enquêtes et d'Analyses pour la Sécurité de l'Aviation Civile (BEA). Investigators from Denmark, the United States, and Canada joined the investigation. Since the incident occurred over Greenland, the Danish Accident Investigation

FIGURE 11.13 A380 engine failure (public domain photo).

[2] Lee, J. L., Dave, P. (2018), Reuters: *Nigerian firm takes blame for routing Google traffic through China,* https://www.reuters.com/article/us-alphabet-disruption/nigerian-firm-takes-blame-for-routing-google-traffic-through-china-idUSKCN1NI2D9, Reuters Technology News. Accessed November 13, 2018.

Board has jurisdiction over the investigation. On October 12, 2017, the American Federal Aviation Administration (FAA) issued an emergency airworthiness directive (EAD) affecting all Engine Alliance GP7270, GP7272 and GP7277 engines.[3]

11.13.3 Equifax Data Breach

In early March 2017, the Department of Homeland Security sent Equifax and other companies an alert about a critical vulnerability in the Apache Struts VCE-2017-5638 framework. A software patch was released on March 6. On March 15, Equifax's information security department ran scans that should have identified systems that were vulnerable to the Apache Struts issue but the scans did not identify any Apache Struts vulnerability. As a result, Equifax did not apply the patch to all its servers using the Apache Struts framework.

As early as May 13, Equifax was attacked. Equifax's security department observed suspicious network traffic associated with the consumer dispute website until July. The Equifax breach exposed sensitive data for as many as 143 million U.S. consumers. It was accomplished by exploiting a vulnerability in a vendor patch released more than two months earlier.

Equifax is not alone. Security researcher Troy Hunt says that more than 772 million email addresses and almost 22 million passwords have been exposed during a series of data breaches dating back to 2008.[4] Table 11.1 lists some of the biggest data breaches between 2006 and 2018. All the breaches can be tracked back to lack of software maintenance.

TABLE 11.1
Largest Data Breaches, 2006–2018

Year	Organization	Number of Accounts Compromised
2018	NASA	Unknown at time of publication
	Google	52.5 million
2017	Equifax	143 million
2016	Adult Friend Finder	412.2 million
2015	Anthem	78.8 million
2014	eBay	145 million
	JP Morgan Chase	76 million
	Home Depot	56 million
2013	Yahoo	3 billion
	Target Stores	110 million
	Adobe	38 million
2012	U.S. Office of Personnel Management	22 million
2011	Sony's PlayStation Network	77 million
	RSA Security	40 million
	JPL	150 only
2008	Heartland Payment Systems	134 million
2006	TJX Companies, Incorporated	94 million

[3] Jamieson, A. and Calabrese, E. (2017). BBC: *Investigation Launched Into 'Serious' Airbus A380 Engine Failure*, https://www.nbcnews.com/storyline/airplane-mode/investigation-launched-serious-airbus-a380-engine-failure-n806301. Accessed December 12, 2018.

[4] Hunt, T. (2019). https://www.troyhunt.com/the-773-million-record-collection-1-data-reach/, accessed January 17, 2019.

FIGURE 11.14 A400M (public domain photo).

11.13.4 A400M Software Vulnerability

An Airbus-built cargo and troop carrier crashed near Seville during a test flight in May 2015 after three out of four engines froze minutes after take-off. The crash killed four of the six crew members. Data needed to run three engines had been accidentally erased when the software installation initially failed, and those files were never restored in the subsequent uploading process. Pilots had no warning that there was a problem until the engines failed. Engine-makers had warned Airbus and the European Aviation Safety Agency (EASA) in October 2014 that software installation errors could lead to a loss of engine data. Installation errors meant that technicians might not receive warnings before take-off that a problem had occurred (Figure 11.14).

Misconfigured engines could only be operated at maximum power or idle. At take-off, the engines were frozen with the power at maximum. This caused the transporter to fly higher and faster. Trying to obey an order from controllers to stay at 457 m, the crew reduced thrust. This action placed the engines in idle mode. The crash is seen by some safety experts as an example of how failures, though rare, can occur in increasingly complex aircraft systems when several apparently minor weaknesses line up together to produce a serious risk. It was a CM implementation failure.[5]

11.14 POOR RISK MANAGEMENT DECISIONS

11.14.1 Ford Pinto

The Ford Pinto of 1971 through 1976 was Ford Motor Company's attempt to regain part of its small-car segment market share being taken by fuel-efficient imports (Figure 11.15). The Ford Pinto/Mercury Bobcat was greenlighted in 1968 and by August of that year the program was underway.

FIGURE 11.15 Ford Pinto (public domain photo).

[5] Helper, T. (2017): Reuters, Airbus knew of software vulnerability before A400M crash. https://www.reuters.com/article/us-airbus-a400m/airbus-knew-of-software-vulnerability-before-a400m-crash-idUSKBN1D819P, Reuters Business News. Accessed November 12, 2018.

Deep into the development cycle, it was discovered that positioning the fuel tank behind the rear axle and in front of the rear bumper was a design flaw. Rear impacts during slow speed tests showed the tanks were easily punctured by a bracket and differential bolts. In addition, the filler neck would tear away from the sheet metal tank. The fuel contained in the tanks would empty in less than two minutes, causing a serious fire risk.

The design could be fixed by leveraging the tank design from the Ford Capri and placing the tank above the rear axle. Ford performed a cost–benefit analysis and determined the design fix would cost €98,690,000 (engineering, test, production delay, and €9.61 per vehicle). If nothing was done, the risk team determined the liabilities would only cost €42,790,000 to €43,670,000. Full production started in 1970 with no change to the gas tank design. The €9.61 per vehicle pushed the price over the €1746.69 price tag cited in the design requirements.

An estimated 500 deaths and 400 other injuries were later linked to the faulty design. A 1979 landmark case, *Indiana vs. Ford Motor Co.*, made the automaker the first U.S. corporation indicted and prosecuted on criminal homicide charges. In 1978, Ford recalled 1.5 million Ford Pintos and 30,000 Mercury Bobcats following an investigation by the National Highway Traffic Safety Administration. There is no way to determine the actual cost impact to Ford as more than 100 lawsuits were filed and many were settled out of court. Later in 1978, General Motors recalled 320,000 of its 1976 and 1977 Chevettes for similar fuel tank issues.

11.15 INADEQUATE SUBSYSTEM TO SUBSYSTEM INTERFACE CONTROL

11.15.1 Chevy Monza

In an article published January 10, 1979, by the *New York Times*, it was reported that on January 9, General Motors admitted that the Chevrolet Monza 2-plus-2 262 V-8 engine sometimes had to be lifted as much as 1.27 cm to change the number three spark plug in tune-ups. The engine design placed the number three plug very close to the steering column. Cars tested at the GM proving ground did not appear to exhibit this issue and it was believed that there was a design change prior to production (Figure 11.16). Using today's digital twin modeling tools, the interface between the 262 V-8 engines and the steering column would have been identified during design rather than in the operations and maintenance cycle.

11.16 LACK OF DATA MANAGEMENT

11.16.1 In 2001, U.S. Congress Created November 31 by Mistake

Many of us learn the saying, "Thirty days hath September, April, June, and November. All the rest have 31, except for February, which has 28." Some of us have perhaps not learned it as well as we should. In 2001,[6] the U.S. Capitol Historical Society printed 650,000 copies of the "We, the People" calendars with 31 days in November. The December dates were advanced one day to account for

FIGURE 11.16 Chevy Monza (public domain photo).

[6] Wolf, B. ABC News (2002): *Congress Accidentally Creates Nov. 31*, ABC News https://abcnews.go.com/Entertainment/WolfFiles/story?id=91981&page=1. Accessed January 15, 2019.

FIGURE 11.17 Lion Boeing 737 MAX 8 (Photo by PK-REN from Jakarta, Indonesia - Lion Air Boeing 737-MAX8; @CGK 2018, CC BY-SA 2.0, https://commons.wikimedia.org/w/index.php?curid=73958203).

the extra day added to November. Most of the production run was distributed by elected officials to their constituents. When the mistake was pointed out by calendar recipients, the society compensated lawmakers and others who ordered the calendars by reducing the price from €6.07 to €2.58.

11.17 LACK OF CHANGE COMMUNICATION

11.17.1 Lion Air Flight 610

On October 29, 2018, Lion Air flight 610, a Boeing 737 Max 8, crashed into the Java Sea, killing all 189 passengers and crew aboard (Figure 11.17). The Federal Aviation Administration confirmed the crash was the first fatal accident involving any type of Boeing 737 Max. The aircraft included an automated system known as the Maneuvering Characteristics Augmentation System (MCAS), a new addition on the 737 Max 8 and 9 models that can pitch the plane's nose down without pilot input when sensor data indicates the possibility of a stall. A lack of pilot information and training on the flight control system is believed to have contributed to the crash. The pilots tried 26 times to counteract the MCAS nose down system before the plane crashed, but failed to follow a documented procedure for countering incorrect activation of the automated safety system. A little over a week after the crash, Boeing put out a bulletin advising airline operators on how to deal with erroneous sensor information that would lead to "uncommanded nose down" maneuvers, while the FAA ordered flight manuals to be updated with the process to follow in such a situation.

Another Boeing 737 Max 8, being flown by Ethiopian Airlines, crashed on March 10, 2019, under apparently similar circumstances, killing 157 people on board. After this second disaster in less than six months, many countries grounded the popular 737 Max until the issue could be investigated and fixed.

11.18 FAILURE TO UNDERSTAND ALL REQUIREMENTS

11.18.1 The U.S. Air Force Expeditionary Combat Support System (ECSS)

The U.S. Air Force Expeditionary Combat Support System (ECSS) project started in 2004. The stated purpose was to create a single engineering resource planning (ERP) system to modernize the Air Force global supply chain infrastructure. The new system would make obsolete more than 240 outdated systems. The ERP implementation would also meet a federal mandate to have an auditable financial record by 2017. The sophistication of the global system was complex. For some reason, firm fixed price contracts were used in an environment where risks were considerable due to the developmental nature of the ECSS charter when cost contracts were warranted. The failure to understand all requirements associated with the systems it would replace—along with ineffective governance, personnel churn, inconsistent acquisition guidance, and the disruptive nature of the change to a single ERP system—doomed the project. When the ECSS acquisition reached a cost of €911,699,350,000 (28 times more than any similar ERP implementation), the program was terminated.[7]

11.18.2 Inadequate Training of Those Performing CM Functions

CM implementation does not rest on a single department, or on the shoulders of a single individual or a single team of individuals. There is a symbiosis between all functional elements that collectively accomplishes CM implementation. Generally, functional implementation responsibilities are defined in guidance and requirements documents.

As products become more complicated and the boundaries between hardware, firmware, software, and biomechanical interfaces blur, CM implementation will become an increasingly critical component in every market sector. CM professionals will take on an increasingly significant role when orchestrating CM implementation across the company infrastructure. This includes defining boundaries/trigger points associated with product planning and management, configuration identification, configuration change management, configuration status accounting and configuration verification, and audit.

Regardless of how well trained the staff of any department called "configuration management" may be, it is incumbent on company management to train everyone at the firm in what their CM activities are and how their department or function relates to the larger sense of CM implementation. The errors identified in this chapter, when viewed retrospectively, stem from a general lack of understanding of how and what they were doing related to the ultimate CM implementation goal.

11.18.3 What Is the Point of Processing an Engineering Order?

A company manufacturing a series of assemblies with unique system architecture was making a block update to the design of the power distribution system. The power distribution system was tested using a specialized test set and with an operating system at V 5.1.

The engineer responsible for the software design (Fred) was being moved to a sister facility, and the development of the new operating system was transitioned to a highly skilled associate (George). Three years later, George was let go due to downsizing. His computer hard drive was reformatted for use by another person.

One year after George was let go, a new update to the power distribution system was in development and the operating system on the test set required a compatible upgrade. Software released in the product data/life cycle management (PD/LM) system was still at V 5.1. Software resident in the test equipment was at V 10.5. No documentation existed for any of the changes between V 5.1 and V 10.5. A set of compact disks were found in a box next to the test set containing the operating system V 6.0 through V 10.4.

The company released each of the versions found in a box in the PD/LM system to capture the unreleased design changes. ECOs for each version had to be *administratively* changed so the release dates matched the file date on the CDs. This was done so the date hierarchy in the PD/LM system was correct. Working with the software CM function, V 10.5 was downloaded from the test set so it could also be released in the PD/LM system. Eventually, software V 11.0 was released.

---

[7] Charette, R. N. (2013). The U.S. Air Force Explains Its $1 Billion ECSS Bonfire. 2013-12-06. IEEE Spectrum. https://spectrum.ieee.org/riskfactor/aerospace/military/the-us-air-force-explains-its-billion-ecss-bonfire. Accessed December 23, 2018.

The research and backtracking effort cost the company 300 hours. During the investigation, one engineer sitting nearby remembered that George had an engineering notebook. This was eventually found and the entry days after George was assigned the task read, "I don't see the point of processing EO for software used on test sets. Test software shouldn't be put under configuration change control."

11.18.4 Where Is That Written Down?

All design activities had been finished and the first set of final assemblies built and turned over to newly qualified test conductors. They were tasked with performing a 100,000-cycle life test. Initial results were encouraging, and preliminary results were provided by email to the program team each night. They simply stated, "All is nominal." No specifics were ever included. Every six hours, 1,000 cycles were run. Every six hours, testing was halted and units disassembled, inspected, and reassembled. This was the daily shift routine during the three-shift, 34-day test cycle.

Expected test results and parts wear had been projected for major assemblies with articulated interfaces based on design and stress analyses. Expected wear tables had been included as exhibits to the test procedures. Columns had been inserted so test conductors could write in the actual results at the end of each six-hour test run. What follows is the interaction between the head of systems test (HST) and the test conductors (TC).

HST: We need you to distribute the plot of test results against expectations.

TC: Do what?

HST: We would like to see the actuals graphed against the predictions. We need to adjust our digital twin models to correlate with the observed data and have the back-up available to support functional audit activities.

TC: Like that's ever going to happen! When the units passed the life test we shredded the as-run test procedure!

The life test had to be re-run, requiring the fabrication of a new set of final assemblies and another 34-day test cycle. The impact was over €720,500. CM implementation was modified to require inclusion of the objective of the test in all procedures.

11.18.5 No Records to Combat Environmental Violation Allegations

In 1999, a major manufacturing facility was being sued by a small community located downstream who claimed that factory pollution of the groundwater was the cause of the high degree of a rare disease in their community when compared to surrounding communities. The facility had been in operation for over 50 years and had been on the environmental watch lists of several proactive environmental organizations.

Prior to the enactment of country, state, or local environmental regulations, the company's stance on environmental issues had been rather lax. In 1960, it drilled test wells along the perimeter of its holdings and ran extensive testing for all chemicals and materials used at the plant. These tests were negative. Not being sufficiently satisfied with the due diligence, the company also drilled wells along the perimeter of each building that either housed or used chemicals for manufacturing. Results were positive for contaminants associated with electroplating, anodizing, and beryllium milling and machining activities. Contamination was also found in its wastewater processing facility as well as in the sewage pipes that tied into the municipal wastewater disposal system. The wells also showed trace contaminants from an upslope source but, as these were not from anything used by the company, no mitigation efforts associated with those pollutants were made. No one at the company in 1960 and involved in this situation was still employed there in 1999.

Extensive mitigation efforts in 1999 included discontinuing all on-site manufacturing of items from beryllium and changing over the electroplating and anodizing processes to what were then considered state-of-the-art eco-friendly methods and materials. Wastewater processing facility upgrades and new piping connecting the facility to the municipal wastewater disposal system were installed. Testing continued until the date of the lawsuit and no other contamination from materials used at the plant from 1960 to 1999 was found.

As the lawsuit proceeded, all environmental records were subpoenaed by the prosecution but only the records from the last five years were located. Local medical authorities were informed, and environmental organizations accused the company of obstruction. As it turned out, all the company's records older than five years had been destroyed in an uncoordinated cost-cutting effort by the facilities department. Eventually, the company that performed the testing and mitigation activities was identified, and fortunately was able to locate the records for use by both the defense and the prosecution. The records proved that the contamination source was not the company in question but rather an old factory located 20 km higher in the terrain, which had been acquired by the city years earlier.

The government was satisfied that the company was a good environmental citizen and started extensive contamination mitigation at the old factory, which used to produce enriched radioactive materials for the country's nuclear efforts. Had there not been a solid link between supply chain management and CM, the company that performed the mitigation could not have been contacted. CM had the records of the report numbers and a copy of the contract in its PD/LM database.

After the lawsuit, CM was tasked with ensuring that all records of environmental activities be located, scanned, and attached to the CM records associated with the vendor in the PD/LM system.

12 Test, Inspection, and Evaluation Master Plan Organized

Ensuring product quality is not accomplished solely through testing and verification activities. Testing is but a fraction of the techniques that are at an organization's disposal to improve its development quality. Good planning of the product incarnations—that is, a phased and incremental delivery of the feature content—makes it possible for an organization to employ tests, inspections, analysis, and demonstrations as tools for creating a competitive advantage. To really improve (and prove product quality and safety), a more comprehensive approach is required.

The Test, Inspection, and Evaluation Master Plan Organized (TIEMPO) was developed by Jon M. Quigley and is included for those not familiar with it. The TIEMPO adds an extra method to the Test and Evaluation Master Plan (TEMP)[1] to support product quality. TIEMPO expands the concept of a TEMP by focusing on staged deliveries, with each product/process release being a superset of the previous release. Each package is well defined; likewise, the test, inspection, and evaluation demands are well defined for all iterations. Ultimately, the planned product releases are coordinated with evaluation methods for each delivery. Under TIEMPO, inspections include:

- Iterative software package contents
- Iterative hardware packages
- Software reviews
- Design reviews
  - Mechanical
  - Embedded product
  - Design failure mode and effects analysis (DFMEA)
  - Process failure mode and effects analysis (PFMEA)
  - Schematic reviews
  - Software requirements specification reviews
  - Systems requirements reviews
  - Functional requirements reviews
- Prototype part inspections
- Production line process (designed)
- Production line reviews
- Project documentation
  - Schedule plans
  - Budget plans
  - Risk management plans

12.1 PHILOSOPHY OF THE MASTER PLAN

At its core, TIEMPO assists in coordinating product functionality improvement. Each package has a defined set of contents, against which our entire battery of quality safeguarding techniques is deployed. This approach of defined builds of moderate size and constant critique has a very nimble

[1] Specified in IEEE 1220 and MIL-STD-499B [Draft]. MIL-STD-499A was canceled and MIL-STD-499B was never released.

character, allowing for readily obtainable reviews of the product. TIEMPO reduces risk by developing superset releases where each subset remains relatively untouched. Most defects will reside in the modified portion of the design. The previously developed part of the product or process becomes a subset proven to be defect free. Should the previous iteration contain unresolved defects, the opportunity exists between iterations to correct those defects. Frequent critical reviews are used to guide design maturity and to find faults. Frequent testing facilitates quality and reliability improvements, generating data used to assess the product readiness level in anticipation of product launch.

12.2 BENEFITS OF THE MASTER PLAN

Experience suggests the following benefits from using TIEMPO.

- Well-planned functional growth in iterative software and hardware packages.
- Ability to prepare for test (known build content), inspection, and evaluation activities based on clearly identified packages.
- Linking test, inspection, and evaluations to design iterations (eliminate testing or inspecting items that are not there).
- Reduced risk.
- Identification of all activities to safeguard the quality—even before material availability and testing can take place.
- Ease of stakeholder assessment, including customer access for review of product evolution and appraisal activities.

An overview of one approach experience also indicates that at least 15% of the time associated with downstream troubleshooting is wasted in unsuccessful searches for data, simply due to lack of meaningful information associations with the developmental process baselines. The TIEMPO approach eliminates this waste, ferreting out issues earlier in the process and allowing more dollars for upfront product refinement. TIEMPO need not be restricted only to phase-oriented product development, but any incremental and iterative approach will be the beneficiary of a constant critique, including entrepreneurial activities. The way each of these pieces fit together is discussed below.

12.2.1 System Definition

The TIEMPO document and processes begin with the system definition. Linked to the CM implementation, TIEMPO describes the iterative product development configurations that build up to final product functionality. In other words, each incarnation of the product was incorporated in the next package delivery. The plan describes functionality growth as development matures from little content to the final product. Each package has incremental feature content and bug fixes from the previous iteration. By defining the process upfront, testing, inspection, and evaluation activities are linked by a digital thread not only to an iteration but to specific attributes of that iteration, and they capture it in an associative data map in the PD/LM system. In the example of testing, specific test cases to be conducted are defined by mapping the product instantiation with the specifications and ultimately to test cases. As a result, a planned road map of the product development is generated. As things change, the TIEMPO document is updated, reviewed, and approved by all stakeholders, and released in the PD/LM system.

12.2.2 Test (Verification)

Test as a verification method consists of activities determining whether the product meets the technical requirements. If an incremental and iterative testing approach is applied, prototype parts are constantly compared against requirements. The parts will initially be made using subtractive or

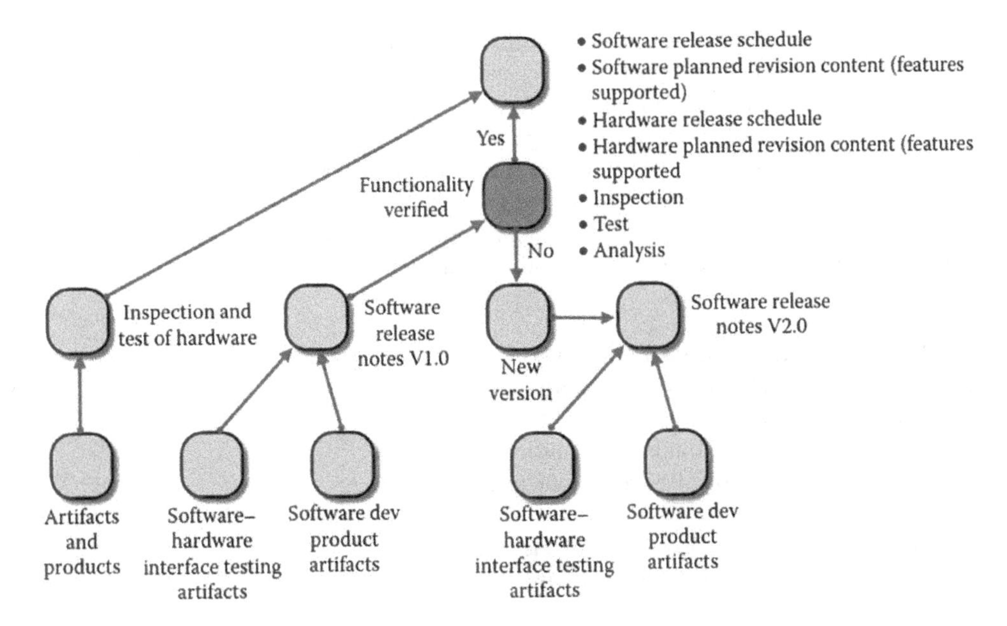

FIGURE 12.1 Testing relationships in a typical software–hardware release plan.

additive manufactured prototypes. As development progresses toward the final design, the prototypes will be replaced with production parts. Prototype parts may not have the same durability as production parts, but they should represent the shape and feature content of the production part. Prototypes reduce risk and cost by allowing us to learn as much as we can along the way about the future safety and quality of the resultant product. It is obvious how testing fits into TIEMPO, however, there are some less obvious opportunities to apply TIEMPO. Can inspection as a form of verification be used? Is the testing scope correct? Will testing stress the product in a way not intended? The TIEMPO can verify software structure long before it is executable. The TIEMPO activity refines the testing scope, test cases, and proposed testing to those needed to meet allocated requirements. It also allows introduction of exploratory-based testing. Testing relationships in a typical software/hardware release plan are shown in Figure 12.1.

12.2.2.1 Reliability Testing

In the case of reliability testing, the probable quality behavior of the product or process over some duration must be tested. Finding failures in the field is a costly proposition, with returned parts often costing the producer to absorb from profit five to ten times the sales price, not to mention the intangible cost of customer dissatisfaction. For reliability testing using combined Weibull and Bayesian analytical techniques, small sample sizes are used when a configuration baseline has been established, and larger sample sizes when there is not a configuration baseline. Physical models are used for accelerated testing to compute probable product life. The TIEMPO document, along with what specific packages (hardware/software) are used to perform this activity, specifies the test methodology. Development and reliability testing are linked together via a digital thread in the PD/LM system.

12.3 SO, WHEN DO WE START TESTING?

Many may believe that it is not possible to test a product without some measure of hardware or software samples available. It is possible to test if we have developed simulators to allow us to explore the product possibilities in advance of the material or software. This requires accurate software models as well as simulation capability. To ensure accurate software models, we will run tests between our software model results and real-world results to determine the gap and make necessary adjustments to the software model. We may even use these tools to develop our requirements if sophisticated enough. These activities reduce the risk and cost of the end design because we have already performed some evaluation of the design proposal. As prototype parts become available, testing on these parts is done alone or in concert with our simulators. If we have staged or planned our function packages delivered via TIEMPO, we will test an incrementally improving product.

There are four stages to software model development. These have been found useful in identifying and resolving issues that affect safety, reliability, and quality.

- Model in the loop
  - Software simulates the environment.
  - Provides nominal functional capabilities evaluation.
- Software in the loop
  - Model is run against simulated hardware.
  - Issues with compiled software architecture are revealed.
- Processor in the loop
  - Model is run against software embedded in a processor.
  - Issues with compiled software and target processor are revealed.
- Hardware in the loop
  - Model is run against hardware instead of simulated hardware.

12.3.1 Types of Tests During Development

When we get into the heavy lifting of the product or service testing, we have a variety of methods in our arsenal. At this stage, we are trying to uncover any product maladies which impact neither us nor our customer. We will use approaches such as

- Compliance testing (testing to specifications)
- Extreme testing (what does it take to destroy and how does the product fail?)
- Multi-stimuli or combinatorial testing
- Stochastic (randomized exploratory)

12.4 REVIEWS

The goal of reviews is to find problems in our effort as early as we can. There may be plenty of assumptions that are not documented or voiced in the creation of these products. The act of reviewing can ferret out the erroneous, deleterious ones, allowing us to adjust. We can employ a variety of review techniques on our project and product, such as

- Concept reviews
- Product requirement reviews
- Specification reviews
- System design reviews
- Software design reviews
- Hardware design reviews
- Bill of materials review
- Project and product pricing review
- Test plan reviews
- Test case reviews
- Prototype inspections
- Technical and users' manuals
- Failure mode effects analysis (discussed further)

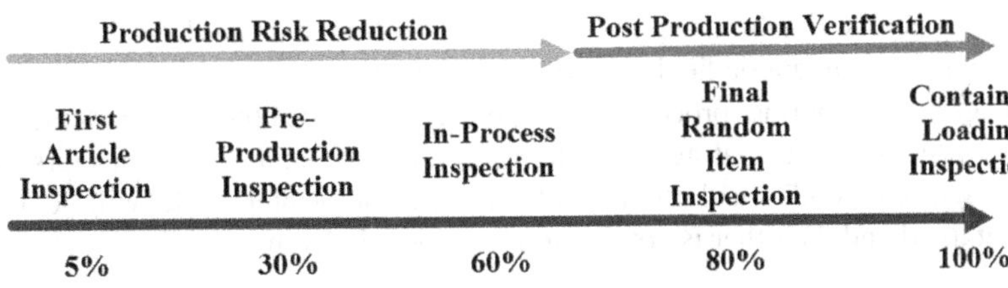

Production Risk Reduction → Post Production Verification →

First Article Inspection	Pre-Production Inspection	In-Process Inspection	Final Random Item Inspection	Container Loading Inspection
5%	30%	60%	80%	100%

FIGURE 12.2 Inspection timeline.

12.5 INSPECTIONS (VERIFICATION)

Inspections are an element of the verification of the finished product against the design. Testing begins after first article inspection is complete. "First article" may imply the first development unit, first production unit, or both (Figure 12.2).

We have listed five common inspection types below:

- First article inspection
  - Ensures engineering design requirements and finish specifications are properly understood and implemented.
- Pre-production inspection
  - Raw materials/components kitted for use, shop order task flows, material data safety sheets, etc.
- In-process (during production) inspection
  - Also called DUPRO or DPI.
  - May include dimensioning and tolerancing verification on the first units produced.
  - Issues found here may be reworked or require redesign.
- Final random inspection
  - Also known as pre-shipment inspection.
  - Takes place after 100% of the production run is finished and not more than 80% is packaged for shipment.
  - Often combined with DUPRO/DPI.
- Container loading inspection
  - Also called container loading supervision.
  - Ensures that all products to be shipped are in the shipping container.

With the advent of additive manufacturing technologies, the traditional use of a dimensioning and tolerancing laboratory for mechanical inspections is no longer possible. The traditional method uses coordinate measuring machines (CMMs) to measure points on the part to determine if it conforms to requirements. This presupposes the manufactured item can be mounted on the CMM for measurement. Measurements of an additive manufactured item must be done with the part being created in situ. Some additive methods are resulting in hybrid manufacturing, combining additive and subtractive capabilities within the same machine.[2]

When directed energy deposition (DED) and powder bed fusion (PBF) are involved, inspection of the complex geometries for residual stress, fatigue, or any other physical defect in the properties of the AM component must be accomplished using nondestructive methods. General dimensioning and tolerancing (GDT) and dimensional inspection are done in the model or continuously during the manufacturing process, not after. This may include using X-ray computed tomography or micro CT

[2] (2018): Inspecting Complex Geometries for 3D-Printed Parts, 2018-10-31, Additive Manufacturing Blog post https://www.additivemanufacturing.media/articles/inspecting-complex-geometries-for-3d-printed-parts. Accessed January 22, 2019.

scanning. There is no one finishing technology for 3D printing. Part hardness may also be different on 3D parts printed in a magnetic field vs. those that are not.

As technologies come online, processes need to evolve, as does the CM implementation associated with them. Be cautioned that items made using the same material on printers manufactured by competitors will not be identical. Items made on the same printer model will be different if one printer is maintained and the other is not due to metrology difference.

12.6 DEMONSTRATION (VERIFICATION)

Demonstrating that the product meets requirements is an element of verification. If the prototype part is of sufficient durability and risk or severity due to malfunction is low, its capabilities may be demonstrated to the customer. Their feedback is used to guide design refinement. There are other ways to employ demonstration. A run-at-rate production demonstration to prove the company can meet anticipated quotas may be performed. Demonstrations in one instance met with questions from the customer when issues surfaced. The small ICBM program concept required self-contained hard mobile launchers to cross varied terrain. During demonstration at the supplier's facility, one of the transporter tracks came off as the vehicle was traversing a slope at a 45-degree angle. The customer became upset because that operation was not a requirement. The supplier politely informed the customer that they had their own minimal performance requirements and would never build a tracked vehicle that wasn't able to perform in that mode (Figure 12.3).

12.7 ANALYSIS (VERIFICATION)

Often, it is not possible to prove that a requirement has been met through test, inspection, or demonstration. Sometimes this is caused by the inability to replicate the requirement environment. Designing a structure that will withstand a 1,000-year flood is an example (Figure 12.4). A 1,000-year flood has the probability of 0.1% that it will occur in any given year. Another example is the predicted failure rate of the final product based on known failure rates of its components.

FIGURE 12.3 Hard mobile launcher (U.S. Air Force photo).

FIGURE 12.4 Los Angeles River, March 2, 1938 (public domain photo).

12.8 DFMEA AND PFMEA

DFMEA and the PFMEA techniques employed by the automotive industry can be applied to any industry. These tools represent a formal and well-structured review of the product and the production processes. The method forces teams to consider the failure mechanism and the impact. If a historical record has been maintained, programs can take advantage of that record or even previous failure modes effects analysis (FMEA) exercises. There are two advantages; the first is the prioritization of severity. The severity is a calculated number known as the "risk priority number" (RPN) and is the result of the product of:

- Severity (ranked 1–10)
- Probability (ranked 1–10)
- Detectability (ranked 1–10)

The larger the resulting RPN is, the higher the severity. Highest severity items are prioritized first. The second advantage fits with the testing portion of the TIEMPO. The FMEA approach links testing to those identified areas of risk as well.

12.9 PRODUCT DEVELOPMENT PHASES

Usually there are industry-specific product development processes captured during the product life cycle.

- Concept
- System level
- Preliminary design
- Critical design
- Test readiness
- Production readiness
- Product launch

Expectations are to perform test, inspection, and analysis in each phase (Figure 12.5). The design aspects will apply to process design just as much as to product or service design.

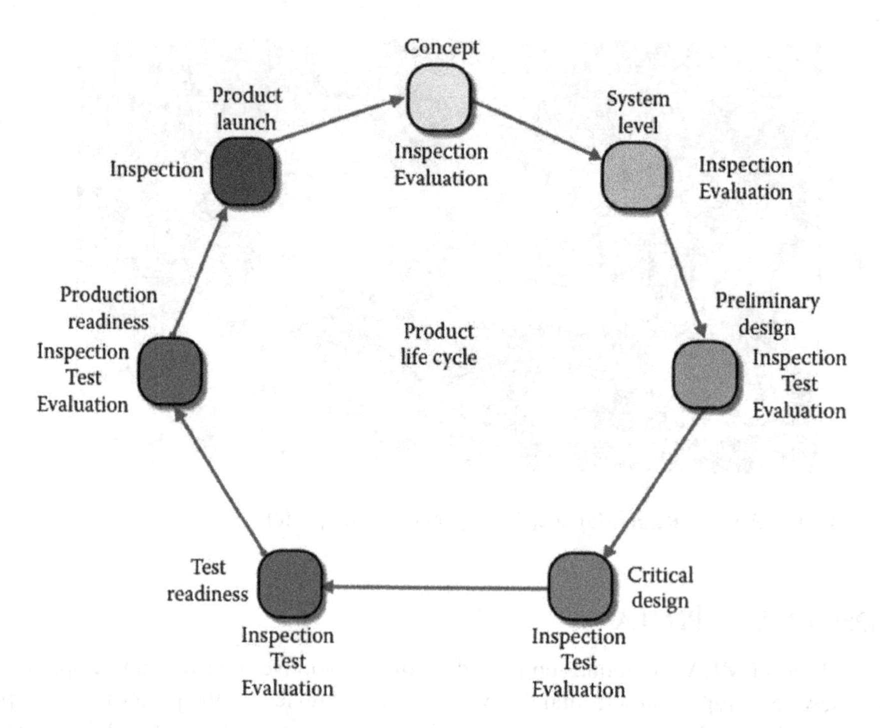

FIGURE 12.5 Test, inspection, and evaluation by phase.

12.10 SUMMATION

Test, inspection, analysis, and demonstration are instrumental to the successful launch of a new product. This is also true for a major modification of a previously released product or service. Test, inspection, analysis, and demonstration are necessary. Without them, verification and validation that the product or process meets requirements and is properly documented fail. The automotive approach has been modified and used in the food and drug industry as the hazard analysis and critical control point system. Critical control points are often inspections for temperature, cleanliness, and other industry-specific requirements.

The following is an outline for a TIEMPO document:[3]

1. Systems Introduction.
   1.1. Mission Description.
   1.2. System Threat Assessment.
   1.3. Minimum Acceptable Operational Performance Requirements.
   1.4. System Description.
   1.5. Inspection Objectives.
   1.6. Critical Technical Parameters.
2. Integrated Test, Inspection, and Evaluation Program Summary.
   2.1. Inspection Schedule.
   2.2. Integrated Test Program Schedule.
   2.3. System Evaluation Schedule.
   2.4. Management.

[3] Pries, K. J., and Quigley, J. M. (2012). *Total Quality Management for Project Management.* Auerbach Publications Taylor & Francis Group, Boca Raton, FL. Reprinted with permission from the Publisher.

3. Developmental Test and Evaluation Outline.
   - 3.1. Simulation.
   - 3.2. Developmental Test and Evaluation Overview.
   - 3.3. Component Test and Evaluation.
   - 3.4. Subsystem Test and Evaluation.
   - 3.5. Developmental Test and Evaluation to Date.
   - 3.6. Future Developmental Test and Evaluation.
   - 3.7. Live-Fire Test and Evaluation.
4. Inspection.
   - 4.1. Inspection Philosophy.
   - 4.2. Inspection Areas (documents, code, models, physical parts, material).
   - 4.3. Inspection of Models (model characteristics, model vs. real world).
   - 4.4. Inspection Material (physical parts).
   - 4.5. Appearance Criteria.
   - 4.6. Prototype Inspections.
   - 4.7. Post-Development Test Part Inspections.
   - 4.8. Inspection Documentation.
   - 4.9. Inspections Software.
   - 4.10. Design Reviews.
5. Operational Test and Evaluation Outline.
   - 5.1. Operational Test and Evaluation Overview.
   - 5.2. Operational Test and Evaluation to Date.
   - 5.3. Features/Function Delivery.
   - 5.4. Future Operational Test and Evaluation.
6. Test and Evaluation Resource Summary.
   - 6.1. Test Articles.
   - 6.2. Test Sites and Instrumentation.
   - 6.3. Test Support.
   - 6.4. Inspection (Requirements and Design Documentation) Resource Requirements.
   - 6.5. Inspection (Source Code) Resource Requirements.
   - 6.6. Inspection Prototype Resource Requirements.
   - 6.7. Threat Systems/Simulators.
   - 6.8. Test Targets and Expendables.
   - 6.9. Operational Force Test Support.
   - 6.10. Simulations, Models, and Test Beds.
   - 6.11. Special Requirements.
   - 6.12. T&E Funding Requirements.
   - 6.13. Manpower/Personnel Training.

13 Assessing and Mitigating Risk

Risks associated with programs meeting requirements and schedules are unavoidable, and so it is important to plan accordingly. One way to do this is by doing a risk assessment, which recognizes, quantifies, and tracks each risk until it is mitigated or overcome by events. Risk assessment starts with and continues through all phases of the program life cycle. Assessment and mitigation of risks is a function that is accomplished at the program/systems level. Aspects of risk assessment flow into the evaluation of every change to functional requirements, into allocation of the changes, as well as into implementation of the changes. Section 7.12.1 presented the 90/90 Rule: 90% of the decisions that drive program costs are made in the first 90 days of a program.

The risk assessment should:

- Identify any potential hazards to the program.
  - Natural
  - Technological
  - Market sector
  - Supply chain
  - Cyber-attack
  - etc.
- Assess the risk probability and magnitude of the impact.
  - People
  - Property
  - Systems
  - Equipment
  - GTS
  - Business operations
  - Reputation
  - Contractual/regulatory
  - Environmental
  - etc.
- Identify the vulnerability and specific impacts.
  - Business interruption
  - Loss of customer base
  - Economic loss
  - Environmental mitigation
  - Contractual penalties
  - Regulatory fines

Some risk areas include:

- Market viability
  - Is the investment to bring a product to market offset by potential profits? (For example, what is the return on investment?)
  - How long is the market likely to remain viable?
- Product market adaptation
  - Can a product developed for one market be modified for a different market with different geopolitical/societal norms?

- Competition
  - Is competition direct or indirect? (Hertz concentrated on city rentals, while AVIS concentrated on airport rentals.)
- Schedule
  - Is the schedule attainable and does it include sufficient scheduling margin?
    - "Green light" schedules that assume no or minimal delays are a precursor for failure.
- Contractual/regulatory
  - Are all contractual and regulatory requirements understood?
  - Are all non-disclosure, trade mark, and intellectual property rights agreements in place?
- Distribution/sales/post-sale support
  - Single-string like Apple and automotive initial sales?
  - Multi-string like PCs?

Each area of risk may branch into multiple specific risks, so the severity (weight) of each risk can be established. The likelihood of the risk manifesting itself can be determined. Finally, how the risk is handled can be set using predetermined handling categories, in risk impact weighting.

- Risk categories need to be weighed according to likelihood of occurrence and impact (Table 13.1).
  - Generally done on a scale of 1–5.
  - Scales of 1–10 can be used if greater risk granularity is desired.
- Some risks are more easily tolerated than others.
  - Safety/environmental risks are of greater concern than product development risks.
- Impact weight rating establishes the risk posture for the program.

TABLE 13.1
Sample Likelihood of Impact

Likelihood	Market Viability	Product Market Adaptation	Direct Competition	Indirect Competition
5 Very High	$P_{MV} = 100\%$	$P_{PMA} = 100\%$	$P_{DC} = 100\%$	$P_{IDC} = 100\%$
4 High	$75\% < P_{MV} \le 100\%$	$75\% < P_{PMA} \le 100\%$	$80\% < P_{DC} \le 100\%$	$80\% < P_{IDC} \le 100\%$
3 Moderate	$25\% < P_{MV} \le 50\%$	$30\% < P_{PMA} \le 50\%$	$50\% < P_{DC} \le 80\%$	$50\% < P_{IDC} \le 80\%$
2 Low	$12.5\% < P_{MV} \le 25\%$	$10\% < P_{PMA} \le 30\%$	$25\% < P_{DC} \le 50\%$	$25\% < P_{IDC} \le 50\%$
1 Very Low	$5\% < P_{MV} \le 12.5\%$	$5\% < P_{PMA} \le 10\%$	$0\% < P_{DC} \le 25\%$	$0\% < P_{IDC} \le 25\%$
Likelihood	**Team Collaboration**	**Product Evolution**	**Safety Environmental**	**Technical**
5 Very High	$P_{TC} = 100\%$	$0\% < P_{PE} \le 12.5\%$	$P_{SE} = 100\%$	$P_{TE} > 50\%$
4 High	$85\% < P_{TC} \le 100\%$	$12.5\% < P_{PE} \le 25\%$	$10\% < P_{SE} \le 100\%$	$25\% < P_{TE} \le 50\%$
3 Moderate	$65\% < P_{TC} \le 85\%$	$25\% < P_{PE} \le 50\%$	$1\% < P_{SE} \le 10\%$	$15\% < P_{TE} \le 25\%$
2 Low	$20\% < P_{TC} \le 65\%$	$50\% < P_{PE} \le 95\%$	$0.1\% < P_{SE} \le 1\%$	$2\% < P_{TE} \le 15\%$
1 Very Low	$5\% < P_{TC} \le 20\%$	$95\% < P_{PE} \le 75\%$	$0.01\% < P_{SE} \le 0.1\%$	$0.1\% < P_{TE} \le 2\%$
Likelihood	**Cost**	**Schedule**	**Contractual Regulatory**	**Distribution, Sales, Support**
5 Very High	$P_C > 80\%$	$P_S > 75\%$	$P_{CR} > 95\%$	$P_{DSS} > 90\%$
4 High	$50\% < P_{CS} \le 80\%$	$50\% < P_{CS} \le 75\%$	$70\% < P_{CR} \le 95\%$	$70\% < P_{DSS} \le 90\%$
3 Moderate	$25\% < P_{CS} \le 50\%$	$25\% < P_{CS} \le 50\%$	$50\% < P_{CR} \le 70\%$	$40\% < P_{DSS} \le 70\%$
2 Low	$10\% < P_{CS} \le 25\%$	$10\% < P_{CS} \le 25\%$	$25\% < P_{CR} \le 50\%$	$15\% < P_{DSS} \le 40\%$
1 Very Low	$2\% < P_{CS} \le 10\%$	$0.2\% < P_{CS} \le 10\%$	$5\% < P_{CR} \le 25\%$	$5\% < P_{DSS} \le 15\%$

TABLE 13.2
Sample Impact Scoring

Risk Domain	1 Very Low	2 Low	3 Moderate	4 High	5 Very High
Market Viability	Negligible or no impact	The market looks to be viable for 4 years	The market looks to be viable for 3 years	The market looks to be viable for 18 months	The market may only remain viable for 6 Months
Product Market Adaptation	Product is viable in all markets	Minor design impact to design to make product viable for all markets	Moderate design impact to design to make product viable for all markets	Major design impact to design to make product viable for all markets	Product will not be viable in all markets
Direct Competition	Negligible	Competing in a 25% to 49% saturated market	Competing in a 50% to 79% saturated market	Competing in an 80% to 99% saturated market	Competing in a fully saturated market
Indirect Competition	Negligible	Indirect competition is a 25% to 49% saturated market	Indirect competition is a 50% to 79% saturated market	Indirect competition is an 80% to 99% saturated market	Indirect competition is a fully saturated market
Team Collaboration	Team collaboration failure is less than 25% assured	Team collaboration failure is 25% to 50% assured	Team collaboration failure is 50% to 79% assured	Team collaboration failure is 80% to 99% assured	Team collaboration failure is assured
Product Evolution	There is a greater than 95% possibility the product design can be leveraged forward	There is a 51% to 95% possibility the product design can be leveraged forward	There is a 26% to 50% possibility the product design can be leveraged forward	There is a 12.6% to 25% possibility the product design can be leveraged forward	There is less than a 12.5% possibility the product design can be leveraged forward
Safety/Evironmental	Negligible	Could cause the need for only minor first aid treatment	May cause minor injory or occupational illness or property damage	May cause severe injury or occupational illness or major property damage	May cause death or permanetly disabling injury and destruction of property
Schedule	Negligible or no schedule impact	Minor impact to schedule milestones; accommodates within reserves; no impact to critical path, <1-week slip	Impact to schedule milestones; accommodates within reserves; moderate impact to critical path > 1-week slip	Major impact to schedule milestones; major impact to critical path >1-month slip	Cannot meet schedule and program milestones; >3-month slip
Cost	<2% increase over allocated and negligible impact on reserve	Between 2% and 5% increase over allocated budgets and can handle with project reserves	Between 5% and 7% increase over allocated budgets and cannot handle with project reserves	Between 7% and 10% increase over allocated budgets and/or exceeds proper project reserves	>10% increase over allocated budgets and/or can't handle with project reserves
Contract Regulatory	Negligible or no impact	Minor impact due to contractual or regulatory requirements; accommodates within reserves; no impact to critical path, <3-month slip	Impact due to contractual or regulatory requirements; accommodates within reserves; moderate impact to critical path 1 to 3-month slip	Major impact due to contractual or regulatory requirements; major impact to critical path 3 to 6-month slip	Cannot meet contractual or regulatory requirements; >6-month slip
Distribution, Sales, Support	<2% risk of Distribution, Sales and Support related issues	Between 2% and 5% decrease in return on investment due to Distribution, Sales and Support related issues	Between 5% and 10% decrease in return on investment due to Distribution, Sales and Support related issues	Between 10% and 20% decrease in return on investment due to Distribution, Sales and Support related issues	>20% decrease in return on investment due to Distribution, Sales and Support related issues

There are several handling categories for risks and opportunities.

- Watch
  - There is not enough information available to establish a mitigation/implementation plan.
- Mitigate/implement
  - One or more paths forward are available to mitigate a risk or take advantage of an opportunity and need to be explored.
- Retire
  - The risk has been mitigated or the opportunity has been approved and implemented.

Once the likelihood of a risk occurring and the risk impact scoring (Table 13.2) have been defined, they are multiplied together to obtain a "summary score." If the summary score is below eight, the risk is watched to see if the likelihood of the risk occurring is changing and the impact that has on the risk impact and the summary score. If the summary score is eight or above, the risk requires that a mitigation/implementation plan be generated. A risk is retired only if it has been mitigated or if it stops being a risk due to some other reason.

Risks should always be viewed as opportunities.

Sample Risk/Opportunity List Column Headers (Table 13.3)

- Risk Identification Number
- Statement of Risk/Opportunity
- Assigned To
- Creation Date
- Likelihood of Impact
- Probability %
- Impact Category – Team Collaboration
- Impact Score
- Impact Date
- Impact Schedule (weeks)
- Summary Score

$$\text{Summary score} = \text{Likelihood of impact} \times \text{Impact score}$$

EQUATION 13.1 Summary score calculation.

- Financial Exposure
- Handling Category
- Mitigation Next Step
- Full Risk Step-Down Plan (Plan Number or NA)
- Status/Notes

Assume that a new program is starting. The development team is comprised of collaboration partners. Collaborators include firms in the following countries: Australia, Canada, France, Germany, Hungary, Iceland, India, Italy, Ireland, Japan, Lithuania, Netherlands, Norway, New Zealand, Portugal, Singapore, South Korea, Turkey, Vietnam, and South Africa.

One of the risks identified is commonality of units, terms, and dates. The assignment to develop the first risk item falls to a junior level engineer named Kreja VanFranken. Kreja does her best to create the first risk item using the Sample Risk/Opportunity List.

The above item is boarded at a risk/opportunity review meeting, where several things were noticed as being inconsistent with an international effort.

TABLE 13.3
Risk/Opportunity List

Risk #	Risk/ Opportunity	Assigned to	Created	Likelihood of Impact	Probability %	Impact Category
1	Commonality of units, terms, and dates	Kreja VanFranken	04-04-18	4	20%	Team Collaboration
	Impact Score	**Impact Date**	**Impact Schedule**	**Summary Score: (= Likelihood of Impact × Impact Score)**	**Financial Exposure**	**Handling Category**
	4	20-5-18	12 weeks	16	12 weeks × (€200,000/week) = €2,400,000	Mitigate/ implement
	Mitigation Next Step	**Due Date**	**Full Risk Step Down Plan (Doc Number or NA)**	**Status/Notes**		
	Create program technical dictionary	Draft due 11-05-18	Document Number TBD	Kick-off meeting with partners scheduled 9-04-18 at 8:30 am Formal review in PDM system scheduled 3 weeks later.		

- Date format appears to be European (DD-MM-YY) and there is no note indicating that the date format was agreed to by all partners/stakeholders; nor is there agreement on what time zone should be used.
  - Creation Date: 04-04-18.
  - Impact Date: 20-05-18.
  - Due Date: Draft due 11-05-18.
  - Status/Notes: Kick-off meeting with partners/stakeholders scheduled 09-04-18 at 8:30 am.
  - The date the risk item status was provided isn't identified.
- No document number was identified for the program technical dictionary.
  - Mitigation Next Step: Create program technical dictionary.
  - Full Risk Step Down Plan (Plan Number or NA) –Document Number TBD.
    - Document number should have been pre-assigned.

In Chapter 11, multiple examples were provided concerning "When Things Go Wrong" that can all be traced back to risks that were not identified and mitigated. Not all risks can be identified at the start of a program. EarthWatch, LLC (later renamed DigitalGlobe, LLC) was formed to provide high resolution commercial images of the Earth. It assessed and mitigated risks based on existing technologies. The business model relied on downloaded satellite data being sent by overnight express to Longmont, Colorado, for data analysis and reduction, image cataloging, and distribution of purchased images. As the program evolved, so did the technological base. By the time the first images were transmitted from the QuickBird-2 satellite, Internet capabilities were robust enough that data could be transmitted to Longmont using 2/3/4G networks. The original risk assessment was overcome by these events, necessitating a new business model and assessment of new risks associated with it.

Risk/opportunity list column headers should include:

- Need for change/variance
  - A clear statement of the need for the change or variance.
- Change classification (major, minor, critical)

- Affected CIs, systems, assemblies, components, suppliers, and configuration information.
- Scope of the change
  - Technical and operational impacts (performance, test, qualification, operation, maintenance, servicing, operation and maintenance training, repair parts, support and test equipment).
  - Risk and consequences of the change not being approved.
- The effectivity (serial numbers of impacted items or the starting lot number or date when the change will roll into the production line) of the change
- Implementation plan and schedules for retrofitting products
  - Rework.
  - Scrap.
  - Using existing inventory.
  - Restricted use.
- Requested change/variance approval date

Appendix 1: Related Configuration Management Standards

We are indebted to the CM PIC organization for providing this list of CM standards for publication. Standards are being canceled and replaced on a regular basis so there is no guarantee that the standards listed below are the current version.

Aerospace

- AS9100C, Advanced Quality System
- DO 178B, Software Considerations in Airborne Systems and Equipment

Alliance for Telecommunications Industry Solutions (ATIS)

- ATIS 0300250, Operations, Administration, Management & Provisioning (OAM&P)—Extension to Generic Network Model for Interfaces between Operations Systems and Network Elements to Support Configuration Management—Analog and Narrowband ISDN Customer Service Provisioning

American National Standards Institute (ANSI)

- ANSI INCITS TR-47, Information Technology—Fiber Channel-Simplified Configuration and Management Specification (FC-SCM)

American Nuclear Society (ANS)

- ANSI/ANS-3.2-1982, Administrative Controls & Quality Assurance for the Operational Phase of Nuclear Power Plants

American Society for Quality (ASQ)

- ANSI/ASQ Q9004-2009, Quality Management Standards—Guidelines for CM Draft Standard for Facilities, 5/8/93; never finished

American Society of Mechanical Engineers (ASME)

- ASME/NQA-1, Quality Assurance Program Requirements for Nuclear Facilities
- ASME/NQA-1B-2011, Quality Assurance Requirements for Nuclear Facilities Applications
- ASME/NQA-2A-1990, addenda to ASME NQA-2-1989 Edition, Quality Assurance Requirements for Nuclear Facility Application
- ANSI/N45.2.11, Quality Assurance Requirements for the Design of Nuclear Power Plants
- ANSI/N18.7, Administrative Controls & Quality Assurance for the Operational Phase of Nuclear Power Plants
- ANSI/N45.2.9-1974, Requirements for the Collection, Storage, & Maintenance of Quality Assurance Records for Nuclear Power Plants
- ANSI/N45.2.13, Quality Assurance Requirements for the Control of Procurement of Items & Services for Nuclear Power Plants
- ASME Y14.100, Government/Industry Drawing Practices
- ASME Y14.24 Types & Applications of Engineering Drawings

- ASME Y14.34, Parts Lists, Data Lists & Index Lists
- ASME Y14.35, Revision of Engineering Drawings & Associated Documents
- ASME NQA Committee, Task Group—CM Draft, January 1991

Australia

- AS/NZS 3907:1996, Quality Management—Guidelines for Configuration Management
- Australian Army Configuration Management Manual (CMMAN) Army CM Manual; version 3.0
- DI(G) LOG 08-4, Configuration Management of Systems and Equipment
- DI(A) SUP 24-2, Configuration Management Policy within Army [currently being updated for issue as DI(A) LOG XX-X, Configuration Management Policy for Capabilities in the Land Environment]
- DI(A) LOG 1-33, Integrated Logistic Support and the Army Material Process (currently being updated)
- The Army Specification Manual (SPECMAN)
- EMEI Workshop A 850, Modifications, Trial Modifications, and Local Modifications to Equipment
- MINE WARFARE-STD-499B, Systems Engineering

Automotive Industry Action Group (AIAG) (North America Automotive Industry)

- QS9000, Quality System Requirements (replaced by ISO/TS16949)
- ISO/TS 16949, Quality management systems particular requirements for the application of ISO 9001:2000 for automotive production and relevant service part organizations.

Bell Canada

- Trillium, Model for Telecom Product Development & Support Process Capability

Bell Communications Research (BELLCORE)

- TR/OPT/000209, Guidelines for Product Change Notice Was Superseded (see Telcordia)

British Standards Institute (BSI)

- BS 6488, CM of Computer-Based Systems
- BSI PD ISO/IEC TR 18018, Information Technology—Systems and Software Engineering—guide for configuration management tool capabilities
- BS EN 46001, Application of EN 29001 (BS5750: Part 1) to manufacture medical devices
- BS5515:1984, British Code of Practice for Documentation of Computer-Based Systems
- BS 7799, Information Security Management
- BS 15000-1 IT Service Management defines the requirements for an organization to deliver managed services of an acceptable quality for its customers. *Note:* replaced by ISO 20000-1
- BS 15000-2 IT Service Management best practices for Service Management processes. *Note:* replaced by ISO 20000-2
- BSI PAS 55:2008, Asset Management
- BS EN 13290-5:2001, Space Project Management. General requirements configuration management

Canada—Department of National Defence (DND) Standards

- C-05-002-001/AG-00, Aerospace Engineering Change Proposal Procedures
- D-01-000-200/SF-001, Joint Electronics Type Designation System (JETS)

- D-01-002-007/SG-001, Requirements for the Preparation of CM Plans
- D-01-002-007/SG-002, Requirements for Configuration Identification
- D-01-002-007/SG-004, Requirements for Configuration Status Accounting
- D-01-002-007/SG-006, Requirements for the Selection of Configuration Items
- D-01-100-215/SF-000, Specification for Preparation of Material Change Notices
- D-01-400-001/SG-000, Engineering Drawing Practices
- D-01-400-002/SF-000, Drawings, Engineering, & Associated Lists
- D-02-002-001/SG-001, Identification Marking of Canadian Military Property
- D-02-006-008/SG-001, The Design Change, Deviation, & Waiver Procedure

Chrysler/Ford/General Motors

- QS-9000, Quality System Requirements

Code of Federal Regulations (CFR)—U.S. and Other Governments

- Clinger–Cohen Act (IT)
- Sarbanes–Oxley—"Define and establish controls ..." is the heart of the Sarbanes–Oxley Act
- Title 21 CFR Part 820, Quality System Medical Devices FDA
- Title 10 CFR Part 830 122, Quality Assurance Criteria DOE
- Title 14 CFR Chapter I (FAA), Part 21 Certification Procedures for Products and Parts
- Title 48 CFR, Federal Acquisition Regulations
- Title 48 CFR 2210 Specifications, Standards, and Other Purchase Descriptions
- Title 48 CFR 1 Part 46 Quality Assurance

Commercial U.S. Airlines Air Transportation Association (ATA)

In 2000, ATA Spec 100 and ATA Spec 2100 were incorporated into ATA iSpec 2200: Information Standards for Aviation Maintenance. ATA Spec 100 and Spec 2100 will not be updated beyond the 1999 revision level.

- ATA 100
- ATA 200
- ATA 2100
- ATA 2200

Department of Defense (DOD) (U.S.) Superseded/Canceled

- AMCR 11-12, Total Decision-Making Process
- AMCR 11-26, CM (Army, 1965)
- ANA Bulletin NO. 390, Engineering Change Proposal (Army, Navy, and Air Force)
- ANA Bulletin 391, Engineering Change Proposal
- ANA Bulletin 445, Engineering Changes to Weapons, Systems, Equipments, & Facilities (1963)
- AFSCM 375-1, CM During the Development & Acquisition Phases (Air Force Systems Command, 1962)
- AFSCM 375-3, System Management (1964)
- AFSCM 375-4, System Program Management Procedures (1966)
- AFSCM 375-5, Systems Engineering Management Procedures (1966)
- AFSCM 375-7, Configuration Management for Systems, Equipment, Munitions, and Computer Programs (1971)
- AFWAMAN33-2 Air Force Weather Agency Consolidated Network Configuration Management Plan (2003)

- AMC Instruction 33-105 Enterprise Configuration Management
- AR 70-37, Joint DOD Service Agency Regulation Configuration Management
- Army Regulation 25-6, Configuration Management for Automated Information Systems
- ASWSPO 5200.4 (Navy, 1965)
- BuWeps Instruction 5200.20, Processing Engineering Change Proposals (NAVY)
- CMI, CM Instructions, Air Force Systems Command, Space Systems Division (1963):
    - CMI No. 1, Facility Engineering Change Proposal procedures (1964)
    - CMI No. 2, Engineering Change Proposal Procedures (1964)
    - CMI No. 3, Specification Maintenance (1964)
    - CMI No. 4, Configuration Change Implementation (1964)
    - CMI No. 5, Configuration Accounting Procedures (1964)
    - CMI No. 7, Configuration Control Board (1964)
    - CMI No. 9, First Article Configuration Inspection (1964)
- COMDTINST 4130.6A, Coast Guard Configuration Management Policy
- COMDTINST M4130.8, Coast Guard Configuration Management for Acquisitions and Major Modifications
- COMDTINST M4130.9, Coast Guard Configuration Management for Sustainment
- DOD 5000.2-M, Defense Manual—Defense Acquisition Management Documents & Reports
- DOD 5000.19, Policies for the Management & Control of Information Requirements
- DOD 5010.12, Management of Technical Data
- DOD 5010.19, DOD CM Program
- DOD 5010.21, CM Implementation Guidance
- DOD 8000 Series, Policies & Procedures for Automated Information Systems
- DOD-D-1000, Drawing, Engineering, & Associated Lists
- DOD-HDBK-287, A Tailoring Guide for DOD-STD-2167A
- DOD-STD-2167, Defense System Software Development
- DOD-STD-7935, DOD Automated Information System Documentation Standards
- M200 (1962), Standardization Policies, Procedures, & Instructions, Defense Standardization Manual, then replaced by 4120.3-M in 1966
- MIL-STD-12, Abbreviations for Use on Drawings, Specs, Standards, & Technical Documents, will eventually be replaced by ANSI Y14.38
- MIL-STD-100, Engineering Drawing Practices
- MIL-STD-454, Standard General Requirements for Electronic Equipment
- MIL-STD-480B, Configuration Control—Engineering Changes, Deviations, & Waivers
- MIL-STD-481, Configuration Control—Short Form
- MIL-STD-482, Configuration Status Accounting Data Elements & Related Features
- MIL-STD-483, CM Practices for Systems, Equipment, Munitions, & Computer Programs
- MIL-STD-490, Specification Practices
- MIL-STD-498, Software Design & Development (replaces DOD-STD-2167, DOD-STD-7935, & DOD-STD-1703)
- MIL-STD-499, Systems Engineering
- MIL-STD-999, Certification of CM/DM Process (DRAFT)
- MIL-STD-1456, CM Plan
- MIL-STD-1521, Technical Reviews & Audits for Systems, Equipments, & Computer Software (Appendixes G, H, & I were superseded by MIL-STD-973)
- MIL-STD-1679, Software Development (1978)
- MIL-STD-2549, Configuration Status Accounting (canceled 9/30/2000). However, Army is using AMC-STD-2549A until EIA 836 is published.
- MIL-STD-3046, DOD Interim Standard Practice Configuration Management (Army, 2013)
- MIL-D-70327, Drawings & Data Lists
- MIL-Q-9858, Quality Program Requirements

- MIL-S-52779, Software Quality Program Requirements (1974)
- MIL-T-47500/1-6, Technical Data Package, General Specification for
- MIL-T-31000, Technical Data Packages, General Specification for
- NAVAIRINST 4000.15, Management of Technical Data & Information
- NAVMAT INSTR 4130.1 (Navy, 1967)
- NAVMATINSTR 4131.1, Configuration Management (Navy, 1967)
- NAVMATINSTR 5000.6, Configuration Management (Navy, 1966)
- NAVSEA 0900-LP-080-2010, Software Configuration Control Procedures Manual (Navy, 1975)
- (U)-E-759/ESD, Software Quality Assurance Plan (Air Force, 1980)
- OPNAVINST 4130.1, Configuration Management of Software in Surface Ship Combat Systems (Navy, 1975)

DOD Active or canceled but still specified in existing contracts

- US COE EC 11-2-173, USCCE Manpower Civil Program Civilian Air Force Configuration and Management
- US COE ER 15-1-33, Automation Configuration Management Boards
- AFPD 21-4, Engineering Data
- AFI 21-401, Engineering Data Storage, Distribution, & Control
- AFI 21-402, Engineering Drawing System
- AFI 21-403, Acquiring Engineering Data
- AFMCPAM 63-104, IWSM Configuration Management Implementation Guide (2000)
- AFWAMAN 33-2, Air Force Weather Agency Consolidated Network Configuration Management Plan (2003)
- AMCI33-105 Configuration Management October 15, 2000
- DID Guide, HQ AFMC/EN DID Guide
- DOD 5000.1, Defense Acquisition
- DOD 5000.2, Defense Acquisition Management Policies & Procedures
- DOD 5010.12-M, Procedure for the Acquisition and Management of Technical Data
- DOD 5010.12-L, Acquisition Management System & Data Requirement List (AMSDL)
- DOD Cataloging Handbook H6, Federal Item Identification Guides for Supply Cataloging
- DOD Cataloging Handbook H7, Manufacturers Part & Drawing Numbering Systems for Use in the Federal Cataloging System
- DODISS, Department of Defense Index of Specifications & Standards
- DOD-STD-1700, Data Management Program
- DOD-STD-2168, Defense System Software Quality Program
- MIL-HDBK-59, Computer Aided Acquisition & Logistics Support (CALS) Program Implementation Guide (CALS is now known as "Continuous Acquisition & Life-Cycle Support")
- MIL-HDBK-61, Configuration Management
- MIL-STD-109, Quality Assurance Terms & Definitions
- MIL-STD-1168, Lot Numbering of Ammunition
- MIL-STD-130, Identification Marking of U.S. Military Property
- MIL-HDBK-245, Preparation of Statement of Work
- MIL-STD-280, Definition of Item levels, Item Exchangeability, Models, & Related Terms
- MIL-HDBK-454, Standard General Requirements for Electronic Equipment
- MIL-STD-881, Work Breakdown Structure for Defense Material Items
- MIL-STD-961, Military Specifications & Associated Documents, Preparation of
- MIL-STD-969, Specifications
- MIL-STD-973, CM Notice 3 (canceled 9/30/2000)

- MIL-STD-974, CITIS (Contractor Integrated Technical Information Service, which is being transitioned to a non-governmental standard)
- MIL-STD-1309, Definitions of Terms for Test, Measurement, & Diagnostic Equipment
- MIL-STD-1465, CM of Armaments, Munitions, & Chemical Production Modernization
- MIL-STD-1520, Corrective Action & Disposition System for Non-conforming Material
- MIL-STD-1767, Procedures for Quality Assurance & Configuration Control of ICBM Weapon System Technical Publications & Data
- MIL-STD-1840, Automated Interchange of Technical Information
- MIL-STD-2084, General Requirements for Maintainability of Avionics & Electronic Systems & Equipment
- MIL-I-8500, Interchangeability & Replaceability of Component Parts for Aerospace Vehicles
- MIL-S-83490, Specification, Types, & Forms
- SMC-S-002, Configuration Management (Space and Missile Command)
- USAFAI33-114 Managing Software Configuration and Controlling Data in the Cadet
- Administrative Management Information System (CAMIS)

DOD, Draft Documents

- MIL-STD-CNI, Coding, Numbering, & Identification
- SD-15, Performance Specification Guide

DOD Initiatives

- AD-A278-102, Blueprint for Change (regarding the use on commercial standards, obtain through NTIS)

Department of Energy (DOE) (U.S.)

- DOE Order 430.1, Life Cycle Asset Management
- DOE Guide G-830-120, Implementation Guide for 10CFR Part 830.120, Quality Assurance
- DOE Order 4330.4A, Maintenance Management Program
- DOE Order 4700.1, Project Management System (will be phased out)
- DOE Order 5480.19, Conduct of Operations Requirements for DOE Facilities
- DOE Order 5700.6C, Quality Assurance
- DOE Order 6430.1A, General Design Criteria
- DOE-STD-1073-93 Parts 1 & 2, Guide for Operational CM Program
- NPO 006-100, DOE Office of New Production Reactors CM Plan
- DOE 5480.CM, Operational CM Program (see DOE-STD-1073)
- DOE Draft, No Number, CM for Non-Nuclear Facilities

Department of Transportation, Federal Highway Administration (DOT/FHA) (U.S.)

- Configuration Management for Transportation Management Systems Handbook, FHWA-OP-04-013. http://ops.fhwa.dot.gov/freewaymgmt/publications/cm/handbook/index.htm
- Configuration Management Fact Sheet. http://ops.fhwa.dot.gov/freewaymgmt/publications/cm/factsheet/
- Configuration Management Primer. http://ops.fhwa.dot.gov/freewaymgmt/publications/cm/primer/
- Configuration Management Tri-Fold Brochure. http://ops.fhwa.dot.gov/freewaymgmt/publications/cm/brochure/
- Configuration Management Technical Presentation. http://ops.fhwa.dot.gov/freewaymgmt/publications/cm/presentation/index.htm

Electronic Industries Alliance (EIA), some standards are also listed under SAE International and TechAmerica

- ANSI/EIA 632-1, Draft Process for Engineering a System—Part 1: Process Characteristics
- ANSI/EIA 632-2, Draft Process for Engineering a System—Part 2: Implementation Guidance
- SAE/EIA 649, Configuration Management Standard (now written by SAE International)
- SAE/EIA 649-1, Configuration Management Requirements for Defense Contracts
- SAE/EIA 649-2, Configuration Management Requirements for NASA Enterprises
- CMB 3, Recommendations Concerning CM Audits
- CMB 4-1, CM Definitions for Digital Computer Programs
- CMB 4-2, Configuration Identification for Digital Computer Programs
- CMB 4-3, Computer Software Libraries
- CMB 4-4, Configuration Change Control for Digital Computer Programs
- CMB 5, CM Requirements for Subcontractors/Vendors
- CMB 6-1, Configuration & Data Management References
- CMB 6-2, Configuration & Data Management In-House Training Plan
- CMB 6-3, Configuration Identification
- CMB 6-4, Configuration Control
- CMB 6-5, Textbook for Configuration Status Accounting
- CMB 6-6, Reviews & Configuration Audits
- CMB 6-7, Data Management
- CMB 6-8, Data Management In-House Training Course
- CMB 6-9, Configuration & Data Management Training Course
- CMB 6-10, Education in Configuration & Data management
- CMB 7-1, Electronic Interchange of CM Data
- CMB 7-2, Guideline for Transitioning CM to an Automated Environment
- CMB 7-3, CALS CM SOW & CDRL Guidance
- EGSA 107, Glossary of DOD CM Terminology & Definitions
- EIA/IS 632, System Engineering
- EIA-748, Earned Value Management Systems
- EIA-927, Common Data Schema for Complex Systems
- EIA SP 3537, Processes for Engineering a System
- EIA SP 4202, Online Digital Information Service (ODIS)
- EIA SSP 3764, Standard for Information Technology—Software Life Cycle Processes Software Development Acquirer/Supplier Agreement
- IEEE/EIA 12207.0, Industry Implementation of ISO/IEC 12207 (Standard for Information Technology)
- IEEE/EIA 12207.1, Guide for Information Technology—Software Life Cycle Processes Life Cycle Data
- IEEE/EIA 12207.2, Guide for Information Technology—Software Life Cycle Processes Implementations Considerations
- J-Std-016 (EIA/IEEE Interim Standard), Standard for Information Technology; Software Life Cycle Processes; Software Development; Acquirer/Supplier Agreement
- Systems Engineering EIA-632
- Systems Engineering Capability Model SECM EIA-732

Electric Power Research Institute (EPRI)

- EPRI TR-103586, Guidelines for Optimizing the Engineering Change Process for Nuclear Power Plants, prepared by Cygna Energy Services, Oakland, CA
- EPRI NP-5640, Nuclear Plant Modifications & Design Control: Guidelines for Generic Problem Prevention

- EPRI NP-6295, Guidelines for Quality Records in Electronic Media for Nuclear Facilities
- EPRI NP-3434, Value-Impact Analysis of Selected Safety Modifications to Nuclear Power Plants
- EPRI NP-5618, Enhancing Plant Effectiveness Through Improved Organizational Communication
- EPRI NSAC-121, Guidelines for Performing Safety System Functional Inspections

European Cooperation for Space Standardization (ECSS)

- http://www.ecss.nl/
- ECSS-M-ST-10, Space Project Management—Project Planning and Implementation
- ECSS-M-ST-40C, Space Project Management—Configuration and Information Management
- ECSS-Q-ST-10-09, Space Product Assurance—Non-conformance Control System
- ECSS-Q-ST-20, Space Product Assurance—Quality Assurance

European Committee for Standardization (ECS)

- CEN EN 13290-5, Space Project Management—General Requirements—Part 5: Configuration Management
- EN 13290-6, Space Project Management—General Requirements—Part 6: Information/Documentation Management
- EN14160, Space Engineering—Software
- EN 9200, Programme Management—Guidelines for Project Management Specification

European Computer Manufacturers Institute (ECMI)

- ECMA-TR 47, CM Service Definition

European Community

- JAR-21, Certification Procedures for Aircraft & Related Products & Parts (Draft)

European Defense Standards Reference System (EDSTAR)

- CEN Workshop 10, European Handbook for Defense Procurement Expert Group 13 Life Cycle (Project) Management Final Report
- ECSS-M-40A, Configuration Management
- ECSS-E-10, System Engineering
- ECSS-M-50, Information/Documentation Management
- ECSS-E-40, Space System Software Engineering. http://www.eda.europa.eu/EDSTAR/home.aspx

European Space Agency (ESA)

- PSS-05-01, ESA Software Engineering Standards
- PSS-05-09, Guide to Software Configuration Management
- PSS-05-10, Guide to Software Verification and Validation
- PSS-05-11, Guide to Software Quality Assurance

European Telecommunications Standards Institute (ETSI)—Too many to list

- http://www.etsi.org/standards

Federal Aviation Administration (FAA) (U.S.)

- Title 14 Code of Federal Regulations, Parts 1–59
- FAA-STD-002, Facilities Engineering Drawing Practices
- FAA-STD-005, Preparation of Specification Documents
- FAA-STD-018, Computer Software Quality Program (1977)
- FAA-STD-021, Configuration Management Contractor Requirements
- FAA-STD-058, Standard Practice Facility Configuration Management
- FAA Order 1800.8, National Airspace Systems Configuration Management
- FAA Order 6030.28, National Airspace Systems Configuration Management
- FAA 1100.57, National Engineering Field Support Division Maintenance Program Procedures, Operational Support (AOS)
- FAA 1800.63, National Airspace System (NAS) Deployment Readiness Review (DRR) Program
- FAA 1800.66, National Policy Configuration Management Requirements
- FAA 6032.1, Modifications to Ground Facilities, Systems, and Equipment in the NAS

Federal Highway Administration (FHA) (U.S.)

- Configuration Management for Transportation Management Systems Handbook (FHWA Publication Number: FHWA-OP-04-013) (EDL Document Number: 13885)
- A Guide to Configuration Management for Intelligent Transportation Systems (FHWA)
- Publication Number: FHWA-OP-02-048) (EDL Document Number: 13622)
- Configuration Management for Transportation Management Systems Primer (FHWA Publication Number: FHWA-OP-04-014) (EDL Document Number: 13886)
- Configuration Management for Transportation Management Systems Brochure (FHWA Publication Number: FHWA-OP-04-016) (EDL Document Number: 13888)
- Configuration Management for Transportation Management Systems Fact Sheet (FHWA Publication Number: FHWA-OP-04-017) (EDL Document Number: 13889)
- Configuration Management for Transportation Management Systems Technical Presentation

Food and Drug Administration (FDA) (U.S.)

- FDA 8541-79, Good Manufacturing Practices, Food & Drug Administration (superseded by QSR)
- QSR Quality System Regulation for the Medical Device Industry (QSR) (based on ISO 9001)

France

- NF EN 13290-5 January 2002, Management of Space Projects—General Requirements—Part 5: Configuration Management
- AFNOR NF EN 300291-1, Telecommunications Management Network (TMN)—Functional Specification of Customer Administration (CA) on the Operations System/Network Element (OS/NE) Interface—Part 1: Single Line Configurations (V1.2.1)
- AFNOR NF ETS 300617, Digital Cellular Telecommunications System (Phase 2)—GSM Network Configuration Management

Germany: Deutsches Institut für Normung (DIN)

- DIN EN 13290-5, Aerospace–Space Project Management—General Requirements—Part 5: Configuration Management, German and English versions

- DIN EN 300291-1, Telecommunications Management Network (TMN)—Functional Specification of Customer Administration (CA) on the Operations System/Network Element (OS/NE) Interface—Part 1: Single Line Configurations [Endorsement of the English version EN 300291-1 V 1.2.1 (1999–02) as the German standard]
- DIN EN 300291-2, Telecommunications Management Network (TMN)—Functional Specification of Customer Administration (CA) on the Operations System/Network Element (OS/NE) Interface—Part 2: Multiline Configurations [Endorsement of the English version EN 300291-2 V 1.1.1 (2002–03) as the German standard]
- DIN EN 300376-1, Telecommunications Management Network (TMN)—Q3 Interface at the Access Network (AN) for Configuration Management of V5 Interfaces and Associated User Ports—Part 1: Q3 Interface Specification [Endorsement of the English version EN 300376-1 V 1.2.1 (1999–10) as German standard]
- DIN ETS 300617, Digital cellular telecommunications system (Phase 2)—GSM Network Configuration Management; English version ETS 300617
- DIN EN 300377-1, Telecommunications Management Network (TMN)—Q3 Interface at the Local Exchange (LE) for Configuration Management of V5 Interfaces and Associated Customer Profiles—Part 1: Q3 Interface Specification [Endorsement of the English version EN 300377-1 V 1.2.1 (1999–10) as the German standard]
- DIN ETS 300377-2, Signaling Protocols and Switching (SPS)—Q3 Interface at the Local Exchange (LE) for Configuration Management of V5 Interfaces and Associated Customer Profiles—Part 2: Managed Object Conformance Statement (MOCS) Performance Specification; English version ETS 300377-2:1995
- DIN EN 300820-1, Telecommunications Management Network (TMN)—Asynchronous Transfer Mode (ATM) Management Information Model for the X Interface Between Operation Systems (OSs) of a Virtual Path (VP)/Virtual Channel (VC) Cross-Connected Network—Part 1: Configuration Management [Endorsement of the English version EN 300820-1 V 1.2.1 (2000-11) as the German standard]
- DIN EN 301268, Telecommunications Management Network (TMN)—Linear Multiplex Section Protection Configuration Information Model for the Network Element (NE) View [Endorsement of the English version EN 301268 V 1.1.1 (1999-05) as the German standard]
- VG 95031-1, Modification of Products—Part 1: Procedure According to CPM
- VG 95031-2, Drawing Set—Part 3: Parts List
- VG 95031-3, Drawing Set—Part 3: Changes on Drawings

IpX (Institute for Process Excellence)

- CMII-100E CMII Standard for Enterprise Configuration Management

Institute of Electrical & Electronic Engineers (IEEE)

- IEEE 323, Qualifying Class IE Equipment for Nuclear Power Generating Stations
- IEEE 344, Recommended Practices for Seismic Qualification of Class IE Equipment for Nuclear Power Generating Stations
- IEEE 352, Guide for the General Principles of Reliability Analysis of Nuclear Power Generating Station Safety Systems
- IEEE Std 610 (ANSI), Computer Dictionary
- IEEE Std 730-1989 (ANSI), Software QA Plans
- IEEE Std 828-90 (ANSI), Standard for Software CM Plans
- IEEE Std 830-84 (ANSI), Guide for Software Requirements Specifications
- IEEE Std 1028 (ANSI), Standard for Software Reviews & Audits
- IEEE Std 1042 (ANSI), Guide for Software CM

- IEEE Std 1063 (ANSI), Standard for User Documentation
- IEEE Std P1220 (ANSI), Systems Engineering
- IEEE Std 803-1983, Recommended Practice for Unique Identification in Power Plants & Related Facilities—Principles & Definitions
- IEEE Std 803.1-1992, Recommended Practice for Unique Identification in Power Plants & Related Facilities—Component Function Identifiers
- IEEE Std 804-1983, Recommended Practice for Implementation of Unique Identification System in Power Plants & Related Facilities
- IEEE Std 805-1984, Recommended Practice for System Identification in Nuclear Power Plants & Related Facilities
- IEEE Std 806-1986, Recommended Practice for System Identification in Fossil-Fueled Power Plants & Related Facilities
- IEEE Std 828-2012, IEEE Standard for Configuration Management in Systems and Software Engineering
- IEEE/EIA 12207.0 Industry Implementation of ISO/IEC 12207 (Standard for Information Technology)
- IEEE/EIA 12207.1 Guide for Information technology—Software Life Cycle Processes Life Cycle Data
- IEEE/EIA 12207.2 Guide for Information Technology—Software Life Cycle Processes Implementations Considerations
- J-Std-016 (EIA/IEEE Interim Standard) Standard for Information Technology; Software Life Cycle Processes; Software Development; Acquirer/Supplier Agreement
- IEEE 828-2012-IEEE Standard for Configuration Management in Systems and Software Engineering: Institute of Electrical and Electronics Engineers/16-Mar-2012/71 pages

Institute of Nuclear Power Operations (INPO) (U.S.)

Note: GP—Good Practice Document

- GP MA-304, Control of Vendor Manuals
- GP TS-402, Plant Modification Control Program
- GP TS-407, Computer Software Modification Controls
- GP TS-412, Temporary Modification Control
- GP TS-415, Technical Reviews of Design Changes
- GP TS-41 l, Temporary Lead Shielding
- INPO 85-016, Temporary Modification Control
- INPO 85-031, Guidelines for the Conduct of Technical Support Activities at Nuclear Power Stations
- INPO 86-006, Report on Configuration Management in the Nuclear Utility Industry
- INPO 87-006, Report on Configuration Management in the Nuclear Utility Industry
- INPO 88-009, System & Component labeling
- INPO 88-016, Guidelines for the Conduct of Design Engineering
- INPO 90-009 REV. 1 Guidelines for The Conduct of Design Engineering
- INPO 94-003, A Review of Commercial Nuclear Power Industry Standardization Experience

International Organization for Standardization (ISO)

Note: Standards can be searched at http://www.iso.org/iso/home.html.

- ISO 9000:2000 Series
- ISO 9000:2000, Quality Management Systems—Fundamentals and Vocabulary

Note: This replaces ISO 9001, ISO 9002, and ISO 9003

- ISO 9001:2000, Quality Management Systems—Requirements
- ISO 9004:2000, Quality Management Systems—Guidelines for Performance Improvements
- ISO 19011, Guidelines on Quality and/or Environmental Management Systems Auditing (under development)
- ISO 10005:1995, Quality Management—Guidelines for Quality Plans
- ISO 10006:2003, Quality Management Systems—Guidelines for Quality Management in Projects
- ISO 10007:2003, Guidelines for Configuration Management
- ISO/DIS 10012, Parts 1 and 2, Quality Assurance Requirements for Measuring Equipment
- ISO/DIS 10303-239, Industrial Automation Systems and Integration—Product Data Representation and Exchange—Part 239: Application Protocol: Product Life Cycle Support
- ISO 10013:1995, Guidelines for Developing Quality Manuals
- ISO/TR 10014:1998, Guidelines for Managing the Economics of Quality
- ISO 10015:1999, Quality Management—Guidelines for Training
- ISO/TS 16949:1999, Quality Management Systems Particular Requirements for the Application of ISO 9001:2000 for Automotive Production and Relevant Service Part Organizations. Joint Effort IAOB (International Automotive Oversight Bureau) and ISO
- ISO/IEC 17025, General Requirements for the Competence of Testing and Calibration Laboratories
- ISO 8402, Quality Management & Quality Assurance Vocabulary
- ISO 9000-1, Guidelines for Use of the ISO 9000 Series (replaced by ISO 9000:2000)
- ISO 9000-2, Guidelines for Applying ISO 9000 to Services (replaced by ISO 9000:2000)
- ISO 9000-3, Guidelines for Applying ISO 9000 to Software (replaced by ISO 9000:2000)

Note: TickIT is an ISO 9000 accreditation scheme for software developers and supporters

- ISO 9001 Model for Quality Assurance in Design/Development, Production, Installation, & Servicing. (replaced by ISO 9001:2000)
- ISO 9001:2008, Quality Management Systems—Requirements
- ISO/DIS 10303-239, Industrial Automation Systems and Integration—Product Data Representation and Exchange—Part 239: Application Protocol: Product Life Cycle Support
- ISO 9002, Model for Quality Assurance in Production & Installation. *Note:* Replaced by ISO 9001:2000 (replaced by ISO 9001:2000)
- ISO 9003, Model for Quality Assurance in Final Inspection & Test. (replaced by ISO 9001:2000)
- ISO 9004, Quality Management & Quality System Elements—Guidelines (replaced by ISO 9004:2000)
- ISO 9004-7, Guidelines for CM (Draft) never released under this number. It was released as ISO 10007.
- ISO 10011-1, -2, and -3: Guidelines for Auditing Quality Systems
- ISO 10303-1 and on 1994 Industrial Automation Systems and Integration—Product Data Representation and Exchange. This standard had too many parts to list here. *Note:* Check ISO for corrections and revisions in the 10303 series. The collection appears to be subject to many changes. What follows are all the parts to this standard.
- ISO 12207, Information Technology—Software Life Cycle Processes
- ISO 13485, Quality Systems—Medical Devices—Particular Requirements for the Application of ISO 9001
- ISO 14001, Environmental Management Systems—Specification with Guidance for Use

- ISO 14004, Environmental Management Systems—General Guidelines on Principles, Systems, and Supporting Techniques
- ISO/IEC TR 15504-1 to -8:1998, Information Technology—Software Process Assessment—Parts 1 through 8
- BSI BS ISO/IEC TR 15846, Information Technology—Software Life Cycle Processes—Configuration Management
- ISO 20000-1: 2005, IT Service Management
- ISO 20000-2: 2005, IT Service Management Code of Practice, describing specific best practices for the processes within ISO 20000-1. Also, there are industry specific variations of ISO 9000: see canceled QS9000 (automotive), TL9000 (Telecommunications), QSRs (FDA), and AS9000 (Aerospace) elsewhere in this guide

International Versions

- AENORUNE-ENISO 10007, Quality Management Systems—Guidelines for Configuration Management (Spanish)
- AFNOR FD ISO 10007, Quality Management Systems—Guidelines for Configuration Management (French)
- AENOR UNE-ISO 10007, Quality Management Systems—Guidelines for Configuration Management (Spanish)
- UNI ISO 10007, Quality Management Systems—Guidelines for Configuration Management (Italian)
- CSACAN/CSA-ISO 10007:03, Quality Management Systems—Guidelines for Configuration Management (Canadian)
- TSE TS EN ISO 10007, Quality Management Systems—Guidelines for Configuration Management (Turkish)
- SNV SN EN ISO 10007, Quality Management Systems—Guidelines for Configuration Management (ISO 10007:1995); Trilingual version (English, German, and French)

Information Technology Infrastructure Library (ITIL)

- ITIL Framework—best practice in the provision of IT service. See http://www.ogc.gov.uk/index.asp?id=1000364

International Telecommunications Union (ITU)

- ITU-T Q.824.5, Stage 2 and Stage 3 Description for the Q3 Interface—Customer Administration: Configuration Management of V5 Interface Environment and Associated Customer Profiles—Series Q: Switching and Signaling Specifications of Signaling System
- ITU-R S.1252, Network Management—Payload Configuration Object Class Definitions for Satellite System Network Elements Forming Part of SDG Transport Networks in the Fixed-Satellite Service
- ITU-T J.705, IPTV Client Provisioning, Activation, Configuration and Management Interface Definition
- ITU-T X.792, Configuration Audit Support Function for ITU-T Applications—Series X: Data Networks and Open System Communications OSI Management—Management Functions and ODMA Functions

National Aeronautics and Space Administration (NASA) (U.S.) Superseded

- NPC 500-1 (or NHB 8040.2), Apollo CM Manual (released 1964) & MSC Supplement #1 (1965)
- PC-093, Maintenance & Configuration Control Requirements, NASA Pioneer Program (1965)

NASA Active

- NASA GPR 1410.2, Configuration Management (GSFC)
- NASA-LLIS-2596, Lessons Learned—Management Principles Employed in Configuration Management and Control in the X-38 Program
- NASA MPR 8040.1, Configuration Management, MSFC Programs/Projects
- NASA MWI 8040.1, Configuration Management Plan, MSFC Programs/Projects
- NASA MWI 8040.7, Configuration Management Audits, MSFC Programs/Projects
- NASA-STD 0005, NASA Configuration Management Standard
- NASA Software Configuration Management Guidebook (1995) from Software Assurance Technology Center
- NASA-STD-2201-93, Software Assurance Standard www.hq.nasa.gov/office/codeq/doctree/canceled/220193.pdf
- GMI 8040.1A CM (for satellite or ground system projects)
- JSC 30000 includes CM Requirements (for space station)
- JSC 31010 CM Requirements
- JSC 31043 CM Handbook
- KHB8040.2B CM Handbook
- KHB 8040.4 Payloads CM Handbook
- KPD 8040.6B CM Plan, National Space Transport System
- MM8040.5C CM Accounting & Reporting System
- MM 8040.12 Contractor CM Requirements
- MM8040.13A Change Integration & Tracking System
- MM1 8040.15 CM Objectives, Policies & Responsibilities
- MMI 8040.15B CM
- MSFC-PROC-1875 Contractor CM Plan Review Procedure
- MSFC-PROC-1916 CM Audit Procedures for MSFC Programs/Projects
- NSTS 07700, Volume IV, Configuration Requirements, Level II Program Definition & Requirements
- SSP 30000 Program Definition & Requirements Document
- Configuration Management Requirements, Space Station Project Office, October 29, 1990

National Institute of Standards and Technology (NIST) (U.S.)

- NIST 800-53 Recommended Security Controls for Federal Information Systems

National Technical Information Service (NTIS) (U.S.)

- ADA076542 CM
- ADA083205 Software CM
- NEI (Nuclear Energy Institute)

North Atlantic Treaty Organization (NATO)

- ACMP-1: Requirements for Preparation of CM Plans
- ACMP-2: Requirements for Identification
- ACMP-3: Requirements for Configuration Control
- ACMP-4: Requirements for Configuration Status Accounting
- ACMP-5: Requirements for Configuration Audits
- ACMP-6: NATO Configuration Management Terms and Definitions
- ACMP-7: Guidance on Application of ACMPs 1-6
- ACMP-2009: (DRAFT) NATO Guidance on Configuration Management
- ACMP-2100: (DRAFT) NATO Contractual Configuration Management Requirements

- AQAP-1: NATO Requirements for an Industrial Quality Control System
- AQAP-13: Software Quality Control Requirements
- AQAP-160: NATO Integrated Quality Requirements for Software Throughout the Life Cycle
- AQAP-2110: Requirements for Design, Development, and Production
- STANAG 4159: NATO Material Configuration Management Policy and Procedures for Multinational Projects
- STANAG 4427: Introduction to Allied Configuration Management

Norway

- Norwegian Defence Acquisition Regulation (ARF)
- ISO 10007 with a contractual adaption replaced in 2014 by NATO ACMP 2100 supplier's quality management systems shall comply with Allied Quality Assurance Requirements—AQAPs

Nuclear Information & Records Management Association (NIRMA)

- CM 1.0-2000 (R2006) DRAFT Configuration Management of Nuclear Facilities
- PP02-1989, Position Paper on CM
- PP03-1992, Position Paper for Implementing a CM Enhancement Program for a Nuclear Facility
- PP04-1994, Position Paper for CM Information Systems
- TG14-1992, Support of Design Basis Information Needs
- TG19-1996, CM of Nuclear Facilities
- TG20-1996, Drawing Management—Principles & Processes

Nuclear Regulatory Commission (NRC) (U.S.)

- 1.28, QA Program Requirements (Design & Construction)
- 1.33, QA program Requirements (Operations)
- 1.33, Rev 2, QA Requirements for the Design of Nuclear Power Plants
- 1.64, Quality Assurance Requirements for the Design of Nuclear Power Plants
- 1.88, Collection, Storage, & Maintenance of Power Plant Quality Assurance Records
- 1.123, QA Requirements for Control of Procurement of Items & Services for Nuclear Power Plants
- 1.152, Criteria for Programmable Digital Computer System Software in Safety-Related Systems of Nuclear Power Plants
- NUREG BR 0167, Software Quality Assurance Programs & Guidelines
- NUREG 1000, Generic Implications of ATWS Events at the Salem Nuclear Power Plant
- NUREG/CR 1397, An Assessment of Design Control Practices & Design Reconstitution Programs in the Nuclear Power Industry
- NUREG CR 4640, Handbook of Software Quality Assurance Techniques Applicable to the Nuclear Industry
- NUREG/CR- 5147, Fundamental Attributes of a Practical CM Program for Nuclear Plant Design Control
- NUMARC (Nuclear Management & Resources Council)
- NUMARC 90-12, Design Basis Program Guidelines, October 1990

Nuclear Science Advisory Committee (NSAC) (U.S.)

- NSAC-105, Guidelines for Design & Procedure Changes in Nuclear Power Plants

Occupational Safety & Health Administration (OSHA) (U.S. Department of Labor)

- OSHA 1910.119, Process Safety Management of Highly Hazardous Chemicals
- OSHA Standards for the Construction Industry (20 CFR Part 1926)
- OSHA Standards for General Industry (29 CFR Part 1910)

Quality Excellence for Suppliers of Telecommunications (QUEST)

A grouping of telecommunications service providers and suppliers founded by members of the Bell family.

- TL9000:2001. *Note:* Based on ISO 9000 and others

Radio Technical Commission for Aeronautics (RTCA)

- RTCA/DO-254, Design Assurance Guidance for Airborne Electronic Hardware
- RTCA/d0-178B, Software Considerations in Airborne Systems and Equipment Certification

SAE International

- AS9000, Aerospace Basic Quality System Standard (aerospace version of ISO 9000)
- AS9100D: Quality Systems Aerospace—Model for Quality Assurance in Design, Development, Production, Installation, and Servicing
- SAE/EIA 649, Configuration Management Standard

Simple Protocol for Independent Computing Environments (SPICE)

- Software Process Improvement and Capability Determination, various documents incorporated into ISO/IEC TR 15504:1998

Software Engineering Institute (SEI)

- http://www.sei.cmu.edu
- Capability Maturity Model® Integration (CMMI®), Continuous
- Capability Maturity Model® Integration (CMMI®), Staged
- CMMI® for Development, Version 1.2 Configuration Management

Spain

- AENOR UNE-EN 13290-5, Space Project Management—General Requirements—Part 5: Configuration Management
- AENOR UNE 135460-1-1, Road Equipment. Traffic Control Centers. Part 1-1: Remote Stations, Services Management. Communications and Configuration Services
- AENOR UNE 73101, Configuration Management in Nuclear Power Plants

TechAmerica – Now SAE

- SAE/EIA 649, Configuration Management
- GEIA-HB-649, Configuration Management Handbook
- GEIA-859A, Data Management
- TECHAMERICA CMB 4-1A, Configuration Management Definitions for Digital Computer Programs (withdrawn)
- TECHAMERICA CMB 5-A, Configuration Management Requirements for Subcontractors/Vendors
- TECHAMERICA CMB 6-10, Education in Configuration and Data Management

- TECHAMERICA CMB 6-1C, Configuration and Data Management References
- TECHAMERICA CMB 6-2, Configuration and Data Management In-House Training Plan
- TECHAMERICA CMB 6-9, Configuration and Data Management Training Course
- TECHAMERICA CMB 7-1, Electronic Interchange of Configuration Management Data
- TECHAMERICA CMB 7-2, Guideline for Transitioning Configuration Management to an Automated Environment
- TECHAMERICA CMB 7-3, CALS Configuration Management SOW and CDRL Guidance
- TECHAMERICA GEIA-TB-0002, System Configuration Management Implementation Template (Oriented for a U.S. Military Contract Environment)

Telcordia Technologies (now part of Telefonaktiebolaget LM Ericsson)

Also see http://www.ericsson.com/ourportfolio/telcordia_landingpage.

- GR-209, Generic Requirements for Product Change Notices (replaces TR-OPT-000209 Issue 2)
- GR-454, Generic Requirements for Supplier-Provided Documentation
- COMMON LANGUAGE Equipment Identification Codes

Telecommunications Industry Forum

- The Alliance for Telecommunications Industry Solutions, in cooperation with the Uniform Code Council (UCC):
  - TCIF-97-001, Item Interchangeability Guideline
  - TCIF-97-009, U.P.C. Implementation Guidelines—Product Identification Coding Schemes Guidelines

Telecommunications Management Network (TMN)

- AFNOR NF EN 300376-1, Q3 Interface to the Access Network (AN) for Configuration Management of V5 Interfaces and Associated Users Ports—Part 1: Q3 Interface Specification

U.K. Civil Aviation Authority

- Issue 94, British Civil Airworthiness Requirements, Section A

U.K. Ministry of Defense

- 00-22, The Identification & Marking of Programmable Items
- AVP 38, Configuration Control, Section 3
- MODUK DEF STAN 02-28, Configuration Management Nuclear Submarines in Service Support
- MOD Def Stan 05-57, Configuration Management Policy and Procedures for Defense Material
- NES 41, Requirements for Configuration Management and Ship Fit Definitions

Appendix 2: Acronyms

Acronym	Definition
2D	Two dimensional
3D	Three dimensional
3di	Three dimensional interactive
3P	Policies, procedures, and practices
5G	Fifth generation cellular technologies
6G	Sixth generation cellular technologies
7G	Seventh generation cellular technologies
AC	Actual cost
ACD	Allocated configuration documentation
ACDM	Association for Clinical Data Management; Association for Configuration and Data Management
ACWP	Actual cost of work performed
AECA	Arms Export Control Act (U.S.)
AFE	Authorization for expenditure
AL	Artwork list
AMC	Army material command (U.S.)
AMCR	AMC regulation (U.S.)
ANA	Army, Navy, Airforce (U.S.)
AR	Army regulation (U.S.); Augmented reality
ASME	American Society of Mechanical Engineers
B&P	Bid and proposal
BAC	Budget at completion
BAE	British Aerospace
BCE	Before common era
BCP	Bridging control protocol
BCWP	Budgeted cost of work performed
BCWS	Budgeted cost of work scheduled
BGP	Border gateway protocol
BI	Business intelligence
BOD	Bill of documents
BOM	Bill of materials
BYO	Bring your own
C_2H_2	Acetylene
CA	Contract award
CaC_2	Calcium carbide
CAD	Computer-aided design
$Ca(OH)_2$	Calcium hydroxide
CCB	Configuration control board
CCDM	Certified configuration data manager
C&DM	Configuration and data management
CCM	Coordinate measuring machines
CCCP	Сою́з Сове́тских Социалисти́ческих Респу́блик
CDR	Critical design review
CER	Cost estimating relationship
CFE	Customer-furnished equipment

CFO	Chief financial officer
CGMA	Current good manufacturing practices (U.S.)
CTFL	Certified tester – foundation level
CI	Configuration item or configured item
CII	Configuration item identifier or configured item identifier
CLIN	Contract line item number
CLM	Contract life cycle management
cm	Centimeter
CM	Configuration management
CMDB	Configuration management data base
CMMI	Capability Maturity Model® Integration®
CMPIC	CM Process Improvement Center
CNC	Computer numerical control
COC	Certificate of conformance
COGS	Cost of goods sold
CORE	Controlled requirements expression
COTS	Commercial off-the-shelf
CPAF	Cost plus award fee
CPFF	Cost plus fixed fee
CPIF	Cost plus incentive fee
CPU	Central processing unit
CS	Cost sharing
CSA	Configuration status accounting
CSCI	Computer software configuration item or Computer software configured item
CT	Computed tomography
CV	Cost variance
DCAA	Defense Contract Audit Agency (U.S.)
DD	Defense document
DDoS	Distributed denial of service
DDTC	Directorate of Defense Trade and Controls
DED	Directed energy deposition
DEF-CON	Defence condition
DEFSTAN	Defence standard (UK)
DevOps	Development and operations
DFAR	DOD FAR Supplement (U.S.)
DFMEA	Design failure mode and effects analysis
DIL	Deliverable items list
DIY	Do-it-yourself
DL	Data list
DM	Data management
DNA	Deoxyribonucleic acid
DND	Department of National Defence (Canada)
DOD	Department of Defense (U.S.)
DOE	Department of Energy (U.S.)
DOORS	Dynamic object-oriented requirements system
dpi	Dots per inch
DPI	During process inspection
DR	Disaster recovery
DRL	Data requirements listing/data requirements list
DRP	Disaster recovery plan
DUPRO	During process (inspection)

DVD	Digital versatile disc
E. coli	*Escherichia coli*
EAC	Estimate at completion
EAD	Emergency airworthiness directive
EAO	Everything-at-once
EAR	Export administration regulations
EASA	European Aviation Safety Agency
EB	exabyte
EBIT	Earnings before interest and taxes
ECO	Engineering change order
ECSS	Expeditionary combat support system
ECP	Engineering change proposal
ECR	Engineering change request
ECU	Electronic control unit
EDM	Electrical discharge machines
EIA	Electronic industries alliance
ELIN	Exhibit line item number
ESA	European Space Agency
ERP	Engineering resource planning
ETC	Estimate to complete
FAA	Federal Aviation Administration
FAR	Federal acquisition regulations (U.S.)
FCA	Functional configuration audit
FCD	Functional configuration documentation
FDA	Food and Drug Administration (U.S.)
FDD	Feature driven development
FDIC	Federal Deposit Insurance Corporation (U.S.)
FDN	Family identification number
FFP	Firm fixed price
FIFO	First-in first-out
FMEA	Failure modes effects analysis
FOB	Freight on board
FOIA	Freedom of Information Act (U.S.)
FOUO	For official use only (U.S.)
FPAF	Fixed price award fee
FPGA	Field-programmable gate array
FPIF	Fixed price incentive firm (target)
FQR	Formal qualification review
FSA	Functional systems audit
FSMA	Food Safety Modernization Act (U.S.)
FSIS	Food Safety and Inspection Service (U.S.)
FSLIC	Federal Savings and Loan Insurance Corporation (U.S.)
FSR	Fixed price with cost redetermination
FSRD	Functional service request disposition
FTC	Federal Trade Commission (U.S.)
G&A	General and administrative
GCM	Gross combined mass
GDT	General dimensioning and tolerancing
GFE	Government furnished equipment (U.S.)
GIDEP	Government and industry data exchange program
GMC	General Motors Corporation

GSFC	Goddard Space Flight Center (U.S. NASA)
GTS	Global technology solutions
GVCS	Global village construction set
H_2O	Water
HDBK	Handbook
HQI	Hardware quality instruction
IABG	Industrieanlagen-Betriebsgesellschaft mbH
IBIS	Issue-based information system
IBM	International Business Machines
IC	Integrated circuit
ICBM	Intercontinental ballistic missile
ICD	Interface control drawing; International Classification of Diseases
ICT	Information and communication technologies
ICWG	Interface control working group
ID	Identification; Identifier
IEEE	Institute of Electrical & Electronic Engineers
IL	Index list
IP	Intellectual property
IR&D	Independent research and development
IS-IS	Intermediate system-to-intermediate system
ISO	International Organization for Standardization; International Organization for Standards
ISP	Internet service provider
I&T	Integration and test
IT	Information technologies
ITAR	International traffic in arms regulations (U.S.)
ITIL	Information technology infrastructure library
JAD	Joint application design/Joint application development
JEM-EF	Japanese experiment module—exposed facility
JPL	Jet Propulsion Laboratory
JSD	Jackson system development
KCl	Potassium chloride
kg	Kilogram
km	Kilometer
LAN	Local area network
LLC	Limited liability corporation
LM	Lunar module
LSA	Link state advertisements
LTMPF	Low temperature microgravity physics facility
m	Meter
MAR	Mission assurance requirements
MBE	Model based engineering
MCAS	Maneuvering characteristics augmentation system
MIRR	Material inspection and receiving report
mm	Millimeter
MMU	Manned maneuvering unit
MoSSEC	Modeling and simulation information in a collaborative system
MRB	Material review board
MRI	Magnetic resonance imaging
MTBF	Mean time between failures
MX	Missile X

NA	Not applicable
NANOG	North American Network Operators Group
NASA	National Aeronautics and Space Administration (U.S.)
NATO	North Atlantic Treaty Organization
NAVAIR	Naval Air Systems Command (U.S.)
NCSC	National Cyber Security Centre (UK)
NDIA	National Defense Industrial Association (U.S.)
NOAA	National Oceanic and Atmospheric Administration (U.S.)
NOR	Notice of revision
NSWC	Naval Surface Warfare Center (U.S.)
O&M	Operations and maintenance
OEM	Original equipment manufacturer
OH	Overhead
OS	Operating system
OSPF	Open shortest path first
PBF	Powder bed fusion
PBP	Piece-by-piece
PC	Personal computer
PCA	Physical configuration audit
PCD	Product configuration documentation
PD&D	Product design and development
PDF	Portable data format
PD/LM	Product data/life cycle management
PDM	Product data management
PDR	Preliminary design review
PFMEA	Process failure mode and effects analysis
PL	Parts list
PLIN	Provisioning list line item number
PLM	Product life cycle management
PMP	Project management professional
PPAP	Production part approval process
PROM	Programmable read-only memory
PRR	Production readiness review
PWA	Printed wiring assembly
QFD	Quality function deployment
QR	Quick response
RDECOM	Research Development and Engineering Command
REE	Requirements engineering environment
RFID	Radio frequency identification
Rh	Rhesus
RIP	Routing information protocol
RTC	Resolution Trust Corporation (U.S.)
SBRE	Scenario-based requirements elicitation
SCN	Specification change notice (U.S.)
SFR	System functional review
SME	Subject matter expert
SMM	Solar maximum mission
SOW	Statement of work
SpaceX	Space Exploration Technologies
SQI	Software quality instruction
SQL	Structured query language

SRR	System requirements review
SSC	Single-string approach based on criticality
SSM	Soft systems methodology
STANAG	Standardization Agreement (NATO)
STS	Space Transportation System (U.S.)
SUMO	Superconducting microwave oscillator
SV	Schedule variance
SVA	Systems verification audit
TAA	Technical assistance agreement
TBD	To be determined
TBN	To be negotiated
TBR	To be reviewed
TBS	To be specified
TCP	Technical assistance agreement
TCTO	Time compliance technical order (U.S.)
TDP	Technical data package
TEMP	Test and Evaluation Master Plan (U.S. DOD)
TIEMPO	Test, Inspection & Evaluation Master Plan Organized
TO	Task order
TPAD	Trunnion pin acquisition device
TPC	Total program cost
TRR	Test readiness review
TTCP	Technology transfer control plan
TTR	Tabletop review
TU/e	Technical University of Eindhoven
UART	Universal asynchronous receiver/transmitter
UCC	Uniform commercial code (U.S.)
USB	Universal serial bus
USML	U.S. munitions list
U.S.S.R.	Union of Soviet Socialist Republics
USTM	User skills task match
UTC	Coordinated universal time
VHS	Video home system
VP	Vice president
VR	Virtual reality
VTC	Value to the company
VW	Volkswagen
WA	Work authorization
WAN	Wide area network
WAWF	Wide area workflow
WAWF RR	Wide area workflow receiving report
WL	Wire list
XP	Extreme programming

Appendix 3: Bibliography

Adami, C., Bryson, D. M., Ofria, C., & Pennock R. T. (Eds.) (2012). *Artificial life XIII: Proceedings of the 13th International Conference on the Synthesis and Simulation of Living Systems.* 11–18. MIT Press. July. Retrieved October 1, 2014, from http://mitpress.mit.edu/sites/default/files/titles/free_-download/9780262310505_Artificial_Life_13.pdf

Alighieri, D. (d.1321). *The divine comedy: Complete the vision of paradise, purgatory and hell.* Translated by Rev. H. F. Cary. Salt Lake City, UT: Project Gutenberg eBook. November 30, 2012. http://www.gutenberg.org/files/8800/8800-h/8800-h.htm

Althin, T. K. (1948). *C. E. Johansson, 1864–1943: The master of measurement.* Stockholm, Sweden: Nordish-Rotogravyr.

American Marketing Association. (2004, October). Definition of marketing. Retrieved March 14, 2014, from https://www.ama.org, https://www.ama.org/AboutAMA/Pages/Definition-of-Marketing.aspx

Angier, N. (2009, March 30). The biggest of puzzles brought down to size. *The New York Times.*

Association for Clinical Data Management (ACDM) established 1988, https://www.acdmglobal.org/

Bannatyne, G. (1896). The Bannatyne manuscript. Retrieved May 20, 2014, from https://openlibrary.org/books/OL7034966M/The_Bannatyne_manuscript

Barkley, B. (2008). *Project management in new product development.* New York: McGraw-Hill.

Barrett, R., Haar, S., & Whitestone, R. (1997, April 25). Routing snafu causes internet outage. *Interactive Week.*

Baumann, P. (2007, April 24). What's the difference between overhead and G&A? Retrieved October 3, 2013, from http://www.theasbc.org, http://www.theasbc.org/news/15315/The-ASBC-Community-News-Whats-the-difference-between-Overhead-and-GA.htm

Beck, K. et al. (2001). *Principles behind the Agile Manifesto.* Washington, DC: Agile Alliance.

Berczuk, S., Appleton, B., & Cowham, R. (2005, November 30). An Agile perspective on branching and merging. Retrieved November 22, 2013, from http://www.cmcrossroads.com/article/agile-perspective-branching-and-merging

Bidault, F., Despres, C., & Butler, C. (1998). *Leveraged innovation: Unlocking the innovation potential of strategic supply.* Houndmills: Macmillan Press Ltd.

Blaustein, J. (2013). What is the purpose of CM? Judith Blaustein of Blaustein and Associates. https://www.linkedin.com/groups/What-is-purpose-Configuration-Management-4393863.S.263480123?trk=groups_most_popular-0-b-ttl&goback=%2Egmp_4393863

Boole, G. (1854). *An investigation of the laws of thought.* Palm Springs, CA: Wexford College Press.

Bordes, F. (1968). *The old stone age.* Translated from French by J. E. Anderson. London: Weidenfeld & Nicolson.

Bottemiller, H. (2012, October 8). 2.5 million pounds of recalled Canadian beef entered U.S. *Food Safety News.* Retrieved November 26, 2013, from http://www.foodsafetynews.com/, http://www.foodsafetynews.com/2012/10/2-5-million-pounds-of-recalled-canadian-beef-entered-u-s/#.U9FDCPldV8E

Bowen, L. R. (2007). *Advanced configuration management.* Boulder, CO: Larry R. Bowen.

Brewster, S. (2013). Elon Musk shapes a 3D virtual rocket part with his hands. Accessed April 22, 2014, from https://gigaom.com/2013/09/05/elon-musk-shapes-a-3d-virtual-rocket-part-with-his-hands/

Brown, H. T. (2010). *Five hundred and seven mechanical movements*, 18th ed. Mendham, NJ: Brown and Seward republished The Astragal Press.

Butler, J. G. (1997). *Concepts in new media.* Tucson, AZ: University of Arizona.

Butler, J. G. (1998). *A history of information technology and systems.* Tucson, AZ: University of Arizona.

Bye, E. (Director). (1991). *Red Dwarf, Series IV,* "Meltdown." [Motion Picture]. London: BBC Broadcasting House.

Calantonea, R. J., Kimb, D., Schmidtc, J. B., & Cavusgil, S. T. (2006). The influence of internal and external firm factors on international product adaptation strategy and export performance: A three-country comparison. *Journal of Business Research*, Vol. 59, No. 2, 176–185.

Centers for Disease Control and Prevention. (2012). *International classification of diseases—Tenth revision-clinical modification (ICD-10-CM).* Atlanta, GA: U.S. Government.

Charette, R. N. (2013, December 6). The U.S. Air Force explains its $1 billion ECSS bonfire. *IEEE Spectrum.* Retrieved December 24, 2013, from https://spectrum.ieee.org/riskfactor/aerospace/military/the-us-air-force-explains-its-billion-ecss-bonfire

Choudhury, G. (2005). Prioritized treatment of specific OSPF Version 2 packets and congestion avoidance, network working group, the Internet Society. Retrieved July 24, 2014, from http://www.rfc-base.org/rfc-4222.html

Christel, M., & Kang, K. C. (1992). Issues in requirements elicitation, technical report CMU/SEI-92-TR-012, ESC-92-TR-012. Pittsburgh, PA: Software Engineering Institute, Carnegie Mellon® University.

Christiansen, B. (2018, December 3). The use of AI and VR In maintenance management. *Engineering.Com.* https://www.engineering.com/AdvancedManufacturing/ArticleID/18100/The-Use-of-AI-and-VR-In-Maintenance-Management.aspx

CMStat (2018, August). Hardware configuration items in As-X configuration management by aerospace & defense contractors (part 2 of 2). CMstat, 3960 Howard Hughes Parkway, Suite 500, Las Vegas, NV 89169. https://cmstat.com/hardware-configuration-items-in-as-x-configuration-management-by-aerospace-defense-contractors-part-2-of-2/

Conover, E. (2018, March 7). Some meteorites contain superconducting bits. *ScienceNews.* Retrieved from https://www.sciencenews.org/article/some-meteorites-contain-superconducting-bits

Coonrod, J. (2010, November 19). What is outgassing and when does it matter? Retrieved July 22, 2014, from Rog Blog: http://mwexpert.typepad.com/rog_blog/2010/11/19/

Cooper, S. B. (2014). Autonomous system routing protocol. Ehow.com. Retrieved July 24, 2014, from http://www.ehow.com/facts_7260499_autonomous-system-routing-protocol.html#ixzz2iAnG97ad

Copeland, L. (2001, December 3). Extreme programming. Retrieved July 12, 2013, from http://www.computerworld.com/s/article/66192/Extreme_Programming.

Cusumano, M. A., Mylonadis, Y., & Rosenbloom, R. S. (1991, March 25). Strategic maneuvering and mass-market dynamics: The triumph of VHS over Beta. WP# BPS-3266-91. Retrieved October 20, 2013, from http://dspace.mit.edu/bitstream/handle/1721.1/2343/SWP-3266-23735195.pdf

Czinkota, M. R., & Ronkainen, I. A. (2003). *International marketing.* South-Western College.

Dahlqvist, A. P. (2001). *CM i ett produktperspektiv—ställer hårdare krav på integration av SCM och PDM.* Stockholm, Sweden: The Association of Swedish Engineering Industries.

DCAA, U.S. (2013). *Cost accounting standards, Defense Contract Audit Agency (Chapter 8-400 Section 4).* U.S. Government.

Demmin, A. (1911). An illustrated history of arms and armour: From the earliest period to the present time. Translated by C. C. Black. London: M.A.G. Bell and Sons from California Digital Library. Accessed on June 22, 2014, from https://archive.org/details/illustratedhisto00demmrich.

Diringer, D., & Minns, E. (2010). *The alphabet: A key to the history of mankind.* Whitefish, MT: Kessinger Publishing.

DOD, U.S. (1985). *MIL-STD-490a, Military Standard Specification Practices.* Washington, DC: U.S. Department of Defense. U.S. Government, Pentagon.

DOD, U.S. (1987). *DOD-STD-100, Engineering Drawing Practices, Revision D.* Washington, DC: U.S. Government, Pentagon.

DOD, U.S. (1992). *MIL-STD-973, Configuration Management.* Washington, DC: U.S. Government, Pentagon.

DOD, U.S. (2001). *MIL-HDBK-61a, Military Handbook: Configuration Management Guidance.* Washington, DC: U.S. Government, Pentagon.

DOD, U.S. (2005). *Dictionary of Military and Associated Terms.* Washington, DC: U.S. Government, Pentagon.

DOD, U.S. (2013). *The Defense Standardization Program Journal.* Retrieved December 26, 2013, from http://www.dsp.dla.mil/, http://www.dsp.dla.mil/APP_UIL/displayPage.aspx?action = content&accounttype = displayHTML&contentid = 75

DOD, U.S. (2018). *MIL-STD-31000B, U.S. Department of Defense Standard Practice: Technical Data Packages.* Washington, DC: U.S. Government, Pentagon.

Doyle, A. C. (1903, December). The adventure of the dancing men. *The Strand Magazine.*

Dragland, Å. (n.d.). Big data—for better or worse. Retrieved September 28, 2013, from http://www.sintef.no/home/Press-Room/Research-News/Big-Data—for-better-or-worse/

Drucker, P. F. (1967). *The effective executive.* New York, NY: Harper/Collins Publishers.

Durak, D., et al. (2018). *Advances in aeronautical informatics technologies towards flight 4.0.* Cham, Switzerland: Springer International Publishing.

EIA. (2003). *EIA-632, Processes for Engineering a System.* EIA.

EMC Corporation. (2012). *IDC digital universe study: Big data, bigger digital shadows and biggest growth in the Far East.* Colorado Springs, CO: EMC Corporation.

Erl, T., Puttini, R., & Mahmood, Z. (2013). *Cloud computing: Concepts, technology & architecture.* Upper Saddle River, NJ: Prentice Hall.

Euler, L. (1735). *Seven bridges of Königsberg*. St. Petersburg, Russia: Imperial Russian Academy of Sciences.

FDA, U.S. (2010). *Guidance for Industry Standards for Securing the Drug Supply Chain—Standardized Numerical Identification for Prescription Drug Packages: Final Guidance*. Silver Spring, MD: U.S. Department of Health and Human Services Food and Drug Administration, U.S. Government.

Federal Government, U.S. (2013). *U.S. Federal Acquisition Regulations 52.249.X, Termination for Default sub-part (a)(1)ii and (a)(1)iii*. U.S. Government.

Fletcher, L., Kaiser, J., Johnson, C., Shea, C., & Cole, B. (2009). Configuration management. http://www.dcs.gla.ac.uk/~johnson/papers/Wingman/Config_Management.pdf

Ford, H., & Crowther, S. (2005). *My life and work*. Salt Lake City, UT: Project Gutenberg eBook. p. 72. Chapter IV.

Forging Industry Association. (2007). Long lifecycle plus high performance makes forged components lowest cost. Retrieved July 25, 2014, from www.forging.org

Francis, S. (2018). FACC joins aerospace industry association SPACE. *Composites world* e-zine. Accessed December 31, 2018, from https://www.compositesworld.com/news/facc-joins-aerospace-industry-association-space

Fujimoto, T. (1999). *The evolution of a manufacturing system at Toyota*. New York: Oxford University Press.

Galbraith, J. R., & Kazanjian, R. K. (1986). *Strategy implementation structure, systems and process*. St. Paul, MN: West Publishing Company.

Gardner, J. (2012, December 19). IT stability and business innovation are not enemies. Retrieved December 20, 2012, from http://www.enterprisecioforum.com/en/blogs/jimgardner/it-stability-and-business-innovation-are.

Garwood, D. (2004). *Bills of material for a lean enterprise*. Marietta, GA: Dogwood Publishing Company.

Ghose, S., & Lee, T. S. (2010). *Intelligent transportation systems*. Boca Raton, FL: CRC Press/Taylor & Francis Group.

Gladwell, M. (2005). *Blink: The power of thinking without thinking*. New York: Little, Brown and Company.

Goldratt, E. M., & Cox, J. (1992). *The goal*. Great Barrington, MA: North River Press.

Goodwin, R. M. (1967). *A growth cycle: Socialism, capitalism and economic growth*. Feinstein, C. H. (Ed.). Chapter 4. Cambridge: Cambridge University Press.

Gopwani, J. (2010, March 11). Subaru builds on loyal customer base. *USA Today*.

Gorzelany, J. (2012, December 29). Biggest auto recalls of 2012 (and why they haven't affected new car sales). *Forbes*. Retrieved November 26, 2013, from http://www.forbes.com/sites/jimgorzelany/2012/12/29/biggest-auto-recalls-of-2012/

Grant, C. C. (2007). A look from yesterday to tomorrow on the building of our safety infrastructure, National Fire Protection Association. Presented at NIST Centennial Standards Symposium, National Institute of Standards and Technology Special Publication 974, 164pp. Boulder, CO: NIST. Retrieved March 7, 2001, from https://www.govinfo.gov/content/pkg/GOVPUB-C13-992cf96b57768b2553927fc04850fd99/pdf/GOVPUB-C13-992cf96b57768b2553927fc04850fd99.pdf

Grodzinsky, F. S., Miller, K., & Wolf, M. J. (n.d.). The ethical implications of the messenger's haircut: Steganography in the digital age. Retrieved September 20, 2013, from http://biblioteca.clacso.edu.ar/ar/libros/raec/ethicomp5/docs/htm_papers/25Grodzinsky,%20Frances%20S.htm.

Groseclose, A. R. (2010). *Estimating of forging die wear and cost*. Columbus, OH: The Ohio State University.

Grush, L. (2018, Oct 22) NASA lost a rover and other space artifacts due to sloppy management, report says. *The Verge*. Retrieved from https://www.theverge.com/2018/10/22/18009414/nasa-lunar-rover-historical-artifacts-lost-space-shuttle

Hammant, P. (2007, April 27). Introducing branching by abstraction. Retrieved July 23, 2014, from http://paulhammant.com/blog/branch_by_abstraction.html#!

Harland, D. M., & Lorenz, R. D. (2005). *Space systems failures: Disasters and rescues of satellites, rockets and space probes*. London: Praxis Publishing Ltd.

Hass, A. M. J. (2003), *Configuration management principles and practice*. Boston, MA: Addison-Wesley Pearson Education, Inc.

Haughey, D. (2013, July 9). *Project management body of knowledge*. Retrieved August 4, 2013, from http://www.projectsmart.co.uk/pmbok.php

Hebden, R. (1995). *Effecting organizational change*. New York: IBM.

Helper, T. (2017): Reuters, Airbus knew of software vulnerability before A400M crash. Reuters Business News. Retrieved from https://www.reuters.com/article/us-airbus-a400m/airbus-knew-of-software-vulnerability-before-a400m-crash-idUSKBN1D819P

Herodotus of Halicarnassus. (2003). *The histories, book 5*, Translated by A. De Selincourt. London: Penguin Group.

Heyerdahl, T. (1950). *The Kon-Tiki expedition: By raft across the south seas.* Skokie, IL: Rand McNally & Comp.

Heyerdahl, T. (1971). *The Ra expeditions.* New York, NY: Doubleday.

Hunt, T. (2019). The 773 million record "Collection #1" data breach. Retrieved from https://www.troyhunt.com/the-773-million-record-collection-1-data-reach/

IASB. (2010, September). *Conceptual framework for financial reporting 2010.* Retrieved April 12, 2014, from http://www.iasplus.com/en/standards/other/framework

IBM. (2013, September 22). What is big data?—Bringing big data to the enterprise. Retrieved August 12, 2014, from http://www-01.ibm.com/software/au/data/bigdata/

IEEE. (2005). *IEEE-1220-2005, IEEE Standard for Application and Management of the Systems Engineering Process. Piscataway,* NJ: IEEE.

IEEE. (2009). *IEEE-P1233, IEEE Guide for Developing System Requirements Specifications.* Piscataway, NJ: IEEE.

IEEE. (2012). *IEEE-828-2012—IEEE Standard for Configuration Management in Systems and Software Engineering.* Piscataway, NJ: IEEE.

ISO. (2002). *ISO 10303-21:2002, Industrial Automation Systems and Integration—Product Data Representation and Exchange—Part 21: Implementation Methods: Clear Text Encoding of the Exchange Structure.* Geneva, Switzerland: ISO.

ISO. (2003). *ISO 10007:2003, International Standard—Quality Management System—Guidelines for Configuration Management.* Geneva, Switzerland: ISO.

ISO. (2003). *ISO 1006:2003, Quality Management Systems—Guidelines for Quality Management in Projects.* Geneva, Switzerland: ISO.

ISO. (2008). *ISO 9001:2008, Quality Management Systems—Requirements.* Geneva, Switzerland: ISO.

ISO. (2008). *ISO/IEC-15288:2008, Systems Engineering: System Life Cycle Processes.* Geneva, Switzerland: ISO.

ISO. (2008). *ISO/IEEE 12207:2008, Systems and Software Engineering—Software Life Cycles.* Geneva, Switzerland: ISO.

Jaeger, E. (1999). *Wildwood Wisdom.* Bolinas, CA: Shelter Publications, Inc.

Jain, S. C. (1989). Standardization of international marketing strategy: Some research hypotheses. *Journal of Marketing, 53*(1), 70–79.

Jamieson, A., & Calabrese, E. (2017). BBC: Investigation launched into 'serious' Airbus A380 engine failure. Retrieved from https://www.nbcnews.com/storyline/airplane-mode/investigation-launched-serious-airbus-a380-engine-failure-n806301

Jones, C. (2011). *Evaluating ten software development methodologies.* Narragansett, RI: Namcook Analytics LLC (Casper Jones & Associates LLC). Retrieved June 14, 2014, from http://namcookanalytics.com/evaluating-ten-software-development-methodologies/

Keller, J. (2018). Navy orders diminishing manufacturing sources (DMS) FPGAs to keep F-35s and other military aircraft flying. *Military and Aerospace Electronics.* Retrieved from https://www.military-aerospace.com/articles/2018/11/diminishing-manufacturing-sources-dms-fpgas-military-aircraft.html?cmpid=enl_mae_weekly_2018-11-21&pwhid=4255a1f5be8f3980ba2f34063c0cb109b8319d5b33b4ad8f1ac4c718d2e3fed61b376465447d983e7f0e5204c306467c323993fa2563ea30d31f3da86303db7f&eid=417926188&bid=2306576

Kent, K., & Ng, V. (2018). Smartphone data tracking is more than creepy—here's why you should be worried. The Conversation. Retrieved February 28, 2019, from http://theconversation.com/smartphone-data-tracking-is-more-than-creepy-heres-why-you-should-be-worried-91110

Kirvan, P. (2009). *IT disaster recovery plan template v 1.0.* Retrieved October 27, 2013, from http://searchdisasterrecovery.techtarget.com/, http://searchdisasterrecovery.techtarget.com/Risk-assessments-in-disaster-recovery-planning-A-free-IT-risk-assessment-template-and-guide

Labovitz, C., Ahuja, A., & Jahanian, F. (1998). *Experimental study of internet stability and wide-area backbone failures.* Ann Arbor, MI: University of Michigan.

Lamichhaney, S., Han, F., Webster, M. T., Andersson, L. B., Rosemary Grant, B. R., & Grant, P. R. (2016, November 23) Report: Rapid hybrid speciation in Darwin's finches. *Science.* Retrieved from http://science.sciencemag.org/content/early/2017/11/20/science.aao4593

Langer, A. M., & Yorks, L. (2013). *Strategic IT: Best practices for managers and executives (CIO).* Hoboken, NJ: John Wiley & Sons.

Lee, G. (2007). So you want to be a systems engineer. Video presentation. Pasadena, CA: Jet Propulsion Laboratory.

Lee, J. L., & Dave, P. (2018, January 13). Nigerian firm takes blame for routing Google traffic through China. Reuters Technology News. Retrieved from https://www.reuters.com/article/us-alphabet-disruption/nigerian-firm-takes-blame-for-routing-google-traffic-through-china-idUSKCN1NI2D9

Leibniz, G. W. (1703). Explication de l'arithmétique binaire VII.223. Retrieved August 22, 2014, from www.leibniz-translations.com.

Leverage, IOT for All (2018). Overview of indoor tracking. Retrieved December 17, 2018, from https://www.iotforall.com/overview-indoor-tracking-technology/?utm_source=newsletter&utm_campaign=IFA_Newsletter_Dec20_2018

Lindsey, B. (2017). Intercontinental ballistic missiles (ICBM): 50 years of sustainment. Presentation at CMPIC seminars, workshops, and training event, August 27, 2017, Rosen Centre Hotel. 9840 International Drive, Orlando, Florida 32819.

Lipson, H., & Kurman, M. (2013). *Fabricated: The new world of 3D printing.* Hoboken, NJ: John Wiley & Sons.

Llewellyn, P., Geilinger, S., & Field, A. (Directors). (1996). *Two fat ladies.* [Motion Picture]. London: BBC Broadcasting House.

Longfellow, H. W. (2013, September 28). *Paul Revere's ride.* Retrieved September 6, 2014, from http://poetry.eserver.org/paul-revere.html

Lönnrot, E. (2008). *The Kalevala: An epic poem after oral tradition.* New York, NY: Oxford World's Classics.

Lorenz, E. U. (2015, March). Predictability: Does the flap of a butterfly's wings in Brazil set off a tornado in Texas? *Indiana Academy of Science Classics, 20*(3), 260–263.

Lotka, A. J. (1925). *Elements of physical biology.* Baltimore, MD: Lippincott Williams & Wilkins.

Luckerson, V. (2013, November 11). Bring your own tech: Why personal devises are the future of work. *TIME Magazine U.S. Edition.*

Maher, M. W., Stickney, C. P., & Weil, R. L. (2011). *Managerial accounting: An introduction to concepts, methods, and uses.* Independence, KY: Cengage Learning.

McTiernan, J. (Director). (1992). *Medicine man.* [Motion Picture].

Miles, S. (2018, March 31). It's official: Termites are just cockroaches with a fancy social life. *Science News,* 193(6), 7. Retrieved from https://www.sciencenews.org/article/itsofficial-termites-are-just-cockroaches?utm_source=editorspicks030418&utm_medium=email&utm_campaign=Editors_Picks

Miller, E. R. (1960–1972). Conversations with Earl Ray "E. R." Miller Sr. (K.L. Robertson, interviewer).

Miller, G. A. (1956). The magical number seven, plus or minus two: Some limits on our capacity for processing information. *The Psychological Review, 63,* 81–97.

Mobley, K. (2003). Major asset life cycles can be managed effectively with configuration management. Plant Services. Retrieved from https://www.plantservices.com/articles/2003/182/

Morris, E. (2007). *From horse power to horsepower.* Berkeley, CA: University of California Transportation Center—ACCESS number 30.

NASA. (2004). *NOAA N-Prime Mishap Investigation Final Report.* NASA.

NOAA, U.S. (2014, July). Magnetic field calculators. Retrieved July 2, 2014, from http://www.ngdc.noaa.gov, http://www.ngdc.noaa.gov/geomag-web/#declination

Noction. (2013, September 20). Routing anomalies—their origins and how do they affect end users. Retrieved July 24, 2014, from http://www.noction.com/, http://www.noction.com/blog/how_routing_anomalies_occur

Nokia Solutions. (2014). Solutions. Retrieved June 10, 2014, from http://nsn.com/, http://nsn.com/portfolio/solutions

Open Security Architecture. (2006, January). Definitions overview. Retrieved July 10, 2011, from http://www.opensecurityarchitecture.org/cms/index.php, http://www.opensecurityarchitecture.org/cms/definitions

Open Source Ecology. (2014). Open source blueprints for civilization build yourself. Retrieved July 25, 2014, from http://opensourceecology.org/, http://opensourceecology.org/#sthash.BtrJ7jWL.dpuf.

Ouellette, N. T. (2019, January 4). Flowing crowds, *Science, 363*(6422), 27–28. DOI: 10.1126/science.aav9869

Paley, W. (1743–1805). *Natural theology.* New York: American Tract Society. Retrieved October 15, 2013, from https://archive.org/details/naturaltheology00pale.

Perlman, R. (1983). Fault-tolerant broadcasting of routing information. *Computer Networks, 7,* 395–405.

Poe, E. A. (1904). *The purloined letter (from the works of Edgar Allan Poe—Cameo Edition).* New York, NY: Funk & Wagnalls Company.

Pries, K. H., & Quigley, J. M. (2009). *Project management of complex and embedded systems.* Boca Raton, FL: Auerbach Publications/Taylor & Francis Group.

Pries, K. H., & Quigley, J. M. (2012). *Total quality management for project management.* Boca Raton, FL: Auerbach Publications/Taylor & Francis Group.

Proctor, K. S. (2011). *Optimizing and assessing information technology: Improving business process execution*. Hoboken, NJ: John Wiley & Sons.

Racskó, P. (2010). Information technology, telecommunication and economic stability, cybernetics and information technologies. *Bulgarian Academy of Sciences, 10*(4) (Sofia), 5.

Raj, A. (2014). *Best practices in contract management: Strategies for optimizing business relationships*. Retrieved April 14, 2014, from http://www.academia.edu/, http://www.academia.edu/5102810/HR_Assgt

Read, R. C. (1965). *Tangrams: 330 puzzles*. Mineola, NY: Dover Recreational Math, Dover Publications.

Redman, T. C. (2008). *Data driven: Profiting from your most important business asset*. Boston, MA: Harvard Business Review Press.

Reuter, D. (2008). Ralph: A visible/infrared imager for the new horizons Pluto/Kuiper belt mission. *Space Science Review, 140*(1–4), 129–154.

Rhoton, J., Clercq, J. D., & Graves, D. (2013). *Cloud computing protected: Security assessment handbook*. Washington, DC: Recursive, Limited.

Rieger, B. (2013). *The people's car: A global history of the Volkswagen Beetle*. Boston, MA: Harvard University Press.

Roe, J. W. (1916). *English and American tool builders*. New Haven, CT: Yale University Press.

Ross, L., & Nieberding, J. (2010). *Space system development: Lessons learned workshop*. Bay Village, OH: Aerospace Engineering Associates LLC.

Rubin, S. M. (1994). *Computer aids for VLSI design*. Retrieved May 19, 2014, from http://www.rulabinsky.com, http://www.rulabinsky.com/cavd/text/chap05-4.html.

Rzepka, W. E. (1989). A requirements engineering testbed: Concept, status, and first results. *Proceedings of the 22nd Annual Hawaii International Conference on System Sciences*, pp. 339–347. Kailua-Kona, HI, January 3–6.

SAE. (2005). *GEIA-HB-649, SAE Bulletin: Implementation guide for configuration management*. SAE International. Warrendale, Pennsylvania.

SAE. (2012). *GEIA-859a, SAE: Data Management*. SAE International. Warrendale, Pennsylvania.

SAE. (2018). *SAE/EIA 649-C, SAE Standard: Configuration Management Standard*. SAE International. Warrendale, Pennsylvania.

Samaras, T. T., & Czerwinski, F. L. (1971). *Fundamentals of configuration management*. New York, NY: Wiley-Interscience, a division of John Wiley & Sons, Inc.

Schneier, B. (2018): *Click here to kill everybody*. London, England: W. W. Norton & Company.

Schumann, R. G. (1980). *Specifications, Issue 13, Government Contracts Monograph Government Contracts Program*. Washington, DC: The George Washington University.

Scott, R. (Director). (1982). *Blade runner*. [Motion Picture]. Burbank, CA: Warner Brothers Studios.

Sexton, D. (2018). F-35 maintenance team explores VR, AR for training needs. Retrieved December 21, 2018, from https://www.arnold.af.mil/News/Article-Display/Article/1719912/f-35-maintenance-team-explores-vr-ar-for-training-needs/

Seglie, D., Menicucci, M., & Collado, H. (2018, February 23) Neanderthal cave art. *Science, 359*(6378), 912–915. Retrieved from http://cesmap.it/neanderart2018-international-conference-report-from-science-23-feb-2018-vol-359-issue-6378-pp-912-915-u-th-dating-of-carbonate-crusts-reveals-neandertal-origin-of-iberian-cave-art/

Severin, T. (2005). *The Brendan voyage*. Dublin, Ireland: Gill & Macmillan Ltd.

Smith, M. A. (2006). *Internet routing instability*. Three papers presented by Michael A. Smith. Tallahassee, FL: Florida State University. www.cs.fsu.edu/~xyuan/.../michael_instability.ppt

Software Engineering Institute. (2010). CMMI® for Development, Version 1.3, CMU/SEI-2010-TR-033 Retrieved July 10, 2014. http://www.sei.cmu.edu/reports/10tr033.pdf

Sole-Smith, V. (2013). *How to protect kids from lead toys*. Retrieved from http://www.goodhousekeeping.com/health/womens-health/poisonous-lead-toys-0907

Stevenson, R. (Director). (1968). *The Love Bug special edition DVD (2003)*. [Motion Picture].

Subramaniam, M., & Hewett, K. (2004). Balancing standardization and adaptation for product-performance in international markets: Testing the influence of headquarters-subsidiary contact and cooperation. *Management International Review, 44*(2), 171–194.

Talley, S. (Director). (2011). *Secrets of the dead: China's terracotta warriors*. [Motion Picture]. Burbank, CA: The Walt Disney Company.

Taylor, G. B. (2008). *Al-Razi's "book of secrets": The practical laboratory in the medieval Islamic world*. Fullerton, CA: California State University.

Terracotta Warriors. (2009, June 15). *National Geographic News*. Retrieved February 22, 2014, from http://news.nationalgeographic.com/news/2009/06/090615-terracotta-excavate-video-ap.html

Turner, V. (2014, December). The EMC digital universe study—with research and analysis by IDC. EMC. Retrieved from http://www.emc.com/leadership/digital-universe/index.htm

United Technologies Corporation. (1998, November). Independent research & development and bid & proposal costs. Retrieved November 3, 2013, from www.utc.com/StaticFiles/UTC/StaticFiles/proposals_english.pdf

Volkswagen. (n.d.). VOLKSWAGEN brand history. *Autoevolution*. Retrieved July 23, 2014, from http://www.autoevolution.com/volkswagen/history/

Voragine, de J. (2013). *The golden legend*. Translated by W. Caxton. Seattle, WA: CreateSpace Independent Publishing Platform.

Walbank, F. W. (1957). *A historical commentary on Polybius*, 3 vols. Oxford: 1999 Special Edition Clarendon Press.

Warriner, K. (2013, December 19). Food Seminars International, Food Safety Culture: When food safety systems are not enough. Retrieved December 19, 2013, from http://foodseminars.net/product.sc;jsessionid=9A15A4DB9773831A41A5B6D46E006C1F.m1plqscsfapp03?productId=163&categoryId=1

Wasson, J. (2008). Configuration management for the 21st century. Retrieved November 23, 2013, from https://www.cmpic.com/whitepapers/whitepapercm21.pdf

Watts, F. B. (2000), *Engineering documentation control handbook*. Park Ridge, U.S.A.: Noyes Publications.

Wessel, D. (2013). What is a configuration item and which consequences does the selection of CIs have to your organization? Consultant on NATO standardization to OSD, ATL. LinkedIn NATO CM Symposium Group moderated by Dirk Wessel started on July 1, 2013.

Wolf, B. (2002). Congress accidentally creates Nov. 31. ABC News. Retrieved from https://abcnews.go.com/Entertainment/WolfFiles/story?id=91981&page=1

World Intellectual Property Organization (2014). *Intellectual property definitions*. Retrieved September 4, 2014, from http://www.wipo.int/export/sites/www/freepublications/en/intproperty/450/wipo_pub_450.pdf

Yost, P. (Director). (2012). *Secrets of the Viking sword—NOVA*. [Motion Picture]. Boston, MA.

Zhewen, L. (1993). *China's imperial tombs and mausoleums*. Beijing, China: Foreign Languages Press.

Zulueta, F. D. (1946). *The institutes of Gaius*. Gloucestershire: Clarendon Press.

Appendix 4: Key Performance Indicators

Many thanks to A. Larry Gurule for providing this compilation.

Accounting/Financial Measures

1. Percent of late reports
2. Percent of errors in reports
3. Errors in input to Information Services
4. Errors reported by outside auditors
5. Percent of input errors detected
6. Number of complaints by users
7. Number of hours spent per week correcting or changing documents
8. Number of complaints about inefficiencies or use of excessive paper
9. Amount of time spent appraising/correcting input errors
10. Payroll processing time
11. Percent of errors in payroll
12. Amount of time to prepare and send a bill
13. Length of time billed and not received
14. Number of final accounting jobs rerun
15. Number of equipment sales miscoded
16. Amount of intracompany accounting bill-back activity
17. Time spent correcting erroneous inputs
18. Number of open items
19. Percent of deviations from cash plan
20. Percent discrepancy in Material Review Board (MRB) and line scrap reports
21. Travel expense accounts processed in three days
22. Percent of advances outstanding
23. Percent data entry errors in accounts payable and general ledger
24. Credit turnaround time
25. Machine billing turnaround time
26. Percent of shipments requiring more than one attempt to invoice
27. Number of untimely supplier invoices processed
28. Average number of days from receipt to processing
29. Percent error in budget predictions
30. Computer rerun time due to input errors
31. Computer program change cost
32. Percent of financial reports delivered on schedule
33. Number of record errors per employee
34. Percent of error free vouchers
35. Percent of bills paid so company gets price break
36. Percent of errors in checks
37. Entry errors per week
38. Number of errors found by outside auditors
39. Number of errors in financial reports
40. Percent of errors in travel advancement records
41. Percent of errors in expense accounts detected by auditors

Engineering

1. Percent of drafting errors per print
2. Percent of prints released on schedule
3. Percent of errors in cost estimates
4. Number of times a print is changed
5. Number of off specs approved
6. Simulation accuracy
7. Accuracy of list of advance materials
8. Cost of input errors to the computer
9. How well the product meets customer expectations
10. Field performance of product
11. Percent of error free designs
12. Percent of errors found during design review
13. Percent of repeat problems corrected
14. Time taken to correct a problem
15. Time required to make an engineering change
16. Cost of engineering changes per month
17. Percent of reports that have errors in them
18. Number of errors in data recording per month
19. Percent of evaluations that meet engineering objectives
20. Percent of special quotations that are successful
21. Percent of test plans that are changed
22. Percent of meetings starting on schedule
23. Cost of spare parts after warranty
24. Number of meetings held per quarter where quality and defect prevention were the main subjects
25. Person months per released print
26. Percent of total problems found by diagnostics as released
27. Customer cost per life of output delivered
28. Number of problems that were also encountered in previous products
29. Cycle time to correct a customer problem
30. Number of errors in publications reported from the plant and field
31. Number of error free products that pass independent evaluation
32. Number of missed shipments of prototypes
33. Number of unsuccessful preanalyses
34. Number of off specs accepted
35. Percent of requests for engineering action open for more than two weeks
36. Number of days late to preanalysis
37. Number of restarts of evaluations and tests
38. Effectiveness of regression tests
39. Number of days for the release cycle
40. Percent of corrective action schedules missed
41. Percent of bills of material that are released in error

Facilities Measurements

1. Percent of facilities on schedule
2. Percent of manufacturing time lost due to bad layouts
3. Percent of error in time estimates
4. Percent of error in purchase requests

5. Hours lost due to equipment downtown
6. Scrap and rework due to calibration errors
7. Repeat call hours for the same problem
8. Change to layout
9. Percent deviation from budget
10. Maintenance cost versus equipment cost
11. Percent variation to cost estimates
12. Number of unscheduled maintenance calls
13. Number of hours used on unscheduled maintenance
14. Number of hours used on scheduled maintenance
15. Percent of equipment maintained on schedule
16. Percent of equipment overdue for calibration
17. Accuracy of assets report
18. Percent of total floor space devoted to storage
19. Number of industrial design completions past due
20. Number of mechanical/functional errors in industrial design artwork
21. Number of errors found after construction had been accepted by the company
22. Percent of engineering action requests accepted

Forecasting Measurements

1. Number of upward pricing revisions per year
2. Number of project plans that meet schedule, price, and quality
3. Percent error in sales forecasts
4. Number of forecasting assumption errors
5. Number of changes in product schedules

Health and Safety

1. Percent of clearance errors
2. Amount of time taken to get clearance
3. Percent of security violations
4. Percent of documents classified incorrectly
5. Number of security violations per audit
6. Percent of audits conducted on schedule
7. Percent of safety equipment checked per schedule
8. Number of safety problems identified by design analysis versus actual safety problems encountered
9. Safety accidents per 100,000 hours worked
10. Safety violations by department
11. Number of safety suggestions
12. Percent of sensitive parts located

Information Systems

1. Keypunch errors per day
2. Input correction on CRT
3. Reruns caused by operator error
4. Percent of reports delivered on schedule
5. Errors per thousand lines of code
6. Number of changes after the program is coded
7. Percent of time required to debug programs

8. Rework costs resulting from computer program
9. Number of cost estimates revised
10. Percent error in forecast
11. Percent error in lines of code required
12. Number of coding errors found during formal testing
13. Number of test case errors
14. Number of test case runs before success
15. Number of revisions to plan
16. Number of documentation errors
17. Number of revisions to program objectives
18. Number of errors found after formal test
19. Number of error free programs delivered to customer
20. Number of process step errors before a correct package is ready
21. Number of revisions to checkpoint plan
22. Number of changes to customer requirements
23. Percent of programs not flow-diagrammed
24. Percent of customer problems not corrected per schedule
25. Percent of problems uncovered before design release
26. Percent change in customer satisfaction survey
27. Percent of defect-free artwork
28. System availability
29. Terminal response time
30. Mean time between system initial program loadings (IPL)
31. Mean time between system repairs
32. Time before help calls are answered

Legal

1. Response time on request for legal opinion
2. Time to prepare patent claims
3. Percent of cases lost

Management Measurements

1. Number of security violations per year
2. Percent variation from budget
3. Percent of target dates missed
4. Percent of personnel turnover rate
5. Percent increase in output per employee
6. Percent absenteeism
7. Percent error in planning estimates
8. Percent of output delivered on schedule
9. Percent of employees promoted to better jobs
10. Department morale index
11. Percent of meetings that start on schedule
12. Percent of employee time spent on first time output
13. Number of job improvement ideas per employee
14. Dollars saved per employee due to new ideas and/or methods
15. Ratio of direct to indirect employees
16. Increased percent of marketshare
17. Return on investment

18. Percent of appraisals done on schedule
19. Percent of changes required to project equipment
20. Normal appraisal distribution
21. Percent of employee output that is measured
22. Number of grievances per month
23. Number of open doors per month
24. Percent of professional employees active in professional societies
25. Percent of managers active in community activities
26. Number of security violations per month
27. Percent of time program plans are met
28. Improvement in opinion surveys
29. Percent of employees who can detect and repair their own errors
30. Percent of delinquent suggestions
31. Percent of documents that require two management signatures
32. Percent error in personnel records
33. Percent of time cards signed by managers that have errors on them
34. Percent of employees taking higher education
35. Number of reports of damaged equipment and property
36. Warranty costs
37. Scrap and rework costs
38. Cost due to poor quality
39. Number of employees dropping out of classes
40. Number of decisions made by higher level management than required by procedures
41. Improvement in customer satisfaction survey
42. Volumes actual versus plan
43. Revenue actual versus plan
44. Number of formal reviews before plans are approved
45. Number of procedures with fewer than three acronyms and abbreviations
46. Percent of procedures less than 10 pages
47. Percent of employees active in improvement teams
48. Number of hours per year of career and skill development training per employee
49. Number of user complaints per month
50. Number of variances in capital spending
51. Percent revenue/expense ratio below plan
52. Percent of executive interviews with employees
53. Percent of departments with disaster recovery plans
54. Percent of appraisals with quality as a line item that makes up more than 30 percent of the evaluation
55. Percent of employees with development plans
56. Revenue generated over the strategic period
57. Number of iterations of the strategic plan
58. Number of employees participating in cost effectiveness
59. Data integrity
60. Result of peer reviews
61. Number of tasks for which actual time exceeded estimated time

Manufacturing and Test Engineering

1. Percent of process operation where sigma limit is within engineering specification
2. Percent of tools that fail certification
3. Percent of tools that are reworked due to design errors

4. Number of process changes per operation due to errors
5. In process yields
6. Percent error in manufacturing costs
7. Time required to solve a problem
8. Number of delays because process instructions are wrong or not available
9. Labor utilization index
10. Percent error in test equipment and tooling budget
11. Number of errors in operator training documentation
12. Percent of errors that escape the operator's detection
13. Percent of testers that fail certification
14. Percent error in yield projections
15. Percent error in quality of output product
16. Asset utilization
17. Percent of designed experiments needing revision
18. Percent of changes to process specifications during process design review
19. Percent of equipment ready for production on schedule
20. Percent of meetings starting on schedule
21. Percent of drafting errors found by checkers
22. Percent of manufacturing used to screen products
23. Number of problems that the test equipment cannot detect during manufacturing cycle
24. Percent correlation between testers
25. Number of waivers to manufacturing procedures
26. Percent of tools and test equipment delivered on schedule
27. Percent of tools and test equipment on change1 level control
28. Percent of functional test coverage of products
29. Percent of projected cost reductions missed
30. Percent of action plan schedules missed
31. Equipment utilization

Marketing Measurements

1. Percent of proposals submitted ahead of schedule
2. Cost of sales per total costs
3. Percent error in market forecasts
4. Percent of proposals accepted
5. Percent of quota attained
6. Response time to customer inquiries
7. Inquiries per $10,000 of advertisement
8. Number of new customers
9. Percent of repeat orders
10. Percent of time customer expectations are identified
11. Sales made per call
12. Number of errors in orders
13. Ratio of marketing expenses to sales
14. Number of new business opportunities identified
15. Errors per contract
16. Percent of time customer expectation changes are identified before impact
17. Man hours per $10,000 sales
18. Percent of reduction in residual inventory
19. Percent of customers called as promised
20. Percent of meetings starting on schedule

21. Percent of changed orders
22. Number of complimentary letters
23. Percent of phone numbers correctly dialed
24. Time required to turn in travel expense accounts
25. Number of revisions to market requirement statements per month
26. Percent of bids returned on schedule
27. Percent of customer letters answered in two weeks
28. Number of complaint reports received
29. Percent of complaint reports answered in three days

Personnel Measurements

1. Percent of employees who leave during the first year
2. Number of days to answer suggestions
3. Number of suggestions resubmitted and approved
4. Personnel cost per employee
5. Cost per new employee
6. Turnover rate due to poor performance
7. Number of grievances per month
8. Percent of employment requests filled on schedule
9. Number of days to fill an employment request
10. Management evaluation of management education courses
11. Time taken to process an application
12. Average time a visitor spends in lobby
13. Time required to get security clearance
14. Time taken to process insurance claims
15. Percent of employees participating in company activities
16. Opinion survey ratings
17. Percent of complaints about salary
18. Percent of personnel problems handled by employees' managers
19. Percent of employees participating in voluntary health screening
20. Percent of offers accepted
21. Percent of retirees contacted yearly by phone
22. Percent of training classes evaluated excellent
23. Percent deviation to resource plan
24. Wait time in medical department
25. Number of days taken to respond to applicant
26. Percent of promotions and management changes publicized
27. Percent of error free newsletters

Process/Industrial Engineering

1. Percent of facilities on schedule
2. Percent of manufacturing time lost due to bad layouts
3. Percent of error in time estimates
4. Percent of error in purchase requests
5. Hours lost due to equipment downtime
6. Scrap and rework due to calibration errors
7. Repeat call hours for the same problems
8. Changes to layout
9. Percent deviation from budget

10. Maintenance cost to equipment cost
11. Percent variation to cost estimates
12. Number of unscheduled maintenance calls
13. Number of hours used on unscheduled maintenance
14. Number of hours used on scheduled maintenance
15. Percent of equipment maintained on schedule
16. Percent of equipment overdue for calibration
17. Accuracy of assets report
18. Percent of total floor space devoted to storage
19. Number of industrial design completions past due
20. Number of mechanical/functional errors in industrial design artwork
21. Number of errors found after construction had been accepted by the company
22. Percent of engineering action requests accepted

Production Control

1. Percent of late deliveries
2. Percent of errors in stocking
3. Number of items exceeding shelf life
4. Percent of manufacturing completed on schedule
5. Time required to incorporate engineering changes
6. Percent of errors in purchase requisitions
7. Percent of products that meet customer orders
8. Inventory turnover rate
9. Time duration when line is down due to assembly shortage
10. Percent of time parts are not in stock when ordered from common parts crib
11. Time spent in shipment of product
12. Cost of rush shipments
13. Spare parts availability in crib
14. Percent of errors in work in progress records versus audit data
15. Cost of inventory spoilage
16. Number of bill of lading errors not caught in shipping

Purchasing and Materials Management

1. Percent of discount orders by consolidating
2. Number of errors per purchase order
3. Number of orders received with no purchase order
4. Number of routing and rate errors per shipment
5. Percent of supplies delivered on schedule
6. Percent decrease in parts costs
7. Expediters per direct employees
8. Number of items on the hot list
9. Percent of suppliers with 100 percent lot acceptance for one year
10. Stock costs
11. Labor hours per $10,000 purchases
12. Purchase order cycle time
13. Number of times per year line stopped due to lack of supplier parts
14. Supplier parts scrapped due to engineering changes
15. Percent of parts with two or more suppliers
16. Average time to fill emergency orders

17. Average time to replace rejected lots with good parts
18. Parts cost per total costs
19. Percent of lots received on line late
20. Actual purchased materials cost per budgeted cost
21. Time taken to answer customer complaints
22. Percent of phone calls dialed correctly
23. Percent of purchase orders resumed due to errors or incomplete description
24. Percent of defect free supplier model parts
25. Percent of projected cost reductions missed
26. Time required to process equipment purchase orders
27. Cost of rush shipments
28. Number of items billed but not received

Quality Assurance

1. Percent error in reliability projections
2. Percent of product that meets customer expectations
3. Time taken to answer customer complaints
4. Number of customer complaints
5. Number of errors detected during design and process reviews
6. Percent of employees active in professional societies
7. Number of audits performed on schedule
8. Percent of quality assurance personnel to total personnel
9. Percent of quality inspectors to manufacturing directs
10. Percent of quality engineers to product and manufacturing engineers
11. Number of engineering changes after design review
12. Number of process changes after process qualification
13. Number of errors in reports
14. Time taken to correct a problem
15. Cost of scrap and rework that was not created at the rejected operation
16. Percent of suppliers at 100 percent lot acceptance for one year
17. Percent of lots going directly to stock
18. Percent of problems identified in the field
19. Variations between inspectors doing the same job
20. Percent of reports published on schedule
21. Number of complaints from manufacturing management
22. Percent of field returns correctly analyzed
23. Time taken to identify and solve problems
24. Percent of laboratory services not completed on schedule
25. Percent of improvement in early detection of major design errors
26. Percent of errors in defect records
27. Number of reject orders not disposed in five days
28. Number of customer calls to report errors
29. Level of customer surveys
30. Number of committed supplier plans in place
31. Percent of correlated test results with suppliers
32. Receiving inspection cycle time
33. Number of requests for corrective action being processed
34. Time required to process a request for corrective action
35. Number of off specs approved
36. Percent of part numbers going directly to stock

37. Number of manufacturing interruptions caused by supplier parts
38. Percent error in predicting customer performance
39. Percent product cost related to appraisal, scrap, and rework
40. Percent of skip lot inspection
41. Percent of qualified suppliers
42. Number of problems identified in process

Shipping Measurements

1. Complaints on shipping damage
2. Percent of parts not packed to required specifications
3. Percent of output that meets customer orders and engineering specifications
4. Scrap and rework cost
5. Number of suggestions per employee
6. Percent of jobs that meet cost
7. Percent of jobs that meet schedule
8. Percent of defect-free product at measurement operations
9. Percent of employees trained to do the job they are working on
10. Number of accidents per month
11. Performance against standards
12. Percent of utilities left improperly running at end of shift
13. Percent of unplanned overtime
14. Number of security violations per month
15. Percent of time log book filled out correctly
16. Time and/or claiming errors per week
17. Time between errors at each operation
18. Errors per 100,000 solder connections
19. Labor Utilization Index
20. Percent of operators certified to do their job
21. Percent of shipping errors
22. Defects during warranty period
23. Replacement parts defect rates
24. Percent of products defective at final test
25. Percent of control charts maintained correctly
26. Percent of invalid test data
27. Percent of shipments below plan
28. Percent of daily reports in by 7:00 a.m.
29. Percent of late shipments
30. Percent of products error free at final test

Appendix 5: Publications contributed to by Value Transformation, LLC.

Publications authored or contributed to by Value Transformation LLC associates include:

Value Transformation LLC associates have extensive cradle-to-grave knowledge and experience in business and product development areas.

Value Transformation LLC also provides consulting services, seminars, and webinars as well as in-person training and distance learning approaches that will heal the wounds of processes gone wild. We do not abandon you while your team acquires the skills it needs for your enterprise to succeed.

Visit us at http://www.valuetransform.com or http://valuetransform.com/testimonial to learn more.

You can contact Value Transformation LLC by sending an email to Jon.Quigley@valuetransform.com.

The sooner you reach out, the sooner you will start to see the transformation you need to compete in today's marketplace.

Index

Note: Locators in *italics* represent figures and **bold** indicate tables in the text.

D

E

F

G

H

J

K

L

M

S